Plains Vizcachas

Luciano Luis Rasia
Claudio Gustavo Barbeito • Francisco Acuña
Editors

Plains Vizcachas

Biology and Evolution of a Peculiar Neotropical Caviomorph Rodent

Springer

Editors
Luciano Luis Rasia
División Paleontología Vertebrados
Museo de La Plata
La Plata, Argentina

Claudio Gustavo Barbeito
Facultad de Ciencias Veterinarias
National University of La Plata
La Plata, Argentina

Francisco Acuña
Facultad de Ciencias Veterinarias
National University of La Plata
La Plata, Argentina

ISBN 978-3-031-49486-4 ISBN 978-3-031-49487-1 (eBook)
https://doi.org/10.1007/978-3-031-49487-1

This Springer imprint is published by the registered company Springer Nature Switzerland AG
The registered company address is: Gewerbestrasse 11, 6330 Cham, Switzerland

Paper in this product is recyclable.

Preface

More than 200 years have passed since the plains vizcacha (*Lagostomus maximus*) was first mentioned by nineteenth century naturalists such as Felix de Azara, Charles Darwin and William Henry Hudson, yet many aspects of the biology of this Neotropical rodent remain unknown. Moreover, considered an agricultural pest and hunted for its skin and meat, the species' range has been considerably reduced as a result of the expansion of agriculture and cattle frontier. Throughout the twentieth century, studies by Augusto Llanos, Jorge Crespo, Barbara Weir, Lyn Branch and John Jackson, among others, contributed greatly to our knowledge of the ecology, growth and reproduction of the plains vizcacha.

Recent studies have revealed unique aspects of the morphology, physiology and evolutionary history of the plains vizcacha, highlighting the need for a meeting point for all these lines of research. The interaction with researchers from a wide range of disciplines has encouraged the edition of this book.

La Plata, Argentina

Luciano Luis Rasia
Claudio Gustavo Barbeito
Francisco Acuña

Contents

Contributors

Agustín M. Abba Centro de Estudios Parasitológicos y de Vectores CEPAVE (CCT La Plata CONICET-UNLP), La Plata, Buenos Aires, Argentina

Francisco Acuña Laboratorio de Histología y Embriología Descriptiva, Experimental y Comparada, Facultad de Ciencias Veterinarias, Universidad Nacional de La Plata, La Plata, Argentina

Consejo Nacional de Investigaciones Científicas y Técnicas (CONICET), La Plata, Argentina

Claudia I. Aguilera Merlo Área de Morfología, Facultad de Química, Bioquímica y Farmacia, Universidad Nacional de San Luis, San Luis, De San Luis, Argentina

Alicia Álvarez CONICET, La Plata, Argentina

Laboratorio de Paleontología de Vertebrados, Instituto de Ecorregiones Andinas (INECOA; CONICET-UNJu), Instituto de Geología y Minería (UNJu), San Salvador de Jujuy, Jujuy, Argentina

María Clara Álvarez INCUAPA, UNCPBA-CONICET. Universidad Nacional del Centro de la Provincia de Buenos Aires (UNCPBA), Facultad de Ciencias Sociales, Olavarría, Buenos Aires, Argentina

Claudio Gustavo Barbeito Laboratorio de Histología y Embriología Descriptiva, Experimental y Comparada, Facultad de Ciencias Veterinarias, Universidad Nacional de La Plata, La Plata, Argentina

Consejo Nacional de Investigaciones Científicas y Técnicas (CONICET), La Plata, Argentina

Fabricio I. Busolini Área de Morfología, Facultad de Química, Bioquímica y Farmacia, Universidad Nacional de San Luis, San Luis, San Luis, Argentina

Victoria Canova Centro de Estudios Parasitológicos y de Vectores CEPAVE (CCT La Plata CONICET-UNLP), La Plata, Buenos Aires, Argentina

Eduardo M. Chaves Área de Morfología, Facultad de Química, Bioquímica y Farmacia, Universidad Nacional de San Luis, San Luis, San Luis, Argentina

Stefanía Cohen Fisheries Research and Development Institute (INIDEP), Mar del Plata, Argentina

Institute of Marine and Coastal Research (IIMyC), FCEyN, CONICET-UNMdP, Mar del Plata, Argentina

Cecilia B. Contarde Área Biología del Comportamiento, Instituto de Investigaciones Biológicas y Tecnológicas, Universidad Nacional de Córdoba, CONICET, Córdoba, Argentina

Albana A. M. Cruceño Área de Morfología, Facultad de Química, Bioquímica y Farmacia, Universidad Nacional de San Luis, San Luis, San Luis, Argentina

Alcira Ofelia Díaz Department of Biology, School of Exact and Natural Sciences (FCEyN), National University of Mar del Plata (UNMdP), Mar del Plata, Argentina

Institute of Marine and Coastal Research (IIMyC), FCEyN, CONICET-UNMdP, Mar del Plata, Argentina

Verónica Berta Dorfman Centro de Estudios Biomédicos Básicos, Aplicados y Desarrollo (CEBBAD), Ciudad Autónoma de Buenos Aires, Argentina

Consejo Nacional de Investigaciones Científicas y Técnicas (CONICET), Ciudad Autónoma de Buenos Aires, Argentina

Marcos D. Ercoli CONICET, La Plata, Argentina

Laboratorio de Paleontología de Vertebrados, Instituto de Ecorregiones Andinas (INECOA; CONICET-UNJu). Instituto de Geología y Minería (UNJu), CONICET, San Salvador de Jujuy, Jujuy, Argentina

Verónica P. Filippa Área de Morfología, Facultad de Química, Bioquímica y Farmacia, Universidad Nacional de San Luis, CONICET, San Luis, San Luis, Argentina

Consejo Nacional de Investigaciones Científicas y Técnicas (CONICET), San Luis, Argentina

Mirta Alicia Flamini Laboratory of Histology and Descriptive, Experimental and Comparative Embryology, Department of Basic Sciences, School of Veterinary Sciences, National University of La Plata, La Plata, Argentina

Carlos A. Galliari Centro de Estudios Parasitológicos y de Vectores (CEPAVE-CONICET-UNLP), La Plata, Buenos Aires, Argentina

María Constanza Gariboldi Centro de Estudios Biomédicos Básicos, Aplicados y Desarrollo (CEBBAD), Ciudad Autónoma de Buenos Aires, Argentina

María A. Gutiérrez INCUAPA, UNCPBA-CONICET. Universidad Nacional del Centro de la Provincia de Buenos Aires (UNCPBA), Facultad de Ciencias Sociales, Olavarría, Buenos Aires, Argentina

Diego A. Guzmán Área Biología del Comportamiento, Instituto de Investigaciones Biológicas y Tecnológicas, Universidad Nacional de Córdoba, CONICET, Córdoba, Argentina

Patricia Hadler Departamento de Geologia, Universidade Federal de Santa Catarina, Florianópolis, SC, Brazil

Julia Halperin Centro de Estudios Biomédicos Básicos, Aplicados y Desarrollo (CEBBAD), Universidad Maimónides, CONICET, Ciudad Autónoma de Buenos Aires, Argentina

Consejo Nacional de Investigaciones Científicas y Técnicas (CONICET), Buenos Aires, Argentina

Pablo Ignacio Felipe Inserra Centro de Estudios Biomédicos Básicos, Aplicados y Desarrollo (CEBBAD), Universidad Maimónides, Ciudad Autónoma de Buenos Aires, Argentina

Consejo Nacional de Investigaciones Científicas y Técnicas (CONICET), Buenos Aires, Argentina

Cristian A. Kaufmann INCUAPA, UNCPBA-CONICET. Universidad Nacional del Centro de la Provincia de Buenos Aires (UNCPBA), Facultad de Ciencias Sociales, Olavarría, Buenos Aires, Argentina

Noelia Paola Leopardo Centro de Estudios Biomédicos Básicos, Aplicados y Desarrollo (CEBBAD), Universidad Maimónides, Ciudad Autónoma de Buenos Aires, Argentina

Consejo Nacional de Investigaciones Científicas y Técnicas (CONICET), Ciudad Autónoma de Buenos Aires, Argentina

Agustina Massigoge INCUAPA, UNCPBA-CONICET. Universidad Nacional del Centro de la Provincia de Buenos Aires (UNCPBA), Facultad de Ciencias Sociales, Olavarría, Buenos Aires, Argentina

María Angélica Miglino Departamento de Cirugía, Facultad de Medicina Veterinaria y Zootecnia, Universidad de San Pablo, San Pablo, San Pablo, Brazil

Fabian H. Mohamed Área de Morfología, Facultad de Química, Bioquímica y Farmacia, Universidad Nacional de San Luis, San Luis, De San Luis, Argentina

Consejo Nacional de Investigaciones Científicas y Técnicas (CONICET), San Luis, Argentina

Graciela T. Navone Centro de Estudios Parasitológicos y de Vectores CEPAVE (CCT La Plata CONICET-UNLP), La Plata, Buenos Aires, Argentina

Ulyses F. J. Pardiñas Instituto de Diversidad y Evolución Austral (IDEAus-CONICET), Puerto Madryn, Argentina

Instituto Nacional de Biodiversidad (INABIO), Quito, Ecuador

Enrique Leo Portiansky Consejo Nacional de Investigaciones Científicas y Técnicas (CONICET), La Plata, Argentina

Laboratorio de Análisis de Imágenes, Facultad de Ciencias Veterinarias, Universidad Nacional de La Plata, La Plata, Argentina

Luciano Luis Rasia División Paleontología Vertebrados, Museo de La Plata, La Plata, Argentina

CONICET, La Plata, Argentina

María del Rosario Robles Centro de Estudios Parasitológicos y de Vectores CEPAVE (CCT La Plata CONICET-UNLP), La Plata, Buenos Aires, Argentina

Juan Manuel Rodríguez INCUAPA, UNCPBA-CONICET. Universidad Nacional del Centro de la Provincia de Buenos Aires (UNCPBA), Facultad de Ciencias Sociales, Olavarría, Buenos Aires, Argentina

Alejandro Raúl Schmidt Centro de Estudios Biomédicos Básicos, Aplicados y Desarrollo (CEBBAD), Universidad Maimónides, CONICET, Ciudad Autónoma de Buenos Aires, Argentina

Consejo Nacional de Investigaciones Científicas y Técnicas (CONICET), Ciudad Autónoma de Buenos Aires, Argentina

María Florencia Tano de la Hoz Department of Biology, School of Exact and Natural Sciences (FCEyN), National University of Mar del Plata (UNMdP), Mar del Plata, Argentina

Institute of Materials Science and Technology Research (INTEMA) UNMdP–CONICET, Applied Electrochemical Division, Mar del Plata, Argentina

Juan Ignacio Túnez Grupo de Investigación en Ecología Molecular, Instituto de Ecología y Desarrollo Sustentable (INEDES CONICET-UNLu), Luján, Buenos Aires, Argentina

Departamento de Ciencias Básicas, Universidad Nacional de Luján, Luján, Buenos Aires, Argentina

Alfredo Daniel Vitullo Centro de Estudios Biomédicos Básicos, Aplicados y Desarrollo (CEBBAD), Universidad Maimónides, CONICET, Ciudad Autónoma de Buenos Aires, Argentina

Consejo Nacional de Investigaciones Científicas y Técnicas (CONICET), Ciudad Autónoma de Buenos Aires, Argentina

Part I
Evolutionary History

Chapter 1
Two Hundred Years of Studies on Plains Vizcachas

Claudio Gustavo Barbeito, Francisco Acuña, Mirta Alicia Flamini, and Luciano Luis Rasia

Abstract The plains vizcacha, *Lagostomus maximus*, is a caviomorph rodent native to southern South America, with gregarious and burrowing habits and a marked sexual dimorphism. It was the subject of myths and legends of the native peoples and "*Criollos*" (early inhabitants of Argentina after the Spanish conquest); it was also hunted for its meat and skin and considered a pest to agriculture. It has been of interest to naturalists for more than 200 years, with some controversy about its taxonomic identity during most of the nineteenth century. From the late nineteenth century to the middle twentieth century, there were many studies about plains vizcachas distribution, behavior, and external anatomy. In the late twentieth century, there were significant advances in the knowledge of its reproductive system. The increase of studies on the plains vizcacha in recent times, regarding its conservation, population genetics, reproductive biology, zooarcheology, and paleontology, has highlighted the need for an update of the knowledge on *Lagostomus maximus*.

Keywords History · South America · Mammals · Systematics · Biodiversity

C. G. Barbeito (✉) · F. Acuña
Laboratorio de Histología y Embriología Descriptiva, Experimental y Comparada, Facultad de Ciencias Veterinarias, Universidad Nacional de La Plata, La Plata, Argentina

CONICET, La Plata, Argentina
e-mail: barbeito@fcv.unlp.edu.ar; facuna@fcv.unlp.edu.ar

M. A. Flamini
Laboratorio de Histología y Embriología Descriptiva, Experimental y Comparada, Facultad de Ciencias Veterinarias, Universidad Nacional de La Plata, La Plata, Argentina

L. L. Rasia
CONICET, La Plata, Argentina

División Paleontología Vertebrados, Museo de La Plata, La Plata, Argentina
e-mail: lucianorasia@fcnym.unlp.edu.ar

L. L. Rasia et al. (eds.), *Plains Vizcachas*, https://doi.org/10.1007/978-3-031-49487-1_1

1.1 Popular Knowledge and Mythology of *Lagostomus maximus*

The plains vizcacha, *Lagostomus maximus* (Table 1.1), is a characteristic member of the fauna of the Pampas in southern South America, being relevant in ancient and present culture (Fig. 1.1).

Long before naturalists began to study it with scientific criteria, the plains vizcacha was always the subject of legends and tales among native peoples. For example, some legends narrated that this rodent arose from the transformation of four children who entered a cave to play (Palermo 1983). Other legends say that vizcachas were wealthy children that asked for credit and never paid, so they had to hide in a cave, becoming beasts (Vidal de Battini 1984). Subsequently, the "*Criollos*" (inhabitants of the Argentine pampas) generated stories related to the habit of this animal

Table 1.1 Common names and systematic classification of *Lagostomus maximus*

Common names: Plains vizcacha, plains viscacha, Argentine plains viscacha, South American viscacha, pampas vizcacha, vizcacha, viscacha.
Systematic classification:
Order RODENTIA Bowdich 1821
Suborder HYSTRICOMORPHA Brandt 1855
Infraorder HYSTRICOGNATHI Tullberg 1899
Parvorder CAVIOMORPHA Wood and Patterson (in Wood 1955)
Superfamily CHINCHILLOIDEA Bennett 1833
Family CHINCHILLIDAE Bennett 1833
Subfamily LAGOSTOMINAE Wiegmann 1835
Genus *Lagostomus* Brookes 1828
Species *Lagostomus maximus* (Desmarest 1817)

Fig. 1.1 Statue (left) and bas-relief (right) of *Lagostomus maximus*, made by Argentinian sculptor Máximo Carlos Maldonado (1900–1980), located in the Museo de La Plata and Paseo del Bosque (La Plata, Argentina) respectively

of taking different things to their caves (*vizcacheras*), and the term vizcacha came to be used to designate people who accumulate things. Beyond the mythical and literary aspects, the vizcacha has had a close relationship with humans for centuries. The native peoples who fed on their meat hunted them, flooding the *vizcacheras* and thus forcing them out. These groups also used their leather, for example, to make bags. Subsequently, the Europeans who had arrived in America also began to consume their meat, especially in areas and times when beef was scarce. The use of its leather was limited until 1887 when a technique for its tanning was developed. However, their skin was always less appreciated than that of other animals. Its meat, on the other hand, continues to be eaten and sold, specially prepared in preserves. In 1953, it was declared a pest for agriculture due to the damage it caused to crops. One of the most controversial aspects of the relationship between vizcachas and human beings was always the possibility that the horses would roll by inserting their legs into the *vizcachera*, causing severe injuries to the animals and riders (Palermo 1983).

1.2 Naturalists of the Nineteenth Century and the First Studies

One of the first mentions of the species by a naturalist was that of Felix de Azara in 1801, but the first exhaustive description of the plains vizcacha (*Lagostomus maximus*) was carried out by Blainville in his "Mémoire sur une nouvelle espèce de gerboise, et illustration de ce genre," who named it as *Dipus maximus*, a giant gerbil. The animal supposedly came from Java ("Nouvelle-Hollande" in the original French) but was later considered as not formally published or not available (see Geoffroi 1820; Jackson et al. 1996).

In 1817, the naturalist, doctor, and writer Anselme Gaëtan Desmarest (1784–1838) mentioned the description of *Dipus maximus* made by de Blainville, becoming the first valid publication of the name (e.g., Jackson et al. 1996).

Later, Brookes (1828) created *Lagostomus trichodactylus*, a name to replace *Dipus maximus*, given that he considered that the species does not belong to the genus *Dipus*.

Numerous genera including several spellings of the common name were used referring to the plains vizcacha in the early nineteenth century (e.g., Schinz 1824 [1829], 1825; Smith 1842), such as *Viscaccia*, *Viscacia*, and *Vizcacia*, but none of these names are considered valid (see Spotorno and Patton 2015 for a review).

Argentinian paleontologist Florentino Ameghino (Ameghino 1889) synonymized with *Lagostomus trichodactylus* the extinct species †*Lagostomus angustidens*, named by Burmeister (1866), and †*Lagostomus fossilis*, named by Gervais and Ameghino (1880); treating them as variants or subspecies of the extant vizcacha.

Charles Darwin, in his "Diary of a Naturalist Around the World," considers the vizcacha to be the most salient feature of the zoology of the Pampas region. The

English naturalist describes its diet, consisting especially of roots, its habitat in the *vizcacheras*, and its distribution, which does not include the lands located south of the Río Negro (below 42 South latitude). He also recounts the habit of this species, which consists of taking different objects to its burrow and tells an anecdote of a lost watch that was recovered from inside a *vizcachera* (Darwin 1839). Charles Darwin mentions the vizcacha again in his masterpiece "The Origin of Species" based on what was written by George Waterhouse (who made the description of the mammal specimens that Darwin brought back on his trip around the world), considering the vizcacha as the rodent most similar to marsupials and the proof that rodents are the group of eutherians closest to metatherians or that they had a common ancestor (Darwin 1859). The Argentine-English writer and naturalist William Henry Hudson studied ecological aspects of the plains vizcacha, based on field observations (Hudson 1872), and dedicated a chapter of his book *A Naturalist in El Plata* (Hudson 1892) where he opposed Darwin's vision because he considered the vizcacha as very advanced. In the aforementioned chapter, Hudson, with his excellent prose, also described many aspects of the biology of the vizcacha, the *vizcacheras*, and their relationship with humans. Both Darwin and Hudson used the name given to it at the time, *Lagostomus trichodactylus*, referring to the presence of hair on the hind autopodia. The work of Ambrosetti (1893) used the same denomination and gave the name of the taxon to Brookes 1828. In this last work, the use of vizcacha leather to create clothing stood out.

1.3 Expanding the Knowledge During the First Decades of the Twentieth Century

In 1910, Thomas proposed the subspecies *Lagostomus maximus immolis* based on specimens from Northwestern Argentina, being the first use of the combination *Lagostomus maximus*. He also described †*Lagostomus crassus* from Peru as a species recently extinct (see Chap. 6). In 1914, Hollister described the subspecies *petilidens* after studying specimens from the southern part of the habitat of this species. The specimens from the central part of the species territory, not included in these two subspecies, were classified in the subspecies *maximus* (Llanos and Crespo 1952). Nevertheless, the lack of clear differences between specimens of the three subspecies led to the use of a single species without mentioning subspecies (e.g., Jackson et al. 1996).

Pocock (1922) studied anatomical differences between genera of Chinchillidae, establishing two separate subfamilies, Chinchillinae (including *Chinchilla* and *Lagidium*) and Lagostominae (which included *Lagostomus maximus*).

Fortuna (1941) used *Lagostomus maximus maximus* for the plains vizcacha in his publication, where he describes the importance of the species as a pest and some ethological and reproductive aspects. In 1949, the German-Argentine zoologist and histologist Miguel Fernández published a study (Fernandez 1949) with an exhaustive description of the *vizcacheras*, including plans of the same. Curiously, Fernández maintains the name *Lagostomus trichodactylus* Brookes 1828. *L. maximus* is always used in later works, although the authorship is sometimes granted to Blanville and other times to Desmarest. For example, in the classic book by Grasse (1955) *L. maximus* Desmarest 1817 is used, as in the classic article by Llanos and Crespo (1952).

Llanos and Crespo (1952) carried out a 3-year study of vizcachas from the province of Entre Ríos in Argentina. In this work, they include many male and female specimens of different ages. They measure and weigh animals, perform craniometric studies, and weigh various organs. They describe their habits and the characteristics of the *vizcahera*. They make a brief description of their reproductive organs, but it is interesting that they describe for the first time the spontaneous and constant embryonic resorption that occurs in the species with the survival and birth of two offspring, the most caudal. Although Llanos and Crespo collect samples to perform histological techniques on the ovaries and testicles, they do not make an exhaustive description of them, so the polyovulation that characterizes the species could not be discovered.

1.4 Barbara Weir and the Recognition of Unprecedented Reproductive Strategies

The English zoologist Barbara Weir, supervised by Idwal Wyn Rowlands, presents in 1968 her dissertation "Some aspects of reproduction in hystricomorph rodents" and begins studies on the reproductive system of different South American rodents that will include *L. maximus* (Carter et al. 2023). Weir manages to capture vizcachas in Argentina and takes them to England where she manages to reproduce them. From the colony that it breeds, she carries out numerous studies, especially on its reproduction, in which she describes the unique shape of the ovary among eutherians, polyovulation, polyimplantation, and embryonic death (Roberts and Weir 1973). These studies are the basis of the numerous works that analyze the reproduction of the vizcacha in the last years of the twentieth century and up to the present (Chaps. 11, 12, 13, 15, 16, 17, 18 of this book). Shortly after, at the Universidad de Cuyo Argentina, a group of researchers, including doctors Piezzi, Scardapane, and Dominguez, began their studies on the endocrine system of the vizcacha and its relationship with the photoperiod, beginning with a study on the pineal gland (Guzmán et al. 1977).

1.5 Future Prospects

In recent years, there has been a notable increase in studies on the plains vizcacha, with very different approaches, such as conservation (Renison et al. 2023), population genetics (Gariboldi et al. 2019), reproductive biology (Flamini et al. 2011;

Inserra et al. 2013; Acuña et al. 2021), zooarcheology (e.g., Quintana and Mazzanti 2011), and paleontology (e.g., Rasia 2016; Rafuse et al. 2019). These studies were the basis for the remaining chapters of this book, representing an update of the knowledge on *Lagostomus maximus*.

References

Acuña F, Barbeito CG, Portiansky EL, Miglino MA, Flamini MA (2021) Prenatal development in *Lagostomus maximus* (Rodentia, Chinchillidae): a unique case among eutherian mammals of physiological embryonic death. J Morphol 282(5):720–732

Ambrosetti JB (1893) Notas biológicas sobre las vizcachas (*Lagostomus trichodactylus* Brookes). Rev Jard Zool Bs As 1:43–46

Ameghino F (1889) Contribución al conocimiento de los mamíferos fósiles de la República Argentina. Bol Acad Nac Cienc Córdoba 6:1–127

Bennett ET (1833) Remarks on the family Chinchillidae and on a new genus referrible [sic.] to it. Proc Zool Soc London 1:57–60

Bowdich TE (1821) An analysis of the natural classifications of Mammalia for the use of students and travelers. Smith, Paris, p 115

Brandt JF (1855) Beitrage zur nahern Kenntniss der Saugethiere Russlands. Kaiserl Academ d Wiss, p 365

Brookes J (1828) A new genus of the order Rodentia. Trans Linn Soc 16:96–105

Burmeister G (1866) Lista de los mamíferos fósiles del terreno diluviano. Anal Mus Púb Buenos Aires 1:121–232

Carter AM, Acuña F, Barbeito CG (2023) I. W. Rowlands, Barbara J. Weir and the biology of the hystricomorph rodents. Reproduction 166(1):R1–R13

Darwin C (1839) Journal and remarks. 1832-1836. In: Narrative of the surveying voyages of H.M.S. Adventure and Beagle between the years 1826 and 1836, describing their examination of the southern shores of South America, and the Beagle's circumnavigation of the globe, vol III. Henry Colburn, London

Darwin C (1859) On the origin of species. John Murray, London, p 502

Desmarest AG (1817) Gerboise premiere espéce: la grande gerboise *Dipus maximus* Blainv. Nouveau Dictionnaire d'Histoire Naturelle 13:117–119

Fernandez M (1949) Sobre la vizcacha (*Lagostomus trichodactylus* Brookes) sus vivendas y su protección. Bol Acad Nac Cienc Córdoba 3:347–378

Flamini MA, Portiansky EL, Favaron PO, Martins DS, Ambrósio CE, Mess AM, Miglino MA, Barbeito CG (2011) Chorioallantoic and yolk sac placentation in the plains viscacha (*Lagostomus maximus*) – a caviomorph rodent with natural polyovulation. Placenta 32(12):963–968

Fortuna J (1941) Consideraciones sobre la vizcacha de La Pampa y apéndice desde el aspecto económico. CEP, Santa Rosa, La Pampa

Gariboldi MC, Inserra PIF, Lucero S, Failla M, Perez SI, Vitullo AD (2019) Unexpected low genetic variation in the South American hystricognath rodent *Lagostomus maximus* (Rodentia: Chinchillidae). PLoS One 14(9):e0221559

Geoffroi M (1820) Grande Gerboise, *Dipus maximus*, Blainv. In: Levrault FG (ed) Dictionnaire des Sciences Naturelles (1816-1830), vol 18. F. G. Levrault/Le Normant, Strasbourg/Paris, pp 471–472

Gervais H, Ameghino F (1880) Los mamíferos fósiles de América del Sud. Savy e Igon, Paris-Buenos Aires, pp XI + 1–XI 225

Grasse PP (1955) Traité de zoologie: Anatomie, systématique, biologie, T 17. fasc. 2. Mammifères. Masson, Paris

Guzmán JA, Piezzi RS, Pelzer LE, Scardapane L, Domínguez S (1977) Hydroxyindole-O-methyl transferase (HIOMT) and serotonin in the pineal gland of the vizcacha (*Lagostomus maximus maximus*). Acta Physiol Lat Am 27(1):31–36. PMID: 616171
Hollister H (1914) Four new neotropical rodents. Proc Biol Soc Wash 27:57–60
Hudson WH (1872) On the habits of the vizcacha (*Lagostomus trichodactylus*). Proc Zool Soc London 1872:822–833
Hudson WH (1892) The naturalist in La Plata. D. Appleton and Company, New York
Inserra PI, Leopardo NP, Willis MA, Freysselinard AL, Vitullo AD (2013) Quantification of healthy and atretic germ cells and follicles in the developing and post-natal ovary of the South American plains viscacha, *Lagostomus maximus*: evidence of continuous rise of the germinal reserve. Reproduction 147(2):199–209
Jackson JE, Branch LC, Villareal D (1996) *Lagostomus maximus*. Mamm Species 543:1–6
Llanos AC, Crespo JA (1952) Ecología de la vizcacha (*Lagostomus maximus maximus* Blainv.) en el nordeste de la Provincia de Entre Ríos. Rev Investig Agrícol 6:289–378
Palermo MA (1983) Ficha antropológica. In: La vizcacha. Fauna Argentina 30. Centro Editor de América Latina, Buenos Aires, pp 23–25
Pocock RI (1922) On the external characters of some hystricomorph rodents. Proc Zool Soc London 92(2):365–427
Quintana CA, Mazzanti DL (2011) Las vizcachas pampeanas (*Lagostomus maximus*, Rodentia) en la subsistencia indígena del Holoceno tardío de las Sierras de Tandilia Oriental (Argentina). Lat Am Antiq 22(2):253–270
Rafuse DJ, Kaufmann CA, Gutiérrez MA, González ME, Scheifler NA, Álvarez MC, Massigoge A (2019) Taphonomy of modern communal burrow systems of the Plains vizcacha (*Lagostomus maximus*, Chinchillidae) in the Pampas region of Argentina: implications for the fossil record. Hist Biol 31(5):517–534
Rasia LL (2016) Los Chinchillidae (Rodentia, Caviomorpha) fósiles de la República Argentina: sistemática, historia evolutiva y biogeográfica, significado bioestratigráfico y paleoambiental. PhD Dissertation, Universidad Nacional de La Plata, La Plata, Argentina
Renison D, Cingolani AM, Contarde C, Guzmán D (2023) Asistiendo a la reintroducción de vizcachas (*Lagostomus maximus*): ¿Cómo aumentar el área de pastoreo seguro? Ecol Austral 33:20–29
Roberts CM, Weir BJ (1973) Implantation in the plains viscacha, *Lagostomus maximus*. J Reprod Fertil 33:299–307
Schinz HR (1824 [1829]) Naturgeschichte und Abbildungen der Säugethiere. Brodtmanns lithographischer Kunstanstalt, Zurich, 417 pp
Schinz HR (1825) Das Thierreich eingetheilt nach dem Bau der Thiere als Grundlage iher Naturgeschichte und der vergleichenden Anatomie von dem Herrn Ritter von Cuvier, vol 4. Zoophyten. Gotta'schen, Stuttgart, 429 pp
Smith CH (1842) Mammalia. Introduction to mammals. In: Jardine W (ed) The naturalist's library, vol 13. W. H. Lizars, Edinburgh, p 313
Spotorno AE, Patton JL (2015) Superfamily Chinchilloidea Bennett, 1833. In: Patton JL, Pardiñas UFJ, D'Elía G (eds) Mammals of South America. 2. Rodents. The University of Chicago Press, Chicago, pp 762–783
Thomas O (1910) A collection of mammals from eastern Buenos Ayres, with descriptions of related new mammals from other localities. Ann Mag Nat Hist Lond Ser 8(5):239–247
Tullberg T (1899) Ueber das System der Nagethiere: eine phylogenetische Studie. Nova Acta Regiae Soc Sci Upsal 3:1–514
Vidal de Battini BE (1984) Cuentos y Leyendas Populares de la Argentina, VII. Ediciones Culturales Argentinas, Buenos Aires
Wiegmann AFA (1835) Eining Bermerkungen über das Chinchilla, vom Herausgeber. Arch Naturgesch 2:204–214
Wood AE (1955) A revised classification of the rodents. J Mammal 36:165–187

Chapter 2
Pan-Chinchillidae, a Once Diverse Clade

Luciano Luis Rasia

Abstract Pan-Chinchillidae is a clade of caviomorph rodents that includes extant chinchillas (*Chinchilla* spp.) and mountain vizcachas (*Lagidium* spp.) within Chinchillinae, and the plains vizcacha (*Lagostomus maximus*) as the sole living member of Lagostominae. Despite its relatively low diversity in recent times, the past diversity of the group was higher (8 extinct genera and nearly 20 extinct species) and with an ancient fossil record starting in the Early Oligocene. Stem chinchillids were recorded since the Early Oligocene to the Early Miocene in Argentina, Bolivia, and Chile and were represented by the extinct genera †*Eoviscaccia*, †*Garridomys*, †*Loncolicu*, †*Saremmys*, and †*Incamys*. Crown Chinchillidae are recorded from the Early to Middle Miocene, including both extant subfamilies, Chinchillinae (†*Miochinchilla*) and Lagostominae (†*Prolagostomus* and †*Pliolagostomus*). Among extant genera, *Lagostomus* has been recorded since the late Middle to Late Miocene, while *Lagidium* has been recorded since the Pleistocene and *Chinchilla* since the Holocene. Five main Diversification Events (DE) can be recognized in the evolutionary history of pan-Chinchillidae. The First DE (late Eocene/Early Oligocene) gave origin to the group. The Second DE (late Early to early Late Oligocene) and Third DE (earliest Early Miocene) are related to the diversification of stem chinchillids. The Fourth DE (late Early Miocene) represents the rising of crown Chinchillidae. First to Fourth DEs match with peaks of maximum cooling, followed by progressive warming intervals. The Fifth DE (Late Miocene) is marked by the origin of the extant genus *Lagostomus* and was probably related to a progressive cooling and the expansion of more arid and open environments in southern South America. The appearance of euhypsodonty in crown chinchillids during the Fourth DE, contrasting with previous protohypsodont stem chinchillids, may be related to the increase of volcanic activity in nearly all the Andean chain since the Late Oligocene. The maximum estimated body mass (MEBM) for the pan-chinchillids is quite stable until the Late Miocene. The MEBM of pan-chinchillids from the Early Oligocene to the Late Miocene was between 300

L. L. Rasia (✉)
División Paleontología Vertebrados, Museo de La Plata, La Plata, Argentina
e-mail: lucianorasia@fcnym.unlp.edu.ar

L. L. Rasia et al. (eds.), *Plains Vizcachas*, https://doi.org/10.1007/978-3-031-49487-1_2

and 1300 g. After the Late Miocene, the MEBM increased notably, reaching more than 5000 g in the Pliocene, 8000 g in the Late Pleistocene-Holocene, and ~9000 g with the extant plains vizcacha, *Lagostomus maximus*. The increase of the MEBM for the pan-Chinchillidae in the latest Late Miocene coincides with the global expansion of grasslands and the Age of the Southern Plains.

Keywords Chinchillidae · Evolution · Fossil record · Climate change · Cenozoic

2.1 Introduction

Caviomorph rodents represent a diverse clade with an ancient fossil record (e.g., Antoine et al. 2012; Arnal et al. 2020; Frailey and Campbell 2004). Four superfamilies are recognized among caviomorphs: Erethizontoidea, Cavioidea, Octodontoidea, and Chinchilloidea, and these clades were already differentiated since the late Eocene (e.g., Arnal et al. 2020). Chinchilloidea is one of the less diverse caviomorph groups in recent times, along with Erethizontoidea (e.g., Mares and Ojeda 1982; Patton et al. 2015), but the fossil record and evolutionary history of chinchilloids are remarkably rich (e.g., Rasia et al. 2021). Chinchilloids include the extant families Chinchillidae and Dinomyidae and the extinct Neoepiblemidae and Cephalomyidae (e.g., Bondesio et al. 1975; Vucetich 1985; Rasia and Candela 2018, 2019a; Rasia et al. 2021).

Modern chinchillids include several species of chinchillas (*Chinchilla* spp.) and mountain vizcachas (*Lagidium* spp.), grouped in the subfamily Chinchillinae, and the plains vizcacha (*Lagostomus maximus*) as the sole living representative of the subfamily Lagostominae (e.g., Jackson et al. 1996; Spotorno and Patton 2015; Spotorno and Valladares Faúndez 2016).

Pan-Chinchillidae (see Bertrand et al. 2012; Rasia et al. 2021) includes several extinct genera recorded in the Early Oligocene to Early Miocene of South America (e.g., Bertrand et al. 2012; Hoffstetter and Lavocat 1970; Vucetich 1989), conforming to the stem Chinchillidae (Rasia et al. 2021; see Fig. 2.1). Since the Early Miocene, the crown Chinchillidae have been recorded in South America, including the extant subfamilies Chinchillinae and Lagostominae (e.g., Ameghino 1887; Flynn et al. 2008; Kramarz 2002; Croft et al. 2021).

In this chapter, the evolutionary history of the pan-Chinchillidae is reviewed in the context of geotectonic and environmental changes that eventually lead to its present impoverished diversity.

2.2 Material and Methods

For the temporal framework (see Fig. 2.1), the South American Land Mammals Ages (SALMAs) system (e.g., Marshall et al. 1983; Pascual and Ortiz Jaureguizar 1990; Flynn and Swisher 1995) is used for the Oligocene-Middle Miocene time

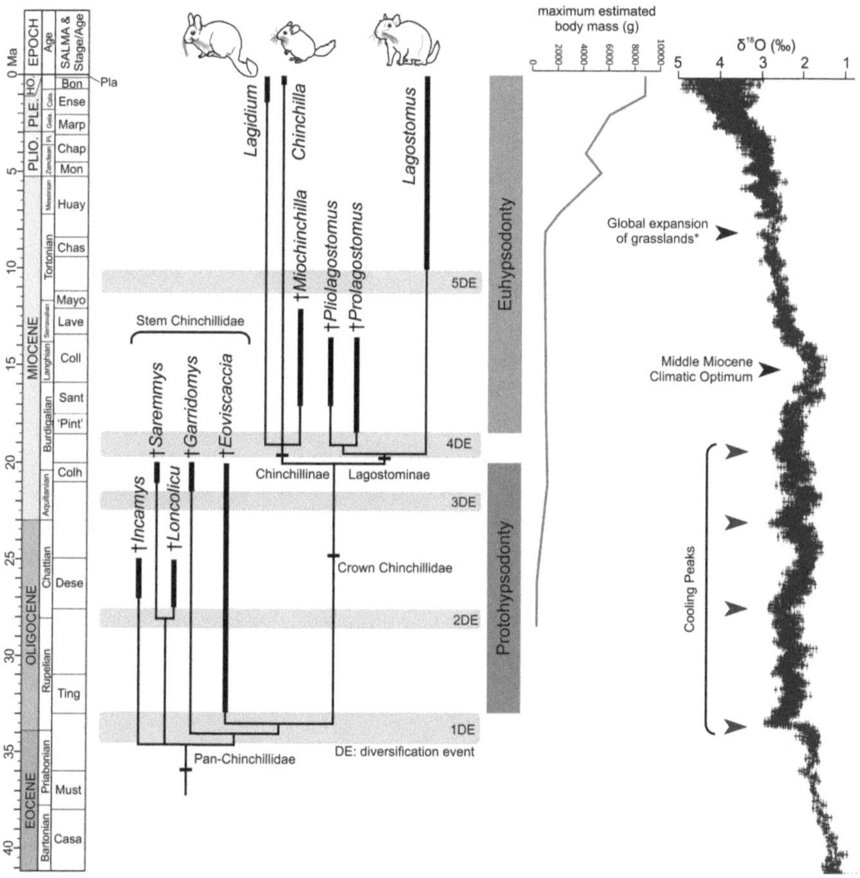

Fig. 2.1 Phylogenetic relationships of pan-Chinchillidae (modified from Rasia et al. (2021) and Croft et al. (2021)) showing changes in hypsodonty and body mass. Isotopic curve modified from Zachos et al. (2008). *Arakaki et al. (2011); Latorre et al. (1997)

period and the Stages/Ages system (e.g., 1995a, b, c, 1999, 2001, 2005) for the Late Miocene-Holocene (see Chap. 3 for details).

The cheek teeth of pan-chinchillids are sidewall hypsodont with balanced wear (Koenigswald 2011) but have a different degree of hypsodonty along their evolutionary history. Here, the degree of hypsodonty is divided into protohypsodont (high crowned but rooted) and euhypsodont (continuous growth, lacking roots), following Mones (1982), despite there are different degrees of protohypsodonty among stem chinchillids (e.g., Patterson and Wood 1982; Vucetich 1989; Vucetich et al. 2015; Busker et al. 2019).

For the body mass estimation, the equation of Croft (2000) was used, which is based on the length of the upper or lower first molar. Here is used only the length of upper molars, given that lower molars returned abnormally high values in living chinchillids (see Rasia 2016).

Fossil material deposited in the following institutions was revised: Museo de La Plata (MLP), La Plata, Argentina; Museo Argentino de Ciencias Naturales "Bernardino Rivadavia" (MACN), Ciudad de Buenos Aires, Argentina; Museo Provincial "Antonio Serrano" (MAS), Paraná, Argentina; and Centro de Investigación Científica y Transferencia Tecnología y Producción (CICYTTP), Diamante, Argentina.

2.3 Fossil Record

2.3.1 Origin and Diversification of the Stem Chinchillids

The oldest of the stem Chinchillidae is †*Eoviscaccia frassinettii* from the Early Oligocene (Tinguirirican SALMA) of Chile (Fig. 2.2; Bertrand et al. 2012).

In the Late Oligocene (Deseadan SALMA), †*Eoviscaccia boliviana* is recognized in the Lacayani Fauna in Bolivia (Fig. 2.2; Vucetich 1989), †*E. australis* is recognized in Deseadan levels of the Sarmiento Formation in Cabeza Blanca (Chubut Province, Argentina; Fig. 2.2; Vucetich 1989), †*Eoviscaccia* sp. is recorded in the Fray Bentos Formation at María Grande and Ávalos creeks (Corrientes Province, Argentina; Fig. 2.2; Bond et al. 1998), †*Incamys bolivianus* is recognized in Bolivia (Salla-Luribay; Fig. 2.2; Hoffstetter and Lavocat 1970; Patterson and Wood 1982), in Deseadan levels of the Sarmiento Formation in Cabeza Blanca (Chubut Province, Argentina; Fig. 2.2; Busker and Dozo 2017) and in the Agua de la Piedra Formation in Quebrada Fiera (Mendoza Province, Argentina; Fig. 2.2; Candela et al. 2021), †*Incamys meniorum* and †*Loncolicu tretos* are recorded in Deseadan levels of the Sarmiento Formation in Cabeza Blanca (Chubut, Argentina; Fig. 2.2; Vucetich et al. 2015).

By the earliest Early Miocene (Colhuehuapian SALMA), stem chinchillids are only recorded in Argentina. †*Eoviscaccia australis* is recorded in the Sarmiento Formation near Gaiman (Chubut Province; Fig. 2.2; Kramarz 2001; Vucetich et al. 2015), the Cerro Bandera Formation in Neuquén Province (Sierra del Portezuelo Norte and Cerro Bayo Mesa; Fig. 2.2; Kramarz et al. 2005), and Colhuehuapian levels at Gran Barranca (Río Negro Province; Fig. 2.2; Vucetich et al. 2010). †*Eoviscaccia* sp. is recorded in the Filo Morado Member of the Sierra Negra Formation in Neuquén Province (Sierra de Huantraicó-Sierra Negra and cerro Villegas; Fig. 2.2; Garrido et al. 2012), †*Saremmys ligcura* is recorded in the Trelew Member of the Sarmiento Formation in Chubut Province (Bryn Gwyn and Gran Barranca; Fig. 2.2; Busker et al. 2019), and †*Garridomys curunuquem* is recognized in the Cerro Bandera Formation in Neuquén Province (NE slope of Barda Negra; Fig. 2.2; Kramarz et al. 2013).

Fig. 2.2 Map of South America showing fossil localities. (1) Tinguiririca; (2) Lacayani; (3) Cabeza Blanca; (4) María Grande and Ávalos creeks; (5) Salla-Luribay; (6) Quebrada Fiera; (7) Gaiman; (8) Sierra del Portezuelo Norte/Cerro Bayo Mesa; (9) Gran Barranca; (10) Sierra de Huantraicó-Sierra Negra and cerro Villegas; (11) Bryn Gwyn; (12) Barda Negra; (13) Chucal; (14) Quebrada Honda; (15) Nazareno; (16) Pinturas river; (17) Santa Cruz river; (18) Collón Curá river; (19) Senguerr river; (20) Laguna del Laja; (21) Ingeniero Jacobacci; (22) Cerro Boleadoras; (23) Las Hornillas/El Horcajo; (24) Cañadón del Tordillo; (25) Pilcaniyeu Viejo; (26) Arroyo Chasicó; (27) Ayacucho; (28) Casa del Diablo; (29) Mylodon Cave, (30) Guitarrero Cave Complex; (31) Tebenquiche Chico; (32) Lago Sofía 1 and 4 caves

2.3.2 Rise of Crown Chinchillidae

In the late Early Miocene, both subfamilies comprising the crown Chinchillidae were already distinguishable, with extinct genera of Chinchillinae and Lagostominae.

The Chinchillinae are represented in the Early to Middle Miocene of Chile and Bolivia (e.g., Flynn et al. 2002; Catena and Croft 2020; Croft et al. 2021). †*Miochinchilla surirense* is recognized in the Chucal Formation (Santacrucian SALMA), in northern Chile (Chucal; Fig. 2.2), and †*M. plurinacionalis* is recorded in the Honda Group (Laventan SALMA) in southern Bolivia (Quebrada Honda; Fig. 2.2), and in the Nazareno Formation (Colloncuran? SALMA) in south-central Bolivia (Nazareno; Fig. 2.2; Croft et al. 2021).

Lagostominae are represented in the Early to Middle Miocene by two genera, †*Prolagostomus* (recorded in Argentina, Bolivia, and Chile) and †*Pliolagostomus* (recognized only in Argentina). The oldest record of †*Prolagostomus* came from the Pinturas Formation ("Pinturan" SALMA) in the Pinturas river (Santa Cruz Province, Argentina; Fig. 2.2) and was identified as †*Prolagostomus* sp. but considering it a different, more primitive species than other known ones (Kramarz 2002). Several species of *Prolagostomus* were originally described from the Santa Cruz Formation (Santacrucian SALMA) in the Santa Cruz river (Santa Cruz Province, Argentina; Fig. 2.2; Ameghino 1887, 1889; Scott 1905), but recent works only consider †*Prolagostomus pusillus* as valid (e.g., Rasia 2016; Arnal et al. 2019). †*Prolagostomus pusillus* was also recorded in the Collón Curá Formation (Colloncuran SALMA) at the Collón Curá river (Neuquén Province; Argentina; Fig. 2.2), in the Río Frías Formation at the Senguerr river (Chubut Province; Argentina; Fig. 2.2; Vucetich 1984), and in the Pampa Castillo fauna (Santacrucian SALMA) in southern Chile (Laguna del Laja; Fig. 2.2; McGrath et al. 2022). †*Prolagostomus rosendoi* was described for the Collón Curá Formation (Colloncuran SALMA) near Ingeniero Jacobacci (Río Negro Province, Argentina; Fig. 2.2; Vucetich 1984). Other records of †*Prolagostomus* in Argentina include the Carrere Formation (Early Miocene) in Sierra de Huantraicó (Neuquén Province, Argentina; Fig. 2.2; Garrido et al. 2009), Cerro Boleadoras Formation (Early Miocene) in Santa Cruz Province (Cerro Boleadoras; Fig. 2.2; Vucetich 1994), Chinches Formation (Santacrucian SALMA) in San Juan Province (between Las Hornillas and El Horcajo; Fig. 2.2; López et al. 2011), and Collón Curá Formation in Neuquén Province (Cañadón del Tordillo, Fig. 2.2; Vucetich et al. 1993). Furthermore, †*Prolagostomus* was reported for the Cura-Mallín Formation (Early-Middle Miocene) in Chile (Flynn et al. 2008) and in Bolivia for the Nazareno Formation (Colloncuran SALMA; Fig. 2.2; see Marshall and Sempere 1991) and Quebrada Honda (Fig. 2.2; e.g., Frailey 1981; Croft et al. 2011).

†*Pliolagostomus notatus* was described for the Santa Cruz Formation (Santacrucian SALMA) in the Santa Cruz river (Santa Cruz Province, Argentina; Fig. 2.2; Ameghino 1887; Rasia and Candela 2019b), and †*Pliolagostomus friasensis* for the Collon Curá Formation (Colloncuran SALMA) in the Collón Curá river

(Neuquén Province, Argentina; Fig. 2.2) and Ingeniero Jacobacci and Pilcaniyeu Viejo (Río Negro Province, Argentina; Fig. 2.2; Vucetich 1984).

2.3.3 Record of Modern Chinchillids

The oldest records of the extant genus *Lagostomus* correspond to late Middle-Late Miocene records of *Lagostomus* sp. from La Pampa Province, Argentina (Montalvo et al. 2023). The oldest species of the genus is †*L. telenkechanum*, from Late Miocene (Chasicoan Stage/Age) sediments of the Chasicó Formation at Arroyo Chasicó (Buenos Aires Province, Argentina; Fig. 2.2; Rasia and Candela 2017a), and Huachipampa Formation in San Juan Province, Argentina (Rasia et al. 2020). By the Huayquerian Stage/Age (Late Miocene), *Lagostomus* was widely spread in southern South America, with records in Bolivia, Argentina, and Uruguay (e.g., Rovereto 1914; Francis and Mones 1965; Marshall and Sempere 1991). During the Pliocene (Montermosan to Chapadmalalan stages/ages), there are records of *Lagostomus* in Argentina, Bolivia, and Uruguay (e.g., Francis and Mones 1966; Marshall and Sempere 1991; Rasia 2021; Rasia and Candela 2013, 2017b). In the Pleistocene to Holocene, *Lagostomus* is recognized in Argentina, Bolivia, Uruguay, and southern Brazil (e.g., Ameghino 1889; Marshall and Sempere 1991; Kerber et al. 2011; Ubilla and Rinderknecht 2016). For a detailed account of the fossil record of *Lagostomus*, see Chap. 3.

Among chinchillines, the extant genus *Lagidium* is recorded since the Pleistocene in caves of the Altiplano in Peru (e.g., Marshall et al. 1984; Walton 1997), including the Ayacucho valley (Fig. 2.2; e.g., Pujos and Salas 2004) and Casa del Diablo cave (Fig. 2.2; Villavicencio and Werdelin 2018), and in the Mylodon Cave in southern Chile (Fig. 2.2; Pardiñas et al. 2021), with Holocene records in Peru (Guitarrero Cave complex; Fig. 2.2; recognized as *Lagidium peruanum*; see Wing 1980), Argentina (Tebenquiche Chico, Catamarca Province; Fig. 2.2; Ortiz et al. 2012), and southern Chile (Lago Sofía 1 and 4 caves; Fig. 2.2; recognized as *Lagidium viscacia*; see Gasparini and Tonni 2022). *Chinchilla* is reported in the Holocene of Argentina as *Chinchilla chinchilla* (Tebenquiche Chico; Fig. 2.2; Ortiz et al. 2012).

2.4 Evolutionary History

2.4.1 Main Events in Pan-Chinchillid Evolution

Five main Diversification Events (DE) can be recognized in the evolutionary history of pan-Chinchillidae (see Fig. 2.1). The First DE would have occurred in the late Eocene/Early Oligocene, giving origin to the group, with the oldest recognized pan-chinchillid in the Early Oligocene, †*Eoviscaccia* (see Rasia et al. 2021). The Second

DE occurred in the late Early to early Late Oligocene and is related to the diversification of stem chinchillids, with the appearance of several taxa such as †*Incamys* and †*Loncolicu* (see Rasia et al. 2021). The third event mentioned by Rasia et al. (2021) represents actually two separate events, the appearance of several taxa of stem chinchillids such as †*Saremmys* and †*Garridomys* in the earliest Early Miocene constitutes the Third DE, and the posterior rising of crown Chinchillidae (after the extinction of stem chinchillids) in the late Early Miocene represents the Fourth DE, with records of †*Miochinchilla*, †*Prolagostomus*, and †*Pliolagostomus*. The Fifth DE is marked by the appearance of the extant genus *Lagostomus* in the Late Miocene.

2.4.2 Hypsodonty

The degree of hypsodonty of pan-chinchillids shows variation along their evolutionary history (see Fig. 2.1; Table 2.1). The oldest members of the group (Early Oligocene-earliest Early Miocene), the stem chinchillids †*Eoviscaccia*, †*Incamys*, †*Saremmys*, †*Loncolicu*, and †*Garridomys*, are protohypsodont. Later (Late Early Miocene to recent times), the crown chinchillids (i.e., Lagostominae+Chinchillinae) developed euhypsodonty.

2.4.3 Body Mass

The maximum estimated body mass (MEBM) for the pan-chinchillids is quite stable since the Early Oligocene until the Late Miocene (see Fig. 2.1; Table 2.1). The body mass of stem chinchillids (Early Oligocene-Early Miocene) was between 300 and 1300 g. During the Early Miocene-Middle Miocene, the earliest representatives of Chinchillinae and Lagostominae weighted between 800 and 1100 g. The body mass of the oldest species of *Lagostomus* (Late Miocene) was between 1000 and 2200 g. After the Late Miocene, the maximum body mass of Lagostominae increases notably, reaching more than 5000 g in the Pliocene and 8000 g in the Late Pleistocene-Holocene. Extant chinchillids reach the maximum body weight with *Lagostomus maximus* (8800 g), with the smaller chinchillines weighing between 450 and 2000 g (Jackson et al. 1996; Spotorno et al. 2004; Spotorno and Valladares Faúndez 2016; Ledesma et al. 2009; Valladares et al. 2018).

Table 2.1 Age, maximum estimated body mass, and degree of hypsodonty of pan-Chinhillidae species

Species	Age	HypDeg	MEBM (g)
Chinchilla chinchilla	Recent	Eu	850[a]
Chinchilla laniger	Recent	Pr	450[a]
†*Eoviscaccia australis*	Colhuehuapian	Pr	317
†*Eoviscaccia australis*	Deseadan	Pr	370
†*Eoviscaccia boliviana*	Deseadan	Pr	326
†*Garridomys curunuquem*	Colhuehuapian	Pr	610
Lagidium ahuacaense	Recent	Eu	2030[a]
Lagidium peruanum	Recent	Eu	1400[a]
Lagidium viscaccia	Recent	Eu	2950[a]
Lagidium wolffsohni	Recent	Eu	3000[a]
†*Lagostomus antiquus*	Huayquerian	Eu	1054
†*Lagostomus compressidens*	Chapadmalalan	Eu	4222
†*Lagostomus euplasius*	Montehermosan	Eu	1687
†*Lagostomus euplasius*	Chapadmalalan	Eu	2179
†*Lagostomus incisus*	Montehermosan	Eu	5435
†*Lagostomus incisus*	Chapadmalalan	Eu	4255
Lagostomus maximus	Recent	Eu	8800[a]
Lagostomus maximus	Ensenadan	Eu	6074
Lagostomus maximus	Platan	Eu	8808
†*Lagostomus pretrichodactyla*	Huayquerian	Eu	2224
†*Lagostomus telenkechanum*	Chasicoan	Eu	1003
†*Loncolicu tretos*	Deseadan	Pr	496
†*Miochinchilla plurinacionalis*	Colloncuran	Eu	1140
†*Miochinchilla surirense*	Santacrucian	Eu	1140
†*Pliolagostomus friasensis*	Colloncuran	Eu	1138
†*Pliolagostomus notatus*	Santacrucian	Eu	361
†*Prolagostomus pusillus*	Santacrucian	Eu	1116
†*Prolagostomus rosendoi*	Colloncuran	Eu	782
†*Saremmys ligcura*	Colhuehuapian	Pr	1293

Eu euhypsodonty, *HypDeg* hypsodonty degree, *MEBM* maximum estimated body mass, *Pr* proto-hypsodonty
[a]Maximum body weight (see text for references)

2.5 Discussion and Conclusions

The first four DE of pan-Chinchillidae match with peaks of maximum cooling, followed by progressive warming intervals (see Fig. 2.1). The Fifth DE, during the Late Miocene, was probably related to a progressive cooling and to the expansion of more arid and open environments in southern South America (Pascual and Ortiz Jaureguizar 1990), a time interval known as "Edad de las Planicies Australes" (Age of the Southern Plains; see Ortiz Jaureguizar 1998; Ortiz Jaureguizar and Cladera 2006).

Euhypsodonty was traditionally linked to grazing habits (e.g., Stebbins 1981; Jacobs et al. 1999), but it has been addressed that it can be related to grit consumption in the diet, not only to grasslands (e.g., Janis 1988; Jardine et al. 2012). The appearance of euhypsodonty in pan-chinchillids during the Early Miocene (which occurred in the Fourth DE), previous to the development of extended grasslands, may be related to the increase of volcanic activity and consequently of volcanic ash. During the Late Oligocene (around 27 Ma), the breakup of the Farallon plate into the Cocos and Nazca plates was a major tectonic event in the evolution of the Andes that resulted in an increase of the volcanic activity along nearly all the Andean chain (e.g., Stern 2004; Trumbull et al. 2006). Grasslands were not widespread in South America before the Late Miocene (e.g., Palazzesi and Barreda 2012; Palazzesi et al. 2014; Barreda and Palazzesi 2021; Catena and Croft 2020); nevertheless, there are records of grass-dominated environments in previous intervals (e.g., Bellosi and González 2010; Cantil et al. 2020).

The drastic increase in the maximum estimated body mass for the pan-Chinchillidae in the latest Late Miocene (Fig. 2.1) coincides with the global expansion of grasslands composed primarily of C4 grasses (e.g., Arakaki et al. 2011; Latorre et al. 1997).

In sum, the diversification, the development of euhypsodonty, and the increase of body mass of pan-chinchillids can be correlated with changes in global temperature and major tectonic events, which in turn affected plant diversity and composition (e.g., Barreda and Palazzesi 2021). Moreover, the concept of pan-Chinchillidae and its taxonomic composition are relatively recent (e.g., Bertrand et al. 2012; Rasia et al. 2021), and there is still a need for testing the phylogenetic position of several Chinchilloidea without a clear position. Future studies would show more accurately the evolutionary history of pan-chinchillids.

Acknowledgments I want to thank the curators (Marcelo Reguero, MLP; Alejo Scarano, MLP; Martin Ezcurra, MACN; Agustin Martinelli, MACN) for allowing access to material under their care. This study is part of the project Red Vizcachera.

References

Ameghino F (1887) Enumeración sistemática de las especies de mamíferos fósiles coleccionados por Carlos Ameghino en los terrenos eocenos de Patagonia austral y depositados en el Museo de La Plata. Bol Mus La Plata 1:1–26

Ameghino F (1889) Contribución al conocimiento de los mamíferos fósiles de la República Argentina. Actas Acad Nac Cienc Córdoba 6:1–1027

Antoine P-O, Marivaux L, Croft DA, Billet G, Ganerød M, Jaramillo C, Martin T, Orliac MJ, Tejada J, Altamirano AJ, Duranthon F, Fanjat G, Rousse S, Salas Gismondi R (2012) Middle Eocene rodents from Peruvian Amazonia reveal the pattern and timing of caviomorph origins and biogeography. Proc Biol Sci 279:1319–1326

Arakaki M, Christin P-A, Nyffeler R, Lendel A, Eggli U, Ogburn RM, Spriggs E, Moore MJ, Edwards EJ (2011) Contemporaneous and recent radiations of the world's major succulent plant lineages. Proc Natl Acad Sci U S A 108:8379–8384

Arnal M, Pérez ME, Deschamps CM (2019) Revision of the Miocene caviomorph rodents from the Río Santa Cruz (Argentinean Patagonia). Publ Electron Paleontol Asoc 19(2):193–229

Arnal M, Kramarz AG, Vucetich MG, Frailey CD, Campbell KE Jr (2020) New Palaeogene caviomorphs (Rodentia, Hystricognathi) from Santa Rosa, Peru: systematics, biochronology, biogeography and early evolutionary trends. Pap Palaeontol 6:193–216

Barreda VD, Palazzesi L (2021) Role of climate and tectonism on the modernization of Patagonian floras: evidence from the fossil record. Glob Planet Chang 204:103556. https://doi.org/10.1016/j.gloplacha.2021.103556

Bellosi E, González M (2010) Paleosols of the middle Cenozoic Sarmiento Formation, central Patagonia. In: Madden R, Carlini A, Vucetich M, Kay R (eds) The paleontology of Gran Barranca: evolution and environmental change through the middle Cenozoic of Patagonia. Cambridge University Press, Cambridge, pp 293–305

Bertrand OC, Flynn JJ, Croft DA, Wyss AR (2012) Two new taxa (Caviomorpha, Rodentia) from the early Oligocene Tinguiririca Fauna (Chile). Am Mus Novit 3750:1–36

Bond M, López G, Reguero MA, ScillatoYané GJ, Vucetich MG (1998) Los mamíferos de la Formación Fray Bentos (Edad Mamífero Deseadense, Oligoceno superior?) de las provincias de Corrientes y Entre Ríos. In: Paleógeno de América del Sur y de la Península Antártica, Publicación Especial 5. Asociación Paleontológica Argentina, Buenos Aires, pp 41–50

Bondesio P, Pascual R, Vucetich MG (1975) Los Neoepiblemidae (Rodentia, Caviomorpha): su caracterización y sus relaciones filogenéticas con los Chinchillidae. Actas del Primer Congreso Argentino de Paleontología y Bioestratigrafía 2:431–447

Busker F, Dozo MT (2017) First confirmed record of *Incamys bolivianus* (Caviomorpha, Chinchilloidea) in the Deseadan of Patagonia (Argentina). Ameghiniana 54:706–712

Busker F, Pérez ME, Dozo MT (2019) A new chinchilloid (Rodentia, Hystricognathi) from the early Miocene of the localities of Bryn Gwyn and Gran Barranca (Patagonia, Argentina). C R Palevol 18(5):525–540

Candela AM, Pérez ME, Rasia LL, Cerdeño E (2021) New late Oligocene caviomorph rodents from Mendoza Province, central-western Argentina. J Vertebr Paleontol 41(2):e1929264. https://doi.org/10.1080/02724634.2021.1929264

Cantil LF, Bellosi ES, Sánchez MV, González MG, Genise JF (2020) The earliest burst of necrophagous dung beetles in South America revealed by the Cenozoic record of *Coprinisphaera*. Lethaia 53:421–438

Catena AM, Croft DA (2020) What are the best modern analogs for ancient South American mammal communities? Evidence from ecological diversity analysis (EDA). Palaeontology 23(1):a03

Cione AL, Tonni EP (1995a) Los estratotipos de los pisos Montehermosense y Chapadmalalense (Plioceno) del esquema cronológico sudamericano. Ameghiniana 32:369–374

Cione AL, Tonni EP (1995b) Chronostratigraphy and land-mammal ages in the Cenozoic of southern South America: principles, practices, and the Uquian problem. J Paleontol 69:135–159

Cione AL, Tonni EP (1995c) Bioestratigrafía y cronología del Cenozoico superior de la región pampeana. In: Alberdi MT, Leone G, Tonni EP (eds) Evolución biológica y climática de la región pampeana durante los últimos cinco millones de años, un ensayo de correlación con el Mediterráneo occidental, Monografias 12. Museo Nacional de Ciencias Naturales, Madrid, pp 47–74

Cione AL, Tonni EP (1999) Biostratigraphy and chronological scale of upper-most Cenozoic in the Pampean area, Argentina. In: Quaternary of South America and Antarctic Peninsula. Balkema, Rotterdam, pp 23–51

Cione AL, Tonni EP (2001) Correlation of Pliocene to Holocene southern South American and European vertebrate-bearing units. Boll Soc Paleontol Ital 40:167–173

Cione AL, Tonni EP (2005) Bioestratigrafía basada en mamíferos del Cenozoico superior de la provincia de Buenos Aires, Argentina. In: de Barrio RE, Etcheverry RO, Caballé MF, Llambías E (eds) Geología y Recursos Minerales de la Provincia de Buenos Aires. Relatorio del XVI

Congreso Geológico Argentino, La Plata. Asociación Geológica Argentina, Buenos Aires, pp 183–200
Croft D (2000) Archaeohyracidae (Mammalia: Notoungulata) from the Tinguiririca Fauna, central Chile, and the evolution and paleoecology of South American mammalian herbivores. PhD Dissertation, University of Chicago, Chicago, USA
Croft DA, Chick JMH, Anaya F (2011) New middle Miocene caviomorph rodents from Quebrada Honda, Bolivia. J Mamm Evol 18:245–268
Croft DA, Flynn JJ, Wyss AR, Charrier R, Anaya F (2021) New chinchillid rodents (Hystricognathi: Caviomorpha) from northern Chile and Bolivia fill a 17-million-year gap in the pan-chinchilline fossil record. J Mamm Evol 28:1205–1236
Flynn JJ, Swisher CC III (1995) Cenozoic South American Land Mammal Ages: Correlation to global geochronologies. In: Berggren WA, Kent DV, Aubry M-P, Hardenbol J (eds) Geochronology, time scales, and global stratigraphic correlation, SEPM (Society for Sedimentary Geology) special publication no. 54. SEPM, Tulsa, pp 317–333
Flynn JJ, Croft DA, Charrier R, Hérail G, Wyss AR (2002) The first Cenozoic mammal fauna from the chilean Altiplano. J Vertebr Paleontol 22:200–206
Flynn JJ, Charrier R, Croft DA, Gans PB, Herriott TM, Wertheim JA, Wyss AR (2008) Chronologic implications of new Miocene mammals from the Cura-Mallín and Trapa Trapa formations, Laguna del Laja area, south central Chile. J S Am Earth Sci 26:412–423
Frailey CD (1981) Studies on the Cenozoic Vertebrata of Bolivia and Peru. PhD Dissertation, University of Kansas, Kansas, USA
Frailey CD, Campbell KE Jr (2004) Paleogene rodents from Amazonian Peru: the Santa Rosa Local Fauna. In: Campbell KE Jr (ed) The Paleogene mammalian fauna of Santa Rosa, Amazonian Peru, Science series 40. Natural History Museum of Los Angeles County, Los Angeles, pp 71–130
Francis JC, Mones A (1965) La presencia de vizcachas [*Lagostomus* (*Lagostomopsis*) *spicatus* (Amegh.)] en la Formación Kiyú, Dto. De San José, R. O. del Uruguay. Rev Fac Humanid Cienc Montevideo 22:155–168
Francis JC, Mones A (1966) Las vizcachas, *Lagostomus* (*Lagostomopsis*) *euplasius* (Amegh.), de la formación Maldonado, departamento de Maldonado, República Oriental del Uruguay. Kraglieviana 1:101–110
Garrido A, Kramarz A, Forasiepi A, Bond M, Chornogubsky L, Martinelli A (2009) Mamíferos fósiles y edad de la Formación Carrere (Provincia del Neuquén). In: XXIV Jornadas Argentinas de Paleontología de Vertebrados. San Rafael, Mendoza
Garrido A, Kramarz A, Forasiepi A, Bond M (2012) Estratigrafía, mamíferos fósiles y edad de las secuencias volcanosedimentarias eoceno-miocenas de la sierra de Huantraico-sierra Negra y cerro Villegas (provincia del Neuquén, Argentina). Andean Geol 39(3):482–510
Gasparini GM, Tonni EP (2022) Quaternary fossil vertebrates of Tierra del Fuego and southernmost Patagonia. Chapter 5. In: Miotti L, Salemme M, Hermo D (eds) Archaeology of Piedra Museo locality: an open window to the early population of Patagonia. Springer, Cham, pp 127–157
Hoffstetter R, Lavocat R (1970) Découverte dans le Déséadien de Bolivie des genres pentalophodontes appuyant les affinités africaines des Rongeurs Caviomorphes. C R Acad Sci Paris 271:172–175
Jackson JE, Branch LC, Villareal D (1996) *Lagostomus maximus*. Mamm Species 543:1–6
Jacobs BF, Kingston JD, Jacobs LL (1999) The origin of grass-dominated ecosystems. Ann Mo Bot Gard 86:590–643
Janis CM (1988) Estimation of tooth volume and hypsodonty indices in ungulate mammals, and the correlation of these factors with dietary preference. In: Russell E, Santoro JP, Signogneau-Russell D (eds) Teeth revisited: proceedings VII international symposium on dental morphology, Mémoires du Muséum national d'Histoire naturelle, Sér. C – Sciences de la Terre 53. Muséum national d'Histoire naturelle, Paris, pp 367–387

Jardine PE, Janis CM, Sahney S, Benton MJ (2012) Grit not grass: concordant patterns of early origin of hypsodonty in Great Plains ungulates and Glires. Palaeogeogr Palaeoclimatol Palaeoecol 365–366:1–10

Kerber L, Pereira Lopes R, Vucetich MG, Ribeiro AM, Pereira J (2011) Chinchillidae and Dolichotinae rodents (Rodentia: Hystricognathi: Caviomorpha) from the late Pleistocene of Southern Brazil. Rev Bras Paleontol 14:229–238

Koenigswald W (2011) Diversity of hypsodont teeth in mammalian dentitions – construction and classification. Palaeontogr Abt A 294(1–3):63–94

Kramarz AG (2001) Registro de *Eoviscaccia* (Rodentia, Chinchillidae) en estratos colhuehuapenses de Patagonia, Argentina. Ameghiniana 38:237–242

Kramarz AG (2002) Roedores chinchilloideos (Hystricognathi) de la Formación Pinturas, Mioceno temprano-medio de la provincia de Santa Cruz, Argentina. Rev Mus Argent Cienc Nat Nueva Ser 4:167–180

Kramarz AG, Garrido A, Forasiepi A, Bond M, Tambussi C (2005) Estratigrafía y vertebrados (Aves y Mammalia) de la Formación Cerro Bandera, Mioceno Temprano de la Provincia del Neuquén, Argentina. Rev Geol Chile 32(2):273–291

Kramarz AG, Vucetich MG, Arnal M (2013) A new early Miocene chinchilloid hystricognath rodent; an approach to the understanding of the early chinchillid dental evolution. J Mamm Evol 20:249–261

Latorre C, Quade J, McIntoch WC (1997) The expansion of C4 grasses and global in the late Miocene: stable isotope evidence from the Americas. Earth Planet Sci Lett 146:83–96

Ledesma KJ, Werner FA, Spotorno AE, Albuja LH (2009) A new species of mountain viscacha (Chinchillidae: *Lagidium* Meyen) from the Ecuadorean Andes. Zootaxa 2126:41–57

López GM, Vucetich MG, Carlini AA, Bond M, Pérez ME, Ciancio MR, Pérez DJ, Arnal M, Olivares AI (2011) New Miocene mammal assamblages from Noegene Manantiales basin, Cordillera frontal, San Juan, Argentina. In: Salfity JA, Marquillas RA (eds) Cenozoic geology of the central Andes of Argentina. SCS Publisher, Salta, pp 211–226

Mares MA, Ojeda RA (1982) Patterns of diversity and adaptation of South American hystricognathous rodents. In: Mares MA, Genoways H (eds) Mammalian biology in South America, Special publication, Pymatuning Laboratory of Ecology. University of Pittsburg, Pittsburg, pp 393–432

Marshall LG, Sempere T (1991) The Eocene to Pleistocene vertebrates of Bolivia and their stratigraphic context: a review. In: Suarez-Soruco R (ed) Fósiles y Facies de Bolivia, Volúmen I-Vertebrados, Revista Técnica de YPFB 12. YPEB, Bolivia, pp 631–652

Marshall LG, Hoffstetter R, Pascual R (1983) Mammals and stratigraphy: geochronology of the continental mammal-bearing Tertiary of South America. Palaeovertebrata, Memoire Extraordinaire 13:1–93

Marshall LG, Berta A, Hoffstetter R, Pascual R, Reig OA, Bombin M, Mones A (1984) Mammals and stratigraphy: geochronology of the continental mammal-bearing Quaternary of South America. Palaeovertebrata, Mémoire Extraordinaire 14:1–76

McGrath AJ, Chick J, Croft DA, Dodson HE, Flynn JJ, Wyss AR (2022) Cavioids, Chinchilloids, and Erethizontoids (Hystricognathi, Rodentia, Mammalia) of the early Miocene Pampa Castillo Fauna, Chile. Am Mus Novit 3984:1–46

Mones A (1982) An equivocal nomenclature: what means hypsodonty? Palaeontol Z 56:107–111

Montalvo CI, Folguera A, Cerdeño E, Verzi DH, Pérez ME, Sostillo R, Tomassini RL (2023) The oldest Andean synorogenic fossiliferous deposits of the Miocene Cerro Azul Formation, central Argentina. J S Am Earth Sci 123:104235. https://doi.org/10.1016/j.jsames.2023.104235

Ortiz Jaureguizar E (1998) Paleoecología y evolución de la fauna de mamíferos de América del Sur durante la "Edad de las Planicies Australes" (Mioceno Superior-Plioceno Superior). Estud Geol 54:161–169

Ortiz Jaureguizar E, Cladera GA (2006) Paleoenvironmental evolution of southern South America during the Cenozoic. J Arid Environ 66:498–532

Ortiz PE, Jayat JP, Nasif NL, Teta P, Haber A (2012) Roedores del Holoceno tardío de la Puna de Atacama, sitio arqueológico Tebenquiche Chico, Catamarca, Argentina. Archaeofauna 21:249–266
Palazzesi L, Barreda V (2012) Fossil pollen records reveal a late rise of open-habitat ecosystems in Patagonia. Nat Commun 3:1–6
Palazzesi L, Barreda VD, Cuitiño JI, Guler MV, Tellería MC, Ventura Santos R (2014) Fossil pollen records indicate that Patagonian desertification was not solely a consequence of Andean uplift. Nat Commun 5:1–8
Pardiñas UFJ, Borrero L, Martin FM, Massone M, Fernández FJ (2021) Fossil rodents in Mylodon Cave as indicators of late Pleistocene–Holocene environmental evolution in southern Chile. Quat Res 105:218–234
Pascual R, Ortiz Jaureguizar E (1990) Evolving climates and mammal faunas in Cenozoic South America. J Hum Evol 19(1–2):23–60
Patterson B, Wood AE (1982) Rodents from the Deseadan Oligocene of Bolivia and the relationships of the Caviomorpha. Bull Mus Comp Zool 149:371–543
Patton JL, Pardiñas UF, D'Elía G (eds) (2015) Mammals of South America, Volume 2: Rodents. University of Chicago Press, Chicago
Pujos F, Salas R (2004) A systematic reassessment and paleogeographic review of fossil Xenarthra from Peru. Bull Inst Fr Etudes Andin 33:331–377
Rasia LL (2016) Los Chinchillidae (Rodentia, Caviomorpha) fósiles de la República Argentina: sistemática, historia evolutiva y biogeográfica, significado bioestratigráfico y paleoambiental. PhD Dissertation, Universidad Nacional de La Plata, La Plata, Argentina
Rasia LL (2021) Reassessing the fossil record of *Lagostomus incisus* Ameghino, 1888 (Rodentia, Caviomorpha) from the late Neogene of southern South America. Publ Electron Paleontol Asoc 21(2):53–61
Rasia LL, Candela AM (2013) Systematic and biostratigraphic significance of a chinchillid rodent from the Pliocene of eastern Argentina. Acta Palaeontol Pol 58:241–254
Rasia LL, Candela AM (2017a) *Lagostomus telenkechanum* sp. nov., a new lagostomine rodent (Caviomorpha, Chinchillidae) from the Arroyo Chasicó Formation (late Miocene; Buenos Aires Province, Argentina). J Vertebr Paleontol 37:e1239205. https://doi.org/10.1080/02724634.2017.1239205
Rasia LL, Candela AM (2017b) Systematic revision of the vizcachas (Rodentia, Caviomorpha, Chinchillidae) from the Chapadmalal Formation, late Pliocene of Buenos Aires Province, Argentina. Ameghiniana 54:50–69
Rasia LL, Candela AM (2018) Reappraisal of the giant caviomorph rodent *Phoberomys burmeisteri* (Ameghino, 1886) from the late Miocene of northeastern Argentina, and the phylogeny and diversity of Neoepiblemidae. Hist Biol 30:486–495
Rasia LL, Candela AM (2019a) Upper molar morphology, homologies and evolutionary patterns of chinchilloid rodents (Mammalia, Caviomorpha). J Anat 234:50–65
Rasia LL, Candela AM (2019b) "*Prolagostomus amplus*" Ameghino, 1894 is a junior synonym of the Patagonian rodent *Pliolagostomus notatus* Ameghino, 1887 (Chinchillidae; Early Miocene, Santa Cruz Formation). Ameghiniana 56:72–77
Rasia LL, Bonini RA, Candela AM (2020) Nuevos registros de *Lagostomus* Brookes (Rodentia, Chinchillidae) en el Mioceno tardío de Argentina y su importancia bioestratigráfica. Andean Geol 47(2):430–445
Rasia LL, Candela AM, Cañón C (2021) Comprehensive total evidence phylogeny of chinchillids (Rodentia, Caviomorpha): cheek teeth anatomy and evolution. J Anat 239(2):405–423
Rovereto C (1914) Los estratos araucanos y sus fósiles. Anales Mus Nac Hist Nat Buenos Aires 25:1–247
Scott WB (1905) Mammalia of the Santa Cruz beds. Vol. 5, Paleontology. Parte 3, Glires. In: Scott WB (ed) Reports of the Princeton University expeditions to Patagonia, 1896–1899. Princeton University, E. Schweizerbart'sche Verlagshandlung (E. Nägele), Stuttgart, pp 384–487

Spotorno AE, Patton JL (2015) Superfamily Chinchilloidea Bennett, 1833. In: Patton JL, Pardiñas UFJ, D'Elía G (eds) Mammals of South America. 2. Rodents. The University of Chicago Press, Chicago, pp 762–783

Spotorno AE, Valladares Faúndez J (2016) Family Chinchillidae. In: Wilson DE, Lacherc TEJ, Mittermeier RA (eds) Handbook of the mammals of the world, Lagomorphs and rodents I, vol 6. Lynx Edicions, Barcelona, pp 462–481

Spotorno AE, Zuleta CA, Valladares JP, Deane AL, Jiménez JE (2004) *Chinchilla laniger*. Mamm Species 758:1–9

Stebbins GL (1981) Coevolution of grasses and herbivores. Ann Mo Bot Gard 68:75–86

Stern CR (2004) Active Andean volcanism: its geologic and tectonic setting. Rev Geol Chile 31(2):161–206

Trumbull RB, Riller U, Oncken O, Scheuber E, Munier K, Hongn F (2006) The time-space distribution of Cenozoic volcanismin the south-central Andes: a new data compilation and some tectonic implications. Chapter 2. In: Oncken O, Chong G, Franz G, Giese P, Götze H-J, Ramos VA, Strecker MR, Wigger P (eds) The Andes: active subduction orogeny. Springer, Berlin, pp 29–43

Ubilla M, Rinderknecht A (2016) *Lagostomus maximus* (Desmarest) (Rodentia, Chinchillidae), the extant plains vizcacha in the late Pleistocene of Uruguay. Alcheringa 40:354–365

Valladares P, Spotorno AE, Cortes MA, Zuleta RC (2018) *Chinchilla chinchilla* (Rodentia: Chinchillidae). Mamm Species 50(960):51–58

Villavicencio NA, Werdelin L (2018) The Casa del Diablo cave (Puno, Peru) and the late Pleistocene demise of megafauna in the Andean Altiplano. Quat Sci Rev 195:21–31

Vucetich MG (1984) Los roedores de la Edad Friasense (Mioceno medio) de Patagonia. Rev Mus La Plata (Nueva Ser). Paleontología 50(8):47–126

Vucetich MG (1985) *Cephalomyopsis hipselodontus* gen. et sp. nov. (Rodentia, Caviomorpha, Cephalomyidae) de la Edad Colhuehuapense (Oligoceno tardío) de Chubut, Argentina. Ameghiniana 22:243–245

Vucetich MG (1989) Rodents (Mammalia) of the Lacayani fauna revisited (Deseadan, Bolivia). Comparison with new Chinchillidae and Cephalomyidae from Argentina. Bull Mus Natl Hist Nat 4° Sér 11(4):233–247

Vucetich MG (1994) La fauna de roedores de la formación Cerro Boleadoras (Mioceno inferior?) en la provincia de Santa Cruz (Argentina). Acta Geol Leopold 39/1(17):365–374

Vucetich MG, Mazzoni MM, Pardiñas UFJ (1993) Los roedores de la formación Collón Curá (Mioceno medio), y la ignimbrita Pilcaniyeu. Canadón del Tordillo, Neuquén. Ameghiniana 30(4):361–381

Vucetich MG, Kramarz AG, Candela AM (2010) Colhuehuapian rodents from Gran Barranca and other Patagonian localities: the state of the art. Chapter 14. In: Madden RH, Carlini AA, Vucetich MG, Kay RF (eds) The paleontology of Gran Barranca evolution and environmental change through the middle Cenozoic of Patagonia. Cambridge University Press, Cambridge, pp 206–219

Vucetich MG, Dozo MT, Arnal M, Pérez ME (2015) New rodents (Mammalia) from the late Oligocene of Cabeza Blanca (Chubut) and the first rodent radiation in Patagonia. Hist Biol 27:236–257

Walton AH (1997) Rodents. Chapter 24. In: Kay RF, Madden RH, Cifelli RL, Flynn JJ (eds) Vertebrate paleontology in the neotropics: the Miocene fauna of La Venta, Colombia. Smithsonian Institution Press, Washington DC, pp 392–409

Wing ES (1980) Faunal remains. In: Lynch TF (ed) Guitarrero cave: early man in the Andes, Studies in archaeology series. Academic Press, New York, pp 149–172

Zachos JC, Dickens GR, Zeebe RE (2008) An early Cenozoic perspective on greenhouse warming and carbon-cycle dynamics. Nature 451:279–283

Chapter 3
Fossil Record of *Lagostomus*: A 10 Million Years Story

Luciano Luis Rasia

Abstract Despite being represented by a single extant species, *Lagostomus maximus*, the fossil record of the genus *Lagostomus* is extremely rich and the oldest of all living caviomorph rodents. It was recognized in the Chasicoan Stage/Age (Late Miocene) of Argentina, with younger records in Argentina, Bolivia, Uruguay, and Brazil. Many extinct species were described in the late nineteenth to early twentieth centuries, most of them based on fragmentary remains. Due to the fossorial habits of extinct vizcachas, they are very frequent in late Cenozoic continental associations, but their remains are not frequently identified to a specific level. Recent revisions have synonymized many of the named extinct species and have identified numerous fossil remains, revealing their biostratigraphic value. In this chapter, the fossil record of *Lagostomus* is revised, with remarks on its biostratigraphy and biogeography.

Keywords *Lagostomus* · Plains vizcacha · Fossil record · Biostratigraphy · Biogeography

3.1 Introduction

Lagostomus Brookes 1828 is represented by a single living species, the plains vizcacha *L. maximus* (Desmarest 1817), but the history of this genus has at least 10 ma, beginning in the late Middle to Late Miocene (Chasicoan Stage/Age, Fig. 3.1; e.g., Bondesio et al. 1980; Rasia and Candela 2017a; Montalvo et al. 2023a), with an abundant fossil record in the late Cenozoic of southern South America (e.g., Ameghino 1889; Kerber et al. 2011; Marshall and Sempere 1991; Ubilla and

L. L. Rasia (✉)
División Paleontología Vertebrados, Museo de La Plata, La Plata, Argentina
e-mail: lucianorasia@fcnym.unlp.edu.ar

L. L. Rasia et al. (eds.), *Plains Vizcachas*, https://doi.org/10.1007/978-3-031-49487-1_3

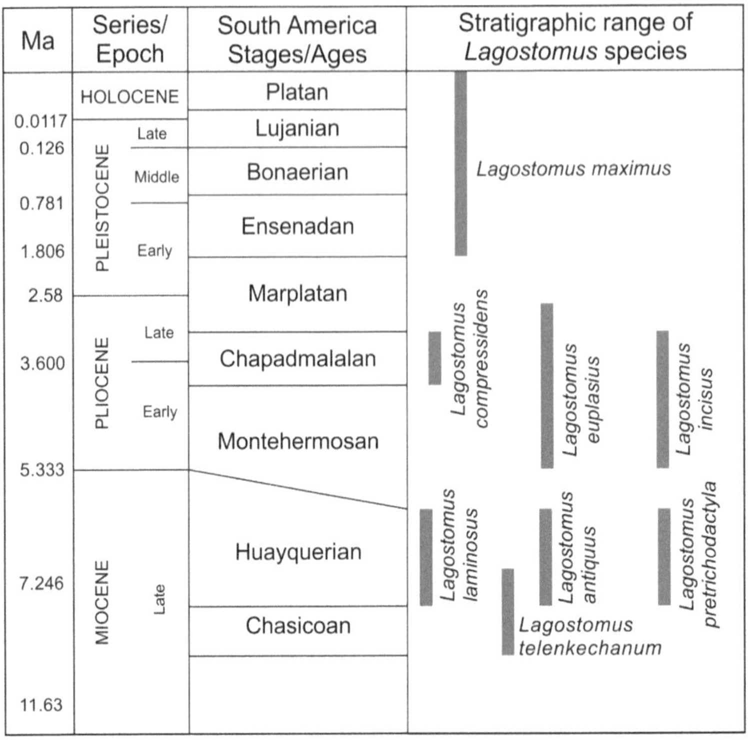

Fig. 3.1 Chronostratigraphic scheme for South America and stratigraphic range of the species of *Lagostomus*. Ma mega annum

Rinderknecht 2016; see Fig. 3.2). The fossorial habits of extinct species of *Lagostomus* (e.g., Elissamburu et al. 2011; Rasia 2016; Cardonatto et al. 2023), similar to the living *L. maximus* (see Weir 1974; Jackson et al. 1996), have favored the preservation of their fossil remains in Cenozoic sediments, being one of the more frequently found caviomorphs.

The longevity of *Lagostomus* is uncommon among extant caviomorph genera, only comparable to that of the echimyid *Myocastor* (Late Miocene; Candela and Noriega 2004) and the caviid *Microcavia* (Early Pliocene; Madozzo-Jaén and Pérez 2016) and also comparable to that of other chinchilloids like †*Eoviscaccia* Vucetich 1989 (Early Oligocene to Early Miocene; see Rasia et al. 2021).

Numerous extinct species of *Lagostomus* were named in the late nineteenth Century (e.g., Bravard 1860; Burmeister 1866; Gervais and Ameghino 1880; Ameghino 1883, 1886, 1888, 1889, 1891; Moreno 1888), but many of them lack proper descriptions, and the holotypes are lost and/or not figured, so only a few of these species have been revised (e.g., Nasif et al. 2013; Rasia 2016; Rasia and Candela 2013, 2017b; Ubilla and Rinderknecht 2016). By the early twentieth century, some researchers (e.g., Ameghino 1908; Rovereto 1914) described new extinct species of *Lagostomus* but placed them in the genus *Viscaccia*, a junior synonym of *Lagostomus* (see Jackson et al. 1996).

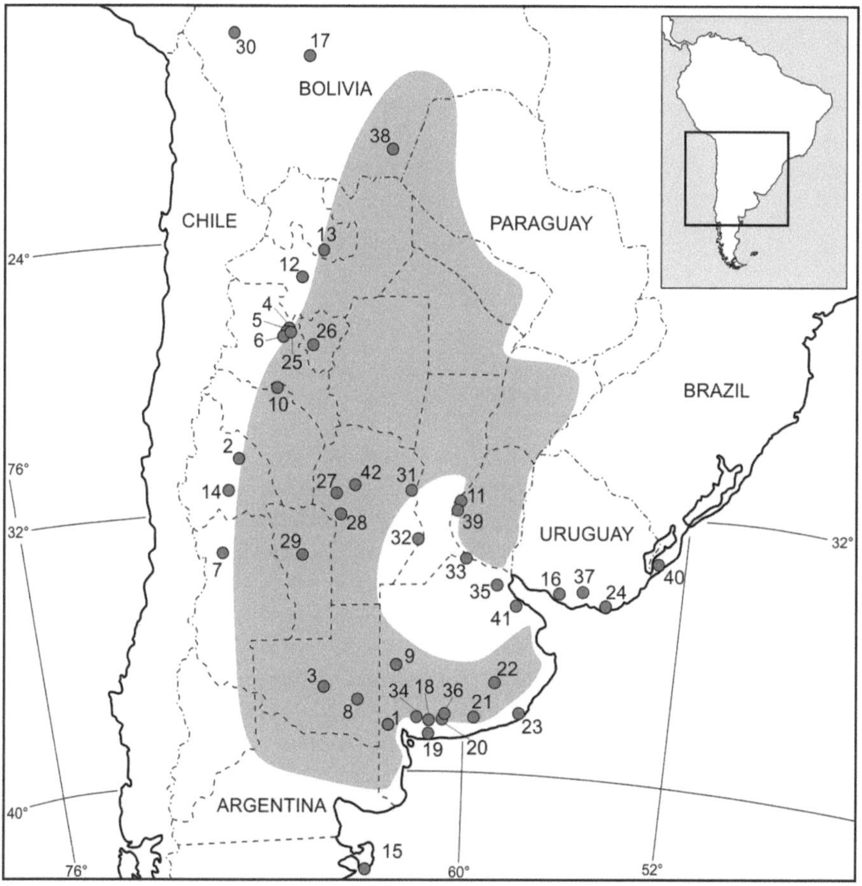

Fig. 3.2 Map showing fossil localities, ordered as they are mentioned in the text. (1) Arroyo Chasicó; (2) Loma del Camello; (3) Cerro La Bota; (4) Santa María Valley; (5) Villavil; (6) Puerta de Corral Quemado; (7) Huayquerías de San Carlos; (8) Salinas Grandes de Hidalgo; (9) Lagunas Encadenadas del Oeste; (10) Sierra de Velasco; (11) Paraná; (12) Tonco Valley; (13) San Salvador de Jujuy; (14) Loma de las Tapias; (15) Valdés Peninsula; (16) San José Department; (17) Muyu Huasi; (18) Las Obscuras; (19) Farola Monte Hermoso; (20) Cascada Grande; (21) Paso Otero; (22) Tandil; (23) Chapadmalal; (24) Maldonado Department; (25) Quebrada del Aspero; (26) El Molino; (27) Nono Valley; (28) Las Caleras; (29) Diaz Nogarol quarry; (30) Umala; (31) San Francisco; (32) Isla Verde; (33) Ramallo; (34) Bajo San José; (35) Río Reconquista; (36) Puente Vanoli; (37) Canelones Department; (38) Ñuapua; (39) Barrenechea Creek; (40) Chui Creek; (41) La Plata; (42) Córdoba. The gray area shows present distribution of *Lagostomus maximus*. (Taken from Jackson et al. 1996)

Later, Kraglievich (1926) erected the subgenus †*Lagostomopsis* Kraglievich 1926 to group all species from the Late Pliocene (Chapadmalalan Stage/Age) and older ones including all post-Chapadmalalan species in the subgenus *Lagostomus*. In a posterior work (Kraglievich 1934), he considered †*Lagostomopsis* as a separate genus. After Kraglievich's contributions (1926, 1934), many authors used †*Lagostomus (Lagostomopsis)* as a subgenus (e.g., Francis and Mones 1965, 1966,

1968; Goin et al. 2000; Rasia and Candela 2013) or †*Lagostomopsis* as a genus (e.g., Pascual 1966, Prado and Cerdeño 1998; Candela 2005; Elissamburu et al. 2011), but recent studies do not support the validity of †*Lagostomopsis* and consider it a synonym of *Lagostomus* (Rasia 2021a, b; Rasia and Candela 2017a, b; Rasia et al. 2020).

In this chapter, the fossil record of *Lagostomus* is revised and briefly summarized, with comments on biogeographic and biostratigraphic aspects.

3.2 Material and Methods

For the stratigraphic framework, the Stages/Ages system proposed by Cione and Tonni (e.g., 1995a, b, c, 1999, 2001, 2005) is used, with successive modifications (Fig. 3.1). The substages/subages of the Chasicoan Stage/Age (i.e., Viveran and Barrancan) are not considered valid, following Zarate et al. (2007). The Lower Huayquerian is not recognized as a valid substage/subage, following Tomassini et al. (2023). The Early Chapadmalalan (sensu; Cione and Tonni 1995a, b) is considered part of the Montehermosan Stage/Age, following Tomassini et al. (2013). The Barrancalobian is not recognized as a valid substage/subage of the Marplatan Stage/Age, following Isla et al. (2015). There are several studies that locally modify some of the biozones of the classic Stages/Ages system (e.g., Verzi et al. 2004; Deschamps 2005; Reguero and Candela 2011; Cruz 2012; Esteban et al. 2014).

Fossil material deposited in the following institutions was revised: Museo de La Plata (MLP), La Plata, Argentina; Museo Argentino de Ciencias Naturales "Bernardino Rivadavia" (MACN), Ciudad de Buenos Aires, Argentina; Museo Provincial "Antonio Serrano" (MAS), Paraná, Argentina; and Centro de Investigación Científica y Transferencia Tecnología y Producción (CICYTTP), Diamante, Argentina.

3.3 Fossil Record of *Lagostomus*

3.3.1 Late Miocene

The oldest record of the genus corresponds to undetermined remains (*Lagostomus* sp.) from late Middle to Late Miocene levels of the Cerro Azul Formation in La Pampa Province, Argentina (Montalvo et al. 2023a).

The oldest recognized species is †*Lagostomus telenkechanum* Rasia and Candela 2017a (Fig. 3.3a, b), from two units referred to the Chasicoan Stage/Age (Late Miocene, Fig. 3.1): the Cerro Azul Formation at Arroyo Chasicó (Buenos Aires Province, Argentina, Fig. 3.2; Rasia and Candela, 2017a; formerly Arroyo Chasicó Formation, see Folguera and Zárate 2009) and from several localities in La Pampa

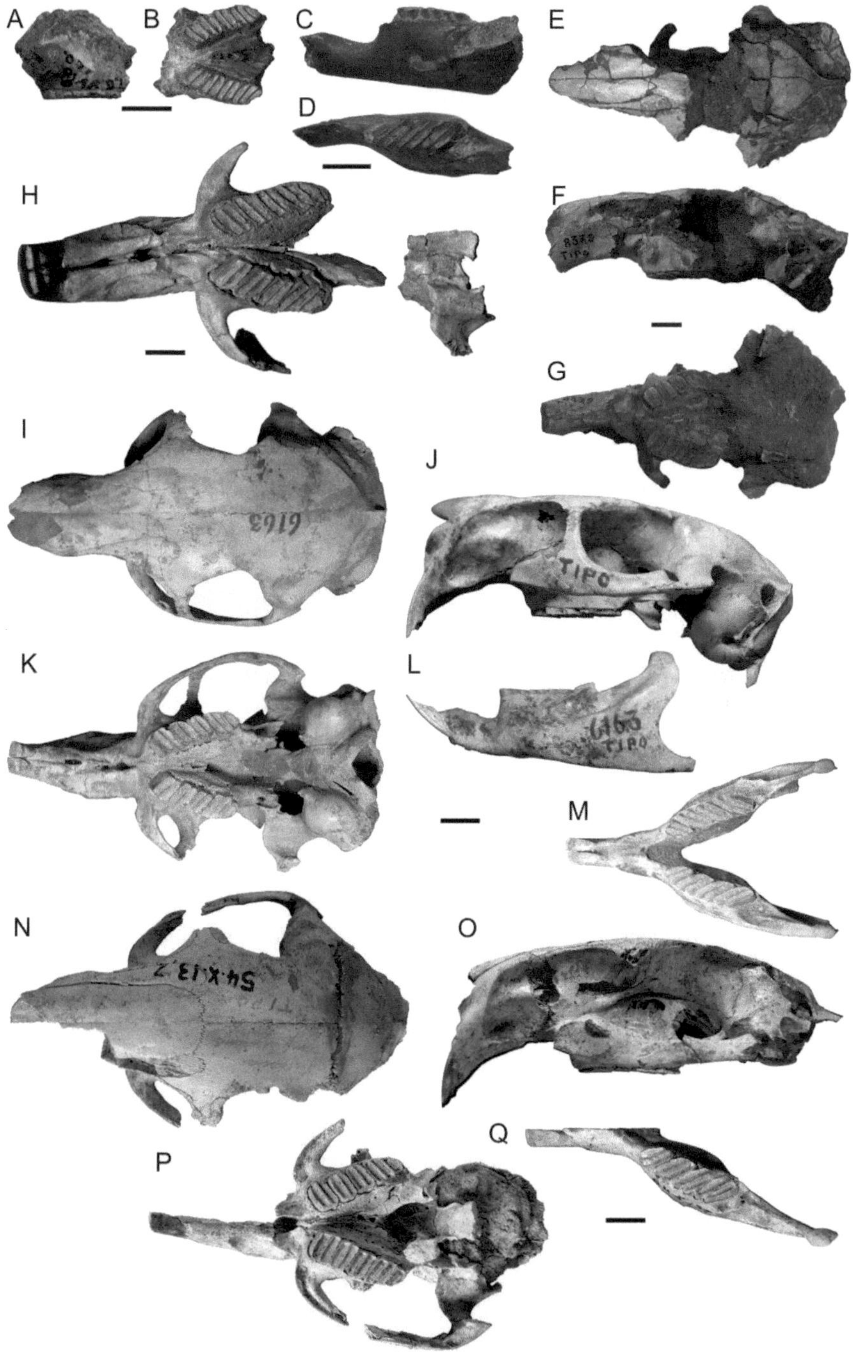

Fig. 3.3 Skull remains of extinct *Lagostomus* species. (**a**, **b**) †*Lagostomus telenkechanum*, MLP 60-VI-18-100 (holotype), palate in lateral (**a**) and ventral (**b**) views; (**c**, **d**) †*Lagostomus antiquus*, MAS-Pv 32 (holotype), left mandible in lateral (**c**) and dorsal (**d**) views; (**e–g**) †*Lagostomus pretrichodactyla*, MACN-Pv 8339 (holotype), skull in dorsal (**e**), left lateral (**f**), and ventral (**g**) views; (**h**) †*Lagostomus incisus*, MACN-A 1112 (holotype), fragmentary skull in ventral view; (**i–m**) †*Lagostomus euplasius*, MACN-Pv 6163 (holotype), skull in dorsal (**i**), left lateral (**j**), and ventral (**k**) views, mandibles in left lateral (**l**) and dorsal (**m**) views; (**n–q**) †*Lagostomus compressidens*, MLP 54-X-13-2 (holotype), skull in dorsal (**n**), left lateral (**o**), and ventral (**p**) views, left mandible in dorsal view (**q**). Scale bar: 1 cm

Province (Montalvo et al. 2023b), and the Huachipampa Formation (Loma del Camello, San Juan Province, Argentina, Fig. 3.2; Rasia et al. 2020). *Lagostomus* sp. was recorded in Chasicoan levels of the Cerro Azul Formation at Cerro La Bota (La Pampa Province, Argentina, Fig. 3.2; Montalvo et al. 2019, 2022).

For the Huayquerian Stage/Age (Late Miocene), †*Lagostomus pretrichodactyla* (Rovereto 1914) (Fig. 3.3e–g) was recorded in the Andalhuala Formation (Santa María Valley, Villavil and Puerta de Corral Quemado, Catamarca Province, Argentina, Fig. 3.2; Rovereto 1914; Marshall and Patterson 1981), the Huayquerías Formation (Huayquerías de San Carlos, Mendoza Province, Argentina, Fig. 3.2; Rovereto 1914; Forasiepi et al. 2014, 2016; Bonini, et al. 2016), and the Cerro Azul Formation (Salinas Grandes de Hidalgo, La Pampa Province, Argentina, Fig. 3.2; Rasia et al. 2020). Also, †*L. pretrichodactyla* was recognized in Late Miocene-Pliocene levels of Buenos Aires Province (Lagunas Encadenadas del Oeste, Fig. 3.2; Bonini et al. 2017), and *Lagostomus* cf. †*L. pretrichodactyla* was reported in the Salicas Formation (Sierra de Velasco, La Rioja Province, Argentina, Fig. 3.2; Tauber 2005).

Recently (Montalvo et al. 2023b), numerous remains of †*Lagostomus telenkechanum* and †*L. pretrichodactyla* have been recorded in association on several Huayquerian localities of the Cerro Azul Formation in La Pampa and Buenos Aires provinces (Argentina), so the stratigraphic range of †*L. telenkechanum* extends to the Huayquerian Stage/Age.

In the "*Conglomerado Osífero*" at the base of the Ituzaingó Formation (Late Miocene, Huayquerian Stage/Age; e.g., Brandoni 2013) near Paraná (Entre Ríos Province, Argentina; Fig. 3.2), two species are recognized, †*Lagostomus antiquus* Ameghino 1883 (Fig. 3.3c, d) and †*L. laminosus* Ameghino 1891 (see Nasif et al. 2013).

Other Late Miocene records of *Lagostomus* sp. from Argentina include the El Jarillal Member of the Chiquimil Formation at Puerta de Corral Quemado and Villavil in Catamarca Province (Fig. 3.2; Herbst et al. 2000; Esteban et al. 2014), the Salicas Formation in La Rioja Province (Brandoni et al. 2012), the Palo Pintado Formation in the Tonco Valley (Salta Province, Fig. 3.2; Zimicz et al. 2018; see also Rasia et al. 2020), the Guanaco Formation in Los Alisos and San Salvador de Jujuy (Jujuy Province, Fig. 3.2; Ercoli et al. 2019, 2021), the Loma de las Tapias Formation in San Juan Province (Fig. 3.2; Contreras and Baraldo 2011), the Cerro Azul Formation in Salinas Grandes de Hidalgo, La Pampa Province (Goin et al. 2000), and the Puerto Madryn Formation in Valdés Peninsula, Chubut Province (Fig. 3.2; Dozo et al. 2002, 2010).

Outside Argentina, Late Miocene records of *Lagostomus* sp. include the Kiyú Lithofacies of the Camacho Formation in San José Department, Uruguay (Fig. 3.2; Francis and Mones 1965; Rasia and Candela 2013), and the Muyu Huasi strata in Bolivia (Fig. 3.2; e.g., Marshall and Sempere 1991).

The ichnotaxa *Maneraichnus pampeanum* was described from Late Miocene sediments of the Cerro Azul Formation in Buenos Aires Province, Argentina (Cardonatto et al. 2023). *Maneraichnus pampeanum* is represented by interconnected burrow systems with ramps and chambers, and due to the architecture of the burrow system, the inferred body mass, and the size of claws and tooth traces, an undetermined species of *Lagostomus* was proposed as its producer (Cardonatto et al. 2023).

3.3.2 Pliocene

Three species of *Lagostomus* are recorded during the Pliocene. †*L. incisus* Ameghino 1888 (Fig. 3.3h) was recognized in several Montehermosan (late Early Pliocene) to Chapadmalalan (early Late Pliocene) units from Buenos Aires Province, Argentina (i.e., unknown unit at Lagunas Encadenadas del Oeste, Saldungaray Formation at Las Obscuras, Monte Hermoso Formation at Farola Monte Hermoso, "Irenean" Fauna at Cascada Grande, Quequén Grande Local Fauna at Paso Otero, Barker Formation near Tandil, and Chapadmalal Formation at Chapadmalal area, Fig. 3.2; Ameghino 1888; Rasia and Candela 2013, 2017b; Rasia 2021b; Bonini et al. 2017), and from the Maldonado Department, Uruguay (San José Member of the Raigón Formation; see Rasia 2021b). †*L. euplasius* (Ameghino 1908) (Fig. 3.3i–m) was recognized in the Monte Hermoso Formation, "Irenean" Fauna, and Chapadmalal Formation in Buenos Aires Province, Argentina (Ameghino 1908; Mones 1980; Rasia and Candela 2017b; Pardiñas et al. 2017) and in the San José Member of the Raigón Formation, Uruguay (Francis and Mones 1966; Rasia 2021b). †*L. compressidens* (Ameghino 1908) (Fig. 3.3n–q) was recorded solely in the Chapadmalal Formation in Buenos Aires Province, Argentina (Chapadmalalan Stage/Age, early Late Pliocene; Ameghino 1908; Rasia and Candela 2017b).

Lagostomus sp. was recognized in the Vorohué Formation (Vorohuean substage/subage; Late Pliocene) in the Chapadmalal area (Buenos Aires Province, Fig. 3.2; Albino and Quintana 1992), but specimens recovered from levels of this unit were identified as †*L. euplasius* (Rasia 2016).

Other Late Miocene to Pliocene records of *Lagostomus* in Argentina include El Cajón Formation (Quebrada del Áspero, Catamarca Province, Fig. 3.2; Parra and Morales 2003), Aconquija Formation (El Molino, Tucumán Province, Fig. 3.2; Nasif et al. 2003; see Armella et al. (2018) for geological context), Brochero Formation (Nono Valley, Córdoba Province, Fig. 3.2; Álvarez and Tauber 2003), Pliocene sediments at Las Caleras (Córdoba Province, Fig. 3.2; Tauber et al. 1997; Tauber 2000), Río Quinto Formation (Diaz Nogarol quarry, San Luis Province, Fig. 3.2; Prado et al. 1998), and Barranca de los Lobos Formation (Chapadmalal area, Buenos Aires Province; Pascual 1966; Vucetich and Verzi 1995; see Isla et al. (2015) for discussion on the age of Barranca de los Lobos Formation).

A lagostomin was recorded in the Umala Formation (Montehermosan-Chapadmalalan stages/ages, Pliocene) in Umala, Bolivia (Marshall and Sempere 1991). It probably corresponds to *Lagostomus*, given that this is the only known lagostomin for the Pliocene.

3.3.3 Pleistocene to Holocene

In the Punta San Andrés Alloformation (Sanandresian Substage/Subage, latest Late Pliocene to Early Pleistocene) in the Chapadmalal area, Buenos Aires Province, there are records of *Lagostomus* sp. (Beilinson and Taglioretti 2013). *Lagostomus* sp. is also recognized in Upper Pleistocene to Holocene sediments of Córdoba Province (San Francisco and Isla Verde, Fig. 3.2; Cruz 2012; Cruz et al. 2009; Luna and Cruz 2014), and Buenos Aires Province (Ramallo, Fig. 3.2; Voglino and Pardiñas 2005).

The fossil record of *Lagostomus maximus* includes several Late Pleistocene to Holocene (Lujanian to Platan Stages/Ages) units of Buenos Aires (e.g., Paso Otero, Bajo San José, Río Reconquista and Puente Vanoli, Fig. 3.2; Prado et al. 1987; Deschamps and Borromei 1992; Lezcano et al. 1992; Tomassini et al. 2017), Santa Fe and Córdoba provinces in Argentina (e.g., Ameghino 1889), Canelones in Uruguay (Francis and Mones 1968; Ubilla and Rinderknecht 2016), and Ñuapua in Bolivia (Fig. 3.2; Marshall and Sempere 1991). Moreover, there are numerous records of *L. maximus* in Late Pleistocene to Holocene archaeological sites of Buenos Aires Province (e.g., Mazzanti and Quintana 1997; Quintana and Mazzanti 2011; Gomez et al. 1999).

There are records of *Lagostomus* cf. *L. maximus* from the Late Pleistocene (Lujanian Stage/Age) of Entre Ríos Province in Argentina (Salto Ander Egg Formation at Barrenechea Creek, Fig. 3.2; Ferrero et al. 2017), Canelones Department in Uruguay (Fig. 3.2; Ubilla and Rinderknecht 2016), and Chui Creek in Brazil (Fig. 3.2; Kerber et al. 2011).

†"*Lagostomus angustidens*" Burmeister 1866 was synonymized with *L. maximus* by Ameghino (1889). †"*Lagostomus cavifrons*" Ameghino 1889, from the Middle to Late Pleistocene of Argentina (Ameghino 1889), was recently synonymized with *L. maximus* (Ubilla and Rinderknecht 2016).

Additionally, the Early to Late Pleistocene (Ensenadan to Bonaerian stages/ages) species, †"*L. minimus*" Ameghino 1889 (Fig. 3.4c, d) (Ensenadan of La Plata, Buenos Aires Province, Fig. 3.2; Ameghino 1889), †"*L. heterogenidens*" Ameghino 1889 (Fig. 3.4e, f) (Bonaerian of Córdoba, Córdoba Province, Fig. 3.2; Ameghino 1889), and †"*L. debilis*" Ameghino 1889 (Fig. 3.4a, b) (Bonaerian-Lujanian of Buenos Aires, Santa Fe and Córdoba provinces; Ameghino 1889), should be treated as synonyms of *L. maximus*, given that they are within the range of size and morphological variability of the living species. This issue was mentioned by Ubilla and Rinderknecht (2016), who considered these species as *Lagostomus* sp. in their study.

Finally, the holotype of †"*L. egenus*" Ameghino 1891 (Fig. 3.4g–i) (Pleistocene of Argentina) is probably a pathological individual of *L. maximus*.

There are records of †"*Lagostomus minimus*" and †"*L. cavifrons*" in the Ensenadan of Buenos Aires (e.g., Ameghino 1889; Rusconi 1929, 1931). Therefore, the fossil record of the extant species *Lagostomus maximus* would be from the Ensenadan Stage/Age (late Early Pleistocene) to recent times.

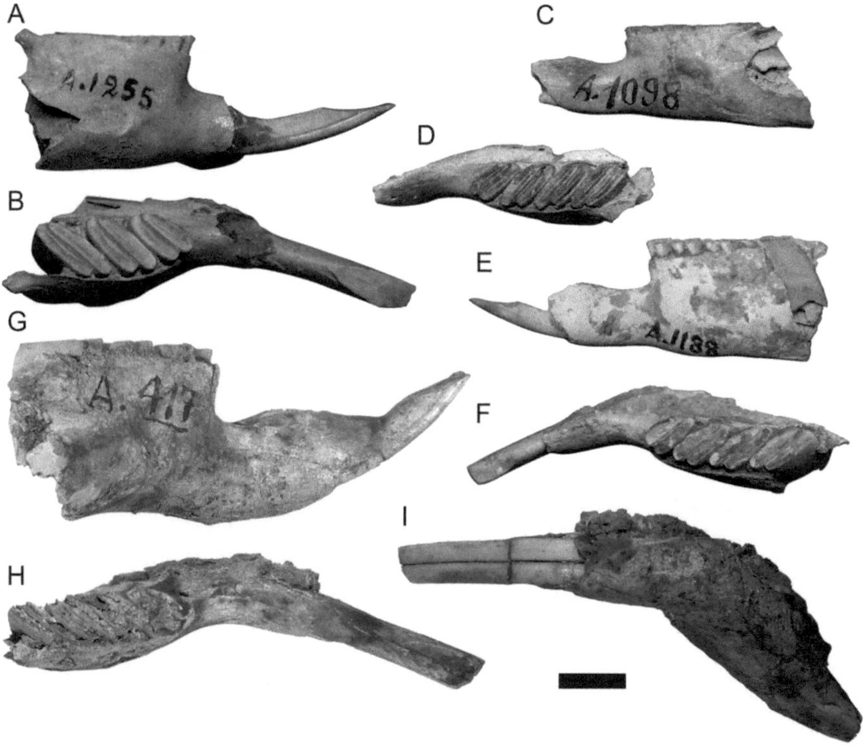

Fig. 3.4 Material referred to *Lagostomus maximus*. (**a**, **b**) MACN-A 1255 (type of †"*L. debilis*"), right mandible in lateral (**a**) and occlusal (**b**) views; (**c**, **d**) MACN-A 1098 (type of †"*L. minimus*"), left mandible in lateral (**c**) and occlusal (**d**) views; (**e**, **f**) MACN-A 1188 (type of †"*L. heterogenidens*"), left mandible in lateral (**e**) and occlusal (**f**) views; (**g–i**) MACN-A 417 (type of †"*L. egenus*"), right mandible in lateral (**g**), occlusal (**h**), and ventral (**i**) views. Scale bar: 1 cm

The species †*Lagostomus brasiliensis* Lund 1841, described from karst deposits in Lagoa Santa, Brazil (Lund 1841), is not valid, given that it does not belong to *Lagostomus*, or even to Lagostominae. As was pointed out by Kerber et al. (2011), the figured specimen (Lund 1841) does not have the occlusal pattern of any known Lagostominae, and the species has not been recognized by posterior reviews of that fauna (e.g., Winge 1888).

3.4 Discussion and Conclusions

3.4.1 Biostratigraphic Comments

It has been addressed that the species of *Lagostomus* have great biostratigraphic significance, not only locally (Rasia and Candela 2013, 2017b) but also in a greater geographic range (Rasia et al. 2020; Rasia 2021a). Despite the species of *Lagostomus* have limited stratigraphic record, some of them encompass two or more stages/ages (see Fig. 3.1), so they do not constitute remarkable biostratigraphic tools as was previously thought (see Montalvo et al. 2023b), but they can be helpful in determining the age of associated faunas or bearing sediments.

During the Chasicoan Stage/Age, only †*Lagostomus telenkechanum* is recorded in Argentina (e.g., Rasia et al. 2020; Montalvo et al. 2023b).

In the Huayquerian Stage/Age, †*Lagostomus pretrichodactyla* is recorded in northwestern and central Argentina (e.g., Rasia et al. 2020). In central Argentina, †*L. telenkechanum* and †*L. pretrichodactyla* have been recorded in association (Montalvo et al. 2023b). †*L. antiquus* and †*L. laminosus* are recorded in northeastern Argentina (e.g., Nasif et al. 2013).

In the Montehermosan Stage/Age, †*Lagostomus incisus* and †*L. euplasius* are recorded in several units (e.g., Rasia and Candela 2017b; Rasia 2021b).

During the Chapadmalalan Stage/Age, †*Lagostomus incisus*, †*L. euplasius,* and †*L. compressidens* are recorded in the Chapadmalal Formation (Buenos Aires Province, Argentina; e.g., Rasia and Candela 2017b), base of the Chapadmalalan Stage/Age.

During the Marplatan Stage/Age, there are records of *Lagostomus* sp. that have not been identified to a specific level and further studies are needed. †*L. euplasius* was recognized in levels of Vorohuean Substage/Subage (Rasia 2016).

In the Ensenadan to Platan stages/ages, the extant *Lagostomus maximus* is recognized. Other species previously described are considered synonyms of *L. maximus* (see above). The stratigraphic range of *L. maximus* encompasses less than 2 ma, but the biostratigraphic value of this species is lower than previously thought, given that it was recognized as a key species of the Platan Stage/Age, with the "Biozone of *Lagostomus maximus*" (see Cione and Tonni 1999, 2005), but now it is considered to comprise four stages/ages.

3.4.2 Biogeography

The past distribution of *Lagostomus* species does not show marked differences with the present distribution of the extant *L. maximus* (see Fig. 3.2), but there are some records that indicate expansions and retractions of the territory occupied by different species of the genus.

During the Chasicoan Stage/Age (Late Miocene), *Lagostomus* species were distributed in the Pampas (Buenos Aires and La Pampa provinces) and Cuyo (San Juan Province).

By the Huayquerian Stage/Age (Late Miocene), *Lagostomus* species reached a wider distribution than the extant *L. maximus*, with records in northern Patagonia in Argentina, southern Uruguay, and central Bolivia.

In the Pliocene (Montehermosan and Chapadmalalan stages/ages), *Lagostomus* species were distributed in central Argentina, southern Uruguay, and western Bolivia, occupying a wider distribution than in recent times (see Fig. 3.2).

The distribution of the extant plains vizcacha, *Lagostomus maximus*, during the Late Pleistocene to Holocene was wider than in recent times, occupying southern Uruguay and probably southern Brazil. In historic times, the distribution of the plains vizcacha has suffered an evident retraction, with virtually no records in some parts of its territory, probably due to agricultural activities, but with high numbers in protected areas (e.g., Jackson et al. 1996; Renison et al. 2023).

Future studies of fossil records of *Lagostomus* from Bolivia, Uruguay, and Brazil would permit to reevaluate the geographic distribution of known extinct species.

Acknowledgments I want to thank the curators (Marcelo Reguero, MLP; Alejo Scarano, MLP; Martin Ezcurra, MACN; Agustin Martinelli, MACN) for allowing access to material under their care. This study is part of the project Red Vizcachera.

References

Albino AM, Quintana C (1992) Los Colubridae (Reptilia: Serpentes) del Chapadmalalense y Uquiense (Plioceno tardío-Pleistoceno temprano?) de la Provincia de Buenos Aires, Argentina. Ameghiniana 29(2):125–133

Álvarez D, Tauber A (2003) Nuevos registros de mamíferos en la Formación Brochero (Mioceno Tardío-Plioceno) del valle de Nono, Córdoba, República Argentina. Ameghiniana 40(4):50R

Ameghino F (1883) Sobre una nueva colección de mamíferos fósiles recogidos por el Profesor Scalabrini en las barrancas del Paraná. Bol Acad Nac Cienc Córdoba 5:112–113

Ameghino F (1886) Contribución al conocimiento de los mamíferos fósiles de los terrenos terciarios antiguos de Paraná. Bol Acad Nac Cienc Córdoba 9:5–228

Ameghino F (1888) Lista de las especies de mamíferos fósiles del Mioceno superior de Monte Hermoso, hasta ahora conocidas. Editorial Coni, Buenos Aires

Ameghino F (1889) Contribución al conocimiento de los mamíferos fósiles de la República Argentina. Actas Acad Nac Cienc Córdoba 6:1–1027

Ameghino F (1891) Mamíferos y aves fósiles argentinos. Especies nuevas, adiciones y correcciones. Rev Argent Hist Nat 1:240–259

Ameghino F (1908) Las Formaciones sedimentarias de la región litoral de Mar del Plata y Chapadmalal. Anales Mus Nac Buenos Aires 10:342–428

Armella MA, García-López DA, Dominguez L (2018) A new species of *Xotodon* (Notoungulata, Toxodontidae) from northwestern Argentina. J Vertebr Paleontol 38(1):e1425882

Beilinson E, Taglioretti M (2013) Análisis icnológico y paleoambiental de la Aloformación Punta San Andrés (Plio-Pleistoceno), Provincia de Buenos Aires, Argentina. Ameghiniana 50(2):153–165

Bondesio P, Laza JH, Scillato Yané GJ, Tonni EP, Vucetich MG (1980) Estado actual del conocimiento de los vertebrados de la Fm. Arroyo Chasicó (Plioceno temprano) de la Prov. de Buenos Aires. Actas II Congreso Argentino de Paleontología y Bioestratigrafía y I Congreso Latinoamericano de Paleontología 3:85–99

Bonini R, Forasiepi AM, Prevosti FJ, Garrido AC, Barbeau DL, Turazzini GF, Echarri S, Pujos F, MacPhee R, Verzi D, Vera B, Cerdeño E, Pérez ME, Rasia LL, Esteban G, De La Fuente MS (2016) Avances en paleontología, estratigrafía y edad de la Formación Huayquerías (Mioceno tardío, Mendoza). Ameghiniana 53(6):R50–R51

Bonini RA, Scanferla AC, Candela AM, Rasia LL, Schmidt GI (2017) Estudio preliminar de una nueva fauna de mamíferos del Neógeno de las Lagunas Encadenadas del Oeste de la Provincia de Buenos Aires. In: Congreso Geológico Argentino, No. 20 y Simposio del Mioceno-Pleistoceno del Centro y Norte de Argentina, No. 4, Actas:19–24. San Miguel de Tucumán

Brandoni D (2013) Los mamíferos continentales del "Mesopotamiense" (Mioceno tardío) de Entre Ríos, Argentina. Diversidad, edad y paleobiogeografía. In: Brandoni D, Noriega JI (eds) El Neógeno de la Mesopotamia Argentina. Asociación Paleontológica Argentina, Publicación Especial 14, pp 179–191
Brandoni D, Schmidt GI, Candela AM, Noriega JI, Brunetto E, Fiorelli LE (2012) Mammals from the Salicas Formation (Late Miocene), La Rioja Province, Northwestern Argentina: paleobiogeography, age, and Paleoenvironment. Ameghiniana 49(3):375–387
Bravard A (1860) Catalogue des espèces d'animaux fossiles recueillies dans l'Amérique du Sud par Auguste Bravard de 1852 à 1860. Unpublished, pp 7
Brookes J (1828) A new genus of the order Rodentia. Trans Linn Soc Lond 16:96–105
Burmeister G (1866) Lista de los mamíferos fósiles del terreno diluviano. Anales Mus Públ Buenos Aires 1:121–232
Candela AM (2005) Los roedores del "Mesopotamiense" (Mioceno tardío, Formación Ituzaingó) de la provincia de Entre Ríos (Argentina). In: Aceñolaza FG (ed) Temas de la Biodiversidad del Litoral fluvial argentino II. INSUGEO, Miscelánea 14, pp 37–48
Candela AM, Noriega JI (2004) Los coipos (Rodentia, Caviomorpha, Myocastoridae) del "Mesopotamiense" (Mioceno tardío; Formación Ituzaingó) de la provincia de Entre Ríos, Argentina. In: Aceñolaza FG (coord) Temas de la Biodiversidad del Litoral fluvial argentino. INSUGEO, Miscelanea, 12, pp 77–82
Cardonatto MC, Feola S, Melchor RN (2023) Neogene communal rodent burrow systems from Central Argentina. Hist Biol. https://doi.org/10.1080/08912963.2023.2228319
Cione AL, Tonni EP (1995a) Los estratotipos de los pisos Montehermosense y Chapadmalalense (Plioceno) del esquema cronológico sudamericano. Ameghiniana 32:369–374
Cione AL, Tonni EP (1995b) Chronostratigraphy and land-mammal ages in the Cenozoic of southern South America: principles, practices, and the Uquian problem. J Paleontol 69:135–159
Cione AL, Tonni EP (1995c) Bioestratigrafía y cronología del Cenozoico superior de la región pampeana. In: Alberdi MT, Leone G, Tonni EP (eds) Evolución biológica y climática de la región pampeana durante los últimos cinco millones de años, un ensayo de correlación con el Mediterráneo occidental. Monografias 12. Museo Nacional de Ciencias Naturales, Madrid, pp 47–74
Cione AL, Tonni EP (1999) Biostratigraphy and chronological scale of upper-most Cenozoic in the Pampean area, Argentina. Quatern South Am Antarctic Peninsula 3:23–51
Cione AL, Tonni EP (2001) Correlation of Pliocene to Holocene southern South American and European vertebrate-bearing units. Boll Soc Paleontol Italiana 40:167–173
Cione AL, Tonni EP (2005) Bioestratigrafía basada en mamíferos del Cenozoico superior de la provincia de Buenos Aires, Argentina. In: de Barrio RE, Etcheverry RO, Caballé MF, Llambías E (eds) Geología y Recursos Minerales de la Provincia de Buenos Aires. Relatorio del XVI Congreso Geológico Argentino, La Plata, pp 183–200
Contreras VH, Baraldo JA (2011) Calibration of the Chasicoan-Huayquerian stages boundary (Neogene), San Juan, western Argentina. In: Salfity JA, Marquillas RA (eds) Cenozoic geology of the Central Andes of Argentina. Instituto del Cenozoico, Universidad Nacional de Salta, pp 111–121
Cruz LE (2012) Biostratigraphy and geochronology of the late Cenozoic of Córdoba Province (Central Argentina). J S Am Earth Sci 42:250–259
Cruz LE, Fernicola JC, Carignano C, Bargo MS (2009) Nueva asociación faunística del Pleistoceno del este de la provincial de Córdoba. Ameghiniana 46(4):17–18R
Deschamps CM (2005) Late Cenozoic mammal biochronostratigraphy in southwestern Buenos Aires province, Argentina. Ameghiniana 42:733–750
Deschamps CM, Borromei AM (1992) La fauna de vertebrados pleistocénicos del Bajo San José (provincia de Buenos Aires, Argentina). Aspectos paleoambientales. Ameghiniana 29:177–183
Desmarest AG (1817) Gerboise première espèce: la grande gerboise *Dipus maximus* Blainv. Nouveau Dictionnaire d'Histoire Naturelle 13:117–119

Dozo MT, Monti A, Bouza P, Vucetich MG, Cione A, Tonni E, Scillato-Yané G (2002) Geología y vertebrados continentales en cercanías de Punta Delgada (Neógeno de Península Valdés, Chubut, Argentina). In: Cabaleri N, Cingolani CA, Linares E, López de Luchi MG, Ostera HA, Panarello HO (eds) Congreso Geológico Argentino, No. 15, Actas 1, pp 536–541

Dozo MT, Bouza P, Monti A, Palazzesi L, Barreda V, Massaferro G, Scasso RA, Tambussi CP (2010) Late Miocene continental biota in Northeastern Patagonia (Península Valdés, Chubut, Argentina). Palaeogeogr Palaeoclimatol Palaeoecol 297:100–109

Elissamburu A, Dondas A, De Santis L (2011) Morfometría de las paleocuevas de la "Fm." Chapadmalal y su asignación a *Actenomys* (Rodentia), *Paedotherium* (Notoungulata) y otros mamíferos fósiles hospedantes. Mastozool Neotrop 18(2):227–238

Ercoli MD, Álvarez A, Santamans C, González Patagua SA, Villalba Ulberich JP, Constantini OE (2019) Los Alisos, a new fossiliferous locality for Guanaco Formation (late Miocene) in Jujuy (Argentina), and a first approach of its paleoecological and biochronology implications. J S Am Earth Sci 93:203–213

Ercoli MD, Álvarez A, Verzi DH, Villalba Ulberich JP, Quiñones SI, Constantini OE, Zurita AE (2021) A new mammalian assemblage for Guanaco Formation (northwestern Argentina), and the description of a new genus of New World porcupine. J S Am Earth Sci 110:103389

Esteban G, Nasif N, Georgieff SM (2014) Cronobioestratigrafía del Mioceno tardío – Plioceno temprano, Puerta de Corral Quemado y Villavil, provincia de Catamarca, Argentina. Acta Geol Lilloana 26(2):165–192

Ferrero BS, Noriega JI, Brunetto E, Nuñez Otaño N (2017) Vertebrate continental assemblage from the Last Interglacial in southern South America (Entre Ríos, Argentina). Biostratigraph Paleoenviron Palaeogeogr Palaeoclimatol Palaeoecol 466:89–99

Folguera A, Zárate M (2009) La sedimentación neógena continental en el sector extrandino de Argentina central. Rev Asoc Geol Arg 64(4):692–712

Forasiepi AM, Prevosti FJ, Vera B, Turazzini GF, Echarri S, Garrido AC, Verzi D, Rasia LL, Schmidt GI, Esteban G, Krapovikas V (2014) The badlands from Mendoza and the Huayquerian Age: insights into the late Miocene. 4th international paleontological congress. Mendoza, 2014 Abstract volume:713

Forasiepi AM, Prevosti FJ, Garrido AC, Vera B, Turazzini GF, Echarri S, Bonini R, Pujos F, MacPhee R, Verzi D, Rasia LL, Schmidt GI, Esteban G, Krapovikas V (2016) Avances en el conocimiento de la fauna de la Formación Huayquerías (Mioceno tardío, Mendoza). Ameghiniana 53(3):R12–R13

Francis JC, Mones A (1965) La presencia de vizcachas [*Lagostomus (Lagostomopsis) spicatus* (Amegh.)] en la Formación Kiyú, Dto. de San José, R. O. del Uruguay. Rev Fac Humanid Cienc Montevideo 22:155–168

Francis JC, Mones A (1966) Las vizcachas, *Lagostomus (Lagostomopsis) euplasius* (Amegh.), de la formación Maldonado, departamento de Maldonado, República Oriental del Uruguay. Kraglieviana 1:101–110

Francis JC, Mones A (1968) Los roedores fósiles de Uruguay. Bol Lab Paleontol Vertebr 1(2):35–64

Gervais H, Ameghino F (1880) Los mamíferos fósiles de América del Sud. Savy e Igon, Paris-Buenos Aires

Goin FJ, Montalvo CI, Visconti G (2000) Los Marsupiales (Mammalia) del Mioceno Superior de la Formación Cerro Azul (Provincia de La Pampa, Argentina). Estud Geol 56(1–2):101–126

Gómez G, Prado JL, Alberdi MT (1999) Micromamíferos del sitio Arroyo Seco 2 (provincia de Buenos Aires, Argentina). Sus implicaciones tafonómicas y paleoambientales. Estud Geol 55:273–281

Herbst R, Anzótegui L, Esteban G, Mautino L, Morton S, Nasif N (2000) Síntesis paleontológica del Mioceno de los valles Calchaquíes del noroeste argentino. In: Aceñolaza F, Herbst R (eds) El Neógeno de Argentina Instituto Superior de Correlación Geológica (INSUGEO), Serie de Correlación Geológica, vol 14, pp 263–287

Isla F, Taglioretti M, Dondas A (2015) Revisión y nuevos aportes sobre la estratigrafía y sedimentología de los acantilados entre Mar de Cobo y Miramar, Provincia de Buenos Aires. Rev Asoc Geol Arg 72:235–250

Jackson JE, Branch LC, Villareal D (1996) Lagostomus maximus. Mamm Species 543:1–6

Kerber L, Pereira Lopes R, Vucetich MG, Ribeiro AM, Pereira J (2011) Chinchillidae and Dolichotinae rodents (Rodentia: Hystricognathi: Caviomorpha) from the late Pleistocene of Southern Brazil. Rev Bras Paleontol 14(3):229–238

Kraglievich L (1926) Sobre el conducto humeral en las vizcachas y paquirucos chapadmalenses con descripción del *Paedotherium imperforatum*. Anales Mus Nac Hist Nat 34:45–51

Kraglievich L (1934) La antigüedad Pliocena de las Faunas de Monte Hermoso y Chapadmalal, deducidas de su comparación con las que le precedieron y sucedieron. Imprenta El Siglo Ilustrado, Montevideo

Lezcano MJ, Reboledo C, Schreiber CE (1992) Bioestratigrafía de los sedimentos de la cuenca alta del Río Reconquista (Pleistoceno tardío, noreste de la Provincia de Buenos Aires, Argentina). Ameghiniana 29:387R

Luna CA, Cruz LE (2014) los mamíferos fósiles del Pleistoceno tardío-Holoceno temprano del sudeste de la provincia de Córdoba y sus implicancias paleoambientales para el centro de Argentina. Rev Bras Paleontol 17(1):69–82

Lund PW (1841) Tillaeg til de to Sidste Afhandlinger over Brasiliens Dyreverden för Sidste Jordomvaeltning. Det Kongelige Danske Videnskabernes Selskabs naturvidenskabelige og Mathematiske Afhandlinger 8:273–296

Madozzo-Jaén MC, Pérez ME (2016) The most ancient caviine rodent (Hystricognathi, Cavioidea) comes from the late Miocene of Northwest Argentina (South America). Hist Biol 29:376–383

Marshall LG, Patterson B (1981) Geology and geochronology of the mammal-bearing Tertiary of the Valle de Santa María and Río Corral Quemado, Catamarca Province, Argentina. Fieldiana Geol (New Ser) 9:1–80

Marshall LG, Sempere T (1991) The Eocene to Pleistocene vertebrates of Bolivia and their stratigraphic context: a review. In: Suárez-Soruco R (ed) Fósiles y Facies de Bolivia, Volumen I-Vertebrados. Revista Técnica YPFB, vol 12(3–4), pp 631–652

Mazzanti DL, Quintana CA (1997) Asociación cultural con fauna extinguida en el sitio arqueológico Cueva Tixi, provincia de Buenos Aires, Argentina. Rev Esp Antropol Am 27:11–21

Mones A (1980) Sobre una colección de vertebrados fósiles de Monte Hermoso (Plioceno superior), Argentina, Con la descripción de una nueva especie de Marmosa (Marsupialia: Didelphidae). Com Paleontol Mus Hist Natl Montevideo 8(1):159–169

Montalvo CI, Tomassini RL, Sostillo R, Cerdeño E, Verzi DH, Visconti G, Folguera A, Schmidt GI (2019) Chasicoan (late Miocene) vertebrate assemblage from Cerro Azul Formation, Central Argentina. Geomorphological and biochronological considerations. J S Am Earth Sci 95:102218

Montalvo CI, Tomassini RL, Cerdeño E, Fernández FJ (2022) A late Miocene mammal assemblage reveals a complex taphonomic history in fluvial deposits of Central Argentina. Hist Biol. https://doi.org/10.1080/08912963.2022.2060100

Montalvo CI, Folguera A, Cerdeño E, Verzi DH, Pérez ME, Sostillo R, Tomassini RL (2023a) The oldest Andean synorogenic fossiliferous deposits of the Miocene Cerro Azul Formation, Central Argentina. J S Am Earth Sci 123:104235. https://doi.org/10.1016/j.jsames.2023.104235

Montalvo CI, Sostillo R, Rasia LL (2023b) New light on the late Middle Miocene–early Pliocene representatives of *Lagostomus* (Chinchillidae, Rodentia) from Cerro Azul Formation, Argentina. J S Am Earth Sci. https://doi.org/10.1016/j.jsames.2023.104481

Moreno FJP (1888) Informe preliminar de los progresos del Museo La Plata durante el primer semestre de 1888, presentado al señor ministro de Obras Públicas de la Provincia de Buenos Aires. Bol Mus La Plata 2:1–35

Nasif N, Esteban G, García J, Falcón C (2003) Nueva localidad con fósiles de vertebrados del Neógeno de la provincia de Tucumán, Noroeste de Argentina. Ameghiniana 40(4):62R–63R

Nasif NL, Candela A, Rasia L, Madozzo-Jaén MC, Bonini R (2013) Actualización del conocimiento de los Roedores del Mioceno tardío de la Mesopotamia Argentina: aspectos sistemáticos, evolutivos y paleobiogeográficos. In: Brandoni D, Noriega J (eds) El Neógeno de la Mesopotamia Argentina, vol 14. Asociación Paleontológica Argentina, Publicación Especial, pp 147–163

Pardiñas UFJ, Prevosti FJ, Voglino D, Cenizo M (2017) A controversial unit within the Argentine neogene: the "Irenean" fauna. Ameghiniana 54:655–680

Parra LE, Morales JA (2003) Descripción e interpretación de la columna estratigráfica tipo de la serie vulcano-sedimentaria neógena del Valle del Cajón (Catamarca, Argentina). Geogaceta 33:55–58

Pascual R (1966) Vertebrata. In: Borrello AV (ed) Paleontografía Bonaerense, Volumen 1. Comisión de Investigaciones Científicas de la Provincia de Buenos Aires, p 1–202

Prado JL, Cerdeño E (1998) Los mamíferos pliocenos de la Fauna Local Quequén Grande (Provincia de Buenos Aires, Argentina). Estud Geol 54:75–83

Prado JL, Menegaz AN, Tonni EP, Salemme MC (1987) Los mamíferos de la Fauna Local Paso Otero (Pleistoceno tardío), Provincia de Buenos Aires. Aspectos paleoambientales y bioestratigráficos. Ameghiniana 24(3–4):217–233

Prado JL, Chiesa J, Tognelli G, Cerdeño E, Strasser E (1998) Los mamíferos de la Formación Río Quinto (Plioceno), Provincia de San Luis (Argentina). Aspectos bioestratigráficos, zoogeográficos y paleoambientales. Estud Geol 54:153–160

Quintana CA, Mazzanti DL (2011) Las vizcachas pampeanas (*Lagostomus maximus*, Rodentia) en la subsistencia indígena del Holoceno tardío de las Sierras de Tandilia Oriental (Argentina). Lat Am Antiq 22(2):253–270

Rasia LL (2016) Los Chinchillidae (Rodentia, Caviomorpha) fósiles de la República Argentina: sistemática, historia evolutiva y biogeográfica, significado bioestratigráfico y paleoambiental. PhD Dissertation, Universidad Nacional de La Plata, La Plata, Argentina

Rasia LL (2021a) What ever happened to *Lagostomopsis*? Assessing the validity of an extinct chinchillid rodent taxa. PE-APA 21(1):R45

Rasia LL (2021b) Reassessing the fossil record of *Lagostomus incisus* Ameghino, 1888 (Rodentia, Caviomorpha) from the late Neogene of southern South America. PE-APA 21(2):53–61

Rasia LL, Candela AM (2013) Systematic and biostratigraphic significance of a chinchillid rodent from the Pliocene of eastern Argentina. Acta Palaeontol Pol 58(2):241–254

Rasia LL, Candela AM (2017a) *Lagostomus telenkechanum* sp. nov., a new lagostomine rodent (Caviomorpha, Chinchillidae) from the Arroyo Chasicó Formation (late Miocene; Buenos Aires Province, Argentina). J Vertebr Paleontol 37(1):e1239205. https://doi.org/10.1080/02724634.2017.1239205

Rasia LL, Candela AM (2017b) Systematic revision of the vizcachas (Rodentia, Caviomorpha, Chinchillidae) from the Chapadmalal Formation, late Pliocene of Buenos Aires Province, Argentina. Ameghiniana 54(1):50–69

Rasia LL, Bonini RA, Candela AM (2020) Nuevos registros de *Lagostomus* Brookes, 1828 (Rodentia, Chinchillidae) en el Mioceno tardío de Argentina y su importancia bioestratigráfica. And Geol 47(2):430–445

Rasia LL, Candela AM, Cañón C (2021) Comprehensive total evidence phylogeny of chinchillids (Rodentia, Caviomorpha): cheek teeth anatomy and evolution. J Anat 239(2):405–423

Reguero MA, Candela AM (2011) Late Cenozoic mammals from the northwest of Argentina. In: Salfity JA, Marquillas RA (eds) Cenozoic geology of the Central Andes of Argentina. SCS, Salta, pp 411–426

Renison D, Cingolani AM, Contarde C, Guzmán D (2023) Asistiendo a la reintroducción de vizcachas (*Lagostomus maximus*): ¿Cómo aumentar el área de pastoreo seguro? Ecol Austral 33:20–29

Rovereto C (1914) Los estratos araucanos y sus fósiles. Anales Mus Nac Hist Nat Buenos Aires 25:1–247

Rusconi C (1929) Enumeración sistemática de las especies fósiles procedentes del piso Ensenadense halladas por el autor. Edición Privada:1–15

Rusconi C (1931) Restos de mamíferos fósiles hallados en las excavaciones del nuevo Mercado de Abastos Proveedor de la Capital Federal. La Ingeniería 684:1–8

Tauber AA (2000) Las Caleras, provincia de Córdoba, primer yacimiento fosilífero de origen cárstico de Argentina. Rev Esp Paleontol 15(2):233–244

Tauber AA (2005) Mamíferos fósiles y edad de la Formación Salicas (Mioceno tardío) de la sierra de Velasco, La Rioja, Argentina. Ameghiniana 42(2):443–460

Tauber AA, Hünicken M, Pensa M (1997) Un singular yacimiento fosilífero del Plioceno de La Calera, Departamento Calamuchita, Córdoba, Argentina. Ameghiniana 34(1):126

Tomassini RL, Montalvo CI, Deschamps CM, Manera T (2013) Biostratigraphy and biochronology of the Monte Hermoso Formation (early Pliocene) at its type locality, Buenos Aires Province, Argentina. J S Am Earth Sci 48:31–42

Tomassini RL, Montalvo CI, Beilinson E, Deschamps CM, Garrone MC, Gasparini GM, Zárate MA, Rabassa J, Ruella A, Tonni EP (2017) Microvertebrates preserved in mammal burrows from the Holocene of the Argentine Pampas: a taphonomic and paleoecological approach. Hist Biol 29(1):63–75

Tomassini R, Montalvo C, Beilinson E, Barasoain D, Schmidt G, Cerdeño E, Zurita A, Bonini R, Miño-Boilini Á, Rasia L, Gasparini G (2023) On the validity of the *Macrochorobates scalabrinii* Biozone (lower Huayquerian stage/age, Late Miocene). Multi-proxy analysis of the enigmatic Arenas Blancas site (Buenos Aires Province, Argentina). And Geol 50(1):22–56

Ubilla M, Rinderknecht A (2016) *Lagostomus maximus* (Desmarest) (Rodentia, Chinchillidae), the extant plains vizcacha in the Late Pleistocene of Uruguay. Alcheringa 40(3):354–365

Verzi DH, Deschamps CM, Tonni EP (2004) Biostratigraphic and palaeoclimatic meaning of the Middle Pleistocene South American rodent *Ctenomys kraglievichi* (Caviomorpha, Octodontidae). Palaeogeogr Palaeoclimatol Palaeoecol 212:315–329

Voglino D, Pardiñas UFJ (2005) Roedores sigmodontinos (Mammalia: Rodentia: Cricetidae) y otros micromamíferos pleistocénicos del norte de la provincia de Buenos Aires (Argentina): reconstrucción paleoambiental para el Ensenadense cuspidal. Ameghiniana 42(1):143–158

Vucetich MG (1989) Rodents (Mammalia) of the Lacayani fauna revisited (Deseadan, Bolivia). Comparison with new Chinchillidae and Cephalomyidae from Argentina. Bull Mus Natl Hist Nat 4° sér 11 sec. c (4):233–247

Vucetich MG, Verzi DH (1995) Los roedores caviomorfos. In: Alberdi MT, Leone G, Tonni EP (eds) Evolución Biológica y Climática de la Región Pampeana durante los últimos cinco millones de años. Un ensayo de correlación con el Mediterráneo occidental. Monografías Museo Nacional de Ciencias Naturales de Madrid, vol 12, pp 213–225

Weir BJ (1974) The tuco-tuco and plains viscacha. In: Rowlands IW, Weir BJ (eds) The biology of Hystricomorph Rodents. Symp Zool Soc London, vol 34, pp 113–128

Winge H (1888) Jordfundne og Nulevende gnavere (Rodentia) fra Lagoa Santa, Minas Geraes, Brasilien. E Museo Lundii 3:1–178

Zárate MA, Schultz PH, Blasi A, Heil C, King J, Hames W (2007) Geology and geochronology of type Chasicoan (late Miocene) mammal-bearing deposits of Buenos Aires (Argentina). J S Am Earth Sci 23:81–90

Zimicz N, Payrola P, Del Papa C (2018) New, Late Miocene mammalian assemblage from the Palo Pintado Formation (Northwestern Argentina). J S Am Earth Sci 81:31–44

Chapter 4
Morphological Variation in the Cranium of *Lagostomus maximus* Along Its Geographic Distribution

Luciano Luis Rasia, Marcos D. Ercoli, and Alicia Álvarez

Abstract Three subspecies of *Lagostomus maximus* have been traditionally recognized (*L. maximus maximus*, *L. maximus immollis*, and *L. maximus petilidens*), but their distributional ranges are uncertain and, although there are brief descriptions of the skull morphology of each subspecies, they are not clearly differentiated. We examined the cranial material of *Lagostomus maximus* from localities encompassing the ranges of the recognized subspecies, grouping them into three regions based on the involved phytogeographic provinces: North (Chaco+Northern Espinal), Pampas (Pampas), and South (Monte+Southern Espinal). Using morphometric variables and different methodological approaches (exploratory methods, between-group Principal Component Analysis, and Discriminant Analysis), we explored and evaluated the presence of morphological variation of the cranium along these three main regions and its possible correspondence with already proposed traits with systematic value. Our analyses do not show remarkable differences between vizcachas of the North, Pampas, and South regions, but some subtle yet statistically significant differences could help to characterize vizcachas from each region. Specimens from the North region have slightly wider heads, wider palates, and longer cheek teeth series than the South and the Pampas regions. Specimens from the South region have narrow palates, shorter cheek teeth series, and smaller skulls. Specimens from the Pampas region have narrow heads and palates, and larger skulls. Some of these differences are in line with already proposed traits that characterize subspecies, and eventually related to environmental factors, but there is a need for larger samples of many of the studied localities for further studies.

L. L. Rasia (✉)
División Paleontología Vertebrados, Museo de La Plata, La Plata, Argentina

CONICET, La Plata, Argentina
e-mail: lucianorasia@fcnym.unlp.edu.ar

M. D. Ercoli · A. Álvarez
CONICET, La Plata, Argentina

Laboratorio de Paleontología de Vertebrados, Instituto de Ecorregiones Andinas (INECOA; CONICET-UNJu), Instituto de Geología y Minería (UNJu),
San Salvador de Jujuy, Jujuy, Argentina

L. L. Rasia et al. (eds.), *Plains Vizcachas*, https://doi.org/10.1007/978-3-031-49487-1_4

Keywords Subspecies · Geographic variation · Skull morphology · Morphometry

4.1 Introduction

Three subspecies of the plains vizcacha, *Lagostomus maximus*, have been traditionally recognized: *L. m. maximus* Desmarest 1817, *L. m. immollis* Thomas 1910, and *L. m. petilidens* Hollister 1914 (e.g., Jackson et al. 1996). The type locality of *Lagostomus maximus maximus* is not designated (see Jackson et al. 1996), the type locality of *L. m. immollis* is Tapia (Tucumán Province, Argentina; Thomas 1910), and the type locality of *L. m. petilidens* is 12.9 km (8 miles in Hollister 1914) north to Carmen de Patagones (Buenos Aires Province, Argentina). Nevertheless, the distributional ranges of these subspecies are uncertain (Llanos and Crespo 1952; Jackson et al. 1996; see Fig. 4.1), and despite the brief descriptions of the skull

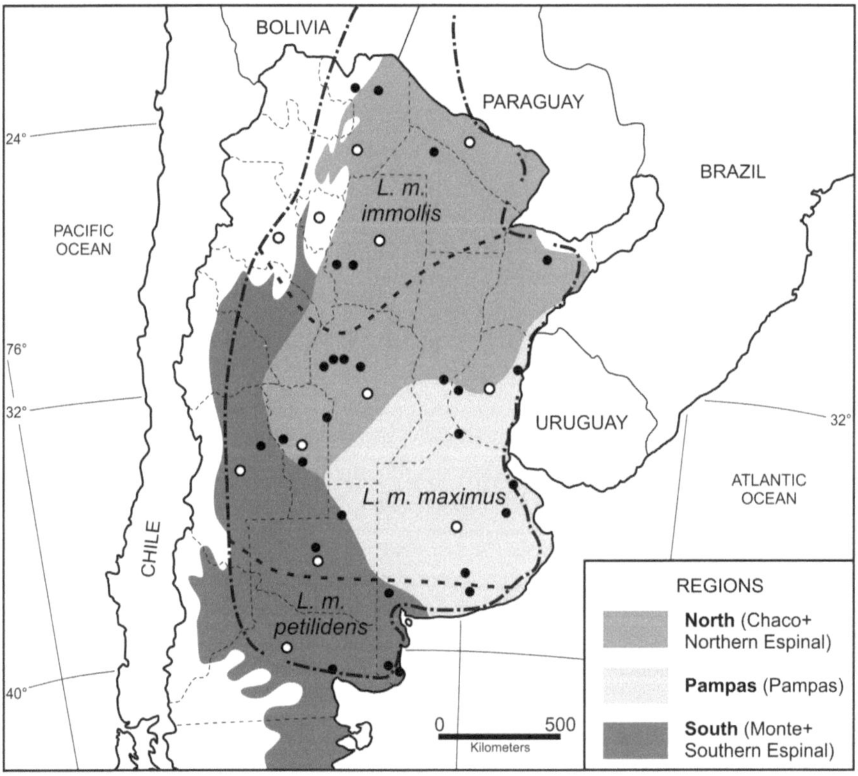

Fig. 4.1 Map of central and northern Argentina showing localities and regions based on phytogeographic provinces (following Cabrera 1976). Black circles: precise locality; white circles: locality indicating only political province; dash-dotted line: distribution limit of *Lagostomus maximus*; dashed line: limits between subspecies (following Jackson et al. 1996)

morphology of each one (Brookes 1828; Thomas 1910; Hollister 1914), they are not clearly differentiated. Thomas (1910) stated that *L. m. immollis* is slightly larger than *L. m. maximus*, with harsher and coarser fur, broader palatal foramina, and larger incisive foramina. According to Hollister (1914), *L. m. petilidens* is smaller than *L. m. maximus*, with smaller and slenderer incisors.

The foundation of subspecies was frequently done without strong support during the early twentieth century (Burt 1954), and it would be the case for plains vizcachas. Moreover, a recent study shows that there is low genetic variation among different populations of *L. maximus* (Gariboldi et al. 2019), reinforcing the possibility that there are no subspecies within extant plains vizcachas.

Among rodents, there are examples of morphological skull variation across wide geographic ranges or climatic gradients (e.g., Alhajeri 2021; Ashrafzadeh et al. 2019; Kageyama et al. 2009; Tabatabaei Yazdi and Adriaens 2011; Zhao et al. 2013), but there are also examples of morphologically conservative and widely distributed species (e.g., Dashti et al. 2022).

We examined and quantified the morphological variation of the cranium of *Lagostomus maximus*, using linear measurements. The specimens come from localities encompassing the geographic ranges of all the recognized subspecies and were grouped into three regions based on phytogeographic provinces (Cabrera 1976). We explored and evaluated the presence of morphological variation of the cranium along these three main regions and its potential systematic value.

4.2 Materials and Methods

4.2.1 Materials and Measurements

We examined 151 skulls of *Lagostomus maximus* deposited in the mammalogical collections of the Museo Argentino de Ciencias Naturales "Bernardino Rivadavia," Buenos Aires, Argentina (MACN), and the Museo de La Plata, La Plata, Argentina (MLP). Only subadult and adult specimens (Stages B and C of Llanos and Crespo (1952); see Rasia (2016) and Rasia et al. (in preparation) for recognition of age stages based on skull anatomy) were included in this study. The sex and locality of all the specimens were determined from collection labels. Many of the specimens do not have sex information ($n = 50$) and some of the localities of the specimens are not precise ($n = 18$), indicating only the political province provenance (Appendix Table 4.7).

Thirteen linear measurements of the cranium were taken in each specimen (Fig. 4.2) with a digital caliper to the nearest 0.01 mm. All specimens were measured by one person (LLR) to minimize error. The measurements are: BAP, breadth of anterior palate; BB, breadth of braincase; BPP, breadth of posterior palate; BZ, zygomatic breadth; HO, occipital height; HR, height of rostrum; LB, length of basicranium; LN, length of nasals; LO, length of orbit; LUD, length of upper diastema;

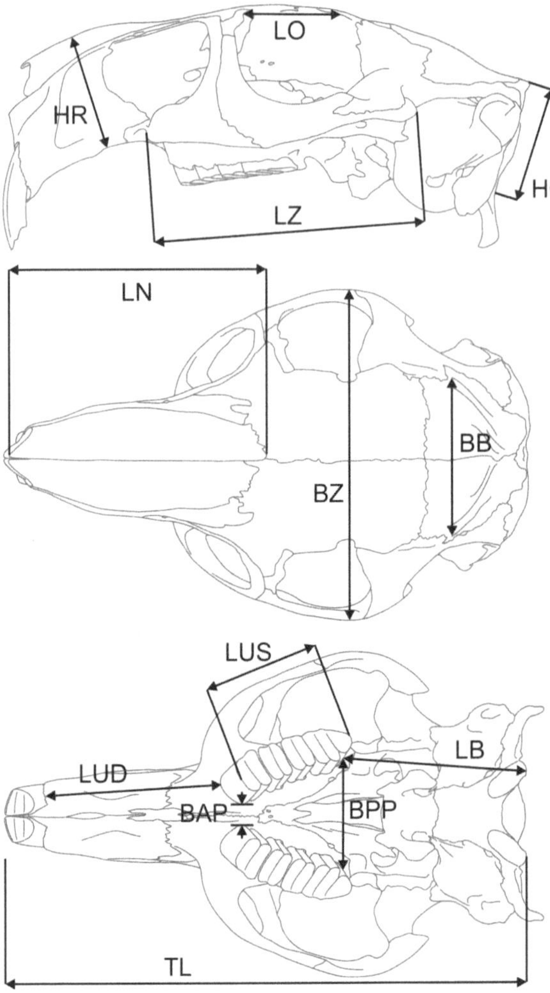

Fig. 4.2 Cranium measurements of *Lagostomus maximus*. BAP, breadth of anterior palate; BB, breadth of braincase; BPP, breadth of posterior palate; BZ, zygomatic breadth; HO, occipital height; HR, height of rostrum; LB, length of basicranium; LN, length of nasals; LO, length of orbit; LUD, length of upper diastema; LUS, length of upper cheek teeth series; LZ, length of zygomatic arch; TL, total length

LUS, length of upper cheek teeth series; LZ, length of zygomatic arch; and TL, total length.

Due to the undefined limits of the distributional ranges of the three subspecies (e.g., Jackson et al. 1996) and considering the results of preliminary exploratory analyses, the specimens were classified using categories based on the phytogeographic provinces proposed by Cabrera (1976). We propose three regions (see Fig. 4.1): Pampas (Pampas phytogeographic province), North (Chaco+Northern Espinal phytogeographic provinces), and South (Monte+Southern Espinal phytogeographic provinces).

4.2.2 Exploratory Methods

Geometric mean values (GM) were computed on the basis of the whole osteological variables and used as a proxy of cranial size. In turn, osteological variables were then standardized by the geometric mean of each specimen (Mosimann variables; e.g., Meachen-Samuels and Van Valkenburgh 2009). This procedure creates new osteological variables in which shape variation without the isometric effect of the size is represented (Appendix Table 4.7). The variation of GM and standardized osteological variables among regions (North, Pampas, and South) was first explored via box plots. Differences between groups for each variable were assessed by Tukey's statistic and informed when significant values were found.

In order to explore the shape variation among crania of vizcachas, we ran a between-groups Principal Component Analysis (bg-PCA; Mitteroecker and Bookstein 2011). Regions were considered as the grouping variable. Specimens without clearly defined phytogeographic province, in marginal geographic situation, or borderline between phytogeographic provinces were not included in this analysis (Appendix Table 4.7). The matrix of standardized variables for each specimen was projected onto the eigenvectors calculated from the variance/covariance matrix among groups. The bg-PCA maximizes morphological differences between selected groups instead of among specimens (as does traditional PCA). The two bg-PCA axes obtained (number of groups minus one) explained the complete variation between the three selected groups and part of the morphological variations among specimens. The bg-PCAs were carried out using the Morpho package (2.9 version; Schlager 2017; Schlager et al. 2022) in R software (4.2.1 version; R Core Team 2022). The among-specimens variation explained by the two bg-PCs was computed by linear regressions using PAST software (4.11 version; Hammer et al. 2001).

4.2.3 Discriminant Analyses

Two discriminant analyses were run using PAST software (4.11 version; Hammer et al. 2001). A first discriminant analysis was done using the standardized osteological variables as predictive variables. A second discriminant analysis was done adding the geometric mean to the standardized osteological variables. Discriminant analyses allow maximization of the segregation between known categories through discriminant axes constructed by the combination of original variables. The three regions were considered as known categories. In order to evaluate the performance of the discriminant functions, the specimens were reassigned to known categories (i.e., constructing the confusion matrices), and error rates for each region were computed. Specimens without clearly defined phytogeographic province were included as unclassified specimens and assigned to regions through the discriminant functions.

4.3 Results

4.3.1 *Boxplots*

Boxplots reveal that any variable alone allows separating the three regions with high confidence, existing high overlap between ranges and nonsignificant differences between regions for most of the variables (Fig. 4.3). Some subtle differences in the cranial shape of vizcachas from different regions can be highlighted. GM indicated that vizcachas from the South tended to be smaller than the North and Pampas ones, but these differences are not statistically supported. BPP partially segregated South from Pampas (Tukey's $Q = 4.413$, $p = 0.007$) and North (Tukey's $Q = 4.654$, $p = 0.005$) vizcachas, with the representatives of the latter group revealing a remarkably restricted variation on this variable. For LUS, North vizcachas frequently presented higher values than South and Pampas ones, particularly segregating North from South (Tukey's $Q = 3.602$, $p = 0.035$). The same applies for BZ, but particularly segregating North from Pampas (Tukey's $Q = 3.715$, $p = 0.028$). Although not significant, BB partially separated the specimens of the South from those of the North and Pampas (Fig. 4.3).

4.3.2 *Between-Group PCA*

The two first axes of the between-group PCA explained 45.54% of the morphological variation of studied specimens (bgPC1 = 20.97%; bgPC2 = 24.57%). High values of bgPC1 are mainly related to high values of LZ, LUD, LUS, BPP, LB, and TL and low values of BB and BAP (Table 4.1). High values of bgPC2 are mainly related to high values of BAP and low values of BZ, LZ, LUD, LB, and TL (Table 4.2). Any group (South, North, or Pampas) was clearly distinguishable. Specimens from Pampas and South are the most clearly separated, with a small overlapping zone. Vizcachas of Pampas region tend to locate on positive (e.g., specimens from La Plata, Juarez, and most of specimens from General Belgrano localities) and central values of bgPC1 (Fig. 4.4). Vizcachas of South region locate on negative (e.g., specimens from Patagones and San Blas localities) and central values of bgPC1. Conversely, specimens from the North have a wide morphological dispersion, encompassing a great portion of the zone occupied by the other two, and in some cases (e.g., specimens from Bialet Masse and some of Pocho localities) reaching extreme negative values on bgPC2 (Fig. 4.4).

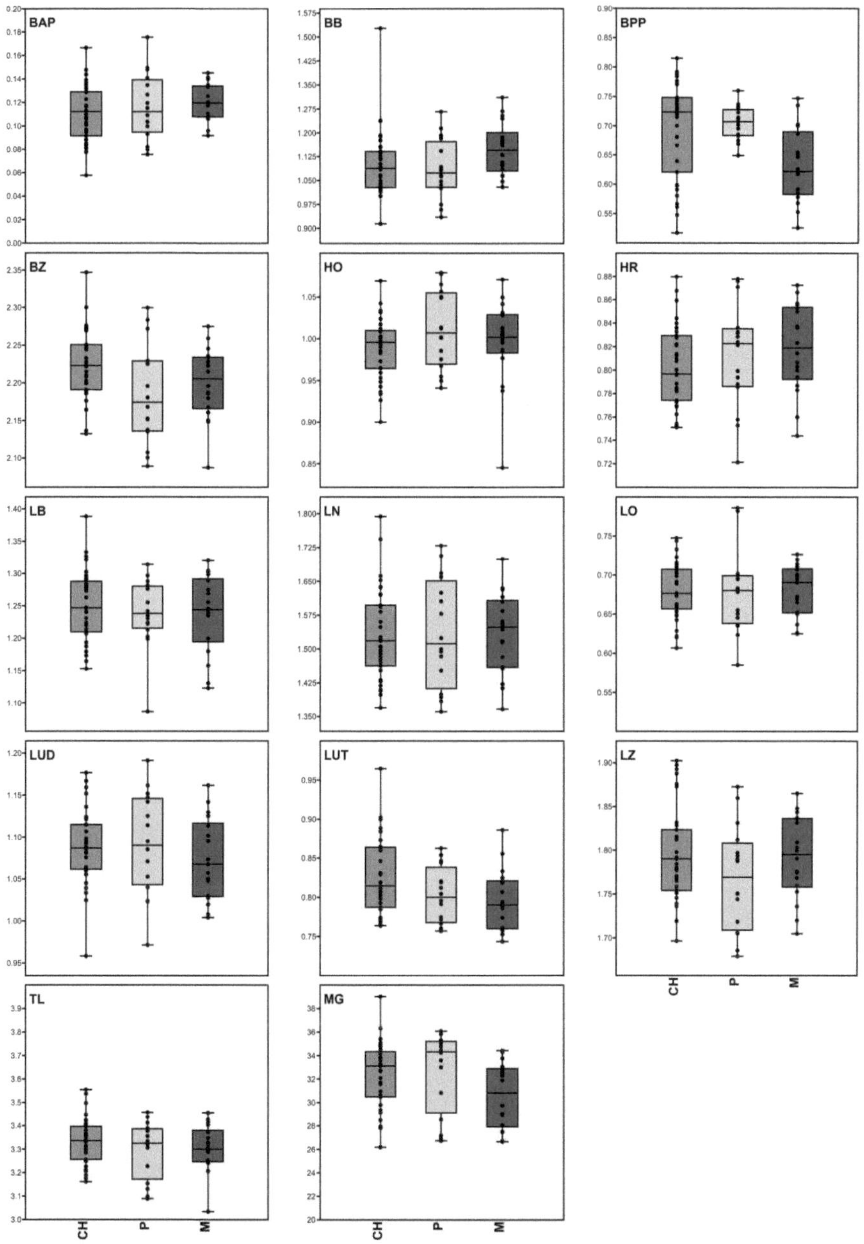

Fig. 4.3 Boxplots of standardized cranial measurements and geometric mean (GM) of analyzed specimens with assigned regions. Gray = North region, light gray = Pampas region, dark gray = South region. Abbreviations as in Fig. 4.2

Table 4.1 Statistics of the multivariate regression between bgPC1 and standardized osteological variables

Variable	Slope	Error	Intercept	Error	*r*	*p*
LN	0.22439	0.13483	1.533	0.011846	0.19925	0.10072
BZ	0.17625	0.069575	2.21	0.0061131	0.29564	0.013651
BB	−0.64032	0.085685	1.102	0.0075286	−0.67424	2.14E-10
HR	−0.0055145	0.052202	0.80961	0.0045866	−0.012905	0.91618
LO	−0.12347	0.050877	0.67932	0.0044702	−0.28425	0.01793
LZ	0.20766	0.06925	1.7885	0.0060845	0.344	0.0038021
HO	−0.0067121	0.059401	0.99631	0.0052192	−0.013803	0.91037
LUD	0.3041	0.057224	1.0852	0.0050279	0.54454	1.32E-06
LUT	0.23553	0.055824	0.81361	0.0049048	0.45817	7.53E-05
BAP	−0.13524	0.027425	0.11426	0.0024097	−0.51603	5.67E-06
BPP	0.51878	0.079921	0.68205	0.0070221	0.62136	1.22E-08
LB	0.20402	0.070983	1.2457	0.0062368	0.3313	0.0054235
TL	0.56156	0.11377	3.3204	0.0099959	0.51641	5.57E-06

Table 4.2 Statistics of the multivariate regression between bgPC2 and standardized osteological variables

Variable	Slope	Error	Intercept	Error	*r*	*p*
LN	−0.29051	0.13301	1.533	0.01168	−0.25781	0.032459
BZ	−0.39114	0.055019	2.21	0.0048314	−0.65572	9.66E-10
BB	−0.042892	0.11597	1.102	0.010184	−0.045138	0.71266
HR	−0.12288	0.050033	0.80961	0.0043935	−0.28738	0.016654
LO	−0.075732	0.052285	0.67932	0.0045912	−0.17425	0.15216
LZ	−0.50185	0.041067	1.7885	0.0036062	−0.83084	1.03E-18
HO	−0.010118	0.059429	0.99631	0.0052185	−0.020795	0.86533
LUD	−0.12962	0.066404	1.0852	0.0058311	−0.23197	0.055124
LUT	−0.27411	0.053173	0.81361	0.0046693	−0.53291	2.43E-06
BAP	0.1923	0.02178	0.11426	0.0019125	0.73335	7.74E-13
BPP	0.13355	0.10075	0.68205	0.0088469	0.15987	0.18947
LB	−0.36851	0.060329	1.2457	0.0052976	−0.59808	5.74E-08
TL	−0.88309	0.077657	3.3204	0.0068192	−0.81161	2.72E-17

4.3.3 Discriminant Analysis

The discriminant function on standardized osteological variables had a modest performance, classifying 63.77% of the total sample correctly (Appendix Table 4.7; Table 4.3; Fig. 4.5a). North was the region with the worst performance; the discriminant function correctly assigned only 18 of 35 representatives. The values for South and Pampas were 13 of 18 vizcachas and 13 of 16 vizcachas, respectively. When the size of vizcachas was considered (i.e., including the geometric mean as an

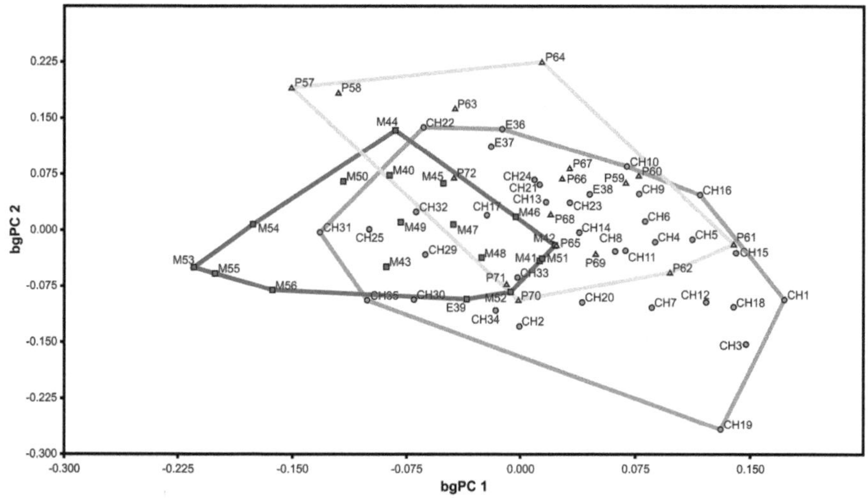

Fig. 4.4 Between-group principal component analysis of the analyzed specimens with assigned regions. Gray = North region, light gray = Pampas region, dark gray = South region. See Appendix Table 4.7 for specimen codes

Table 4.3 Confusion matrix of the discriminant analysis based on standardized osteological variables

	Predicted, CH	Predicted, M	Predicted, P	Predicted, Total	Percentage
CH	18	10	7	35	51.43%
M	4	13	1	18	72.22%
P	3	0	13	16	81.25%
Total	25	23	21	69	63.77%

Last column indicated the percentage of correct reclassification for each region and the total sample

additional predictor variable; Fig. 4.5b), the correct reclassification of specimen increases to 72.46%, by the improvement of the assignment of North (21 of 35) and South (16 of 18) vizcachas, while the proportion of correctly reclassified specimens of Pampas region remained equal (Appendix Table 4.7; Table 4.4).

In agreement with the bg-PCA, the specimens pertaining to the North region were less clearly identified as member of the group by the discriminant functions (Appendix Table 4.7).

The variables that contributed the most to the discrimination of groups Pampas and South are BB and BPP, and secondarily LUD. The variables that contributed the most to the separation of South and North are LUS, BZ, TL, and BPP and secondarily LB. The variables that contributed the most to the separation of Pampas and North are TL, BZ, HO, and LZ (Tables 4.5 and 4.6).

Specimens from peripheral regions of North (Salta, Tucumán, see also San Luis) were assigned to South (Appendix Table 4.7). Specimens from Estancia la Elisa

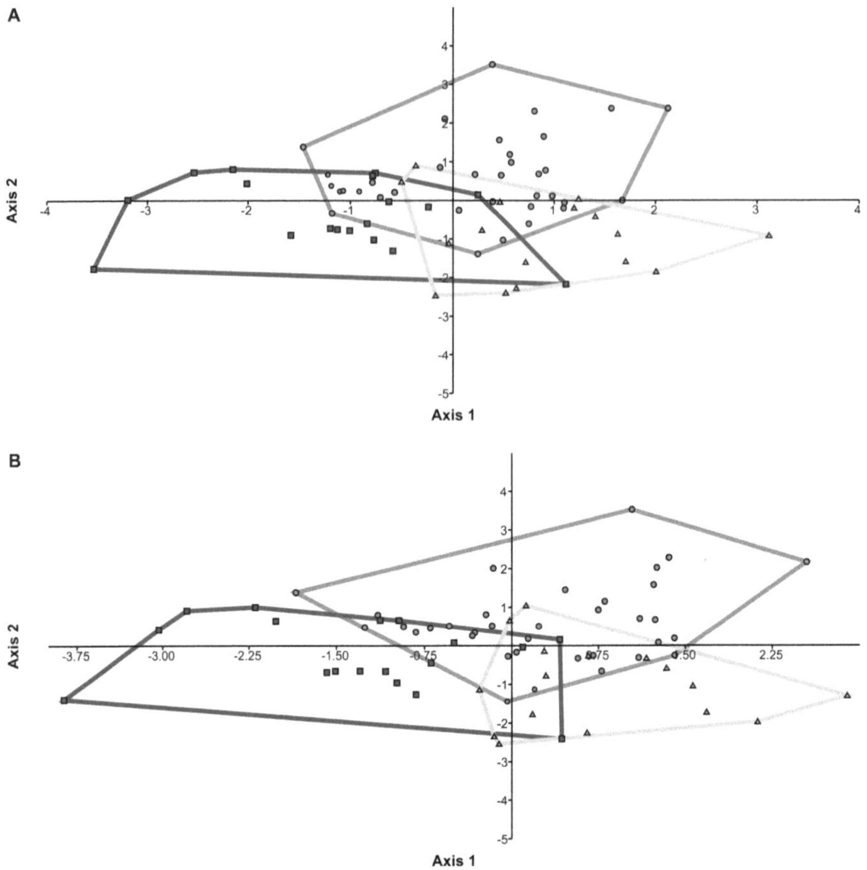

Fig. 4.5 Discriminant analyses of the analyzed specimens with assigned regions. Gray = North region, light gray = Pampas region, dark gray = South region. (**a**) Discriminant analysis based on osteological standardized measurements, (**b**) discriminant analysis based on osteological standardized measurements + geometric mean

Table 4.4 Factor loadings of the discriminant analysis based on standardized osteological variables

	Axis 1	Axis 2
LN	−0.00056	−0.00348
BZ	−0.00046	0.02568
BB	−0.03144	−0.00354
HR	−0.00551	−0.00827
LO	−0.00147	0.00071
LZ	−0.00795	0.01421
HO	0.00462	−0.01230
LUD	0.00885	0.00208
LUT	0.00887	0.01968
BAP	−0.00364	−0.00518
BPP	0.03838	0.01062
LB	0.00274	0.00791
TL	0.00373	0.02551

Table 4.5 Confusion matrix of the discriminant analysis based on standardized osteological variables plus the geometric mean

	Predicted, CH	Predicted, M	Predicted, P	Predicted, Total	Percentage
CH	21	6	8	35	60.00%
M	1	16	1	18	88.89%
P	2	1	13	16	81.25%
Total	24	23	22	69	72.46%

The last column indicates the percentage of correct reclassification for each region and the total sample

Table 4.6 Factor loadings of the discriminant analysis based on standardized osteological variables plus the geometric mean

	Axis 1	Axis 2
MG	1.01260	0.09088
LN	−0.00086	−0.00337
BZ	0.00198	0.02542
BB	−0.02998	0.00057
HR	−0.00597	−0.00746
LO	−0.00132	0.00089
LZ	−0.00616	0.01506
HO	0.00320	−0.01274
LUD	0.00854	0.00091
LUT	0.01022	0.01829
BAP	−0.00392	−0.0047
BPP	0.03719	0.00554
LB	0.00333	0.00746
TL	0.00592	0.02475

(Entre Ríos Province), a locality in a transitional zone between North and Pampas, as well as marginal for the distribution range of the species, have a wide distribution in the values of the osteological variables (Appendix Table 4.7), as well as in the discriminant morphospace, overlapping with the polygons of the three regions and even outside of them (results not shown), resulting in wrong predicted classifications assigning different specimens from this locality to some of the three regions (Appendix Table 4.7).

4.4 Discussion

Based on the analysis of the cranial morphology of *Lagostomus maximus*, it does not seem to be remarkable differences between the three considered regions (North, South, and Pampas), but some subtle differences can be highlighted. The Discriminant Analysis evidences that the combination of variables allows us to separate groups more clearly than the exploratory analyses. So, there is no key variable that permits to identify a geographic provenance, but the overall sum of morphological changes helps to discriminate groups.

Thomas (1910) stated that the northern subspecies *Lagostomus maximus immollis* was larger than the Pampean *L. m. maximus*, also mentioning broader incisive foramina and differences in fur coat. The geometric mean (GM) and the total length of cranium (TL) did not support this statement of size differences between North and Pampas vizcachas, but a relatively wider zygomatic arch (BZ) for North specimens was recovered as a significant difference compared to those of the Pampas region.

Hollister (1914) differentiated the southern subspecies *Lagostomus maximus petilidens* from *L. m. maximus* essentially because of its smaller and more slender incisors, mentioning also smaller cheek teeth, narrower presphenoid, much reduced choanae, among other features. The estimated distribution of *Lagostomus maximus petilidens* is included (see Hollister 1914; Jackson et al. 1996) in the southern portion of the South region here considered (see Fig. 4.1). The posterior breadth of the palate (BPP), can be considered equivalent to the width of the choanae. Specimens from the South region show narrower BPP with respect to the other two regions, and the differences between these pair of groups were statistically supported by our analysis. Additionally, the length of the upper cheek teeth (LUS) indicated a transitional increase from South, Pampas, to North regions, and the differences between South and North were recovered as significant. Both shape traits support differences between South and other regions in agreement with Hollister's (1914) description.

Beyond probable correspondences between some linear measurements and previous proposals in order to distinguish subspecies, the here presented analyses allow us to propose clues to characterize different morphotypes or morphological tendencies along the geographic distribution of vizcachas. The vizcachas from the South region are the most clearly differentiated from the other two regions, particularly from those of the Pampas region. When size and shape were considered together, the discriminant function allowed to recognize South vizcachas as pertaining to this region with an accuracy of 89%, and only few vizcachas of other regions are incorrectly classified as pertaining to South. North vizcachas showed a remarkable shape and size variations, including forms similar to other regions, and some specimens with unique morphologies. In relation to this, the morphological characterization of North vizcachas was a difficult task, and the discriminant function had the worst performance when trying to recognizing them.

In sum, the cranium of vizcachas allows us to partially distinguish morphotypes that could characterize each region, partially supporting and improving the traits described by previous authors to describe subspecies. Vizcachas from the North region have a slightly wider heads and wider palates and possess longer cheek teeth series than those from the South and the Pampas. The specimens from the South region can be more clearly differentiated by their narrow palate and shorter cheek teeth series and tend to possess smaller skulls. Finally, Pampas vizcachas have narrow heads and palates and tend to have larger skulls.

Nevertheless, some localities record particular cases. For example, the specimens of the very well-represented Estancia La Elisa locality (Entre Ríos Province; Appendix Table 4.7) show a wide morphological variation, surely in part related to the wide representation of this locality, but may also be related to its location in a transitional zone, represented by a mosaic of Chaco and riparian forests. Llanos and Crespo (1952) indicate that vizcachas of this locality do not exploit the forested environments, so the reason for this wide morphological variation is not clear.

General morphological trends indicate that specimens from North and Pampas, being the largest vizcachas, may be adapted to environments with high humidity and high availability of food sources, and specimens from South, the smaller ones, may be conditioned by the dryer conditions of central-western Argentina (see Figs. 4.1 and 4.3). This size difference was already noted for specimens from Entre Ríos and San Luis provinces, with marked variation in annual precipitation (see Jackson et al. 1996). Contrary to what is established by Bergmann's Rule (Bergmann 1847), plains vizcachas from the South are smaller than those from the North, as is observed in other fossorial rodents (Medina et al. 2007), given that, underground, temperature would not affect individuals in the same way as aboveground.

Interestingly, some vizcachas that were collected in the Argentinian Northwestern area (e.g., Salta, Tucumán; Appendix Table 4.7) are misclassified as pertaining to the South region. Although not clearly understood, the combination of cranial features of those specimens resembles those of the South, eventually linked to similar ecological demands between Monte (and Southern Espinal) and some specific environments of Yungas. Alternatively, a septentrional extension of Monte phytogeographic province has been proposed in phytogeographic schemes other than that we used (Parodi 1934). Alternatively, these cases can be interpreted as indicating phenotypic convergences, arguing against some of the traits that support the subspecies recognition for *L. maximus*. Further analyses would indicate if variation in the skull morphology may be related to adaptations to different environmental contexts (e.g., variations in diet, tooth-digging strategies, physiology, etc.).

4.5 Conclusions

Our quantitative analyses of cranial measurements of *Lagostomus maximus* do not show remarkable differences between vizcachas of the North, Pampas, and South regions. Nevertheless, some subtle but statistically significant differences, partially supporting previous subspecies definitions, could characterize each region as follows:

- North region: slightly wider head, wider palate, and longer cheek teeth series than South and the Pampas regions.
- South region: more clearly differentiated by its narrow palate, shorter cheek teeth series, and smaller skull.
- Pampas region: narrow head and palate and larger skull.

Some of these differences may be related to environmental factors; for example, larger vizcachas from the North and Pampas may be adapted to higher temperatures and humidity or high availability of food sources, and smaller ones from the South may be conditioned by the lower temperatures and dryer conditions of central-western Argentina.

Acknowledgments We want to thank the curators P. Teta (MACN), S. Lucero (MACN), D. Verzi (MLP), and I. Olivares (MLP) for permitting access to the material under their care. M.D.E. and A.A. thank INECOA-PUE 2017 22920170100027CO.

Appendix

Table 4.7 Studied specimens indicating collection number, sex (female: F, male: M), locality, geometric mean (GM), and standardized measurements (see Fig. 4.2) for each sample

Collection number	Sex	Locality/political province	Code	GM	LN	BZ	BB	HR	LO	LZ	HO	LUD	LUS	BAP	BPP	LB	TL	R	PR
MACN 39.12	?	Córdoba, Bialet	CH1	34.157	1.593	2.246	1.022	0.783	0.653	1.888	0.997	1.115	0.862	0.085	0.786	1.284	3.447	N	N
MACN 39.17	?	Córdoba, Pocho	CH2	29.140	1.506	2.250	1.237	0.770	0.703	1.832	0.953	1.095	0.862	0.083	0.725	1.302	3.396	N	N
MACN 39.21	?	Córdoba, Pocho	CH3	33.211	1.597	2.210	1.034	0.833	0.643	1.898	0.990	1.136	0.864	0.077	0.731	1.388	3.537	N	N
MACN 39.27	?	Córdoba, Pocho	CH4	33.787	1.526	2.244	1.043	0.797	0.660	1.798	1.042	1.081	0.831	0.101	0.724	1.290	3.357	N	N
MACN 39.32	?	Córdoba, Pocho	CH5	35.112	1.519	2.249	1.002	0.784	0.657	1.784	1.010	1.062	0.864	0.113	0.720	1.263	3.336	N	N
MACN 39.35	?	Córdoba, Pocho	CH6	34.778	1.410	2.164	1.527	0.752	0.629	1.736	0.973	1.094	0.793	0.113	0.700	1.221	3.205	N	N
MACN 39.42	?	Córdoba, Pocho	CH7	30.656	1.493	2.276	1.154	0.821	0.662	1.876	0.949	1.086	0.884	0.081	0.788	1.278	3.354	N	N
MACN 39.43	?	Córdoba, Pocho	CH8	27.828	1.471	2.269	1.157	0.828	0.723	1.816	0.996	1.045	0.902	0.081	0.792	1.207	3.284	N	P
MACN 39.49	?	Córdoba, Pocho	CH9	33.232	1.489	2.190	1.050	0.782	0.676	1.812	1.032	1.124	0.846	0.107	0.745	1.180	3.298	N	N
MACN 39.50	?	Córdoba, Pocho	CH10	34.947	1.636	2.199	1.014	0.774	0.712	1.780	1.001	1.055	0.807	0.112	0.742	1.210	3.249	N	N
MACN 39.53	?	Córdoba, Pocho	CH11	29.375	1.497	2.222	1.192	0.751	0.716	1.790	0.900	1.115	0.889	0.088	0.815	1.240	3.296	N	N
MACN 39.57	?	Córdoba, Pocho	CH12	33.329	1.476	2.300	1.088	0.754	0.702	1.829	0.936	1.060	0.864	0.097	0.777	1.293	3.352	N	N
MACN 39.58	?	Córdoba, Pocho	CH13	29.798	1.407	2.233	1.102	0.773	0.674	1.765	0.995	1.025	0.787	0.129	0.729	1.288	3.248	N	N

Collection number	Sex	Locality/political province	Code	GM	LN	BZ	BB	HR	LO	LZ	HO	LUD	LUS	BAP	BPP	LB	TL	R	PR
MACN 39.61	?	Córdoba, Pocho	CH14	33.608	1.483	2.203	1.084	0.814	0.658	1.792	0.960	1.081	0.819	0.117	0.723	1.231	3.339	N	N
MACN 39.62	?	Córdoba, Pocho	CH15	35.430	1.523	2.208	1.023	0.769	0.649	1.813	0.926	1.177	0.798	0.110	0.771	1.282	3.408	N	N
MACN 39.64	?	Córdoba, Pocho	CH16	34.517	1.419	2.186	1.020	0.752	0.622	1.791	0.993	1.075	0.860	0.132	0.773	1.220	3.298	N	N
MACN 39.68	?	Córdoba, Pocho	CH17	27.818	1.399	2.227	1.239	0.810	0.714	1.768	0.993	0.958	0.900	0.095	0.784	1.241	3.209	N	P
MACN 39.80	?	Córdoba, Pocho	CH18	33.870	1.369	2.272	1.040	0.789	0.689	1.893	1.024	1.159	0.815	0.092	0.748	1.327	3.421	N	N
MACN 39.86	?	Córdoba, Pocho	CH19	31.564	1.652	2.347	1.126	0.836	0.699	1.902	1.012	1.098	0.965	0.058	0.723	1.297	3.554	N	N
MACN 39.96	F	Córdoba, Bialet	CH20	27.942	1.560	2.245	1.136	0.796	0.748	1.873	0.997	1.082	0.873	0.078	0.715	1.274	3.368	N	N
MLP Mz 1634	F	Córdoba, El Quebracho	CH21	30.615	1.549	2.132	1.141	0.785	0.657	1.754	1.003	1.121	0.860	0.103	0.735	1.214	3.313	N	P
MLP Mz 565	F	San Luis, A Pencoso	CH22	30.496	1.463	2.189	1.118	0.800	0.677	1.751	1.033	1.098	0.764	0.137	0.681	1.153	3.161	N	P
MACN 28231	M	Santiago del Estero, Choya, Villa La Punta	CH23	34.788	1.638	2.223	1.002	0.840	0.657	1.746	1.034	1.167	0.785	0.117	0.640	1.194	3.357	N	P
MACN 28.264	?	Santiago del Estero	CH24	39.023	1.794	2.136	0.914	0.868	0.663	1.697	0.934	1.152	0.766	0.148	0.561	1.273	3.364	N	P
MACN 19183	?	Santiago del Estero, Loreto, La Florida	CH25	32.107	1.662	2.246	1.122	0.829	0.620	1.754	1.070	1.061	0.811	0.129	0.567	1.188	3.310	N	S

(continued)

Table 4.7 (continued)

Collection number	Sex	Locality/political province	Code	GM	LN	BZ	BB	HR	LO	LZ	HO	LUD	LUS	BAP	BPP	LB	TL	R	PR
MACN 23527	?	Chaco, Dto Guemes, Misión Nueva Pompeya	CH29	33.118	1.620	2.273	1.139	0.859	0.710	1.770	0.986	1.088	0.800	0.101	0.621	1.247	3.299	N	N
MACN 47.392	?	Formosa	CH30	34.335	1.621	2.228	1.067	0.880	0.707	1.778	0.983	1.123	0.803	0.117	0.517	1.280	3.423	N	S
MACN 39.98	F	Córdoba, B Massé	CH31	28.490	1.429	2.271	1.177	0.822	0.744	1.759	1.001	1.039	0.772	0.132	0.591	1.209	3.225	N	S
MACN 39.95	M	Córdoba, Guasta	CH32	32.676	1.431	2.191	1.062	0.785	0.706	1.790	0.965	1.033	0.764	0.166	0.599	1.224	3.256	N	N
MLP Mz 1752	?	Santiago del Estero, Defensa Agrícola	CH33	33.657	1.519	2.272	1.040	0.859	0.648	1.777	1.001	1.107	0.769	0.123	0.597	1.333	3.380	N	S
MLP Mz 1759	M	Santiago del Estero	CH34	32.743	1.744	2.176	1.028	0.844	0.691	1.812	1.003	1.086	0.829	0.106	0.548	1.322	3.497	N	S
MLP Mz 3-XII-02-3	?	Santiago del Estero, Dto Coya, Villa la Punta	CH35	26.179	1.504	2.199	1.189	0.762	0.733	1.824	1.017	1.092	0.807	0.113	0.581	1.247	3.400	N	S
MLP Mz 1659	F	Entre Ríos, Salto	E36	30.964	1.518	2.200	1.093	0.808	0.691	1.719	0.965	1.039	0.771	0.135	0.742	1.165	3.176	N	P
MLP Mz 1657	F	Entre Ríos, Salto	E37	31.739	1.453	2.186	1.117	0.782	0.673	1.739	0.943	1.064	0.774	0.144	0.738	1.173	3.189	N	N
MLP Mz 720	?	Sta Fe, Santo Tome	E38	36.318	1.582	2.215	1.001	0.813	0.607	1.719	1.009	1.087	0.802	0.139	0.666	1.240	3.331	N	N
MLP Mz 2-VI-60-6	?	Buenos Aires, Chasicó	E39	33.772	1.635	2.186	1.029	0.857	0.665	1.865	1.050	1.162	0.789	0.111	0.526	1.321	3.456	S	S
MLP Mz 1649	F	Catamarca, Sinquillin	M40	28.926	1.561	2.149	1.175	0.803	0.727	1.775	0.938	1.050	0.818	0.119	0.687	1.158	3.207	S	S

Collection number	Sex	Locality/political province	Code	GM	LN	BZ	BB	HR	LO	LZ	HO	LUD	LUS	BAP	BPP	LB	TL	R	PR
MLP MZ 1650	F	Catamarca, Sinquillin	M41	29.042	1.517	2.215	1.167	0.807	0.695	1.848	0.995	1.020	0.886	0.092	0.747	1.199	3.321	S	S
MLP Mz 1651	M	Catamarca, Sinquillin	M42	34.437	1.632	2.161	1.047	0.856	0.637	1.803	0.999	1.095	0.855	0.107	0.652	1.239	3.404	S	S
MLP Mz 1729	M	Mendoza, La Paz, Arroyito	M43	27.477	1.422	2.180	1.186	0.794	0.714	1.832	1.071	1.004	0.825	0.108	0.626	1.303	3.347	S	S
MLP Mz 1728	F	Mendoza, La Paz, Arroyito	M44	28.941	1.457	2.087	1.177	0.760	0.701	1.720	1.042	1.047	0.761	0.125	0.700	1.289	3.243	S	P
MLP Mz 1654	M	Bs As, Patagones	M45	32.657	1.413	2.150	1.160	0.744	0.698	1.776	0.977	1.073	0.761	0.133	0.702	1.269	3.248	S	N
MLP Mz 05-IX-97-01	?	Río Negro, San Antonio Oeste	M46	32.298	1.460	2.196	1.130	0.783	0.690	1.753	0.845	1.102	0.744	0.140	0.735	1.242	3.293	S	S
MACN 53.8	F	Catamarca	M47	29.746	1.515	2.228	1.107	0.823	0.669	1.809	0.943	1.057	0.833	0.135	0.654	1.131	3.241	S	S
MACN 53.13	M	Catamarca	M48	32.851	1.700	2.233	1.065	0.872	0.626	1.736	0.985	1.142	0.820	0.119	0.591	1.180	3.414	S	S
MACN 53.70	M	Catamarca	M49	32.525	1.584	2.223	1.097	0.850	0.625	1.760	1.028	1.114	0.753	0.134	0.584	1.235	3.328	S	S
MACN 53.69	M	Catamarca	M50	33.044	1.615	2.168	1.100	0.853	0.650	1.705	1.004	1.068	0.758	0.145	0.568	1.246	3.288	S	S
MACN 12952	M	Río Negro	M51	34.312	1.553	2.275	1.064	0.837	0.691	1.768	1.013	1.125	0.759	0.106	0.647	1.299	3.372	S	N
MACN 20393	?	Mendoza	M52	31.894	1.557	2.223	1.085	0.866	0.673	1.836	1.031	1.130	0.807	0.096	0.618	1.304	3.426	S	S

(continued)

Table 4.7 (continued)

Collection number	Sex	Locality/political province	Code	GM	LN	BZ	BB	HR	LO	LZ	HO	LUD	LUS	BAP	BPP	LB	TL	R	PR
MLP Mz 1617	F	Buenos Aires, Patagones	M53	26.650	1.366	2.245	1.311	0.787	0.720	1.791	0.987	1.008	0.791	0.141	0.578	1.256	3.034	S	S
MLP Mz 1527 (2)	?	Río Negro, Bahía San Blas	M54	27.488	1.605	2.188	1.254	0.836	0.652	1.844	1.006	1.067	0.794	0.118	0.617	1.123	3.251	S	S
MLP Mz 1527 (3)	?	Río Negro, Bahía San Blas	M55	26.696	1.543	2.259	1.268	0.800	0.711	1.800	1.009	1.028	0.774	0.120	0.553	1.275	3.252	S	S
MLP Mz 1527 (4)	?	Río Negro, Bahía San Blas	M56	28.066	1.482	2.235	1.245	0.815	0.707	1.837	0.987	1.030	0.786	0.120	0.585	1.236	3.306	S	S
MLP Mz 19-IX-47-3	F	Bs As, San Nicolas	P57	27.174	1.384	2.107	1.267	0.758	0.683	1.705	1.001	0.972	0.767	0.148	0.737	1.222	3.089	P	P
MLP Mz 1603	F	Bs As, Gral Belgrano	P58	28.542	1.399	2.153	1.214	0.753	0.655	1.719	1.049	1.023	0.757	0.149	0.714	1.202	3.098	P	P
MLP Mz 1601	M	Bs As, Gral Belgrano	P59	35.832	1.625	2.229	0.975	0.821	0.636	1.751	1.057	1.147	0.775	0.119	0.675	1.231	3.329	P	P
MLP Mz 1602	M	Bs As, Gral Belgrano	P60	36.083	1.707	2.272	0.935	0.876	0.696	1.750	1.080	1.125	0.760	0.103	0.669	1.224	3.307	P	P
MLP Mz 1605	M	Bs As, Gral Belgrano	P61	35.024	1.668	2.300	0.958	0.871	0.585	1.788	1.050	1.191	0.804	0.100	0.707	1.235	3.412	P	P
MLP Mz 1604	M	Bs As, Gral Belgrano	P62	35.309	1.729	2.283	1.026	0.878	0.651	1.832	1.078	1.161	0.812	0.075	0.702	1.276	3.437	P	P
MACN 5.46	?	Bs As, La Plata	P63	30.837	1.361	2.152	1.142	0.829	0.700	1.686	0.941	1.085	0.854	0.141	0.733	1.086	3.130	P	P
MACN 5.45	?	Bs As, La Plata	P64	33.010	1.393	2.089	1.047	0.721	0.682	1.679	0.975	1.053	0.759	0.175	0.760	1.213	3.153	P	P
MACN 54.153	M	Bs As, Juarez	P65	33.593	1.500	2.168	1.086	0.786	0.695	1.797	0.968	1.071	0.820	0.115	0.695	1.255	3.378	P	N
MACN 54.151	M	Bs As, Juarez	P66	34.464	1.578	2.181	1.082	0.788	0.635	1.744	1.065	1.142	0.791	0.109	0.707	1.242	3.334	P	P

Collection number	Sex	Locality/political province	Code	GM	LN	BZ	BB	HR	LO	LZ	HO	LUD	LUS	BAP	BPP	LB	TL	R	PR
MACN 54.150	M	Bs As, Juarez	P67	34.774	1.484	2.100	1.036	0.824	0.678	1.790	0.950	1.095	0.772	0.127	0.723	1.277	3.306	P	P
MACN 54.147	M	Bs As, Juarez	P68	34.394	1.660	2.135	1.065	0.836	0.702	1.793	1.014	1.114	0.796	0.093	0.685	1.297	3.389	P	P
MACN 54.146	M	Bs As, Juarez	P69	34.260	1.524	2.137	1.062	0.833	0.623	1.812	1.013	1.152	0.819	0.104	0.683	1.314	3.457	P	P
MACN 36.144	M	Bs As, San Cayetano	P70	26.734	1.494	2.229	1.183	0.799	0.782	1.860	0.986	1.053	0.863	0.080	0.712	1.282	3.356	P	N
MACN 36.145	M	Bs As, San Cayetano	P71	26.934	1.452	2.225	1.192	0.794	0.786	1.873	1.002	1.024	0.847	0.082	0.729	1.288	3.321	P	P
MACN 39.149	?	Bs As, San Cayetano	P72	35.281	1.606	2.196	1.093	0.832	0.645	1.706	0.955	1.040	0.844	0.135	0.649	1.199	3.227	P	N
MACN 14316	M	Salta	CH26	31.605	1.621	2.193	1.070	0.884	0.667	1.836	0.989	1.030	0.786	0.120	0.590	1.273	3.364	?	S
MACN 17.101	?	Salta, Dto San Martin, Embarcacion	CH27	34.738	1.535	2.239	1.096	0.837	0.672	1.765	0.969	1.149	0.791	0.113	0.609	1.289	3.433	?	N
MACN 36.282	F	Salta, Dragones	CH28	27.525	1.569	2.200	1.257	0.845	0.694	1.784	1.056	1.075	0.789	0.126	0.496	1.193	3.300	?	S
MLP Mz 1642	F	San Luis	?	27.209	1.338	2.195	1.272	0.756	0.603	1.731	1.018	1.014	0.773	0.144	0.716	1.294	3.185	?	S
MACN 27136	?	San Luis, Dto Gral Pedernera, cerca de La Punilla	?	31.991	1.440	2.287	1.048	0.829	0.710	1.883	1.093	1.207	0.704	0.107	0.610	1.194	3.617	?	S
MACN 27586	?	Corrientes, Mercedes, RP Iberá	?	36.027	1.738	2.176	0.972	0.847	0.640	1.778	1.037	1.154	0.863	0.109	0.578	1.258	3.439	?	N

(continued)

Table 4.7 (continued)

Collection number	Sex	Locality/political province	Code	GM	LN	BZ	BB	HR	LO	LZ	HO	LUD	LUS	BAP	BPP	LB	TL	R	PR
MLP Mz 2515	?	San Luis, Estancia El Centenario	?	33.631	1.598	2.264	1.026	0.854	0.635	1.830	1.010	1.171	0.839	0.096	0.635	1.275	3.523	?	N
MLP Mz 2516	?	San Luis, Estancia El Centenario	?	34.355	1.592	2.228	0.950	0.848	0.687	1.832	0.981	1.177	0.802	0.120	0.559	1.276	3.505	?	N
MLP Mz 1941	?	San Luis, Dto Gral Pedernera, Ea El Centenario	?	29.669	1.487	2.189	1.158	0.827	0.719	1.728	1.006	1.114	0.822	0.115	0.579	1.251	3.363	?	N
MLP Mz 2363	?	San Luis, Dto Gral Pedernera, Ea El Centenario	?	33.905	1.527	2.240	1.089	0.797	0.702	1.789	1.026	1.113	0.843	0.108	0.618	1.238	3.380	?	N
MACN 17.67	?	Tucumán	?	29.806	1.467	2.194	1.172	0.818	0.687	1.815	0.926	1.057	0.825	0.147	0.558	1.193	3.288	?	S
MACN 48.100	F	Entre Ríos, Ea. La Elisa	?	27.703	1.317	2.134	1.177	0.772	0.693	1.651	0.990	0.986	0.735	0.166	0.776	1.211	3.053	?	P
MACN 48.101	F	Entre Ríos, Ea. La Elisa	?	28.064	1.420	2.134	1.215	0.779	0.714	1.706	0.973	1.045	0.722	0.158	0.691	1.160	3.070	?	P
MACN 48.103	M	Entre Ríos, Ea. La Elisa	?	29.288	1.465	2.106	1.187	0.776	0.627	1.679	1.027	1.033	0.740	0.152	0.715	1.240	3.155	?	P
MACN 48.105	F	Entre Ríos, Ea. La Elisa	?	29.059	1.499	2.190	1.202	0.784	0.650	1.708	0.956	0.982	0.752	0.146	0.725	1.227	3.167	?	S
MACN 48.107	F	Entre Ríos, Ea. La Elisa	?	32.383	1.645	2.186	1.036	0.799	0.667	1.765	0.939	1.079	0.802	0.119	0.743	1.217	3.266	?	S
MACN 48.108	F	Entre Ríos, Ea. La Elisa	?	29.391	1.509	2.024	1.159	0.795	0.640	1.679	1.001	1.028	0.750	0.151	0.724	1.236	3.178	?	P
MACN 48.109	F	Entre Ríos, Ea. La Elisa	?	29.670	1.566	2.163	1.115	0.795	0.676	1.730	0.957	1.058	0.774	0.134	0.696	1.216	3.203	?	S

Collection number	Sex	Locality/political province	Code	GM	LN	BZ	BB	HR	LO	LZ	HO	LUD	LUS	BAP	BPP	LB	TL	R	PR
MACN 48.112	M	Entre Ríos, Ea. La Elisa	?	28.175	1.418	2.087	1.162	0.789	0.635	1.693	0.980	0.988	0.741	0.169	0.750	1.223	3.085	?	P
MACN 48.113	F	Entre Ríos, Ea. La Elisa	?	29.238	1.599	2.165	1.195	0.789	0.696	1.734	0.936	1.045	0.784	0.107	0.785	1.221	3.232	?	P
MACN 48.114	M	Entre Ríos, Ea. La Elisa	?	36.964	1.662	2.232	0.975	0.835	0.607	1.755	0.972	1.104	0.740	0.143	0.688	1.220	3.268	?	S
MACN 48.117	M	Entre Ríos, Ea. La Elisa	?	35.524	1.667	2.223	1.016	0.820	0.611	1.752	0.993	1.142	0.740	0.131	0.672	1.253	3.280	?	S
MACN 48.271	M	Entre Ríos, Ea. La Elisa	?	38.314	1.636	2.151	1.004	0.797	0.624	1.666	0.941	1.117	0.692	0.165	0.724	1.226	3.209	?	P
MACN 48.272	M	Entre Ríos, Ea. La Elisa	?	35.679	1.740	2.238	1.040	0.799	0.626	1.731	0.957	1.150	0.748	0.119	0.726	1.197	3.346	?	N
MACN 48.275	M	Entre Ríos, Ea. La Elisa	?	34.446	1.595	2.151	1.006	0.804	0.622	1.705	0.950	1.130	0.733	0.150	0.710	1.244	3.273	?	P
MACN 48.276	M	Entre Ríos, Ea. La Elisa	?	36.492	1.709	2.190	0.979	0.823	0.579	1.759	0.987	1.128	0.744	0.132	0.662	1.320	3.421	?	S
MACN 48.277	F	Entre Ríos, Ea. La Elisa	?	29.161	1.592	2.132	1.188	0.815	0.667	1.726	0.977	1.077	0.797	0.119	0.696	1.177	3.233	?	S
MACN 48.279	M	Entre Ríos, Ea. La Elisa	?	39.510	1.769	2.238	0.857	0.850	0.621	1.780	0.934	1.163	0.749	0.129	0.695	1.267	3.384	?	N
MACN 48.85	M	Entre Ríos, Ea. La Elisa	?	37.880	1.790	2.180	0.974	0.878	0.625	1.729	0.945	1.161	0.730	0.113	0.705	1.291	3.367	?	P
MACN 48.86	F	Entre Ríos, Ea. La Elisa	?	29.598	1.519	2.150	1.174	0.795	0.686	1.741	0.951	1.038	0.785	0.145	0.667	1.165	3.152	?	S
MACN 48.87	F	Entre Ríos, Ea. La Elisa	?	29.150	1.535	2.195	1.174	0.773	0.674	1.747	0.962	1.041	0.781	0.131	0.719	1.184	3.174	?	S

(continued)

Table 4.7 (continued)

Collection number	Sex	Locality/political province	Code	GM	LN	BZ	BB	HR	LO	LZ	HO	LUD	LUS	BAP	BPP	LB	TL	R	PR
MACN 48.91	M	Entre Ríos, Ea. La Elisa	?	36.424	1.643	2.210	0.970	0.809	0.650	1.715	0.977	1.115	0.741	0.139	0.714	1.203	3.263	?	P
MACN 48.92	M	Entre Ríos, Ea. La Elisa	?	29.222	1.466	2.124	1.176	0.767	0.654	1.659	0.935	1.017	0.720	0.178	0.732	1.196	3.069	?	P
MACN 49.12	F	Entre Ríos, Ea. La Elisa	?	32.849	1.641	2.159	1.050	0.770	0.668	1.749	0.940	1.076	0.804	0.129	0.711	1.220	3.278	?	N
MACN 49.13	M	Entre Ríos, Ea. La Elisa	?	38.499	1.710	2.208	0.960	0.797	0.642	1.742	0.943	1.142	0.743	0.137	0.698	1.209	3.340	?	N
MACN 49.16	F	Entre Ríos, Ea. La Elisa	?	30.039	1.649	2.146	1.116	0.811	0.677	1.747	0.993	1.065	0.795	0.107	0.709	1.269	3.264	?	P
MACN 49.17	F	Entre Ríos, Ea. La Elisa	?	30.147	1.492	2.205	1.126	0.792	0.656	1.733	1.008	1.040	0.782	0.128	0.707	1.237	3.258	?	S
MACN 49.226	M	Entre Ríos, Ea. La Elisa	?	36.893	1.633	2.181	0.947	0.819	0.640	1.771	0.961	1.112	0.745	0.134	0.710	1.283	3.298	?	P
MACN 49.231	M	Entre Ríos, Ea. La Elisa	?	40.868	1.837	2.222	0.924	0.838	0.599	1.772	0.978	1.201	0.740	0.116	0.700	1.258	3.363	?	P
MACN 49.268	F	Entre Ríos, Ea. La Elisa	?	30.969	1.626	2.132	1.153	0.774	0.680	1.774	0.931	1.081	0.817	0.121	0.696	1.190	3.253	?	S
MACN 49.269	F	Entre Ríos, Ea. La Elisa	?	32.600	1.636	2.172	1.068	0.798	0.641	1.792	0.934	1.090	0.823	0.116	0.716	1.237	3.332	?	S
MACN 49.271	F	Entre Ríos, Ea. La Elisa	?	31.365	1.600	2.191	1.095	0.826	0.646	1.789	0.936	1.053	0.815	0.114	0.746	1.213	3.280	?	S
MACN 49.272	M	Entre Ríos, Ea. La Elisa	?	36.303	1.633	2.104	0.954	0.829	0.596	1.736	0.961	1.093	0.745	0.154	0.732	1.239	3.263	?	P
MACN 49.275	F	Entre Ríos, Ea. La Elisa	?	28.979	1.530	2.116	1.132	0.790	0.668	1.748	0.987	1.052	0.790	0.128	0.706	1.235	3.233	?	S

Collection number	Sex	Locality/political province	Code	GM	LN	BZ	BB	HR	LO	LZ	HO	LUD	LUS	BAP	BPP	LB	TL	R	PR
MACN 49.278	M	Entre Ríos, Ea. La Elisa	?	30.423	1.599	2.192	1.107	0.786	0.682	1.769	0.935	1.025	0.831	0.114	0.776	1.198	3.216	?	N
MACN 49.280	M	Entre Ríos, Ea. La Elisa	?	35.531	1.604	2.114	0.988	0.822	0.632	1.743	0.953	1.088	0.749	0.151	0.703	1.236	3.248	?	P
MACN 49.281	F	Entre Ríos, Ea. La Elisa	?	32.290	1.623	2.190	1.104	0.780	0.673	1.779	0.943	1.059	0.823	0.116	0.711	1.227	3.296	?	N
MACN 49.282	F	Entre Ríos, Ea. La Elisa	?	30.705	1.559	2.188	1.113	0.784	0.688	1.735	0.946	1.048	0.782	0.123	0.729	1.239	3.259	?	N
MACN 49.284	M	Entre Ríos, Ea. La Elisa	?	31.794	1.606	2.157	1.061	0.772	0.688	1.775	0.952	1.059	0.800	0.129	0.716	1.208	3.218	?	S
MACN 49.287	F	Entre Ríos, Ea. La Elisa	?	28.917	1.594	2.143	1.096	0.819	0.673	1.776	1.009	1.049	0.841	0.122	0.715	1.074	3.261	?	S
MACN 49.289	F	Entre Ríos, Ea. La Elisa	?	30.975	1.586	2.137	1.145	0.801	0.674	1.780	0.933	1.042	0.826	0.116	0.730	1.213	3.254	?	S
MACN 49.290	M	Entre Ríos, Ea. La Elisa	?	35.109	1.596	2.113	1.012	0.791	0.607	1.746	0.968	1.108	0.750	0.148	0.697	1.285	3.276	?	P
MACN 49.4	M	Entre Ríos, Ea. La Elisa	?	27.044	1.473	2.094	1.234	0.782	0.696	1.721	1.009	1.036	0.745	0.132	0.680	1.243	3.225	?	P
MACN 49.46	F	Entre Ríos, Ea. La Elisa	?	30.885	1.641	2.223	1.098	0.817	0.635	1.753	0.933	1.079	0.749	0.118	0.750	1.251	3.287	?	S
MACN 49.6	M	Entre Ríos, Ea. La Elisa	?	36.388	1.700	2.204	1.011	0.853	0.610	1.773	0.952	1.097	0.788	0.115	0.686	1.316	3.355	?	S
MACN 50.10	M	Entre Ríos, Ea. La Elisa	?	36.467	1.651	2.146	0.920	0.788	0.630	1.732	0.981	1.105	0.761	0.146	0.726	1.245	3.288	?	P
MACN 50.11	M	Entre Ríos, Ea. La Elisa	?	38.929	1.640	2.221	0.998	0.798	0.614	1.699	0.935	1.093	0.717	0.148	0.701	1.313	3.302	?	N

(continued)

Table 4.7 (continued)

Collection number	Sex	Locality/political province	Code	GM	LN	BZ	BB	HR	LO	LZ	HO	LUD	LUS	BAP	BPP	LB	TL	R	PR
MACN 50.12	M	Entre Ríos, Ea. La Elisa	?	36.920	1.685	2.227	0.873	0.855	0.575	1.795	0.990	1.132	0.793	0.130	0.699	1.263	3.381	?	N
MACN 50.13	M	Entre Ríos, Ea. La Elisa	?	38.935	1.754	2.170	0.942	0.816	0.630	1.770	0.951	1.106	0.764	0.125	0.708	1.274	3.370	?	N
MACN 50.14	F	Entre Ríos, Ea. La Elisa	?	32.883	1.649	2.129	1.075	0.789	0.697	1.732	0.924	1.046	0.783	0.139	0.717	1.165	3.158	?	P
MACN 50.15	M	Entre Ríos, Ea. La Elisa	?	37.701	1.676	2.206	0.943	0.805	0.625	1.737	0.964	1.128	0.756	0.127	0.720	1.291	3.379	?	N
MACN 50.19	M	Entre Ríos, Ea. La Elisa	?	32.316	1.555	2.155	1.135	0.781	0.633	1.729	0.988	1.051	0.745	0.137	0.709	1.259	3.242	?	S
MACN 50.20	M	Entre Ríos, Ea. La Elisa	?	31.621	1.539	2.122	1.102	0.769	0.628	1.748	0.973	1.006	0.774	0.144	0.710	1.300	3.262	?	S
MACN 50.21	M	Entre Ríos, Ea. La Elisa	?	38.035	1.605	2.168	0.963	0.786	0.599	1.718	0.975	1.105	0.729	0.162	0.702	1.276	3.238	?	P
MACN 50.22	M	Entre Ríos, Ea. La Elisa	?	31.320	1.587	2.129	1.120	0.813	0.649	1.713	0.981	1.042	0.756	0.131	0.716	1.247	3.238	?	P
MACN 50.24	F	Entre Ríos, Ea. La Elisa	?	27.551	1.438	2.121	1.193	0.765	0.667	1.678	0.981	1.020	0.738	0.168	0.712	1.185	3.067	?	P
MACN 50.25	M	Entre Ríos, Ea. La Elisa	?	28.624	1.492	2.129	1.219	0.775	0.605	1.716	1.002	1.012	0.745	0.151	0.707	1.251	3.174	?	S
MACN 50.26	M	Entre Ríos, Ea. La Elisa	?	29.640	1.491	2.099	1.127	0.775	0.661	1.692	0.990	1.024	0.752	0.159	0.723	1.188	3.144	?	P
MACN 50.8	F	Entre Ríos, Ea. La Elisa	?	31.408	1.616	2.151	1.118	0.797	0.677	1.777	0.982	1.074	0.781	0.115	0.732	1.200	3.230	?	S
MACN 50.9	F	Entre Ríos, Ea. La Elisa	?	31.485	1.571	2.155	1.160	0.786	0.652	1.735	0.936	1.025	0.791	0.136	0.722	1.210	3.186	?	S
MACN 59.6	M	Entre Ríos	?	37.893	1.680	2.185	0.953	0.822	0.628	1.771	0.959	1.144	0.770	0.126	0.705	1.244	3.361	?	N

Collection number	Sex	Locality/political province	Code	GM	LN	BZ	BB	HR	LO	LZ	HO	LUD	LUS	BAP	BPP	LB	TL	R	PR
MLP Mz 25-X-95-2	F	La Pampa, Jagual del Monte	?	29.985	1.472	2.097	1.125	0.765	0.713	1.743	1.002	1.051	0.822	0.124	0.801	1.128	3.123	?	P
MACN 31.301	?	La Pampa, Conhelo	?	31.730	1.577	2.338	0.966	0.837	0.735	1.810	1.015	1.149	0.871	0.101	0.602	1.155	3.517	?	N
MACN 34.721	?	La Pampa, Conhelo	?	34.935	1.685	2.274	1.029	0.838	0.653	1.841	0.986	1.177	0.808	0.094	0.622	1.302	3.525	?	N
MACN 43.5	?	La Pampa, Conhelo	?	33.861	1.668	2.157	0.948	0.874	0.707	1.922	1.032	1.188	0.753	0.096	0.651	1.212	3.518	?	P
MACN 31.302	?	La Pampa, Conhelo	?	33.227	1.413	2.301	1.030	0.788	0.612	1.816	1.037	1.094	0.875	0.133	0.591	1.282	3.400	?	N
MACN 32.2	?	La Pampa, Conhelo	?	32.052	1.560	2.208	1.065	0.851	0.685	1.814	0.991	1.089	0.849	0.118	0.525	1.313	3.466	?	S
MACN 21.5	?	Buenos Aires	?	34.201	1.538	2.229	0.962	0.773	0.685	1.772	1.021	1.163	0.764	0.110	0.738	1.288	3.405	?	N
MACN 3.7	?	Buenos Aires	?	35.727	1.568	2.165	0.971	0.816	0.645	1.812	1.015	1.141	0.793	0.114	0.680	1.303	3.423	?	P
MACN 49.157	M	La Pampa	?	33.790	1.609	2.285	1.100	0.868	0.682	1.854	1.066	1.155	0.823	0.083	0.633	1.232	3.451	?	P
MACN 49.158	F	La Pampa	?	29.908	1.438	2.173	1.181	0.812	0.719	1.774	1.013	1.084	0.833	0.139	0.578	1.111	3.192	?	S
MACN 49.140	M	La Pampa	?	31.019	1.685	2.231	1.140	0.847	0.663	1.744	0.952	1.084	0.816	0.122	0.592	1.190	3.293	?	S

Code used in analyses, region (R), and predicted region (PR) by the discriminant analysis with the best performance (i.e., standardized osteological variables + geometric mean) is also indicated. North (N), Pampas (P), South (S)

References

Alhajeri BH (2021) A geometric morphometric analysis of geographic mandibular variation in the dwarf gerbil *Gerbillus nanus* (Gerbillinae, Rodentia). J Mamm Evol 28:469–480

Ashrafzadeh MR, Madjdzadeh MS, Azarpira M, Shahi T (2019) Intraspecific variation within *Meriones persicus* (Rodentia: Gerbillinae) populations in south central of Iran. J Wildl Biodivers 3:40–51

Bergmann C (1847) Ueber die Verhältnisse der wärmeökonomie der Thiere zu ihrer Grösse. Göttinger Studien, Göttingen 3(1):595–708

Brookes J (1828) A new genus of the order Rodentia. Trans Linn Soc Lond 16:96–105

Burt WH (1954) Mammalogy: the subspecies category in mammals. Syst Zool 3(3):99–104

Cabrera AL (1976) Regiones fitogeográficas argentinas. In: Parodi LR (ed) Enciclopedia Argentina de agricultura y jardinería, vol 2, no 1. Acme, Buenos Aires, pp 1–85

Dashti Z, Alhaddad H, Alhajeri BH (2022) A geometric morphometric analysis of geographic variation in the Cape Short-eared gerbil, *Desmodillus auriculares* (Rodentia: Gerbillinae). Mammalia 86(6):615–631

Desmarest AG (1817) Gerboise premiere espéce: la grande gerboise *Dipus maximus* Blainv. Nouveau Dictionnaire d'Histoire Naturelle 13:117–119

Gariboldi MC, Inserra PIF, Lucero S, Failla M, Perez SI, Vitullo AD (2019) Unexpected low genetic variation in the South American hystricognath rodent *Lagostomus maximus* (Rodentia: Chinchillidae). PLoS One 14(9):e0221559

Hammer Ø, Harper DAT, Ryan PD (2001) PAST: paleontological statistics software package for education and data analysis. Palaeontol Electron 4:9. Available at: https://www.nhm.uio.no/english/research/resources/past/

Hollister H (1914) Four new neotropical rodents. Proc Biol Soc Wash 27:57–60

Jackson JE, Branch LC, Villareal D (1996) *Lagostomus maximus*. Mamm Species 543:1–6

Kageyama M, Motokawa M, Hikida T (2009) Geographic variation in morphological traits of the large Japanese field mouse, *Apodemus speciosus* (Rodentia, Muridae), from the Izu Island group. Jpn Zool Sci 26:266–276

Llanos AC, Crespo JA (1952) Ecología de la vizcacha (*Lagostomus maximus maximus* Blainv.) en el nordeste de la Provincia de Entre Ríos. Rev Investig Agríc 6:289–378

Meachen-Samuels J, Van Valkenburgh B (2009) Forelimb indicators of prey-size preference in the Felidae. J Morphol 270:729–744

Medina AI, Martí DA, Bidau CJ (2007) Subterranean rodents of the genus *Ctenomys* (Caviomorpha, Ctenomyidae) follow the converse to Bergmann's rule. J Biogeogr 34:1439–1454

Mitteroecker P, Bookstein FL (2011) Linear discrimination, ordination, and the visualization of selection gradients in modern morphometrics. Evol Biol 38:100–114

Parodi LR (1934) Las plantas indígenas no alimenticias cultivadas en la Argentina. Rev Arg Agron 1(3):165–212

Rasia LL (2016) Los Chinchillidae (Rodentia, Caviomorpha) fósiles de la República Argentina: sistemática, historia evolutiva y biogeográfica, significado bioestratigráfico y paleoambiental. PhD Dissertation, Universidad Nacional de La Plata, La Plata, Argentina

Rasia LL, Segura V, Flores DA, Candela AM (in preparation) Skull ontogeny and sexual dimorphism of *Lagostomus maximus* (Rodentia, Chinchillidae): cranial features and sutures, and comparison with the extinct *Lagostomus incisus*

R Core Team (2022) R: A language and environment for statistical computing. R Foundation for Statistical Computing, Vienna. https://www.R-project.org

Schlager S (2017) Morpho and Rvcg—shape analysis in R. In: Zheng G, Li S, Szekely G (eds) Statistical shape and deformation analysis. Academic Press, Cambridge, pp 217–256

Schlager S, Jefferis G, Ian D (2022) Morpho: calculations and visualisations related to geometric morphometrics. R-Package Version 4.2.1

Tabatabaei Yazdi F, Adriaens D (2011) Patterns of skull shape variation in *Meriones persicus* (Rodentia: Muridae) in relation to geoclimatical conditions. Iran J Anim Biosyst 7:129–142
Thomas O (1910) A collection of mammals from eastern Buenos Ayres, with descriptions of related new mammals from other localities. Ann Mag Nat Hist Lond Ser 8(5):239–247
Zhao L, Wang Y, Liu N, Liao J (2013) Effects of change in altitude on the auditory bulla of midday gerbil, *Meriones meridianus*. Pak J Zool 45:581–588

Chapter 5
Genetic Variation Across *Lagostomus maximus'* Geographical Distribution in Argentina

María Constanza Gariboldi, Pablo Ignacio Felipe Inserra, Juan Ignacio Túnez, and Alfredo Daniel Vitullo

Abstract Studying the amount and pattern of genetic diversity within and between populations is one of the central aims of population genetics. However, the population genetic structure of many species, including *Lagostomus maximus*, is currently poorly understood. The analysis of the mitochondrial DNA control region reveals low levels of genetic variation in *L. maximus* populations throughout Argentina. Despite the recognition of three morphologically distinct subspecies in Argentina, and the species' tendency for philopatry among females and the average home range of the species of 1.3 ha, a single widely distributed population inhabits Argentina. Metapopulation processes and changes in human population dynamics in the late Holocene may have played a role in shaping the population genetic structure of *L. maximus* in Argentina.

Keywords Mitochondrial DNA · Population structure · Phylogeography

M. C. Gariboldi (✉)
Centro de Estudios Biomédicos Básicos, Aplicados y Desarrollo (CEBBAD), Buenos Aires, Argentina
e-mail: gariboldi.constanza@maimonides.edu

P. I. F. Inserra · A. D. Vitullo
Centro de Estudios Biomédicos Básicos, Aplicados y Desarrollo (CEBBAD), Buenos Aires, Argentina

Consejo Nacional de Investigaciones Científicas y Técnicas (CONICET), La Plata, Buenos Aires, Argentina

J. I. Túnez
Grupo de Investigación en Ecología Molecular, Instituto de Ecología y Desarrollo Sustentable (INEDES CONICET-UNLu), Lujan, Buenos Aires, Argentina

Departamento de Ciencias Básicas, Universidad Nacional de Luján, Lujan, Buenos Aires, Argentina

L. L. Rasia et al. (eds.), *Plains Vizcachas*, https://doi.org/10.1007/978-3-031-49487-1_5

5.1 Introduction

One of the central aims of population genetics is the study of the amount and pattern of genetic variation within and between populations. This variation can be used to understand the ecological and evolutionary history of a given species. However, the population genetic structure of numerous species remains poorly understood, as is the case of the species belonging to the Chinchillidae family, encompassing the genera *Chinchilla*, *Lagidium*, and *Lagostomus*. Species belonging to these genera are found primarily in arid regions of western and southern South America.

Lagostomus maximus inhabits mostly the Pampean along with the adjoining Espinal, Monte, and Chaquenean regions of Argentina. However, it can also be found in the southeastern area of Bolivia and the western regions of Paraguay. Considering that *L. maximus* lives in lowland habitats that include subtropical grasslands, dry thorn scrub, and desert scrub (Patton et al. 2015), it would be expected to observe behavioral, morphological, and/or genetic changes within its distribution in Argentina, as a result to its adaptation to heterogeneous environments (Schmid and Guillaume 2017).

From a morphologic perspective, three subspecies of *L. maximus* are recognized based on variations in coat color, skull morphology, and incisor width (Llanos and Crespo 1952; Jackson et al. 1996). However, the ranges of subspecies are uncertain (Jackson et al. 1996). The three subspecies are identified as *L. m. immollis* Thomas, which is found in Bolivia, Paraguay, and northcentral Argentina; *L. m. maximus* Desmarest, which is found in central Argentina; and *L. m. petilidens* Hollister, which is found in southern Buenos Aires, La Pampa, and Río Negro provinces in Argentina. It is noteworthy that certain characteristics distinguish these subspecies. For example, *L. m. immollis* is slightly larger than *L. m. maximus*, with its skull resembling that of *L. m. maximus*, except for the distinct broadness of the palatal foramina and the presence of distinct incisive foramina. Additionally, the general color of *L. m. immollis* is rather a browner gray than in *L. m. maximus* (but see Thomas 1910). On the other hand, the skull of *L. m. petilidens* is similar in size to that of *L. m. maximus*, but with much smaller, slender, and weak incisors (Thomas 1910; Hollister 1914; Llanos and Crespo 1952).

The primary source of information regarding the behavior and ecology of *L. maximus* in Argentina is derived from a particular area in the province of Entre Ríos, one site in the province of La Pampa and different sites in the province of San Luis (Llanos and Crespo 1952; Giulietti and Jackson 1986; Jackson 1989, 1990; Branch et al. 1993). As far as behavioral perspectives are concerned, the authors are not aware of any discernible differences among putative populations and/or the presently described subspecies in Argentina. *L. maximus* is a highly social species. Social groups consist of one to three adult males, two to four times more females, and immature individuals. These groups reside in communal burrow systems that are known as *vizcacheras* (Llanos and Crespo 1952). All members from a *vizcachera* share a common home range, with an average size in the semiarid scrub of La Pampa, and minimal overlap between neighboring *vizcacheras* (Branch 1993a, b).

While home range boundaries are not defended, all members of a social group protect the *vizcachera* against nonresident individuals of the same sex-age class, and adult females also defend it against juveniles attempting to disperse into it. Aggression is low among members of a *vizcachera*, and cooperative behaviors are frequent (Branch 1993a, b). Females and immature individuals remain in the same social group, but there is a complete turnover of adult males each year (Branch 1993b; Branch et al. 1993). Individuals are active throughout the year, from dusk to dawn. Regarding diet, vizcachas are herbivorous and coprophagous (e.g., Jackson 1985; Kufner et al. 1992; Branch et al. 1994; Kufner and Monge 1998; Puig et al. 1998). During spring to fall, adult males often forage alone, while in winter they forage with females and immature individuals (Branch 1993c).

From a genetic standpoint, the current understanding of the population genetic structure of *L. maximus* is limited due to the fact that only one study has been conducted on this matter (Gariboldi et al. 2019). In terms of other species belonging to the Chinchiliidae family, the only other study that partially addresses this topic was carried out by Walker et al. (2007), in which the functional connectivity of the mountain vizcacha (*Lagidium viscacia*) in northern Patagonia was evaluated by analyzing the correlations between genetic distances and the cost-distance among the sampled areas. Moreover, considering all the species within the Caviomorpha group, there exist only a few studies that address the population genetic structure of these species (e.g., Ojeda 2010; Lopes et al. 2013; Rivera et al. 2016; Ibañez et al. 2021).

5.2 Results and Conclusions

By means of examining a fragment of the control region of mitochondrial DNA (mtDNA) obtained from a group of 90 individuals of *L. maximus* collected in Argentina, Gariboldi et al. (2019) were able to identify nine haplotypes determined by 24 polymorphic sites. It was observed that the most frequent haplotype was H1, as shown in Table 5.1.

Across much of its geographical distribution, *L. maximus* displays limited genetic diversity. This is apparent not only when considering all samples as a single large population (haplotype diversity: $2.30 \times 10^{-1} \pm 5.93 \times 10^{-2}$; nucleotide diversity: $2.3 \times 10^{-3} \pm 1.6 \times 10^{-3}$), but also when examining the three subspecies or regions within subspecies as defined by Gariboldi et al. (2019) (see Fig. 5.1). In all cases, haplotype diversity was below 4.4×10^{-1} and nucleotide diversity was below 5.9×10^{-3}. Notably, the observed genetic diversity values in *L. maximus* were generally lower than those reported in other caviomorph rodents (e.g., Ojeda 2010; Lopes et al. 2013; Rivera et al. 2016; Ibañez et al. 2021).

In addition to exhibiting low values of genetic diversity, Gariboldi et al. (2019) also observed that subspecies, or the defined groups, were not genetically structured. However, the low genetic diversity and genetic differentiation among groups of individuals of this species throughout Argentina can be accounted for if this

Table 5.1 Relative position of variable nucleotides and haplotype frequencies in the mtDNA control region of *L. maximus* from Argentina

Haplotype	*n*	Relative position of variable nucleotide																							
		43	70	71	79	88	114	136	141	149	153	162	201	207	210	216	224	231	249	262	270	304	308	343	346
H1	79	A	A	T	C	T	A	C	G	A	T	C	T	A	T	A	C	T	A	A	C	G	A	A	T
H2	2	G																							
H3	3				T	C	G	T					C		A			C	G		T	A	G	G	C
H4	1			C	T		G		A	G		T		G	A		T	C		G	T	A	G		
H5	1											T													
H6	1															G									
H7	1																		G						
H8	1		G																						
H9	1										C														

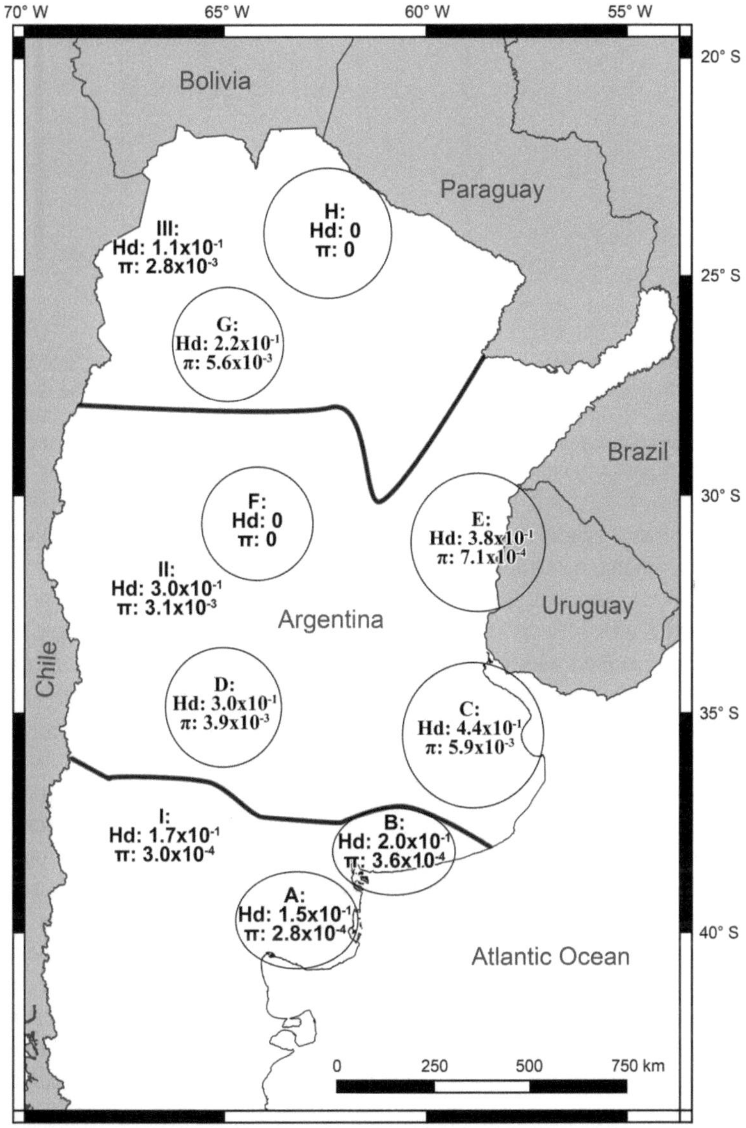

Fig. 5.1 Geographical locations of the sampled sites along the distributional range of *L. maximus* in Argentina and genetic diversity indexes for each subspecies and group. I, II, and III correspond to *L. m. petilidens*, *L. m. maximus*, and *L. m. immollis*, respectively; A–H corresponds to the defined groups. H: haplotype diversity; π: nucleotide diversity. (Figure adapted from Fig. 1 in Gariboldi et al. (2019))

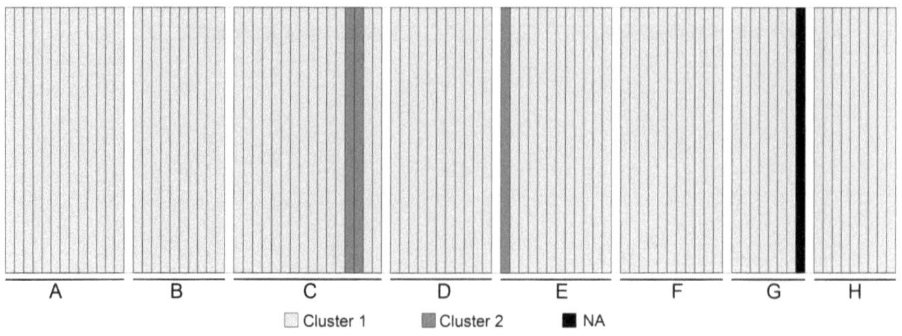

Fig. 5.2 Bayesian assignment analysis of the mtDNA control region sequences. Each vertical bar represents an individual and its associated probability of belonging to one of the two clusters detected. Cluster 1 (light gray), cluster 2 (dark gray), nonassigned individual (black). Abbreviations of sampling sites correspond to those described in Fig. 5.1. The analysis was performed without defining a priori putative populations, using an admixture model

widely distributed species has undergone local extinction processes, followed by subsequent expansion (Grant and Bowen 1998; Pannell and Charlesworth 2000; Tero et al. 2003).Moreover, a Bayesian Analysis of Population Structure performed by using BAPS 6.0 software (Corander et al. 2013) showed that the majority of the samples were grouped within the same cluster (Fig. 5.2). Additionally, through the implementation of phylogenetic analyses, no phylogeographic relationship amongst haplotypes was discovered, with most of the haplotypes being grouped and forming a single well-supported clade (Fig. 5.3). These findings further corroborate the proposition of a single widely distributed population. Furthermore, it is noteworthy that given the tendency of ancient demographic events to allow for greater genetic diversification (Avise 2000), the closeness between the majority of the haplotypes identified throughout Argentina serves to strengthen the hypothesis of a relatively recent expansion event.In analyzing the demographic history of the species, Gariboldi et al. (2019) discovered that the effective population size of the species exhibited a moderate decline in Argentina from 1500 to 300 years before present (YBP), followed by a population expansion from 300 YBP to the present. The changes in the effective population size of *L. maximus* can be attributed to various factors such as changes in human occupation, population density, and subsistence strategy in Argentina (e.g., Barrientos and Perez 2005; Martínez et al. 2013, 2015).In this respect, the archaeological evidence suggests a rapid increase in human population size after 1500 YBP, alongside a rise in human territoriality in several regions of Argentina (Martínez et al. 2013, 2015; Muscio and Lopez 2016; Nores et al. 2017). This increase in human population size triggered heightened exploitation and consumption of *L. maximus* and other small-sized mammals (Quintana and Mazzanti 2011; Stoessel and Martínez 2014; Rafuse et al. 2017). Such an intensified predation pressure could have conceivably affected the population size of small-sized

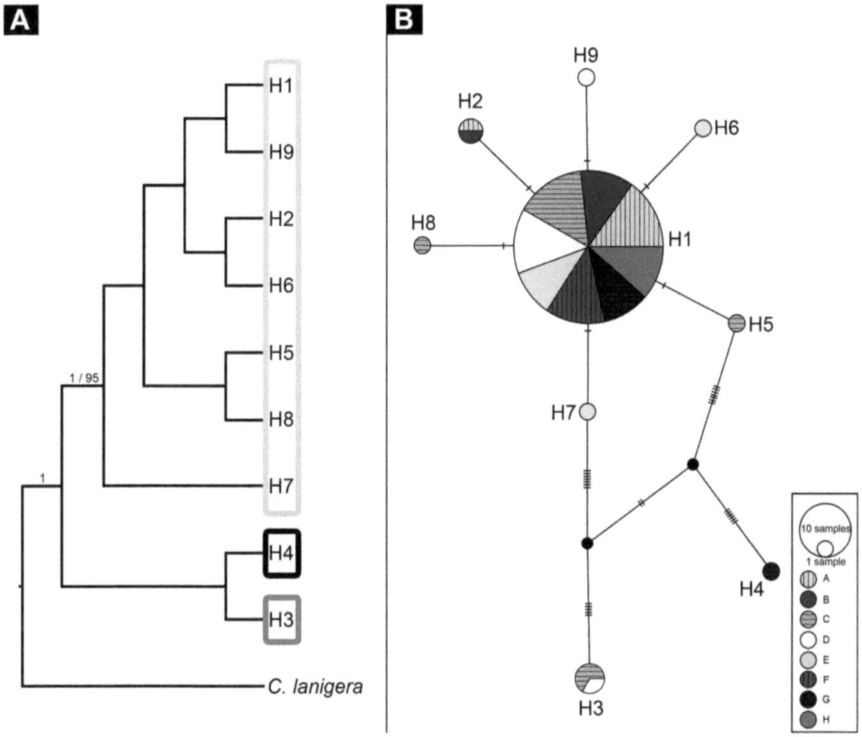

Fig. 5.3 Phylogenetic analyses. (**a**) Cladogram based on the mtDNA control region. *Chinchilla lanigera* (GenBank accession number JX312692) was used as an outgroup. Numbers above branches represent Bayesian posterior probabilities/maximum likelihood bootstrap support values after 1000 iterations. Bootstrap and Bayesian posterior probability values below 70% and 0.95, respectively, are not shown. Clusters obtained from BAPS analysis are highlighted based on Fig. 5.2. (**b**) Median-joining network displaying the mtDNA control region variation. (Figure adapted from Fig. 2 in Gariboldi et al. (2019)). Haplotypes are represented with discs, and shading indicates geographical locations. Mutational steps are indicated with stripes. Abbreviations of sampling sites correspond to those described in Fig. 5.1

mammal species, including *L. maximus*. After that period, the subsequent expansion in the population size of *L. maximus* may be linked to the decline in the indigenous human population during the Hispanic-indigenous contact period and the changes in Pampas and Patagonian subsistence strategies related to horse adoption (Mitchell 2015).

Acknowledgments We are grateful to Fundación Científica Felipe Fiorellino, Museo Argentino de Ciencias Naturales "Bernardino Rivadavia," Mauricio Failla, and Sergio Lucero, who made this work possible. We are indebted to Ezequiel Ibañez for his contributions, which improved the manuscript.

References

Avise JC (2000) Phylogeography: the history and formation of species. Harvard University Press, Cambridge

Barrientos G, Perez SI (2005) Was there a population replacement during the late mid-Holocene in the southeastern Pampas of Argentina? Archaeological evidence and paleoecological basis. Quat Int 132:95–105. https://doi.org/10.1016/j.quaint.2004.07.018

Branch LC (1993a) Social organization and mating system of the plains viscacha. J Zool 229(3):473–491. https://doi.org/10.1111/j.1469-7998.1993.tb02649.x

Branch LC (1993b) Intergroup and intragroup spacing in the plains viscacha, *Lagostomus maximus*. J Mammal 74(4):890–900. https://doi.org/10.2307/1382427

Branch LC (1993c) Seasonal patterns of activity and body mass in the plains vizcacha, *Lagostomus maximus* (family Chinchillidae). Can J Zool 71(5):1041–1045. https://doi.org/10.1139/z93-137

Branch LC, Villarreal D, Fowler GS (1993) Recruitment, dispersal, and group fusion in a declining population of the plains vizcacha (*Lagostomus maximus*; Chinchillidae). J Mammal 74(1):9–20

Branch LC, Villarreal D, Sbriller AP, Sosa RA (1994) Diet selection of the plains vizcacha (*Lagostomus maximus*, family Chinchillidae) in relation to resource abundance in semi-arid scrub. Can J Zool 72:2210–2216. https://doi.org/10.1139/z94-296

Corander J, Marttinen P, Sirén J, Tang J (2013) BAPS: Bayesian analysis of population structure. Manual 6. Department of Mathematics and Statistics, University of Helsinki, Finland

Gariboldi MC, Inserra PIF, Lucero S, Failla M, Perez SI, Vitullo AD (2019) Unexpected low genetic variation in the South American hystricognath rodent *Lagostomus maximus* (Rodentia: Chinchillidae). PLoS One 14(9):e0221559. https://doi.org/10.1371/journal.pone.0221559

Giulietti JD, Jackson JE (1986) Composición anual de la dieta de vizcacha (*Lagostomus maximus*) en pastizales naturales en la pcia. de San Luis, Argentina. Rev Arg Prod Anim 6:229–237

Grant WS, Bowen BW (1998) Shallow population histories in deep evolutionary lineages of marine fishes: insights from sardines and anchovies and lessons for conservation. J Hered 89:415–426. https://doi.org/10.1093/jhered/89.5.415

Hollister N (1914) Four new neotropical rodents. Proc Biol Soc Wash 27:57–60

Ibañez EA, Peralta DM, Obregón G, Nardelli M, Túnez JI (2021) The first assessment of genetic diversity, population structure, and historical population dynamics of *Myocastor coypus* (Rodentia: Echimyidae) in the center of its distribution range. Biol J Linn Soc 134:620–635. https://doi.org/10.1093/biolinnean/blab097

Jackson JE (1985) Ingestión voluntaria y digestibilidad en la vizcacha (*Lagostomus maximus*). Rev Arg Prod Anim 5:113–119

Jackson JE (1989) Reproductive parameters of the plains vizcacha in San Luis Province, Argentina. Vida Silv Neotrop 2:57–62

Jackson JE (1990) Construction of a reliable lens-weight growth curve for plains vizcacha (*Lagostomus maximus*) of unknown age in Argentina. Vida Silv Neotrop 2:56–58

Jackson JE, Branch LC, Villareal D (1996) *Lagostomus maximus*. Mamm Species 543:1–6. https://doi.org/10.2307/3504168

Kufner MB, Monge S (1998) *Lagostomus maximus* (Rodentia, Chinchillidae) diet in areas under human intervention in the Monte Desert, Argentina. Iheringia Ser Zool 10:175–184

Kufner MB, Sbriller AP, Monge S (1992) Relaciones tróficas de una comunidad de herbívoros del desierto del Monte (Argentina) durante la sequía invernal. Iheringia 72:113–119

Llanos AC, Crespo JA (1952) Ecología de la vizcacha (*Lagostomus maximus maximus* Blainv.) en el nordeste de la Provincia de Entre Ríos. Rev Investig Agrar 6:289–378

Lopes C, Ximenes S, Gava A, de Freitas TRO (2013) The role of chromosomal rearrangements and geographical barriers in the divergence of lineages in a South American subterranean rodent (Rodentia: Ctenomyidae: Ctenomys minutus). *Heredity 111*:293–305. https://doi.org/10.1038/hdy.2013.49

Martínez G, Flensborg G, Bayala PD (2013) Chronology and human settlement in northeastern Patagonia (Argentina): patterns of site destruction, intensity of archaeological signal, and population dynamics. Quat Int 301:123–134. https://doi.org/10.1016/j.quaint.2013.03.027

Martínez G, Prates L, Flensborg G, Stoessel L, Alcaráz AP, Bayala P (2015) Radiocarbon trends in the Pampean region (Argentina). Biases and demographic patterns during the final late Pleistocene and Holocene. Quat Int 356:89–110. https://doi.org/10.1016/j.quaint.2014.09.056

Mitchell P (2015) Horse nations: the worldwide impact of the horse on indigenous societies post-1492. Oxford University Press, Oxford

Muscio HJ, Lopez GEJ (2016) Radiocarbon dates and anthropogenic signal in the South-Central Andes (12,500-600 cal. years BP). J Archaeol Sci 65:93–102. https://doi.org/10.1016/j.jas.2015.11.007

Nores R, Fabra M, García A, Demarchi DA (2017) Diversidad genética en restos humanos arqueológicos del sitio El Diquecito (costa sur, Laguna de Mar Chiquita, Provincia de Córdoba). Rev Argent Antropol Biol 19:1–12. https://doi.org/10.17139/raab.2017.0019.01.05

Ojeda AA (2010) Phylogeography and genetic variation in the South American rodent Tympanoctomys barrerae (Rodentia: Octodontidae). J Mammal 91(22):302–313. https://doi.org/10.1644/09-MAMM-A-177.1Pannell JR, Charlesworth B (2000) Effects of metapopulation processes on measures of genetic diversity. Philos Trans R Soc B 355:1851–1864. https://doi.org/10.1098/rstb.2000.0740

Patton JL, Pardiñas UFJ, D'Elías G (2015) Mammals of South America, Volume 2: Rodents. University of Chicago Press, Chicago. https://doi.org/10.7208/chicago/9780226169606.001.0001

Puig S, Videla F, Cona M, Monge S, Roig V (1998) Diet of the vizcacha *Lagostomus maximus* (Rodentia, Chinchillidae), habitat preferences and food availability in northern Patagonia, Argentina. Mammalia 62:191–204. https://doi.org/10.1515/mamm.1998.62.2.191

Quintana CA, Mazzanti DL (2011) Las vizcachas pampeanas (*Lagostomus maximus*, Rodentia) en la subsistencia indígena del Holoceno Tardío de las sierras de Tandilia Oriental (Argentina). Lat Am Antiq 22:253–270. https://doi.org/10.7183/1045-6635.22.2.253

Rafuse DJ, Kaufmann CA, Gutiérrez MA, González ME, Scheifler NA, Álvarez MC, Massigoge A (2017) Taphonomy of modern communal burrow systems of the Plains vizcacha (*Lagostomus maximus*, Chinchillidae) in the Pampas region of Argentina: implications for the fossil record. Hist Biol 31:517–534. https://doi.org/10.1080/08912963.2017.1374957

Rivera DS, Vianna JA, Ebensperger LA, Palma RE (2016) Phylogeography and demographic history of the Andean degu, Octodontomys gliroides (Rodentia: Octodontidae). Zool J Linn Soc 178(2):410–430. https://doi.org/10.1111/zoj.12412

Schmid M, Guillaume F (2017) The role of phenotypic plasticity on population differentiation. Heredity 119:214–225. https://doi.org/10.5061/dryad.418p5

Stoessel L, Martínez G (2014) El proceso de intensificación en la transición pampeano-patagónica oriental. Discusión y perspectivas comparativas con regiones aledañas. Comechingonia 18:65–94. https://doi.org/10.37603/2250.7728.v18.n2.18155

Tero N, Aspi J, Siikamäki P, Jäkäläniemi A, Tuomi J (2003) Genetic structure and gene flow in a metapopulation of an endangered plant species, *Silene tatarica*. Mol Ecol 12:2073–2085. https://doi.org/10.1046/j.1365-294x.2003.01898.x

Thomas O (1910) A collection of mammals from eastern Buenos Ayres, with descriptions of related new mammals from other localities. Ann Mag Nat Hist 5(8):239–247. https://doi.org/10.1080/00222931008692761

Walker RS, Novaro AJ, Branch LC (2007) Functional connectivity defined through cost-distance and genetic analyses: a case study for the rock-dwelling mountain vizcacha (Lagidium viscacia) in Patagonia, Argentina. Landscape Ecol 22:1303–1314. https://doi.org/10.1007/s10980-007-9118-2

Chapter 6
Chancy History of a Supposedly Vanished South American Rodent, *Lagostomus crassus*, with Remarks on Recent Extinctions in Caviomorphs

Patricia Hadler, Carlos A. Galliari, and Ulyses F. J. Pardiñas

Abstract *Lagostomus crassus* is a chinchillid rodent exclusively based in an isolated cranium exhumed near Cusco, Peru, by Jan Kalinowski. Although controversial from the very beginning, the validity of the species persisted for more than a century included in treatises and analyses of recent extinctions. Here, we support the hypothesis advanced in the 1960s by Angel Cabrera: *L. crassus* is a junior synonym of *Lagostomus maximus* and probably represents an animal taken to Peru. The case of *L. crassus* is used here as a trigger to discuss the extinction of caviomorphs that occurred in the South American mainland during the Late Holocene-Anthropocene. Biological extirpations differentially affected this group of rodents in contrast with the resilience of cricetids, the other diverse clade of the order inhabiting the continent. Cases and potential causes are reviewed and combined to explore how several caviomorphs went out in recent times.

Keywords Peru · Cusco · Lagostominae · *Lagostomus maximus* · Holocene · Extinctions

P. Hadler
Departamento de Geologia, Universidade Federal de Santa Catarina, Florianópolis, SC, Brazil
e-mail: patricia.hadler@ufsc.br

C. A. Galliari
Centro de Estudios Parasitológicos y de Vectores (CEPAVE-CONICET-UNLP), Buenos Aires, Argentina

U. F. J. Pardiñas (✉)
Instituto de Diversidad y Evolución Austral (IDEAus-CONICET, Puerto Madryn, Argentina

Instituto Nacional de Biodiversidad (INABIO), Quito, Ecuador
e-mail: ulyses@cenpat-conicet.gob.ar

L. L. Rasia et al. (eds.), *Plains Vizcachas*, https://doi.org/10.1007/978-3-031-49487-1_6

6.1 Introduction

Machu Picchu (also spelled as Machupicchu; Astete and Bastante 2020), including the Urubamba River sacred valley (Peru), is one of the most emblematic archaeological localities of the world and a symbol of the human resistance to European domination in America. Mythical from its inception, shrouded in ambiguity regarding its exact location, and elevated to spectacular status since its dramatic discovery by Hiram Bingham, Machu Picchu remains eternal—capturing the imagination to the extent that it inspired a descriptive opus penned by a young "Che" Guevara (1953). Less widely recognized is the fact that the enigma of Machu Picchu and its environs also encompasses two mammals species. Both shared not only their taxonomic affiliation, since they belong to the speciose group of the caviomorph rodents, but also complex stories spent over a century. One of them, originally called *Abrocoma oblativa* and collected during the initial excavations carried out in the Ciudadela of Machu Picchu (Eaton 1916), is now solved. Thanks to the field and cabinet effort of Emmons (1999), this rodent is today a member of the abrocomid genus *Cuscomys*, still alive in the forest covering the Urubamba Valley (Patton and Emmons 2015; Ochoa et al. 2020). The other, a chinchillid rodent, is the goal of this chapter.

Vizcachas (genus *Lagostomus*) are animals of open, unforested, and mainly flat country. Since the initial detailed descriptions provided by travelers and naturalists, which were based on observations of wild individuals, it has been an established fact. In one of the most detailed studies made on this rodent, written by Hudson (1892), and based on the direct experience of him living in the countryside around Buenos Aires (Argentina) in the last decades of the nineteenth century, he remarked that "the sites of which the vizcacha invariable makes choice to work on, as well as his manner of burrowing adapt him to peculiarly to live and thrive on the open pampas." Because of that, from the very beginning, Thomas (1910) tacitly highlighted the contradictory situation aroused by a *Lagostomus* associated with montane forest in Peru. Despite the fact that he named the taxon as a new species, *Lagostomus crassus*, the "environmental incompatibility" of this finding permeated its entire history. After more than a century, this never-revised Peruvian species is summarily regarded as extinct in more recent works (e.g., Spotorno and Patton 2015; Spotorno and Valladares Faúndez 2016).

Lagostomus crassus is one of the few rodents that disappeared from the South American mainland during historical times (i.e., since 1500 AD). A prerequisite to studying extinctions, one of the most candent biological themes, is taxonomic certainty (MacPhee and Flemming 1999). Therefore, the status of *L. crassus* is not solely a matter of interest per se in the context of the systematics of the genus or the family. It is relevant in the discussion about the differential tendencies to extinctions that apparently show the different groups of Neotropical mammals. The aim of this chapter is not only to review the history of *L. crassus* and their validity but also to take this taxon as an excuse for a shallow dive into the still poorly known universe of recent extinctions affecting rodents in the South American mainland.

6.2 History of *Lagostomus crassus*

The origin of this Peruvian vizcacha combines two passionate men, one as a collector and the other as a describer. The former, the renowned naturalist and traveler Jan Kalinowski (1857–1941), was tasked to work in Peru by Władysław Taczanowski (Shydlovskyy and Zatushevskyy 2015). The other was Oldfield Thomas, the outstanding mammal taxonomist who served at the British Museum (Natural History), London, United Kingdom (BM, thereafter; currently, The Natural History Museum) for more than four decades (Hill 1990). Kalinowski collected numerous animals in Peru, including caviomorphs and cricetids. Some of them have been named after his prolific action, such as *Dasyprocta kalinowskii* or *Thomasomys kalinowskii* (Thomas 1894, 1897). This is not the case with *Lagostomus crassus* as with several others. His relationship with O. Thomas can be traced to 1893, being Kalinowski, the collector of the holotype of *Ichthyomys stolzmanni* (Thomas 1893). Starting in Peru in 1888, he married Maria Villamonte, and both formed a large family at the hacienda La Cadena in the region of Marcapata (Pérez Ruiz 2011). It is not far from the Machu Picchu area where the holotype of *L. crassus* was obtained. This material was described by Thomas (1910, p. 245–246), who briefly indicated, "… the Museum contains the skull of a *Lagostomus* which was obtained in Peru by Mr. Kalinowski, by whom it was found buried in sand… Hab. Santa Ana, District of Cuzco, Peru."

According to Ł. Piechnik, who made a deep study about the South American mammals collected by Polish naturalists, there is no mention of the finding of *L. crassus* in the unpublished notes and memoirs of J. Kalinowski (Ł. Piechnik, email to UFJP in 2017). Although the exact date when the holotype of *L. crassus* was collected is unknown, Piechnik and Kurek (2016, p. 329) conjectured that it was found perhaps in the second half of 1894 and sent, along with other specimens, to Warsaw, from where it was transferred to the BM. Therefore, Thomas (1910) is the single source of original data about this rodent. He described it as new under the name of *L. crassus*, whose etymology (a Latin word meaning fat, gross) seems transparent because Thomas (1910) repeatedly remarked on the large size of the Peruvian vizcacha. Acutely, Thomas (1910, p. 245–246) highlighted "No Viscachas of this genus [*Lagostomus*] are known to live in Peru, and the animal is probably now extinct, but the skull is in no way fossilized…." Thomas (1920) briefly mentioned *L. crassus* when he had the opportunity to study a large collection of recent mammals secured as part of the field activities around Machu Picchu. On the occasion, he highlighted "No specimens were obtained by Mr. Heller of the two rodents described by Mr. Eaton, the osteologist of the party, on skulls found by him in the Inca tombs of Machu Picchu during the 1912 expedition, *Abrocoma oblativa* and *Agouti thomasi*; nor of *Lagostomus crassus* described by myself in 1910 on a skull found in the sand at Santa Ana by Mr. Kalinowski. The first and third of these animals are in all probability now completely extinct, judging by the fact that their relatives are no longer found in this faunal area, but only far south in Chile and Argentina" (Thomas 1920, p. 220).

Until the 1960s, no fresh data was added to the knowledge of *L. crassus*, being this form passingly listed by Tate (1935, p. 368) and by Ellerman (1940, p. 236). Cabrera (1961, p. 561), on the contrary, deserves the merit to advance a hypothesis to explain this clearly extralimital occurrence of a *Lagostomus* in Peru, but also was the first to question the validity of the species, indicating that "O se trata de una forma extinguida, como generalmente se cree, o de un ejemplar de la especie común [*Lagostomus maximus*] llegado, quien sabe por qué conducto, al territorio peruano. Conviene tener en cuenta que las medidas de dicho cráneo, a las que Thomas dio tanta importancia taxonómica, caben dentro de los límites de la especie argentina." Cabrera's opinion on the origin of *L. crassus* was taken by Jackson et al. (1996, p. 5) as follows: "Possibly, this skull [*L. crassus*] was from a *L. maximus* that was taken to Peru." Anderson (1997, p. 477) describing the mammals of Bolivia acutely noted in reference to the type locality of *L. crassus* that "this place is about 1000 km from the nearest known occurrence in southeastern Bolivia of *Lagostomus* of the living species, *L. maximus*."

The different mammal treatises produced during the current century just repeated the basic data about the Peruvian vizcacha with minor modifications and errors. For example, Woods and Kilpatrick (2005, p. 1551) directly listed *L. crassus* as extinct, following the statement of Nowak (1999). The same status for the species was given by Ojeda and Bidau (2013) in a pale UICN assessment. Spotorno and Patton (2015, p. 779 and 781) summarized all the information available for this taxon and concluded "Given the lack of any additional specimens in the modern or archaeological record of such an obvious rodent in the Altiplano of Peru or Bolivia, we agree with Jackson et al. (1996) that the single skull of *L. crassus*, which is at the upper end of the size range of *L. maximus*, probably represents a Plains Viscacha that was taken to Peru." Spotorno and Valladares Faúndez (2016, p. 475 resurrected *L. crassus* as different from *L. maximus* indicating "The Peruvian Plains Viscacha (*Lagostomus crassus*)… was larger and stouter than the Plains Viscacha… it is considered to be extinct."

6.3 Description of *Lagostomus crassus*

6.3.1 Type Locality

It was given as "Santa Ana, District of Cuzco, Peru" (Thomas 1910, p. 246). Directly taken, Santa Ana is a well-known historical place, originally a Jesuitic Mission and then a hacienda settled very near the Urubamba River. According to Stephens and Traylor (1983, p. 198), the locality is "SANTA ANA; Cuzco 1252/7243 (USBGN) A few km SW of Quillabamba on Rio Urubamba (MAPA); 3,480 ft [1,060 m]." However, it can also be interpreted as a general reference since J. Kalinowski largely worked collecting animals, especially birds, across the entire valley of Santa Ana (Chapman 1921; Parker III and O'Neill 1980). In fact, the type locality of *Dasyprocta*

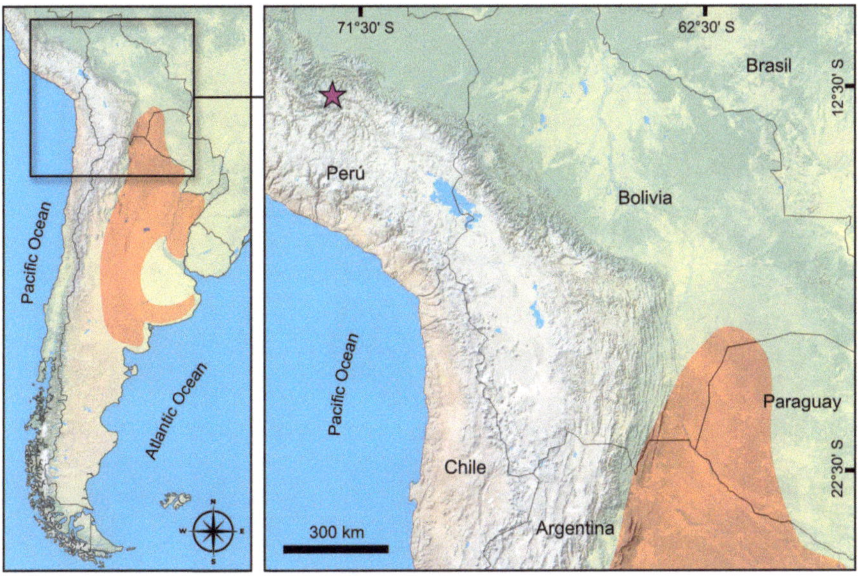

Fig. 6.1 Map of a west-central portion of South America depicting the type locality of *Lagostomus crassus* (marked with a star) and the current approximate distribution of *Lagostomus maximus* (in shading)

kalinowskii was referred to by Thomas (1897, p. 219) as "Idma, Valley of Santa Ana, Cuzco, Peru. Alt. 4600 feet." Idma (12 53′ S, 72 45′ W) is a locality about 7 km west of Santa Ana (see Chapman 1921: map). According to the available data, there is no form to resolve if the place where J. Kalinowski collected the holotype of *L. crassus* was the proper hacienda Santa Ana or another location across the valley of Santa Ana. In addition, the original emplacement where Santa Ana was is currently occupied by a large city, Quillabamba (12 52′ S, 72 41′ W), mostly developed to the west of the Urubamba River (Fig. 6.1). Therefore, the type locality of *L. crassus* must be restricted to Quillabamba, Cusco, Peru, being the reference of Santa Ana just retained with historical purposes.

6.3.2 Holotype

Marked with the number BM 97.10.3.16, the holotype of *L. crassus* is a cranium here illustrated for the first time (Fig. 6.2). It is almost complete, just lacking both incisors and small portions of the pterygoid region, especially the left one and the Eustachian tube of the left auditory capsule. In addition, the anterior faces of both fourth premolars are scraped, as well as fine cracks cross each molariform.

Judged by the moderate development of the sagittal crest (Fig. 6.2c), the cranium probably belongs to a male (cf. discussion in Rasia and Candela 2013; Ubilla and

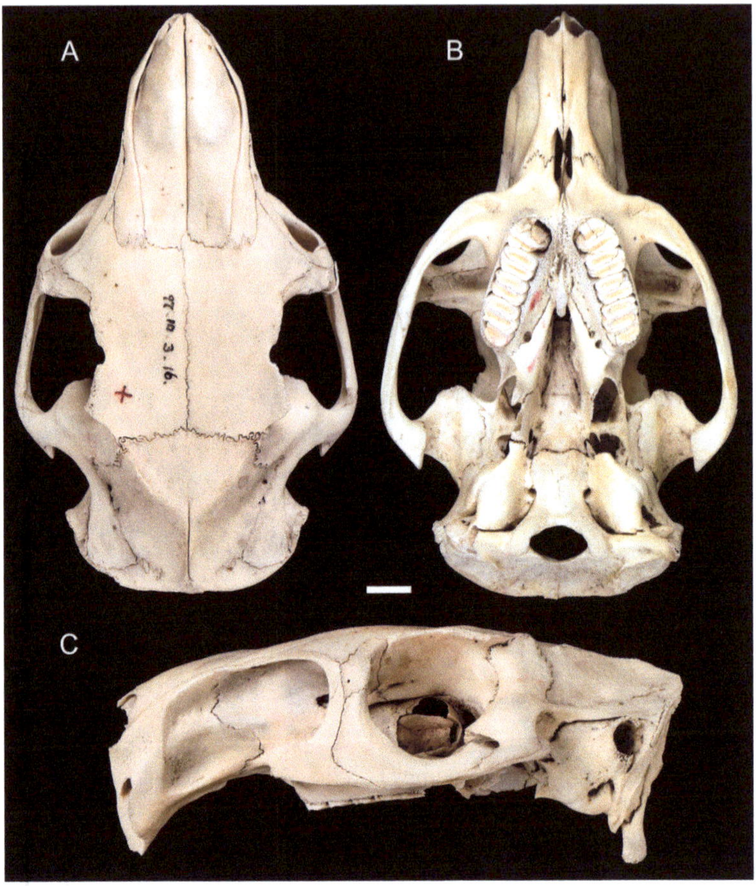

Fig. 6.2 Holotype of *Lagostomus crassus* (BM 97.10.3.16), cranium in (**a**) dorsal, (**b**) ventral, and (**c**) lateral view. Scale bar = 10 mm

Rinderknecht 2016). With respect to the ontogenetic stage, the suture between the basioccipital and the basisphenoid is open while the suture between the latter bone and the presphenoid is closed but marked. This combination suggests that the holotype of *L. crassus* is an adult individual, but not an old one. The material shows several marks and other superficial modifications. Among those more noticeable are a small group of potential cut marks on the left anteorbital bar, a mark surely attributable to the etching of root acids on the left premaxillary near the incisor case, and punctual spots elsewhere apparently constituted by manganese hydroxide. Overall, these bone modifications, as well as the above-described dental damage, reflect a taphonomic history including weathering and burial in a humid soil and/or humid environment. As Thomas (1910) recorded, the cranium was extracted from sand and some grains of sediment are still adhered in some osteological cavities. There is no form to establish if the material was exhumed when building something or when extracting sand from the Urubamba River's extensive banks.

6.4 Taxonomic Status of *Lagostomus crassus*

L. crassus was mostly distinguished from *L. maximus* by its larger size. Thomas (1910, p. 246) described the species indicating "the material difference in size between this skull [*L. crassus*] and that of the Argentine Viscacha [*L. maximus*] is evident on comparing its measures with those of the type of *L. m. immollis*, the largest male *L. maximus* skull in the Museum collection." Cabrera (1961) was the first to passingly dismiss these invoked size-specific differences. Since then, other authors uncritically accepted Cabrera's observation.

To test the supposedly metric differences between both species, we take advantage of the measurements collected by Ubilla and Rinderknecht (2016, Online Supplementary Data 2 and 3). The analyzed database included 28 cranial and upper dentition dimensions (Appendix 6.1) derived from 59 specimens of *L. maximus* (27 males, including the holotype, and 32 females) and the cranium of *L. crassus*. A Principal Component Analysis (PCA) on a covariance matrix was conducted using the software PAST (Hammer et al. 2001) to graphically assess the position in the multivariate space of *L. crassus* relative to *L. maximus*. As there are differences between sexes in the size of the skull in adults, a Discriminant Analysis (DA) and a posteriori classification of cases to infer the sex of the specimen of *L. crassus* were performed. The latter analysis was made using Statistica Software version 8 (Statsoft 2007).

Figure 6.3 shows the dispersion of the measured specimens in the first two axes of the PCA, which accumulate practically 98% of the variance (Appendix 6.1). The loadings of variables in the first axis are all positives and surely correlate with size. The second axis has the high loadings of the variables palatal length and frontal

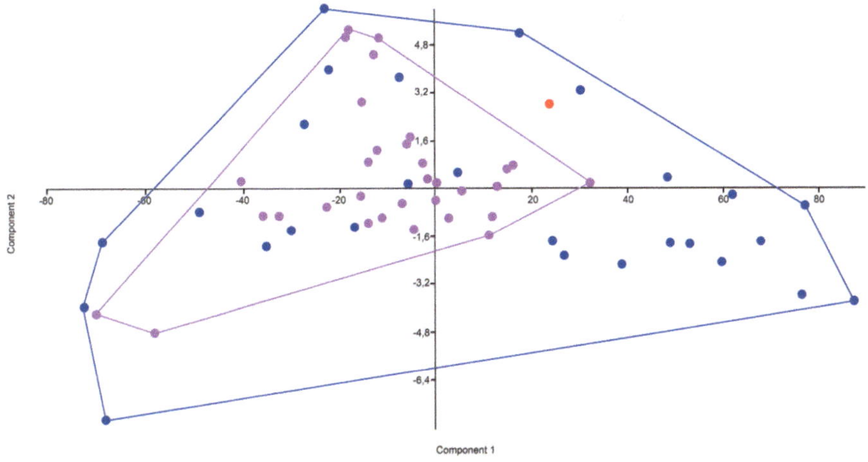

Fig. 6.3 PCA (axis 1 vs. axis 2) of crania of *Lagostomus* showing the position in the multivariate space of *L. crassus* (red point) in respect of males (blue points) and females (pink points) of *L. maximus*

length as positive and nasal length and upper diastema length as negative, positioning *L. crassus* near the specimens of *L. maximus* with the largest palatal and frontal measurements (Appendix 6.1). The analysis of the posterior probabilities of classifications in the DA indicates that of the 28 males (including *L. crassus*), five were classified as females, and of the 32 females, two were classified as males. The probability that *L. crassus* is a male is $p = 0.98$. The incorrect classifications refer to skulls of young females and males whose measurements overlap. The sexual differences are more evident in specimens of larger sizes.

Mio-Pleistocene species in *Lagostomus* have been recently reviewed by Rasia and collaborators (e.g., Rasia and Candela 2013, 2017; Rasia et al. 2020; Rasia 2021). The reported differences among the numerous nominal forms based on fossil material are sometimes subtle and many times imply metrical differences, sometimes related to ontogenetic variation and/or sexual dimorphism (Rasia and Candela 2013). However, there are also more trenchant anatomical features including the obliquity in the implantation of the molariforms with respect to the sagittal plane, the degree of expansion of the palate, etc. (Rasia and Candela 2013). In this context, the high morphological similarity between *L. crassus* and *L. maximus*, including not solely the agreement of measurements but also dental and osteological features, indicates that there is no base to maintain the specific status of the former. We propose that *L. crassus* must be treated as a junior synonym of *L. maximus*. How this isolated specimen reached an extralimital locality such as Santa Ana (Peru) remains a mystery. The nearest living population of this rodent is in Bolivia, about 700 km SSE Santa Ana, where *L. maximus* occupies lowlands (<600 m elevation) in the Chaco (Anderson 1997; Cuéllar and Noss 2003; Tarifa and Yensen 2010). The conjecture advanced by Cabrera (1961), where *L. crassus* represents an animal brought to Peru in an indeterminate past event, seems a suitable explanation (see also Piechnik and Kurek 2016). The Inca Empire extended largely to the south of South America embracing extensive areas currently inhabited by this *L. maximus*. Therefore, it is not disparate to suppose that the specimen then became the holotype of this Peruvian nominal form, which was originally taken as a pet or trade object or even a special gift to someone in Peru. If the marks here highlighted as present on the cranium of *L. crassus* are actually cutting marks, the possibility that the animal served as a food resource seems reinforced.

6.5 Rodent Extinctions in Near Time in South American Mainland

If *L. crassus* is treated as a junior synonym of *L. maximus*, a previously considered extinct species is erased from the list of disappear forms (see Turvey 2009). Nevertheless, several other rodents have been recorded as biologically extirpated during the Late Holocene-Anthropocene in the South American mainland (Appendix 6.2).

At least three of these forms belong to Cricetidae, the most speciose group of rodents living currently on the continent, with a diversity reaching 90 genera and 460 species (Pardiñas et al. 2017a). The single cricetid extinguished during the Holocene is one species of the akodontine *Bibimys*, a genus characterized by three living forms (two of them counting less than ten recording localities; Pardiñas et al. 2017b). *Bibimys massoiai*, recorded for a tropical hilly system in Central-eastern Brazil (Serra da Capivara), apparently disappeared near the Late Holocene (Neves et al. 2020). The other two are also akodontines from subtropical areas: *Juscelinomys candango* and *Gyldenstolpia fronto*. They are known almost exclusively for their type series. The former was described from a handful of animals collected in 1960 during the construction of the city of Brasília (Moojen 1965). The latter is known from a single individual obtained from an interior locality in northeastern Argentina in 1896, plus a few fossil remains of indeterminate Quaternary age collected in Brazil (Pardiñas et al. 2009). *J. candango* is considered extinct according to the UICN (Roach 2019), while *G. fronto* was categorized as critically endangered (Pardiñas et al. 2018).

Recent extinctions in caviomorph are affecting at least three families, Caviidae, Ctenomyidae, and Echimyidae (Appendix 6.2). *Galea tixiensis* was proposed as a fossil form living during the Holocene in the hilly system of Tandil in southeastern Buenos Aires Province (Quintana 2001). Its condition as junior synonym of the living *Galea leucoblephara* was advanced by Teta and Campo (2017). However, these authors failed to study the type of materials and based the conclusions exclusively on the published data for *G. tixiensis*. Therefore, their hypothesis needs further scrutiny. According to Quintana (2001), this *Galea* disappears from the record during historical times. *Ctenomys viarapaensis* was named to distinguish populations of *Ctenomys* largely recorded in an archaeological site in a medium altitude hilly plateau in Córdoba (De Santi et al. 2020). It is very similar to *Ctenomys osvaldoreigi*, a poorly known species currently living in the vicinities of *viarapaensis*' type locality (De Santi et al. 2020, 2021). Echimyids champion holocenic biological extinctions within caviomorphs. At least three species disappeared during the Late Holocene, and more importantly, these extirpations involve two genera, *Dicolpomys* and *Proclinodontomys*. The known records mostly come from southern Brazilian archaeological sites (Roth et al. 2008; Hadler et al. 2008, 2016) but also involve the famous region of Lagoa Santa (Winge 1887) and one Argentinean locality (Verzi et al. 2018).

The first raw contrast between these two major groups of extinguished rodents is their biological significance. While the three lost cricetids represent a single tribe, Akodontini, within the 13 recognized, caviomorphs embrace at least three families. However, this contrast can be viewed as partially spurious since the latter rodents boat a much larger familial diversity than sigmodontines (Patton et al. 2015). Another difference between both groups is the antiquity of these extinctions. At least two of the cricetids disappeared during the last century, while all the caviomorphs apparently went during the Late Holocene but prior to the main period of human impact (from 1500 AD to present).

Late Holocene-Anthropocene rodent extinctions in the South American mainland show some features more or less shared by all the members. They appear geographically associated with the eastern part of tropical and subtropical South America (except *C. viaparensis* and *G. tixiensis*). They are not animals from forested country but probably inhabited open or shrubby environments. Finally, they share a fossorial mode of life (except in *G. tixiensis*). The latter condition deserves more in-depth attention because it was associated with vulnerability. Subterranean rodents share a combination of characteristics linked with their style of life as patchy distribution, limited vagility, territoriality, etc. (e.g., Reig et al. 1990; Nevo 1999; Lacey et al. 2000). Low-range size could be associated with increased risk extinction in rodents (Fourcade and Alhajeri 2023). Small and isolated populations are more vulnerable to local extinction (Lopes and Freitas 2012).

Several factors linked with anthropogenic activities, including the trampling by domestic cattle and also the overall landscape change by the transformation to agroecosystems, are pointed as responsible for biological diversity losses (e.g., Turvey and Fritz 2011; Teta et al. 2014; Harfoot et al. 2021; Roycroft et al. 2021). Habitat fragmentation is the human-induced change most often associated with population declines and increased risk of extinction, even in populations of subterranean rodents (Fernandes et al. 2007; Mapelli and Kittlein 2009; Mapelli et al. 2020). The case of Patagonian *Ctenomys magellanicus* is eloquent about the consequences of human disturbance. The species is supposed to inhabit a large area in the southernmost South American mainland (Lazo-Cancino et al. 2020; Caraballo et al. 2023). However, almost a century ago, Thomas (1929), based on the field data collected by E. Budin, highlighted the local extinction of this *Ctenomys* near Lago Argentino (Santa Cruz Province, see also Barquez and Pardiñas 2019). In addition, the vast populations recorded by Hatcher (1903) and William Greenwood (around the end of the nineteenth century, see Grace Paz and Campbell 2015) near the Andes have vanished (Pardiñas 2013). And the same is true for several occurrences of *C. magellanicus* in the steppic portion of this same province (e.g., Pardiñas et al. 2011). Apparently, in this barren portion of the southern cone, this *Ctenomys* occupies mostly small punctual wetlands ("mallines") because the rest of the landscape is dominated by gravel plains. These special suitable habitats were the primary target of sheep farmers when they began wool exploitation around 1880 (Aagesen 2000). The well-known sheep overload of the Patagonian fields not only caused overgrazing and general desertification but also compaction of the fragile soils and the eradication of hypogeum animals (Mann Fischer 1978).

Patches of fossorial mammals are also a predictable resource for aboriginal groups, and clearly, they could have contributed to local, regional, and, perhaps, total extinctions. De Santi et al. (2020) and Medina et al. (2021) associate the eradication of *C. viaparensis* with human activity implying hunting and consumption. Most of the remains of the large rodents extinguished in the archaeological sites of Rio Grande do Sul are attributable to cooking practices (Fernández et al. 2019). However, there are many contra examples suggesting direct human exploitation is not enough to drive total extinctions. *C. magellanicus* was deeply consumed in the north portion of the Tierra del Fuego Island, been the main protein resource of humans living there since the Late Pleistocene (e.g., Santiago et al. 2016 and the references cited therein). This rodent also suffered the biggest impact of the major sheep overload ever recorded. Finally, this *Ctenomys* was the target of a specific and direct campaign for its eradication via the usage of a large roller dotted with numerous steel points designed to pick the animals when inside their subterranean galleries (P. Braun, comm. pers. to UFJP, 2006). By this latter method, vast portions of the island were "cleaned" of rodents. Today, the populations of these *Ctenomys* are everywhere, being clearly detectable because of large mounds of black soil produced during excavations (Massoia 1967). The case of *Ctenomys sociabilis*, other large-bodied tuco-tuco with a long history of human exploitation (by consumption) in northern Patagonia, is also pointing to resilience (e.g., Tammone et al. 2017).

Recent mammal extinctions in mainland South America are clearly challenging. The few examined with deep attention, including the concourse of recent numerical methods (e.g., niche modeling), are pointing to multivariate factors (e.g., the extinction of the canid *Dusicyon avus*, see Prevosti et al. 2015). In fact, *L. maximus* was almost extirpated from the entire Buenos Aires Province a few decades ago, and this territory was one of its main ranges according to the nineteenth-century travelers (e.g., Darwin 1839; Burmeister 1879). This process of eradication largely contributed to land transformation with the implantation of extensive agroecosystems, but intensive hunting activity was a main component of this regional disappearance (Jackson et al. 1996). Vulnerability associated with the hypogeum mode of life, particularly in those animals living aggregated but patchily distributed, can be viewed in a working hypothesis as a major driver of recent rodent extinctions mediated by human impact.

Acknowledgments The authors are grateful to the editors of this book for the kind invitation to contribute. Special thanks are extended to the following persons: Łukasz Piechnik, who generously contributed two of the pictures of the holotype of *L. crassus*, his book, and his deep knowledge of J. Kalinowski; Roberto Portela Miguez (BM Senior Curator in Charge of Mammals), who favored our access to the holotype and weaving the charge of £75 for the publication of each picture; Phaedra Kokkini, who took the ventral view of the holotype; Stephen Atkinson (BM, Image Sales), who kindly authorized the publication of the photographs of the holotype; Damian Voglino, who made a detailed analysis of the marks on the cranium of *L. crassus* and prepared the Fig. 6.1 with his usual generosity; and Luciano Rasia, who freely shared with us several photographs of *L. maximus*. Economic funds to cover some aspects of this research were derived from PICT Agencia 2020–2068 (to UFJP).

Appendices

Appendix 6.1: PCA of Crania of Lagostomus *Including* L. crassus *and* L. maximus

Twenty-eight cranial and upper dentition measurements from 27 males—including the type of *L. maximus* [NHMUK3.6.6.11], 32 females, and the specimen of *L. crassus* collected by Ubilla and Rinderknecht (2016, Online Supplementary Data 2 and 3) from the series of the following institutions: Museo Argentino de Ciencias Naturales "Bernardino Rivadavia" (MACN), Museo de La Plata (MLP), Museo Nacional de Historia Natural de Montevideo (MNHN), and Natural History Museum of London (BM).

Measurements are: TL, total length; CBL, condylobasilar length; CBaL, condylobasal length; PL, palatal length; NL, nasal length; NW, nasal width; FL, front length; IC, interorbital constriction; PaL, parietal length; PaW, parietal width; BZW, zygomatic width; BMW, mastoid width; PW, palatal width; RW, rostral width; DL, upper diastema length; P4–M3L, upper cheek-teeth length; M1–M3L, upper molar length; ItW, upper incisor transverse width; ApIW, upper anteroposterior incisor width; Hsk, height of skull; P4L, upper P4 length; P4W, upper P4 width; M1, M2, and M3 L, upper molars length; M1, M2, and M3 W, upper molars width.

Eigenvalues of the first three axes of the PCA, showing the percentages of variance explained.

Axis	Eigenvalue	% variance
1	1390.24	97.496
2	7.17867	0.50343
3	5.19378	0.36423

Variable loadings of the 1, 2, and 3 PCA axis.

	Axis 1	Axis 2	Axis 3
TL	0.4871	−0.05611	−0.3128
CBL	0.4251	0.07804	0.06756
CBaL	0.4569	−0.04492	0.0187
PL	0.2432	0.5308	0.07992
NL	0.2824	−0.4584	0.3559
NW	0.09678	−0.1229	−0.08478
FL	0.09924	0.4365	−0.1321
IC	0.1151	−0.1117	−0.1504
PaL	0.07747	−0.1427	−0.6229
PaW	0.08231	0.1616	−0.07717
BZW	0.2813	0.1437	0.05025
BMW	0.1447	−0.12	0.08544

	Axis 1	Axis 2	Axis 3
PW	0.103	0.04252	−0.145
RW	0.1109	−0.01486	−0.1493
DL	0.1644	−0.2978	0.2246
P4-M3L	0.08674	0.1991	0.2679
M1-M3L	0.07173	0.08328	0.2304
ItW	0.02411	−0.02571	0.01297
ApIW	0.02661	−0.0077	0.01052
Hsk	0.1411	−0.1199	−0.03727
P4L	0.02079	0.01585	0.03503
P4W	0.02553	0.1214	0.1012
M1L	0.01603	−0.02081	0.03472
M1W	0.02917	0.1176	0.09235
M2L	0.01788	−0.00635	0.0841
M2W	0.02826	0.02876	0.146
M3L	0.03475	0.04434	0.1422
M3W	0.02786	0.1129	0.1515

Appendix 6.2: Biological Extinctions of Rodents in the South American Mainland During the Holocene-Anthropocene

Species	Geographic provenance	More recent occurrence	Main references
Bibimys massoiai (Cricetidae)	Piauí, Brazil	6890 ± 60 years BP (5920–5606 AD)	Neves et al. (2020)
Gyldenstolpia fronto (Cricetidae)	Minas Gerais, Brazil and Chaco, Argentina	1896 AD	Winge (1887), Gyldenstolpe (1932) and Pardin~as et al. (2009)
Juscelinomys candango (Cricetidae)	Brasília, Distrito Federal, Brazil	1960 AD	Moojen (1965)
Galea tixiensis (Caviidae)	Tandilia, Buenos Aires, Argentina	170 ± 60 years BP (1660–1862 AD)	Quintana (2001) and Teta and Campo (2017)
Ctenomys viarapaensis (Ctenomyidae)	Córdoba, Argentina	Ca. 360 years BP	De Santi et al. (2020) and Medina et al. (2021)
Clyomys riograndensis (Echimyidae)	Rio Grande do Sul, Brazil	Late Holocene (<3000 years BP)	Hadler et al. (2008, 2016)
Dicolpomys fossor (Echimyidae)	Minas Gerais and Rio Grande do Sul, Brazil; Entre Ríos, Argentina	Medieval Warm Period ca. 800–1300 AD	Winge (1887), Hadler et al. (2008, 2016), Roth et al. (2008) and Verzi et al. (2018)
Proclinodontomys mordax (Echimyidae)	Minas Gerais and Rio Grande do Sul, Brazil	Late Holocene (<3000 years BP)	Winge (1887) and Hadler et al. (2008, 2016)

References

Aagesen D (2000) Crisis and conservation at the end of the world: sheep ranching in Argentine Patagonia. Environ Conserv 27:208–215

Anderson S (1997) Mammals of Bolivia: taxonomy and distribution. Bull Am Mus Nat Hist 231:1–652

Astete F, Bastante JM (eds) (2020) Machupicchu. Investigaciones interdisciplinarias, Tomo 1. Dirección Desconcentrada de Cultura de Cusco, Lima

Barquez RM, Pardiñas UF (2019) Un pionero de la colección de mamíferos en Patagonia: el viaje de Emilio Budin en 1927–1928. Mastozool Neotrop 26(1):7–29

Burmeister H (1879) Description physique de la République Argentina d'après des observations personnelles et étrangères. 3 (Animaux vertébrés, 1: Mammifères vivants et éteints). Paul-Emile Coni, Buenos Aires

Cabrera A (1961) Catálogo de los mamíferos de América del Sur. Rev Mus Argent Cienc Nat Bernardino Rivadavia Cienc Zool 4:xxii + 309–732

Caraballo DA, López SL, Botero-Cañola S et al (2023) Filling the gap in distribution ranges and conservation status in *Ctenomys* (Rodentia: Ctenomyidae). J Mammal. https://doi.org/10.1093/jmammal/gyac099

Chapman FM (1921) The distribution of bird life in the Urubamba Valley of Peru. A report on the birds collected by the Yale University-National Geographic Society's expeditions. Smithsonian Institution. U S Natl Mus Bull 117:1–138 + map

Cuéllar E, Noss A (2003) Mamíferos del Chaco y la Chiquitanía de Santa Cruz, Bolivia. Editorial Fundación Amigos de la Naturaleza, Santa Cruz

Darwin C (1839) Narrative of the surveying voyages of His Majesty's ships Adventure and Beagle, between the years 1826 and 1836, describing their examination of the southern shores of South America, and the Beagle's circumnavgation of the globe, vol 3. Journal and Remarks. Henry Colburn, London

De Santi NA, Verzi DH, Olivares AI et al (2020) A new peculiar species of the subterranean rodent *Ctenomys* (Rodentia, Ctenomyidae) from the Holocene of Central Argentina. J S Am Earth Sci 100:102499

De Santi NA, Verzi DH, Olivares AI et al (2021) A new Pleistocene *Ctenomys* and divergence dating of the hyperdiverse South American rodent family Ctenomyidae. J Syst Palaeontol 19(5):377–392

Eaton GF (1916) The collection of osteological material from Machu Picchu. Mem Conn Acad Arts Sci 5:1–96 + 39 plates

Ellerman JR (1940) The families and genera of living rodents. Volume I. Rodents other than Muridae. Trustees of the British Museum (Natural History), London

Emmons LH (1999) A new genus and species of abrocomid rodent from Peru (Rodentia: Abrocomidae). Am Mus Novit 3279:1–14

Fernandes FA, Fernández-Stolz GP, Lopes CM et al (2007) The conservation status of the tuco-tucos, genus *Ctenomys* (Rodentia: Ctenomyidae), in southern Brazil. Braz J Biol 67(4, suppl):839–847

Fernández FJ, Hadler P, Cherem JJ et al (2019) Holocene small mammals hunted by owls and humans in southern Brazil: taphonomic evidence and biological significance. Boreas 48(4):953–965

Fourcade Y, Alhajeri BH (2023) Environmental correlates of body size influence range size and extinction risk: a global study in rodents. Glob Ecol Biogeogr 32(2):206–217

Grace Paz G, Campbell D (eds) (2015) Patagonia bravía. Naturaleza, vidas y aventuras. Memorias originales del baqueano William H. Greenwood. Edición de los autores, Santiago

Guevara E (1953) Machu-picchu, enigma de piedra en América. Suplemento Semanal Siete, 12 de Diciembre de 1953, Panamá, reproducido en Otra vez, diario del segundo viaje por Latinoamérica, Ed. Ocean Sur, 2012, México

Gyldenstolpe N (1932) A new *Scapteromys* from Chaco Austral, Argentine. Ark Zool 24B(1):1–2

Hadler P, Verzi DH, Vucetich MG et al (2008) Caviomorphs (Mammalia, Rodentia) from the Holocene of Rio Grande do Sul state, Brazil: systematics and paleoenvironmental context. Rev Bras Paleontol 11(2):97–116

Hadler P, Cherem JJ, Turbay R et al (2016) Diversidade de pequenos mamíferos (Didelphimorphia e Rodentia) do Holoceno do nordeste do estado do Rio Grande do Sul, Brasil: implicações taxonômicas e paleoambientais. Rev Bras Paleontol 19(1):127–144

Hammer Ø, Harper DAT, Ryan PD (2001) PAST: paleontological statistics software package for education and data analysis. Palaeontol Electron 4(1):1–9

Harfoot MBJ, Johnston A, Balmford A et al (2021) Using the IUCN red list to map threats to terrestrial vertebrates at global scale. Nat Ecol Evol 5:1510–1519

Hatcher JB (1903) Narrative of the expeditions. Geography of Southern Patagonia. In: Reports of the Princeton Expeditions to Patagonia, 1896–1899, vol 1. The University/E. Schweizwerbart'sche Werlagshandlung (E Nägele), Princeton/Stuttgart

Hill JE (1990) A memoir and bibliography of Michael Rogers Oldfield Thomas, FRS. Bull Br Mus Nat Hist (Hist Ser) 18(1):25–113

Hudson WH (1892) Biography of the vizcacha. In: Hudson WH (ed) The Naturalist in La Plata. Chapman & Hall, London, pp 289–313

Jackson JE, Branch LC, Villarreal D (1996) *Lagostomus maximus*. Mamm Species 543:1–6

Lacey EA, Patton JL, Cameron GN (eds) (2000) Life underground: the biology of subterranean rodents. University of Chicago Press, Chicago

Lazo-Cancino D, Rivera R, Paulsen-Cortez K et al (2020) The impacts of climate change on the habitat distribution of the vulnerable Patagonian-Fueguian species *Ctenomys magellanicus* (Rodentia, Ctenomyidae). J Arid Environ 173:104016

Lopes CM, Freitas TR (2012) Human impact in naturally patched small populations: genetic structure and conservation of the burrowing rodent, tuco-tuco (*Ctenomys lami*). J Hered 103(5):672–681

MacPhee RDE, Flemming C (1999) *Requiem Aeternam*. The last five hundred years of mammalian species extinctions. In: MacPhee RDE (ed) Extinctions in near time. Kluwer Academic/Plenum Publisher, New York, pp 333–371

Mann Fischer G (1978) Los pequeños mamíferos de Chile. Gayana Zool 40:1–342

Mapelli FJ, Kittlein MJ (2009) Influence of patch and landscape characteristics on the distribution of the subterranean rodent *Ctenomys porteousi*. Landsc Ecol 24:723–733

Mapelli FJ, Boston ES, Fameli A et al (2020) Fragmenting fragments: landscape genetics of a subterranean rodent (Mammalia, Ctenomyidae) living in a human-impacted wetland. Landsc Ecol 35:1089–1106

Massoia E (1967) Algunos rasgos ecológicos del "tucu tuco" de Magallanes. Hoja Informativa, INTA, CNIA 12:1–12

Medina ME, De Santi NA, Rivero DE et al (2021) Fossorial rodents and applied zooarchaeology to ecosystem conservation in Sierras of Córdoba. Argentina Austral Ecol 46(1):139–147

Moojen J (1965) Novo gênero de Cricetidae do Brasil Central (Glires, Mammalia). Rev Bras Biol 25(3):281–285

Neves SB, Pardiñas UFJ, Hadler P et al (2020) A new fossil cricetid (Rodentia, Sigmodontinae) from northeastern Brazil with remarks on small mammal extinctions in the tropical Quaternary. J Mammal 101(4):1133–1147

Nevo E (1999) Mosaic evolution of subterranean mammals. Regression, progression, and global convergence. Oxford University Press, Oxford

Nowak RM (1999) Walker's mammals of the world, 6th edn. Johns Hopkins University Press, Baltimore

Ochoa JA, Quispe R, Jara Moscoso N et al (2020) Confirmación de la presencia de la enigmática "rata chinchilla arborícola de Machupicchu" *Cuscomys oblativus* (Abrocomidae). Rev Peru Biol 27(2):251–254

Ojeda R, Bidau C (2013) *Lagostomus maximus*. The IUCN Red List of Threatened Species 2013:e. T11170A22190618

Pardiñas UFJ (2013) Localidades típicas de micromamíferos en Patagonia: el viaje de J. Hatcher en las nacientes del río Chico, Santa Cruz, Argentina. Mastozool Neotrop 20(2):413–420
Pardiñas UFJ, D'Elía G, Teta PV (2009) Una introducción a los mayores sigmodontinos vivientes: revisión de *Kunsia* Hershkovitz, 1966 y descripción de un nuevo género (Rodentia: Cricetidae). Arq Mus Nac 66(3–4):509–594. [Number 3–4 of the Arquivos in which this paper appeared is dated "jul/dez 2008" but did not appear until 2009]
Pardiñas UFJ, Teta P, Formoso A et al (2011) Roedores del extremo austral: tafonomía, diversidad y evolución ambiental durante el Holoceno tardío. In: Borrero L, Borrazzo K (eds) Bosques, montañas y cazadores. Investigaciones arqueológicas en Patagonia Meridional. Editorial Dunken, Buenos Aires, pp 61–84
Pardiñas UFJ, Myers P, León-Paniagua L et al (2017a) Family Cricetidae (True hamsters, voles, lemmings and new world rats and mice). In: Wilson DE, Lacher TE Jr, Mittermeier RA (eds) Handbook of the mammals of the world, Rodents II, vol 7. Lynx Edicions, Barcelona, pp 204–279
Pardiñas UFJ, Galliari CA, Voglino D (2017b) Miscellany on *Bibimys* (Rodentia, Sigmodontinae), a unique akodontine cricetid. Mastozool Neotrop 24(1):241–250
Pardiñas U, D'Elía G, Teta P et al (2018) Gyldenstolpia fronto. The IUCN Red List of Threatened Species. 2018:e.T11062A22388720
Parker TA III, O'Neill JP (1980) Notes on little known birds of the upper Urubamba valley, southern Peru. The Auk 97:167–176
Patton JL, Emmons LH (2015) Family Abrocomidae G. S. Miller and Gidley, 1918. In: Patton JL, Pardiñas UFJ, D'Elía G (eds) Mammals of South America, vol 2 – rodents. The University of Chicago Press, Chicago, pp 805–818
Patton JL, Pardiñas UFJ, D'Elía G (eds) (2015) Mammals of South America, vol 2 – rodents. The University of Chicago Press, Chicago
Pérez Ruiz W (2011) Recordando a Celestino Kalinowski. https://lamula.pe/2011/06/25/recordando-a-celestino-kalinowski/wilfredoprezruiz/. Accessed 11 May 2023
Piechnik L, Kurek P (2016) Ssaki Neotropików – Odkryte Przez Polskich Naturalistów. Instytut Botaniki im. W. Szafera Polskiej Akademii Nauk, Kraków
Prevosti FJ, Ramírez MA, Schiaffini M et al (2015) Extinctions in near time: new radiocarbon dates point to a very recent disappearance of the South American fox *Dusicyon avus* (Carnivora: Canidae). Biol J Linn Soc 116(3):704–720
Quintana C (2001) *Galea* (Rodentia, Caviidae) del Pleistoceno Superior y Holoceno de las sierras de Tandilia Oriental, Provincia de Buenos Aires. Argentina Ameghiniana 38(4):399–407
Rasia LL (2021) Reassessing the fossil record of *Lagostomus incisus* Ameghino, 1888 (Rodentia, Caviomorpha) from the late Neogene of southern South America. Publ Electron Asoc Paleontol Argent 21(2):53–61
Rasia LL, Candela AM (2013) Systematic and biostratigraphic significance of a chinchillid rodent from the Pliocene of eastern Argentina. Acta Palaeontol Pol 58(2):241–254
Rasia LL, Candela AM (2017) Systematic revision of the vizcachas (Rodentia, Caviomorpha, Chinchillidae) from the Chapadmalal Formation, Late Pliocene of Buenos Aires Province, Argentina. Ameghiniana 54(1):50–69
Rasia LL, Bonini RA, Candela AM (2020) Nuevos registros de *Lagostomus* (Rodentia, Chinchillidae) en el Mioceno tardío de Argentina y su importancia bioestratigráfica. Andean Geol 47(2):430–445
Reig OA, Busch C, Ortells MO et al (1990) An overview of evolution, systematics, population biology, cytogenetics, molecular biology and speciation in *Ctenomys*. Prog Clin Biol Res 335:71–96
Roach N (2019) *Juscelinomys candango* (amended version of 2019 assessment). The IUCN Red List of Threatened Species 2019. e.T10946A160756258
Roth PRDO, Jung DMH, Christoff AU (2008) Identificação de remanescentes de roedores de um abrigo sob rocha do nordeste do RS: implicações paleoambientais. Rev de Iniciaç Cient ULBRA 7:71–81
Roycroft E, MacDonald AJ, Moritz C et al (2021) Museum genomics reveals the rapid decline and extinction of Australian rodents since European settlement. PNAS 118(27):e2021390118

Santiago F, Salemme M, Pardiñas UFJ (2016) Análisis de restos de roedores del sitio arqueológico Las Vueltas 1, Tierra del Fuego, Argentina. Arqueología 22, Dossier:211–230

Shydlovskyy I, Zatushevskyy A (2015) Jan Kalinowski – outstanding naturalist of Far East and South America. Contribution of Amateur naturalists into biological diversity studies. Proceed. Intern. Scien. Conf. devoted to the 200-th anniversary of Lajos Vagner's birthday, May 14–16, 2015, Beregszasz, p 93–95

Spotorno AE, Patton JL (2015) Superfamily Chinchilloidea Bennett, 1833. In: Patton JL, Pardiñas UFJ, D'Elía G (eds) Mammals of South America, vol2 – rodents. The University of Chicago Press, Chicago, pp 762–783

Spotorno AE, Valladares Faúndez P (2016) Family Chinchillidae (Viscachas and Chinchillas). In: Lacher TE, Mittermeier RA, Wilson DE (eds) Handbook of the mammals of the world, Lagomorphs and rodents I, vol 6. Lynx Edicions, Barcelonap, pp 462–481

Statsoft, Inc (2007) STATISTICA (data analysis software system), version 8.0

Stephens L, Traylor MA (1983) Ornithological gazetteer of Peru. Cambridge, Museum of Comparative Zoology, Harvard University

Tammone MN, Lezcano MJ, Lacey EA et al (2017) Los roedores tuco-tucos (*Ctenomys* sp.) y su relacion con las ocupaciones humanas de cazadores-recolectores en el valle superior del río Limay. Macroscopia:20–25

Tarifa T, Yensen E (2010) Chinchillidae. In: Wallace RB, Gómez H, Porcel Z, Rumiz D (eds) Distribución, ecología y conservación de los mamíferos medianos y grandes de Bolivia. Centro de Ecología Difusión, Fundación Simón I. Patiño, Santa Cruz de la Sierra

Tate GHH (1935) The taxonomy of the genera of Neotropical hystricoid rodents. Bull Am Mus Nat Hist 68:295–447

Teta P, Campo DH (2017) Is *Galea tixiensis* Quintana, 2001 a synonym of *G. leucoblephara* Burmeister, 1861? Therya 8(3):209–216

Teta P, Formoso A, Tammone M et al (2014) Micromamíferos, cambio climático e impacto antrópico: ¿Cuánto han cambiado las comunidades del sur de América del Sur en los últimos 500 años? Therya 5(1):7–38

Thomas O (1893) On some mammals from Central Peru. Proc Zool Soc Lond 1893(Part I):333–341 + 2 plates

Thomas O (1894) Descriptions of some new Neotropical Muridæ. Ann Mag Nat Hist Ser 6 14:346–366

Thomas O (1897) Descriptions of four new South American mammals. Ann Mag Nat Hist Ser 6 20:218–221

Thomas O (1910) A collection of mammals from eastern Buenos Ayres, with descriptions of related new mammals from other localities. Ann Mag Nat Hist Ser 8 5:239–247

Thomas O (1920) Report on the Mammalia collected by Mr. Edmund Heller during the Peruvian expedition of 1915 under the auspices of Yale University and the National Geographic Society. Proc U S Natl Mus 58:217–249

Thomas O (1929) The mammals of Señor Budin's Patagonian expedition, 1927–28. Ann Mag Nat Hist Ser 10 4:35–45

Turvey ST (ed) (2009) Holocene extinctions. Oxford University Press, Oxford

Turvey ST, Fritz SA (2011) The ghosts of mammals past: biological and geographical patterns of global mammalian extinction across the Holocene. Philos Trans R Soc B Biol Sci 366(1577):2564–2576

Ubilla M, Rinderknecht A (2016) *Lagostomus maximus* (Desmarest) (Rodentia, Chinchillidae), the extant plains vizcacha in the Late Pleistocene of Uruguay. Alcheringa 40:354–365

Verzi DH, Olivares AI, Hadler P et al (2018) Occurrence of *Dicolpomys* (Echimyidae) in the late Holocene of Argentina: the most recently extinct South American caviomorph genus. Quat Int 490:123–131

Winge H (1887) Jordfundne og Nulevende gnavere (Rodentia) fra Lagoa Santa, Minas Geraes, Brasilien. E Museo Lundii 1:1–200

Woods CA, Kilpatrick CW (2005) Infraorder Hystricognathi Brandt, 1855. In: Wilson DE, Reeder DM (eds) Mammal species of the world: a taxonomic and geographic reference, 3rd edn. Johns Hopkins University Press, Baltimore, pp 1538–1600

Chapter 7
Use of the Plains Vizcacha (*Lagostomus maximus*) by Indigenous Societies in the Pampas Ecoregion (Argentina)

Agustina Massigoge, Juan Manuel Rodríguez, María Clara Álvarez, María A. Gutiérrez, and Cristian A. Kaufmann

Abstract This chapter reviews the plains vizcacha (*Lagostomus maximus*) bone record from archaeological sites in the Pampas ecoregion. Information from 82 archaeological sites was collected from the literature, standardized, and assembled into a database. This study shows that this rodent is well-represented in the archaeological record. However, its presence does not always relate to human action but also to natural deposition processes. The number of vizcacha remains recovered from archaeological contexts is variable, and findings mainly consist of fragmented cranial and postcranial elements. Evidence shows that the vizcacha was an important complementary resource for indigenous societies that inhabited the Pampas ecoregion from the Late Pleistocene to the Late Holocene. Regionally, the role of this rodent became more significant in the eastern end of the Tandilia Hills during the Late Holocene but remained unchanged in the rest of the Pampas.

Keywords Plains vizcacha · Pampas ecoregion · Archaeology · Hunter-gatherers · Late Pleistocene · Holocene · Economic resource

7.1 Introduction

The plains vizcacha (*Lagostomus maximus* [Desmarest 1817]) is one of the most characteristic mammals of the Pampas ecoregion, and its skeletal remains are frequently recovered in archaeological sites. Despite this, few archaeological studies

Supplementary Information The online version contains supplementary material available at [https://doi.org/10.1007/978-3-031-49487-1_7].

A. Massigoge (✉) · J. M. Rodríguez · M. C. Álvarez · M. A. Gutiérrez · C. A. Kaufmann
INCUAPA, UNCPBA-CONICET. Universidad Nacional del Centro de la Provincia de Buenos Aires (UNCPBA), Facultad de Ciencias Sociales, Olavarría, Buenos Aires, Argentina
e-mail: amassigo@soc.unicen.edu.ar; juanmanuel.rodriguez@soc.unicen.edu.ar; malvarez@soc.unicen.edu.ar; mgutierr@soc.unicen.edu.ar; ckaufman@soc.unicen.edu.ar

L. L. Rasia et al. (eds.), *Plains Vizcachas*, https://doi.org/10.1007/978-3-031-49487-1_7

have focused on evaluating the use of this rodent by indigenous societies that inhabited the region. These societies had a hunter-gatherer way of life from the early peopling at the end of the Pleistocene (ca. 14.000 cal years BP) to the Spanish Conquest in the sixteenth century (Politis 2008). The study of animal remains recovered from archaeological sites, also known as zooarchaeology, allows investigation of the relationships that humans established with other animals in the past and provides information about the environments in which people lived as well as their impact on those environments, among other issues (Gifford-González 2018). The focus of zooarchaeological research in the Pampas has been on the subsistence strategies of the indigenous societies. Current subsistence models propose a generalized regional economy based on the exploitation of a broad spectrum of animal resources during the Late Pleistocene, the Early Holocene and the early part of the Middle Holocene (Miotti and Salemme 1999; Quintana and Mazzanti 2001; Martínez and Gutiérrez 2004; Politis and Messineo 2008; Martínez et al. 2016). Towards the second half of the Middle Holocene (ca. 6000–5000 cal years BP), a shift from this generalist economy to a more specialized one is proposed, characterized by the preferential hunting of artiodactyls, especially guanaco (*Lama guanicoe*) and Pampas deer (*Ozotoceros bezoarticus*) (Quintana and Mazzanti 2001; Martínez and Gutiérrez 2004; Álvarez 2014; Scheifler and Messineo 2016). This specialized economy would have continued during the first part of the Late Holocene until around 2000–1000 cal years BP. From this time onwards, a process of diversification and intensification in subsistence is proposed to have occurred, characterized in some areas of the Pampas by the incorporation of a more significant number of small vertebrates into the diet and more intensive processing of large and small vertebrates (González de Bonaveri 1997; Quintana and Mazzanti 2001, 2014; Martínez and Gutiérrez 2004; Frontini and Bayón 2015). This change in the economy of indigenous societies co-occurred and could be linked to demographic and social changes at a regional level, such as an increase in population density, a reduction in residential mobility, an increase in territoriality, and an intensification of interaction social networks (Mazzanti 2006; Politis 2008; Quintana and Mazzanti 2014; Martínez et al. 2015; Kaufmann et al. 2023).

The archaeological record from the Pampas shows that the indigenous societies exploited the plains vizcacha throughout the occupational history of the region (Martínez and Gutiérrez 2004; Quintana and Mazzanti 2011). Due to the scarcity of its remains in most archaeological contexts, it has generally been considered a complementary or occasional resource, like other small vertebrates (Salemme et al. 2012; Quintana and Mazzanti 2014; Frontini and Bayón 2015). A different situation was recorded in some archaeological sites from the eastern end of the Tandilia Hills with chronologies close to and after 1000 cal years BP, where an abundance of vizcacha remains with evidence of human processing suggests an increased use of this resource as part of the process of diversification and intensification of subsistence above-mentioned (Quintana and Mazzanti 2011, 2014).

This chapter evaluates and discusses the plains vizcacha's role in the Pampean hunter-gatherers' subsistence from the Late Pleistocene to the Late Holocene. From a comprehensive literature overview, we have built a standardized database on the vizcacha bone record from archaeological sites across the ecoregion. Special attention was paid to the number of skeletal remains, the minimal number of individuals, the skeletal and age-class representation, the cultural modifications, and the relative abundance of remains compared to other commonly exploited taxa to discuss the human use of this species over time and space. Also, the incorporation of vizcacha bones into the archaeological sites due to natural causes was examined.

7.2 The Pampas Ecoregion

The Pampas ecoregion constitutes Argentina's most important grassland ecosystem (Fig. 7.1). The general relief is flat or gently undulating. The highest sector of the ecoregion comprises the Ventania and Tandilia Hills (Frengüelli 1946; Matteucci 2012). Steppe grasses predominate, and their species composition varies according to the local climate and soil (Soriano et al. 1992; Matteucci 2012). As for the fauna,

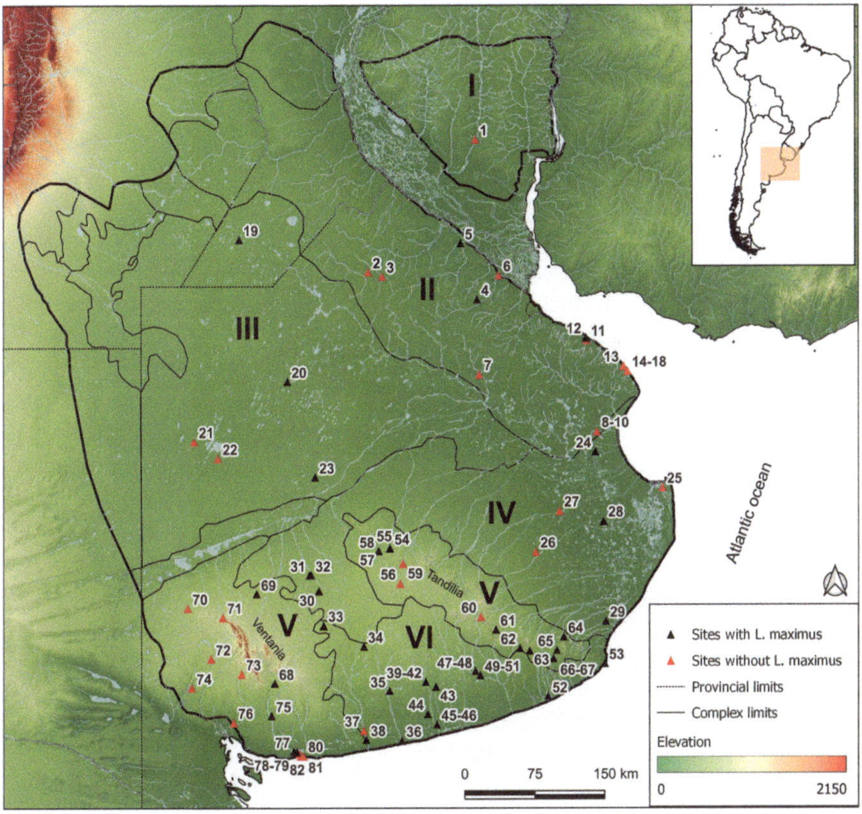

Fig. 7.1 Pampas ecoregion, ecosystem complexes, and archaeological sites surveyed in this work. References: I: Complex Mesopotamic Pampa; II: Complex Rolling Pampa; III: Complex Sandy Pampa; IV: Complex Depressed Pampa; V: Complex Bonaerense Hills; VI: Complex Inter-hill Pampa. Reference numbers for the sites are listed in Table 7.1

this ecoregion largely coincides with the Pampasic Domain of the Guayano-Brazilian subregion of Ringuelet (1961); this domain can be considered an ecotonal zone, which includes an impoverished Guayano-Brazilian fauna, especially in the northern sector, and some fauna elements of the Andean-Patagonian subregion further south (Ringuelet 1955, 1961). The plains vizcacha is one of the most conspicuous mammals (Cabrera and Yepes 1960). Other common rodents are the tuco-tuco (*Ctenomys* sp.) and the cavies (*Cavia* sp. and *Microcavia* sp.). There are also different species of armadillos of the Dasypodidae and Chlamyphoridae families (Vizcaíno et al. 1995). Carnivores include the weasel (*Galictis cuja*), the skunk (*Conepatus chinga*), and different species of canids and felids (Cabrera and Willink 1980). Large native ungulates such as guanaco and Pampas deer are almost absent from the region, although they were abundant in the recent past (Politis et al. 2011). There are also many arboreal, aquatic, and grassland birds, being the greater Rhea (*Rhea americana*) the largest in the ecoregion (Tambussi 1995).

7.3 Materials and Methods

We reviewed published faunal information from archaeological sites in the Pampas ecoregion and built a standardized database, considering journal articles, books or book chapters, conference proceedings, and theses publicly available in libraries or digital repositories. In addition, unpublished information on the sites investigated by the authors of this work was included. Archaeological sites where only test pits or shovel tests were excavated were not included. Archaeological assemblages from the same site but recovered from different contexts (e.g., layers, levels, units, strata, components, and sectors) were entered into the database as discrete analytical units. The database did not include assemblages with no animal bones and those containing only modern bones. Also, sites interpreted by the authors exclusively as human burial settings were not considered for this work.

We employ terms and quantification units commonly used in zooarchaeology. A specimen or bone specimen is defined as "a bone or tooth, or fragment thereof" (Grayson 1984:16). The NISP is the number of identified specimens per taxon, and the MNI is the minimum number of individuals needed to account for all the specimens identified for a particular taxonomic group (Grayson 1984; Lyman 2008).

For each of the assemblages, the following variables were surveyed: site name, geographic location, site location (e.g., open, cave, rock shelter), site function if proposed by the original author(s), radiocarbon age(s), and dated material(s), relative age when absolute age is not available, plains vizcacha NISP, guanaco NISP, Pampas deer NISP, and bibliographic reference(s). Only radiocarbon dates accepted by authors of the original publications were considered. Dates on organic matter from soil samples were only included if they were the only ones available for an analytical unit. Nor were dates on human remains considered because burials are generally intrusive in the deposit. Statistical analyses and calibration of radiocarbon dates were performed using Calib 8.2 software; the SHCal20 calibration curve was used for terrestrial samples, and the Marine20 calibration curve for marine samples (Stuiver and Reimer 1993; Hogg et al. 2020). All ages were calibrated at two sigma.

For the vizcacha remains, the following data were also collected: MNI, skeletal representation, age (based on fusion state), cultural modifications (e.g., cut marks, cultural fracture; Johnson 1985; Fisher 1995), burning (Lyman 1994), and origin of the remains (e.g., cultural, natural, mixed) according to the original publication(s).

To assess intraregional variation in vizcacha exploitation, sites were grouped by ecosystem complexes following Matteucci (2012). All maps were elaborated using QGIS version 3.22.0. It is important to note that the Bonaerense Hills complex was divided into the Ventania and the Tandilia hilly systems since they are different archaeological areas with particular cultural and natural processes.

Temporal trends in the exploitation of the plains vizcacha were evaluated, grouping them according to the Holocene subdivision proposed by Walker et al. (2018): Early Holocene (ca. 11,600–8200 cal years BP), Middle Holocene (ca. 8200–4200 cal years BP), and Late Holocene (starting at ca. 4200 cal years BP). Contexts with a low-temporal resolution that have a calibrated age range greater than 3000 years were not considered for evaluating temporal trends. A Vizcacha Index (VI) was used to assess temporal changes in the relative importance of this rodent in relation to the artiodactyls guanaco and the Pampas deer, the main prey for indigenous societies in most of the Pampas ecoregion through the Holocene. It was calculated using NISP. The formula was modified from the original Artiodactyl Index defined by Szuter and Bayham (1989) and is: [ΣNISP vizcacha/ΣNISP vizcacha, guanaco, and Pampas deer]. The results vary between 0 and 1, with 0 being a diet totally oriented toward artiodactyls and 1 being a diet totally oriented toward vizcachas. The VI was calculated only for Holocene sites in the southwestern portion of the Pampas, where there is a continuous record of human occupations throughout the Holocene. Coastal sites oriented to the exploitation of marine resources were excluded. The index value was zero in contexts where vizcacha remains were reported as of natural origin. The median probability of calibrated dates was calculated for each context to arrange them over time. In contexts with more than one dating, the average radiocarbon age was calculated, and the resulting age was calibrated.

7.4 Results

One hundred fifteen archaeological contexts were surveyed out of 82 sites (Table 7.1). Most of the sites are concentrated in the southwestern portion of the Pampas ecoregion, in the Bonaerense Hills (35%), and in the Inter-hill Pampas (26%) complexes (Fig. 7.1). Most are open-air sites, while in the Tandilia Hills there are two sites in caves (Cueva Tixi and Cueva El Abra) and three in rock shelters (Cerro La China 1, Lobería I, and Amalia 3). Most of the surveyed archaeological contexts are in stratigraphy, although there are a few cases where part or all of the material was recovered on the surface (Laguna El Doce, La Olla, Barrio Las Dunas, El Americano II, and Laguna Sauce Grande II and III).

Table 7.1 Archaeological sites from the Pampas ecoregion included in this study

N°	Site	Context	Age (Cal years BP)[a]	Vizcacha remains	Origin	References
Mesopotamic Pampas						
1.	Laguna del Negro 1	Units II–III	Late Holocene (2108–1633)	–	–	Apolinaire and Castiñeira Latorre (2017) and Bastourre and Apolinaire (2017)
Rolling Pampas						
2.	Hunter	Unit D	Late Holocene (2000–1754)	–	–	Loponte et al. (2010) and Tchilinguirián et al. (2013–2014)
3.	Meguay	Unit D	Late Holocene (1052–929)	–	–	
4.	Cañada de Rocha	Layer 2	Late Holocene (625–336)[b]	Presence	Indet.	Salemme and Miotti (1987), Toledo (2011), Buc and Loponte (2016) and Politis et al. (2019b)
5.	Cañada Honda	Layer 4	Late Holocene (2300–1700)	84	Cultural/ natural	Lanzelotti and Bonaparte (2009), Lanzelotti et al. (2011), Escosteguy and Salemme (2012), Salemme et al. (2012) and Escosteguy and Salemme (2017)
6.	Rio Luján II		Late Holocene (1697–1418)	–	–	Escosteguy and Salemme (2012) and Mucciolo (2013)
7.	Techo Colorado		Late Holocene (1982–141)	–	–	Frère et al. (2016) and Escosteguy et al. (2017)
8.	La Guillerma 1	Unit 1 (A-horizon)	Late Holocene (1280–290)	–	–	González (2005), Escosteguy and Salemme (2012), Salemme et al. (2012) and Frère et al. (2016)
9.	La Guillerma 4	Unit 1 (A-horizon)	Late Holocene (1860–1360)	–	–	González (2005) and Escosteguy et al. (2012)
10.	La Guillerma 5	Unit 1 (A-horizon)	Late Holocene (1450–455)	–	–	González (2005), Escosteguy and Salemme (2012), Salemme et al. (2012) and Frère et al. (2016)

(continued)

Table 7.1 (continued)

N°	Site	Context	Age (Cal years BP)[a]	Vizcacha remains	Origin	References
11.	La Norma		Late Holocene (no dates)	–	–	Brunazzo (1999)
12.	La Maza I	Units I–V	Late Holocene-Post-Hispanic (no dates)	Presence	Natural	Salemme et al. (1985)
13.	Las Marías		Late Holocene (1533–1053)	–	–	Day Pilaría (2018)
14.	San Clemente I		Late Holocene (no dates)	–	–	
15.	San Clemente II		Late Holocene (790–0)	–	–	
16.	San Clemente III		Late Holocene (no dates)	–	–	
17.	San Clemente IV		Late Holocene-Post-Hispanic (489–286)	–	–	
18.	San Clemente VI		Late Holocene (920–690)	–	–	
Sandy Pampas						
19.	Laguna El Doce	Surface assemblage	Middle to Late Holocene (7940–1280)	10	Indet.	Ávila (2011) and Cornaglia Fernández (2013)
20.	Lagunas de los Pampas	Excavated assemblage	Middle Holocene (7932–6447)	29	Cultural	Politis et al. (2012), Álvarez (2018) and Messineo et al. (2018)
21.	Huencú Nazar	Sector A	Late Holocene (3325–2886)	–	–	Scheifler (2020)

(continued)

Table 7.1 (continued)

N°	Site	Context	Age (Cal years BP)[a]	Vizcacha remains	Origin	References
22.	Laguna Cabeza de Buey 2	Unit I	Late Holocene-Post-Hispanic (no dates)	–	–	Messineo and Scheifler (2016) and Scheifler and Messino (2016)
		Unit II	Late Holocene (no dates)	1	Indet.	
		Unit III	Middle Holocene (4824–4449)	1	Indet.	
		Unit IV	Middle Holocene (7692–7512)	–	–	
23.	La Susana 1	Unit II	Late Holocene (no dates)	–	–	Scheifler (2019) and Scheifler et al. (2023)
		Unit III	Early to Middle Holocene (8360–4613)	–	–	
Depressed Pampas						
24.	La Salada	A-horizon	Late Holocene (no dates)	23	Indet.	Aldazabal (2002)
25.	El Divisadero Monte 6		Late Holocene (630–334)	–	–	Aldazabal et al. (2017)
26.	Pessi	Upper levels	Late Holocene (no dates)	–	–	Aldazabal (2002)
		Lower levels	Late Holocene (3330–2120)	–	–	
27.	San Ramón 7	A-horizon	Late Holocene (1177–570)	–	–	Escosteguy et al. (2015) and Frère et al. (2016)
28.	Los Molles	A-horizon/ B-horizon	Late Holocene-Post-Hispanic (no dates)	Presence	Cultural[c]	Aldazabal (2002) and Aldazabal and Eugenio (2013)
29.	Laguna de Sotelo	A-horizon	Late Holocene (no dates)	3	Possible cultural[c]	Eugenio and Pardiñas (1991) and Aldazabal (2002)

(continued)

Table 7.1 (continued)

N°	Site	Context	Age (Cal years BP)[a]	Vizcacha remains	Origin	References
30.	Fortín Necochea 2	Units A–B	Late Holocene-Post-Hispanic (no dates)	6	Cultural[c]	Crivelli Montero et al. (1987–88) and Crivelli Montero et al. (1994)
		Unit C	Middle to Late Holocene (7590–3700)	18	Cultural[c]	
31.	Laguna Muscar 2	Sector 1	Middle Holocene (5740–5470)	16	Natural	Barros et al. (2018a), Bellinzoni (2018), Kaufmann et al. (2019) and this paper
		Sector 2	Middle to Late Holocene (no dates)	58	Natural	
32.	Hangar	Upper levels	Late Holocene (1288–668)	16	Cultural	Barros et al. (2018b) and this paper
Inter-hill Pampas						
33.	Laguna del Trompa	Sector H-Layer I A1 A3	Late Holocene-Post-Hispanic (no dates)	6	Cultural[c]	Crivelli Montero (1991), Eugenio (1991) and Silveira (1991)
		Sector H Layer I B3	Late Holocene (no dates)	1	Cultural[c]	
		Sector J- Layer I A1 A3	Late Holocene-Post-Hispanic (no dates)	–	–	
		Sector J- Layer I B3	Late Holocene (no dates)	–	–	
34.	Laguna Tres Reyes 1	Upper component	Late Holocene (2350–1588)	14	Cultural	Madrid and Barrientos (2000) and Salemme and Madrid (2007)
		Lower level	Late Holocene (no dates)	1	Indet.	

(continued)

Table 7.1 (continued)

N°	Site	Context	Age (Cal years BP)[a]	Vizcacha remains	Origin	References
35.	Arroyo Seco 2	Units X-X/Y	Late Holocene-Post-Hispanic (no dates)	2	Indet.	Rafuse (2013), Politis (2014), Salemme (2014), Politis et al. (2016) and Rafuse (2017)
		Units Y-Y/S-S--S/Z-Z	Late Pleistocene to Middle Holocene (14,150–6400)	34	Possible cultural/ possible natural	
36.	Claromecó 1	Upper component	Late Holocene-Post-Hispanic (no dates)	3	Indet.	Bonomo et al. (2008) and Leon (2014)
		Lower component	Late Holocene (735–576)	26	Cultural	
37.	La Represa Antigua	Level 1	Late Holocene (3353–3067)	–	–	Hoguin and March (2007–2008) and March et al. (2011)
		Level 2	Late Holocene (3822–3489)	–	–	
		Level 3	Late Holocene (2283–1927)	–	–	
38.	Quequén Salado 1	Stratum IV	Late Holocene (921–569)	19	Possible cultural	Madrid et al. (2002) and Leon (2020)
39.	Las Brusquillas 1	Upper levels	Late Holocene-Post-Hispanic (no dates)	2	Natural	Massigoge (2009, 2012)
		Lower levels	Late Holocene (3680–3400)	1	Natural	
40.	Las Brusquillas 2	Unit III	Late Holocene (1880–1430)	10	Indet.	Massigoge (2009, 2011)
41.	Las Brusquillas 3	Unit III	Late Holocene (3230–2720)	22	Possible cultural	Massigoge et al. (2018)

(continued)

Table 7.1 (continued)

N°	Site	Context	Age (Cal years BP)[a]	Vizcacha remains	Origin	References
42.	Las Brusquillas 4	Unit IIIb	Late Holocene (3353–3171)	372	Cultural	Rodríguez (2023) and this paper
43.	Cortaderas	Lower component	Late Holocene (2740–1760)	20	Cultural/ possible natural	Massigoge (2007, 2009) and Massigoge and Pal (2011)
44.	Las Toscas 5	Unit 2	Middle Holocene (5572–4980)	5	Possible natural	Massigoge et al. (2021)
45.	El Guanaco 1	Sector 1-Ap horizon	Early to Late Holocene-Post-Hispanic (no dates)	–	–	Flegenheimer et al. (2010) and Frontini (2012)
		Sector 1-unit 3	Early to late Holocene (no dates)	17	Cultural	
		Sector 1-unit 2	Early to Middle Holocene (10,500–8040)	9	Indet.	
		Sector 2	Middle Holocene (6270–4450)	3	Indet.	
46.	El Guanaco 2		Early to Middle Holocene (10,360–7160)	10	Cultural	Flegenheimer et al. (2010), Frontini (2012) and Frontini and Vecchi (2014)
47.	Zanjón Seco 2		Late Holocene (3366–3076)	3	Indet.	Politis et al. (2001, 2004)
48.	Zanjón Seco 3		Late Holocene (1408–1179)	1	Indet.	Politis (1984), Politis and Beukens (1991) and Martínez (1999)
49.	Paso Otero 3	Ab7	Middle to Late Holocene (5580–2940)[d]	25	Cultural[c]	Martínez et al. (2001), Martínez (2006) and Johnson et al. (2012)
50.	Paso Otero 4	Upper levels	Middle Holocene (7668–4981)[d]	118	Cultural	Gutiérrez et al. (2011), Álvarez (2012), Álvarez et al. (2013) and Martínez and Gutiérrez (2019)
		Lower levels	Early Holocene (9660–9320)	144		

(continued)

Table 7.1 (continued)

N°	Site	Context	Age (Cal years BP)[a]	Vizcacha remains	Origin	References
51.	Paso Otero 5	Puesto Callejón Viejo Paleosol	Late Pleistocene (11,880–11,720)	–	–	Martínez (2001) and Martínez and Gutiérrez (2011)
52.	Nutria Mansa 1	Upper levels	Late Holocene-Post-Hispanic (no dates)	–	–	Bonomo (2005), Bonomo et al. (2009) and Bonomo and Matarrese (2012)
		Lower component	Late Holocene (3480–2520)	113	Cultural[c]	
53.	Alfar		Middle Holocene (6630–6310)	49	Cultural/natural	Bonomo and Leon (2010), Bonomo et al. (2013) and Leon (2014)
Bonaerense Hills (Tandilia Hills)						
54.	El Puente	Upper component	Late Holocene (3160–1840)	–	–	Messineo et al. (2014)
		Lower component	Middle Holocene (6550–6312)	1	Indet.	
55.	Calera	Pit 2	Late Holocene (3350–1887)	17	Indet.	Politis et al. (2005), Kaufmann and Álvarez (2007) and Álvarez (2012)
		Grid N°8	Late Holocene (no dates)	3	Indet.	Rodríguez (2023)
56.	La Barrancosa 1		Late Holocene (1694–1410)	–	–	Messineo (2003, 2008, 2011)
57.	Empalme Querandíes 1		Late Holocene (3382–1830)	63	Cultural	Álvarez (2012) and Messineo et al. (2013)
58.	Campo Laborde	Unit 4	Late Pleistocene (13,240–11,850)	8[e]	Natural	Politis and Messineo (2008), Scheifler et al. (2015), Politis et al. (2019a) and Messineo et al. (2021)
59.	La Moderna 1	Lower component	Early to Middle Holocene (9250–7620)[f]	–	–	Politis and Gutiérrez (1998)

(continued)

Table 7.1 (continued)

N°	Site	Context	Age (Cal years BP)[a]	Vizcacha remains	Origin	References
60.	El Picadero		Middle to Late Holocene (5569–521)	–	–	Colombo (2013)
61.	La Liebre		Late Holocene (1583–1366)	1	Natural	Flegenheimer (1991) and Pupio (1995)
62.	Cerro La China site 1	Unit 3	Late Pleistocene (13,050–12,090)	2	Indet.	Flegenheimer et al. (2015) and Mazzia and Flegenheimer (2021)
63.	Lobería I site 1	Upper component	Late Holocene (899–551)	382	Cultural	Mazzanti et al. (2010) and Quintana and Mazzanti (2011, 2014)
64.	Cueva Tixi	Level 4	Late Holocene (676–555)	180	Cultural	Mazzanti (1997), Quintana (2001), Valverde (2001), Quintana and Mazzanti (2001, 2011, 2014), Martínez et al. (2013) and Mazzanti et al. (2015)
		Level 3	Late Holocene (3640–3240)	4	Cultural[c]	
		Level 2	Middle Holocene (5720–5330)	12	Cultural	
		Level 1	Late Pleistocene to Early Holocene (12,590–11,240)	31	Cultural	
65.	Cueva El Abra	Upper component	Late Holocene (919–741)	468	Cultural	Mazzanti (1999), Quintana et al. (2002), Quintana and Mazzanti (2011, 2014), Mazzanti et al. (2012), Martínez et al. (2013) and Quintana (2016)
		Lower component	Late Pleistocene to Early Holocene (12,610–10,870)	12	Indet.	
66.	Amalia site 3	Unit A	Post-Hispanic (no dates)	–	–	Mazzanti (2007)

(continued)

Table 7.1 (continued)

N°	Site	Context	Age (Cal years BP)[a]	Vizcacha remains	Origin	References
67.	Amalia site 4	Sector 1	Post-Hispanic (no dates)	–	–	Mazzanti (2007) and Quintana and Mazzanti (2011)
		Sector 2	Post-Hispanic (430–0)	19	Cultural	
Bonaerense Hills (Ventania Hills)						
68.	La Toma	Upper component	Late Holocene (960–730)	6	Possible cultural	Álvarez and Salemme (2015)
		Lower levels	Late Holocene (3870–1721)	6	Possible cultural	
69.	Laguna del Indio		Middle to Late Holocene (4825–566)	25	Indet.	Vecchi et al. (2023)
70.	Laguna de Puán 1	Unit B	Late Holocene (no dates)	–	–	Oliva et al. (1991)
71.	La Montaña 1		Late Holocene (998–180)	–	–	Catella (2014)
72.	Los Chilenos 2		Late Holocene (2401–2135)	–	–	Barrientos et al. (2002), Catella (2014) and Morales (2022)
73.	Cerro del Aguila		Late Holocene (3453–3217)	–	–	Frontini et al. (2018)
74.	San Martín 1		Late Holocene (2727–2365)	1	Indet.	Catella (2014) and Morales (2015, 2022)
75.	Paso Mayor Y1S1–2	Upper levels	Late Holocene (2948–2576)	–	–	Bayón et al. (2010) and Tomassini et al. (2019)
		Lower levels	Middle to Late Holocene (6840–3984)	2	Indet.	
76.	Paso Vanoli		Late Holocene (720–550)	–	–	Vecchi et al. (2013)
77.	Puente de Fierro		Late Holocene (2240–1412)	2	Cultural	Frontini and Bayón (2015) and Borella et al. (2017)

(continued)

Table 7.1 (continued)

N°	Site	Context	Age (Cal years BP)[a]	Vizcacha remains	Origin	References
78.	Laguna del Sauce II		Late Holocene (no dates)	1	Indet.	Frontini et al. (2021b)
79.	Laguna del Sauce III		Late Holocene (2846–2348)	–	–	
80.	Barrio Las Dunas		Middle Holocene (7420–6860)	–	–	Bayón et al. (2012), Frontini (2013) and Frontini and Bayón (2017)
81.	La Olla	Sector 1	Middle Holocene (7760–6670)	–	–	Politis et al. (1994), Bayón and Politis (2014) and Leon et al. (2017)
		Sector 2	Middle Holocene (7910–7470)	–	–	
		Sector 3	Middle Holocene (7827–7010)	–	–	
		Sector 4	Middle Holocene (7930–7260)	–	–	
82.	El Americano II		Middle Holocene (7674–7513)	–	–	Frontini and Bayón (2017) and Frontini et al. (2019, 2021a)

[a]Individual ^{14}C dates and their calibrated age ranges are listed in Supplementary Table 7.1
[b]Politis et al. (2019b) argued that the available dates for the site Cañada de Rocha cannot be used to estimate its chronology because the dated bones were taphonomically altered and contaminated
[c]The authors report no evidence to support anthropic use
[d]Radiocarbon dates are exclusively from organic matter in sediments
[e]Minimum Number of Elements (MNE, Lyman 2008: 214)
[f]Politis et al. (2019a) suggest that more radiocarbon analyses using amino acid-based methods are necessary to test the supposedly Holocene age of the extinct megamammals in the site La Moderna

Vizcacha skeletal remains were recovered in 57% of the sites. The Inter-hill Pampas has the highest percentage of sites with vizcacha specimens (90%). This is followed by the Tandilia Hills (71%) and the Depressed and Sandy Pampas (67% and 60%, respectively). In the Ventania Hills, this rodent has only been recorded in 40% of the sites. A lower representation of vizcacha is observed in the northern portion of the Pampas ecoregion: only 18% of the sites in the Rolling Pampas present remains while this taxon is absent from the only site in the Mesopotamic Pampas (Fig. 7.1).

Although the survey was carried out on archaeological sites, only in some cases the presence of vizcacha has been attributed by researchers to human action. Of the 64 archaeological contexts where this rodent has been identified, its existence has been linked to human exploitation with certainty in only 43.8% of them and as a possibility in 9.4% (Table 7.1). For four of these contexts, the researchers have proposed that the bones were deposited by human action and natural processes (mixed origin). In 12.5% of the contexts, the natural incorporation of vizcacha remains has been suggested, either by etho-ecological causes—such as the death of animals in their burrows—or by animal predation. Finally, in 34.3% of the cases, the origin of the specimens could not be determined because either the evidence to assess the cause of deposition is unclear or the authors did not discuss the origin of these remains.

Human utilization has been generally determined by cut marks, cultural fractures, and burn damage (Table 7.2). Cut marks are considered the most robust evidence of human exploitation because their interpretation is less ambiguous than the other two types of modification. Examples of cut marks on vizcacha bones recovered in archaeological sites of the Pampas are shown in Fig. 7.2.

A difference in the number of vizcacha specimens is observed when comparing cultural and natural contexts. Contexts of cultural origin exhibit between 1 and 468 remains (median = 19), while those of natural origin show between 1 and 58 remains (median = 5). Most cultural contexts present less than 50 vizcacha remains, but a few exceed this number, and a small group shows an unusually high quantity. Examples of the latter are Las Brusquillas 4 (372 remains) in the Inter-hill Pampas and the upper components of Lobería I and Cueva El Abra sites in the Tandilia Hills (382 and 468 remains, respectively). The site with the largest number of naturally introduced vizcacha remains is Laguna Muscar 2 in the Depressed Pampas; in this particular case, researchers have argued that the presence of these remains is related to the construction of a *vizcachera* and the natural death of individuals in the site after the human occupation (Kaufmann et al. 2019) (Fig. 7.3).

This chapter also sought to gather information on the representation of age classes in the vizcacha assemblages from different contexts. However, few works provide information on this variable. Only a few mentions indicate the presence of some unfused postcranial elements, such as the innominate, vertebrae, and long bones. Although limited, this information is interesting because it shows that immature individuals are represented in at least 24% of the contexts in which vizcacha was exploited.

Our results show that the natural and cultural assemblages share several characteristics in the representation of skeletal parts (Fig. 7.4). In most assemblages, dense portions or elements, such as the skull (tympanic bullae, teeth), pelvic girdle, and hind limb (femur, tibia, and tarsals), are best represented. In contrast, vertebrae, fibulae, clavicles, carpals, and sternum are present in a few assemblages.

The contexts in which the presence of vizcacha is due to human activity indicate that this rodent was used by hunter-gatherers from the Late Pleistocene to the end of the Late Holocene, even after the Spanish conquest (Table 7.1). When analyzing the spatial distribution of sites where vizcacha was exploited, the results indicate that this resource is more frequently recorded in the southeast of the ecoregion. The percentage of contexts showing exploitation of vizcacha is 46% in the Inter-hill Pampas, 42% in the Depressed Pampas, and 38% in the Tandilia Hills. This

Table 7.2 Sites with evidence of plains vizcacha use in the Pampas ecoregion

		Vizcacha			Specimens with evidence of human use					
Site/context	Site function	NISP	MNI	Immature individuals	Cut marks	Cultural fracture	Burning	Bone tool	Interpretation	References
Cañada Honda	Multiple-activity site/ burial site	84	12	No data	2	7	1	–	Skinning, disarticulation, defleshing, and consumption of meat and bone marrow	Escosteguy and Salemme (2017)
Laguna de los Pampas	Multiple-activity site/ burial site	29	2	No data	1	–	3	–	Consumption	Álvarez (2018)
Los Molles	Multiple-activity camp	No data	No data	No data	–	–	–	–	Consumption	Aldazabal and Eugenio (2013)
Laguna de Sotelo	Multiple-activity camp/burial site	3	1	No data	–	–	–	–	Complementary resource?	Eugenio and Pardiñas (1991) and Aldazabal (2002)
Fortín Necochea 2/ units A–B	Campsite (residential bases)	6	No data	No data	–	–	–	–	Consumption	Crivelli Montero et al. (1987–88) and Crivelli Montero et al. (1994)
Fortín Necochea 2/ unit C	Campsite (residential bases)	18	No data	No data	–	–	–	–	Consumption	
Hangar	Residential or base camps	16	1	No data	1	–	4	–	Consumption	Barros et al. (2018b) and this paper

(continued)

Table 7.2 (continued)

Site/context	Site function	Vizcacha		Immature individuals	Specimens with evidence of human use				Interpretation	References
		NISP	MNI		Cut marks	Cultural fracture	Burning	Bone tool		
Laguna del Trompa/sector H-layer I A1 A3	Campsite (residential bases)	6	No data	No data	–	–	–	–	Consumption	Crivelli Montero (1991), Eugenio (1991) and Silveira (1991)
Laguna del Trompa/sector H layer I B3	Campsite (residential bases)	1	No data	No data	–	–	–	–	Consumption	
Laguna Tres Reyes 1/upper component	Multiple-activity site/ burial site	14	1	No data	Presence	–	–	–	Consumption	Madrid and Barrientos (2000) and Salemme and Madrid (2007)
Arroyo Seco 2/ units Y-Y/S-S-S/Z-Z	Short-term campsites/carcass processing sites (Late Pleistocene)-base camps/burial site (Early to Middle Holocene)	34	2	Presence	–	–	Presence	–	Consumption?	Rafuse (2013), Politis (2014), Salemme (2014), Politis et al. (2016) and Rafuse (2017)
Claromecó 1/ lower component	Short-term camp	26	2	Presence	–	–	4	–	Hunting of several individuals; cooking by roasting the whole animal; consumption	Bonomo et al. (2008) and Leon (2014)
Quequén Salado I/ stratum IV	Base camp	19	2	No data	–	–	Presence	–	Complementary resource?	Madrid et al. (2002) and Leon (2020)

Site/context	Site function	Vizcacha		Immature individuals	Specimens with evidence of human use				Interpretation	References
		NISP	MNI		Cut marks	Cultural fracture	Burning	Bone tool		
Las Brusquillas 3/unit III	Multiple-activity site	22	3	No data	–	–	5	–	Cooking by roasting the whole animal; consumption?	Massigoge et al. (2018)
Las Brusquillas 4/unit IIIb	Multiple-activity site	372	16	Presence	46	2	3	–	Skinning, disarticulation, defleshing, and consumption of meat and bone marrow	Rodríguez (2023)
Cortaderas/ lower component	Residential camp	20	2	No data	1	–	2	–	Disarticulation, roasting, and consumption	Massigoge (2009)
El Guanaco 1/ sector 1- unit 3	Burial pit with a mixture of materials from different occupations	17	4	Presence	1	–	–	–	Processing and consumption	Frontini (2012)
El Guanaco 2	Residential camp/ burial site	10	2	Presence	–	–	3	–	Cooking and consumption	Frontini (2012)
Paso Otero 3/ Ab7	Short-term camps	25	6	No data	–	–	–	–	Processing and consumption	Martínez et al. (2001), Martínez (2006) and Johnson et al. (2012)

(continued)

Table 7.2 (continued)

Site/context	Site function	Vizcacha		Immature individuals	Specimens with evidence of human use				Interpretation	References
		NISP	MNI		Cut marks	Cultural fracture	Burning	Bone tool		
Paso Otero 4/upper levels	Residential and multiple-activity site	118	7	No data	2	–	4	–	Skinning, processing, and consumption	Gutiérrez et al. (2011), Álvarez (2012), Álvarez et al. (2013) and Martínez and Gutiérrez (2019)
Paso Otero 4/lower levels	Residential and multiple-activity site	144	9	No data	14	1	2	–	Skinning, disarticulation, defleshing, and consumption of meat and bone marrow	
Nutria Mansa 1/lower component	Multiple-activity site	113	No data	No data	–	–	–	–	Consumption	Bonomo (2005)
Alfar	Coastal residential camp	49	5	Presence	–	1	2	–	Consumption	Bonomo and Leon (2010) and Leon (2014)
Empalme Querandíes 1	Multiple-activity base camp	63	2	No data	3	–	1	–	Processing and consumption	Álvarez (2012) and Messineo et al. (2013)
Lobería I site 1/upper component	Residential camp	382	13	No data	16[a]	Presence	No data	–	Skinning, disarticulation, defleshing, and consumption of meat and bone marrow	Mazzanti et al. (2010) and Quintana and Mazzanti (2011, 2014)

Site/context	Site function	Vizcacha		Immature individuals	Specimens with evidence of human use				Interpretation	References
		NISP	MNI		Cut marks	Cultural fracture	Burning	Bone tool		
Cueva Tixi/ level 4	Residential camp	180	6	No data	14[a]	Presence	Presence	–	Skinning, disarticulation, defleshing, and consumption of meat and bone marrow	Mazzanti (1997), Quintana (2001), Valverde (2001), Quintana and Mazzanti (2001, 2011, 2014), Martínez et al. (2013) and Mazzanti et al. (2015)
Cueva Tixi/ level 3	Ephemeral occupations during which faunal resources were processed	4	1	No data	–	–	–	–	Consumption	
Cueva Tixi/ level 2	Ephemeral occupations during which faunal resources were processed	12	1	Presence	1	–	–	–	Processing and consumption	
Cueva Tixi/ level 1	Residential camp	31	2	No data	2	–	Presence	–	Defleshing for consumption	
Cueva El Abra/ upper component	Residential camp	468	12	No data	32[a]	Presence	–	1	Skinning, disarticulation, defleshing, and consumption of meat and bone marrow; use of bones as raw material for tools	Quintana and Mazzanti (2011, 2014)

(continued)

Table 7.2 (continued)

Site/context	Site function	Vizcacha		Immature individuals	Specimens with evidence of human use				Interpretation	References
		NISP	MNI		Cut marks	Cultural fracture	Burning	Bone tool		
Amalia site 4/ sector 2	Residential camp	19	2	No data	–	–	Presence	–	Consumption	Mazzanti (2007) and Quintana and Mazzanti (2011)
La Toma/upper component	Base camp	6	1	No data	–	–	–	–	Consumption?	Álvarez and Salemme (2015)
La Toma/lower levels	Multiple-activities camps	6	2	No data	–	–	–	–	Consumption?	
Puente de Fierro	Residential camp	2	1	Presence	2	–	–	–	Disarticulation and consumption	Frontini and Bayón (2015) and Borella et al. (2017)

[a]Calculated by the authors of this work from the percentages provided in the original publication

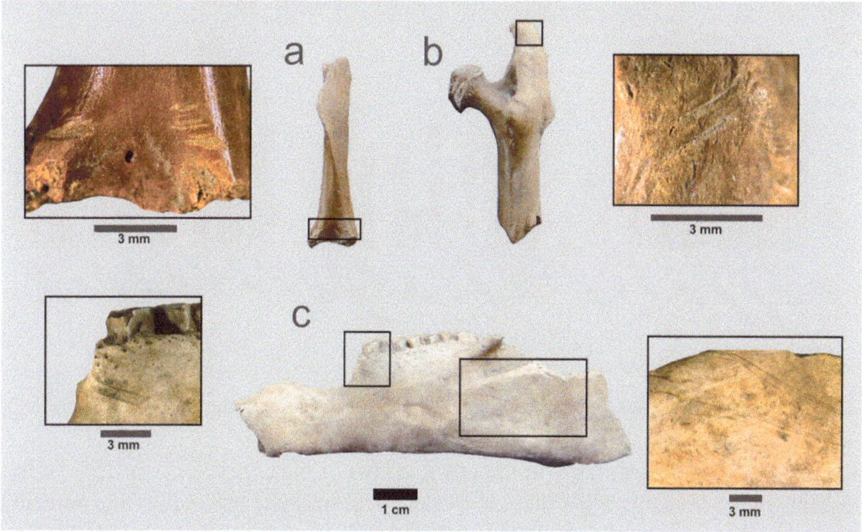

Fig. 7.2 Plains vizcacha bones with cut marks recovered from archaeological sites in the Pampas ecoregion. (**a**) Humerus from Las Brusquillas 4 site; (**b**) femur from Paso Otero 4 site (lower levels); (**c**) Hemimandible from Hangar site

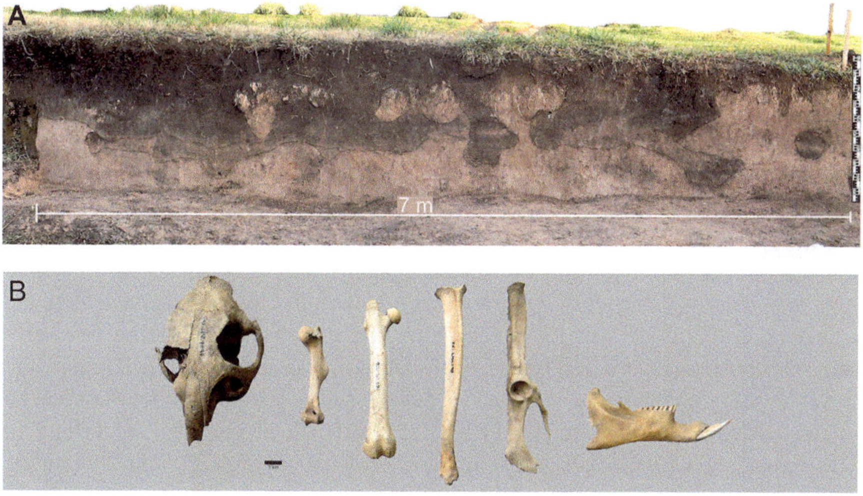

Fig. 7.3 Stratigraphy and materials from the Laguna Muscar 2 site. (**a**) Profile showing the disturbance generated by the *vizcacheras*; (**b**) Skeletal remains of plains vizcacha of natural origin

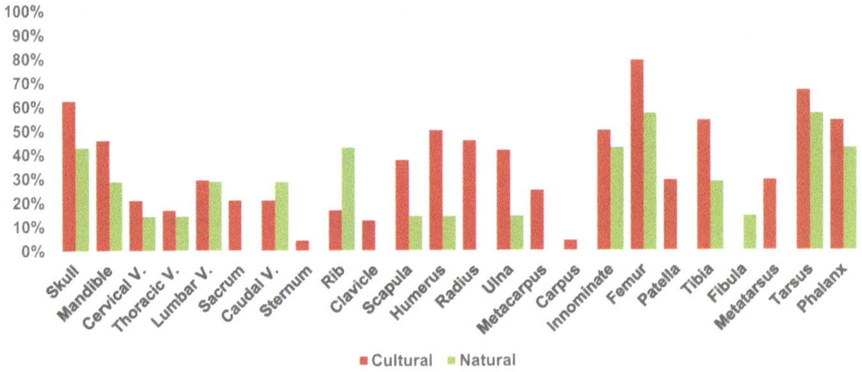

Fig. 7.4 Percentage of sites with plains vizcacha bone specimens of cultural and natural origin

percentage drops considerably in the other sectors of the region: only in 15% of the contexts in the Ventania Hills there is evidence of vizcacha use, while this percentage is even lower in the Sandy Pampas (11%) and in the Rolling Pampas (6%).

No clear temporal trend is observed when evaluating the percentage of contexts with vizcacha exploitation and the number of remains per context for each chronological period. Evidence of exploitation is shown in two out of six Late Pleistocene/Early Holocene contexts (33%), five of 17 Middle Holocene contexts (29%), and 23 out of 79 Late Holocene contexts (29%). In terms of the number of remains per context, a minimum of 31 and a maximum of 144 are observed for the Late Pleistocene/Early Holocene (mean = 87.5; median = 87.5); a minimum of 12 and a maximum of 118 for the Middle Holocene (mean = 46.6; median = 29); and a minimum of 1 and a maximum of 468 for the Late Holocene (mean = 83.3; median = 19).

The earliest record of vizcacha exploitation comes from Level 1 of the Cueva Tixi site, at the eastern end of the Tandilia Hills, with a chronology of 12,590–11,240 cal years BP (Table 7.1; Fig. 7.5). In this context, which was interpreted as a residential camp, 31 vizcacha remains were identified, two of which show cut marks (Table 7.2). Quintana and Mazzanti (2011) suggested that the defleshing and consumption of at least two vizcachas occurred during this early time. The only context dated to the Early Holocene in the Inter-hill Pampas, the Lower levels of the Paso Otero 4 site (Fig. 7.5), shows a high number of vizcacha remains (NISP = 144) corresponding to at least nine individuals. According to Álvarez et al. (2013), the presence of remains showing cut marks, cultural fractures, and burning suggests the occurrence of activities such as skinning, disarticulation, defleshing, and consumption of meat and bone marrow of vizcacha at a residential site (Table 7.2).

There is a larger number of archaeological contexts dated to the Middle Holocene distributed more widely in the ecoregion (Fig. 7.5). Five of these contexts show evidence of the use of vizcacha, including Laguna de los Pampas in the Sandy Pampas; Paso Otero 3, Paso Otero 4 (Upper levels), and the Alfar sites in the Inter-hill Pampas; and the Level 2 of the Cueva Tixi site in Tandilia. A variable number of

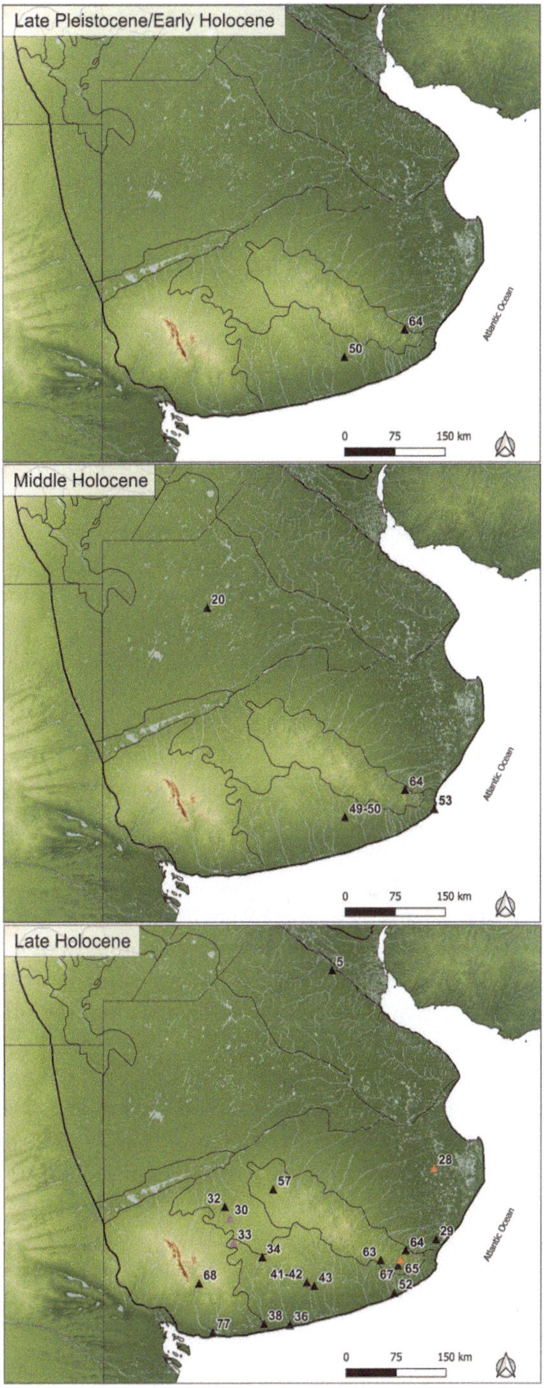

Fig. 7.5 Archaeological contexts with evidence of plains vizcacha use by chronological period. Color references for the Late Holocene map: Black triangles: pre-Hispanic; Purple triangles: pre- and post-Hispanic; Orange: post-Hispanic

vizcacha specimens were recovered at these sites, with a maximum of 118 at the Lower levels of Paso Otero 4 (MNI = 7) and a minimum of 12 at Cueva Tixi (MNI = 1). The difference in the number of remains may be due in part to the functionality of these sites; Paso Otero 4 has been interpreted as a residential and multiple-activity site where diverse animals were consumed (Álvarez et al. 2013), while Level 2 of Cueva Tixi has been interpreted as the result of ephemeral occupations focused on the processing of faunal resources, especially artiodactyls (Quintana and Mazzanti 2001, 2011).

For the Late Holocene, there are considerably more sites distributed throughout the ecoregion, except in its northwestern portion (Fig. 7.5). During this time, the first human occupations are recorded in the Mesopotamic and Rolling Pampas and the eastern portion of the Depressed Pampas. There is also a notable increase of sites in the Inter-Hill Pampas and the Bonaerense Hills, both in Tandilia and Ventania (Table 7.1).

In the northeastern portion of the ecoregion, vizcacha remains have been identified in only 3 of the 17 sites of the Rolling Pampas, with only one showing evidence of human exploitation. This is the Cañada Honda site, a multiple-activity and burial site dated to around 2000 cal BP, where the presence of cut marks, cultural fractures, and burning damage on bones supports the utilization of this rodent. Although abundant vizcacha remains were recovered (NISP = 84; MNI = 12), researchers have suggested that some of these could have been incorporated naturally since a *vizcachera* was observed at the site (Escosteguy and Salemme 2012).

In the Depressed Pampas, vizcacha remains were identified in five of the nine archaeological contexts of the Late Holocene. In four of them, it has been proposed that this rodent was used by hunter-gatherers (Table 7.1). However, evidence of this utilization has only been described for the Hangar site, located in the western portion of this complex. At this site, dated to ca. 1000 cal years BP and interpreted as a base camp, a few vizcacha bones were identified (NISP = 16; MNI = 1), and cut marks and burning damage were observed (Table 7.2).

Information on 23 Late Holocene archaeological contexts is available for the Inter-Hill Pampas. Seventeen present vizcacha remains, and nine have been proposed as having a cultural or possible cultural origin (Table 7.1). The sites with the most substantial evidence of consumption are Las Brusquillas 4, Cortaderas, and Claromecó 1, all interpreted as campsites. The bone record shows notable differences among these sites. While only a few remains were recovered in Cortaderas and Claromecó 1, Las Brusquillas 4 stands out as one of the sites with the highest number of remains in the ecoregion corresponding to a minimum of 16 individuals. Another outstanding feature of this assemblage is that 32% of the specimens correspond to nonfused elements. Evidence of exploitation is abundant and includes bones with cut marks, cultural fractures, and thermal alteration (Table 7.2). Based on this evidence, this rodent's skinning, disarticulation, defleshing, and consumption of meat and bone marrow have been proposed (Rodríguez 2023). A cut marked vizcacha humerus recovered from this site was dated to 3090 ± 20 ^{14}C years BP (PSUAMS-8569; 3353–3171 cal years BP). This is the first and only taxon date for the vizcacha in the Pampas ecoregion for archaeological and paleontological contexts.

The Bonaerense Hills complex also has many contexts dated to the Late Holocene, 13 in Tandilia and 12 in Ventania. The first sector stands out for the high number of contexts with vizcacha, which in most cases show evidence of consumption (6 out of 9). Except for Empalme Querandíes 1, the remaining sites with vizcacha exploitation are grouped in the eastern end of Tandilia. Three of these contexts, interpreted as residential camps, present a large number of remains with abundant evidence of human use: Level 4 of the Cueva Tixi site (NISP = 180; MNI = 6) and the upper components of Lobería I (NISP = 382; MNI = 13) and Cueva El Abra (NISP = 468; MNI = 12) (Table 7.2). These contexts are dated to between ca. 900 and 500 cal years BP. Based on this record, Quintana and Mazzanti (2011) proposed systematic and integral processing of this prey for using the skin, the meat, the bone marrow, and the long bones as raw material for tools. The findings from the site Amalia 4 are also worth mentioning in this sector. This site, with an archaeological context dating to the eighteenth century, shows a continuity in the use of the vizcacha by indigenous societies after the Spanish conquest (Mazzanti 2007; Quintana and Mazzanti 2011).

Markedly different results emerge from analyzing the contexts in the Ventania Hills. A few vizcacha remains were identified in five contexts dated to the Late Holocene. The use of this species was proposed for three of them: the Upper component (960–730 cal years BP) and the Lower levels (3870–1721) of the site La Toma and the site Puente de Fierro (2000 cal years BP), all interpreted as residential camps. No direct evidence of exploitation was identified in the few vizcacha bones recovered in the contexts from La Toma. Still, the possible consumption of this rodent was proposed (Álvarez and Salemme 2015). In Puente de Fierro, the only two bones recovered show cut marks indicating disarticulation and consumption of at least one individual (Frontini and Bayón 2015).

To evaluate temporal changes in the relative importance of the vizcacha in the subsistence strategies of indigenous societies, a Vizcacha Index (VI) was calculated for 71 contexts (Supplementary Table 7.2). Most of them (61.97%) have a VI value of 0, followed by those with values between 0.001 and less than 0.5 (32.39%), and finally, those with values of more than 0.5 (5.63%). Among the latter contexts are Lobería I and Cueva El Abra in the Tandilia Hills and Las Brusquillas 4, Paso Otero 4, and El Guanaco 1 in the Inter-hill Pampas. However, the latter has a small sample and a broad relative chronology.

When evaluating the trend lines for each of the surveyed complexes, some patterns can be observed, except for the Depressed Pampas, for which only two sites are available (Fig. 7.6). For Ventania, there are only scarce remains of vizcacha in a few Late Holocene contexts. For Sandy Pampas, there is a very similar situation, Laguna de los Pampas being the only site with vizcacha, dated to the first half of the Middle Holocene. In the case of Inter-hill Pampas, many contexts with vizcacha remains are recorded. In this case, the values are variable throughout the Holocene, with some peaks in the Middle and the first part of the Late Holocene. The trend line shows a slight decrease toward late times. Finally, for Tandilia, the most marked trend line is observed, with a significant increase in the final Late Holocene.

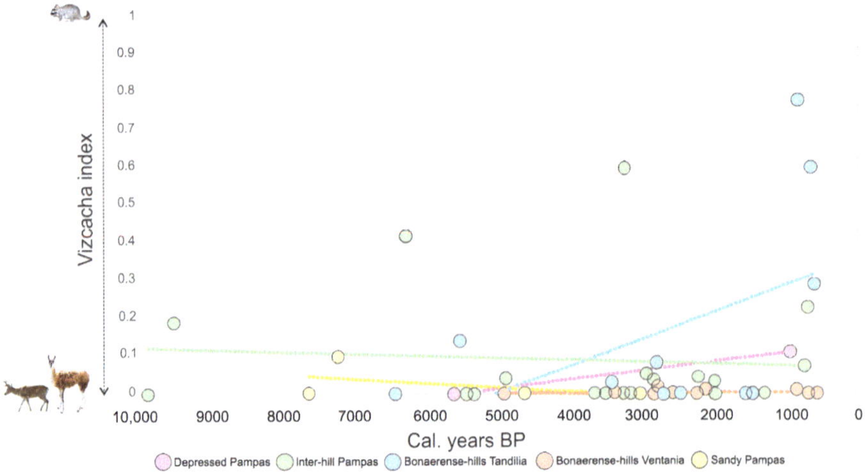

Fig. 7.6 Vizcacha index (VI) and trend lines for archaeological contexts from the southwestern portion of the Pampas

7.5 Discussion

7.5.1 Plains vizcachas in the Pampean Archaeological Record

The plains vizcacha is well-represented in the archaeological record of the Pampas ecoregion. According to the present study, their remains have been identified in more than half of the archaeological sites in the region. Although these sites are widely distributed in the ecoregion, they are concentrated in the south, particularly in the Inter-hill Pampas and the Tandilia Hills (Fig. 7.1). The low representation of vizcacha in the northern portion of the Pampas may be due in part to the reduced number of archaeological sites. Discussing the causes of this low representation of sites exceeds the objectives of this chapter. Still, it is worth mentioning that other researchers have attributed it to a combination of factors, ranging from differences in population dynamics to scientific and taphonomic biases (Politis and Madrid 2001; Martínez et al. 2015).

As this study has shown, human action does not always explain the presence of vizcacha remains at archaeological sites. In about 12% of the contexts, the existence of vizcacha has been attributed exclusively to natural causes, mainly to their death in the burrows. The results of actualistic studies indicate that the vizcachas can be characterized as ecosystemic engineers capable of accumulating, mixing, and fragmenting different materials (Tomassini et al. 2017, 2019; Rafuse et al. 2019; Bellinzoni et al. 2021). To construct their burrows, the vizcachas select elevated areas, which often coincide spatially with the location of the archaeological sites. The construction of *vizcacheras* on archaeological sites contributes to the natural incorporation of vizcacha skeletal remains, as well as those of other animals that occupy these loci (e.g., the burrowing owl, *Athene cunicularia*) and their prey (Pocheville 2015). Moreover, the tunnels and chambers profoundly affect the stratigraphic integrity and cause vertical and horizontal displacements of archaeological materials, altering the original associations and mixing evidence from different human occupations (Kaufmann et al. 2019; Rafuse et al. 2019; Tomassini et al. 2019; Bellinzoni et al. 2021).

The number of vizcacha remains recovered from archaeological contexts is variable. According to the present study, this number is higher in contexts where the presence of vizcacha links to human action. However, we cannot rule out the possibility of a sampling bias due to the low number of contexts where vizcacha is present due to natural causes. Mostly, vizcacha remains appear in considerably lower frequency than those of guanaco. While this difference may be primarily due to indigenous societies' differential selection of these two species, the influence of other factors in the lower representation of vizcacha bones cannot be dismissed. As well recognized by zooarchaeologists, the abundance of prey remains does not directly reflect their importance to past societies. Still, there are also taphonomic and methodological biases that can affect the representation of smaller vertebrates. The taphonomic bias is because different postdepositional processes destroy small vertebrate specimens more easily (e.g., differential destruction by carnivores and weathering) than larger ones (Behrensmeyer 1978; Lyman 1994; Behrensmeyer and Miller 2012). Among the methodological biases, we must mention failures in their recovery during fieldwork due to the small size of some bone remains. Also, methodological biases can be expected during laboratory analysis since many of the remains of small species are part of the material recovered in the sieve, which is only sometimes analyzed. Difficulties stress during the taxonomic identification due to a lack of reference collections and poor knowledge of archaeologists of the anatomy of the species that appear less frequently in the faunal record (Peres 2010; Reitz and Wing 2008).

The types of anatomical elements recovered are variable. Vizcacha findings often consist of cranial and postcranial elements, frequently fragmented. The high variability in the representation of vizcacha skeletal parts shown in actualistic studies makes it difficult to propose a characteristic pattern of naturally deposited assemblages that would be useful to differentiate them from those of human origin. The various vizcacha bone contexts attributed to hunting and human processing analyzed in this work show the presence of remains of the entire skeleton, suggesting that this rodent would have been brought whole to the residential camps. Although all skeletal elements entered the residential camps, the intrinsic characteristics of the bones (e.g., size, shape, bone mineral density), combined with the impact of destructive natural processes, human processing, and archaeological recovery and identification methods, could directly influence the skeletal representation.

To date, there is no specific bone mineral density index for vizcacha that would allow us to evaluate the differential survival of its bones in the face of destructive processes. Beyond the particular values of mineral density characteristic of each bone element, we can generalize that the axial region of the mammalian skeleton, with a more significant proportion of trabecular tissue, would be the least dense portion and the most affected by taphonomic agents (Lyman 1994). Naturalistic and experimental observations have recognized that during the consumption of their prey, carnivores chew and destroy vertebrae, sternum, and ribs in more significant proportion, surviving to a larger extent hard parts of the skull, long bones, and autopodia (Lloveras et al. 2008; Álvarez et al. 2012; Rafuse et al. 2014; Gutiérrez et al. 2016; Cohen and Kibii 2015). Furthermore, variability in human practices of processing, cooking, and preserving the meat of prey may generate differential segmentation, distribution, and discard of anatomical parts within the camp, each with its taphonomic trajectories.

As for the age-class representation of vizcacha in the archaeological record, scarce information is available in the literature as it has yet to be a systematically surveyed variable in the zooarchaeological analyses. Only the presence of unfused postcranial elements is mentioned for a limited number of sites indicating the occurrence of immature individuals. However, the lack of a bone fusion schedule for this species does not allow better resolution age-class assignments to be made. There are only two studies of the ontogenetic development of the vizcacha skeleton: Rasia (2016) defined age classes based on the sutures, size, and shape of the skull, and dental features of the mandible, and Garrone et al. (2021) established ontogenetic stages based on histological changes in selected long bones. On the one hand, and despite their importance, these models of skeletal development are not readily applicable to the study of vizcacha bones recovered from archaeological sites. In the first case, craniums and mandibles are rarely recovered in a good state of completeness. In the second case, because it is a type of study that requires more time and resources and specific training, in addition to destroying the sample. On the other hand, the fusion state is a variable that faunal analysts can easily and quickly record.

Chronological information from archaeological sites containing vizcacha remains resulting from human action shows that the vizcacha bone record covers a time from the Late Pleistocene to the Late Holocene. However, it is important to highlight that direct dates on vizcacha bones are nonexistent in the published literature of the Pampas. The first taxon date for the vizcacha is reported in the present work and was obtained for the Las Brusquillas 4 archaeological site.

7.5.2 *Plains vizcacha as an Economic Resource for Indigenous Societies in the Pampas*

The present study shows that the vizcacha was exploited by the indigenous societies of the Pampas ecoregion from the Late Pleistocene to at least the eighteenth century. According to the reviewed literature, this rodent appears to have been used throughout the region. However, evidence of its exploitation is more frequent in the southeast, particularly in the Inter-hill Pampas and Tandilia, where the percentage of contexts showing vizcacha use is the highest, and some contexts with exceptionally high numbers of remains exist.

The vizcachas were most likely an attractive resource for hunter-gatherers for several reasons: (a) they are widely distributed in the landscape and are available all year round; (b) their location is predictable, as they are always found in or near the *vizcacheras*, which in turn are built in particular sectors of the landscape, may be active for many years, and are highly visible; and (c) they are easy to capture, and being a colonial animal, they can be taken either singly or in groups, with different technologies (Quintana and Mazzanti 2011, 2014; Tomassini et al. 2019).

The variable number of individuals (MNI) in the different contexts in which the vizcacha was exploited suggests that different strategies may have been implemented in its capture. Vizcacha bone assemblages with few individuals could result from single capture events, possibly when they were foraging outside the *vizcacheras*. Because vizcachas are nocturnal, this type of hunting would have required organizing special hunting parties, leaving the camp at dusk and with limited participants. The hunters may have used throwing weapons, but because vizcachas communicate through alarm vocalizations when detecting a predator in their home range and ran into the main *vizcachera* (Branch 1993), only a few individuals could have been taken by this method. Individual hunting could also have been carried out in satellite *vizcacheras*, used by adult males from spring to autumn.

Additionally, as documented in ethnographic, ethnoarchaeological, and historical literature for other small mammals, other procurement methods that did not require nocturnal parties may have been used for individual capture, including spring-pole traps, nets, snares or, in the case of fossorial mammals, prizing them from their burrows with specially constructed skewers (Yellen 1991; Lupo and Schmitt 2002; Andrade and Boschín 2015). Specific observations on vizcacha capture made by Arenas (2003) on indigenous groups in the Central Chaco of northern Argentina in the late twentieth and early twenty-first centuries describe nocturnal hunting carried out using arrows on clear nights. Jackson et al. (1996) also mention hunting vizcachas with fire guns or using wire cages by modern rural settlers. Escosteguy et al. (2023) note that present-day inhabitants of the Rolling and Depressed Pampas use fire and smoke to force the vizcachas to leave the *vizcacheras* to hunt them.

On another note, the presence of vizcacha assemblages with numerous individuals in some of the sites raises the question of whether they represent an accretional accumulation resulting from multiple individual captures or correspond to one or a few mass capture events. Because vizcachas are colonial and up to 30 individuals can congregate in a *vizcachera* during the day, they are a good candidate for mass procurement, which could be conducted during the daytime by all members of a social group, including the youngest. Mass capture of fossorial mammals can include the excavation of their burrows with digging sticks or forcing them out of their burrows by smoking or flooding them (Jones 2006; Andrade and Boschín 2015; Maryañski and Nielsen 2015; Fernández et al. 2017). Regarding vizcacha mass procurement, Arenas (2003) describes capturing on days of heavy rain: flooding the burrows through the excavation of channels that direct the water toward them and killing the animals with a club when they come out. Mass hunting is not selective, and in the particular case of hunting in *vizcacheras*, it would have involved capturing a large part of the individuals of the social group, including adult females and males, as well as offspring.

The information available at the moment does not allow us to confirm that indigenous societies of the Pampas used mass-collecting techniques to procure vizcachas. However, the large number of individuals in some of the assemblages and the presence in one of them of a high number of unfused remains allow us to consider this as a possibility. Detailed zooarchaeological studies are required to estimate the age-class representation in the vizcacha bone assemblages and thus test this hypothesis. This approach will also require knowledge of the age at which the skeletal elements of this species fuse. Given the importance of having a bone fusion schedule to identify vizcacha age classes at archaeological sites, zooarchaeologists should undertake this task. This schedule should consider differences in skeletal maturity in females and males.

Identifying whether vizcacha was mass-collected is very important when assessing the subsistence strategies of indigenous populations. In optimal foraging theory, it has been traditionally assumed that there is a close correspondence between the prey's body size and its energetic return rate (Broughton 1994). However, this would change when large numbers of small animals are taken simultaneously (Steele and Baker 1993; Madsen and Schmitt 1998). Although some authors proposed that returns from these events would still be low because of the high costs of capture and processing (Lupo and Schmitt 2002; Ugan 2005), Jones (2006) mentioned that the costs involved in capturing animals that live in warrens, like the European rabbit, are minimal, because simple techniques such as the use of water or fire could be easily performed. Considering that in each *vizcachera* there is an approximate number of between 10 and 30 individuals, mass-collecting techniques would allow acquiring a critical number of animals yielding a protein return that could be similar o even higher than that obtained from a single larger prey. Numerous *vizcacheras* are placed nearby, increasing this resource's local abundance and return. This characteristic makes the vizcacha an even more attractive resource, especially during times of scarcity of larger animals such as artiodactyls. Another advantage of this resource is that it does not require logistical parties or specialized hunters since it is easy to acquire gregarious animals living in dens.

As was previously mentioned, the diversity of elements recovered in the assemblages, corresponding to the different skeletal regions, suggests that once captured, the vizcachas were transported whole to the camps, as proposed by other researchers (Quintana and Mazzanti 2011; Álvarez 2012; Rodríguez 2023). Considering that the average weight of adult vizcachas is 6 kg, no anatomical part selection strategies would have been implemented to reduce transportation costs, as occurs with large mammals (Binford 1978; Bunn et al. 1988; Yellen 1991, among others).

The vizcacha bone remains from the archaeological sites show different evidence of human processing. The modifications mentioned allow us to identify activities linked to different stages of the processing and consumption sequence. The presence of skinning marks indicates the use of the pelts. In this regard, some records show the traditional and commercial use of vizcacha pelts for making coats, bags, rugs, and bedspreads in modern times (Jackson 1985; Arenas 2003). There is also evidence of thermal alteration, which would be linked to cooking techniques. Some authors mentioned the possibility of cooking by roasting the whole animal (e.g., Leon 2014; Massigoge et al. 2018). Arenas (2003) reports that vizcachas can be roasted in ovens made in wells, on roasters, fried, and cooked in soups or stews. This author also describes that during hunting events in which numerous vizcachas are captured, they are roasted and dehydrated in the fire's heat to allow the animals' deferred consumption (Arenas 2003).

Traces of disarticulation and filleting are also linked to consumption and butchery. This large rodent has white meat with good nutritive value due to its high digestibility coefficient, low-fat fusion point, and high percentage of proteins (Jackson 1985). Another characteristic of the sites is the presence of fresh fractures with associated features indicating bone marrow use. Although vizcacha meat is of good quality, it is poor in fat (Barbarán 2003), an essential nutrient for hunter-gatherer societies (Speth and Spielmann 1983). The scarcity of fat could have led to the implementation of strategies such as the selection of fat prey and the consumption of bone marrow. It is important to mention that the availability of ceramic containers in the Late Holocene may have generated better conditions for fat utilization, especially in animals with lean meat and scarce bone marrow, such as the vizcacha. Finally, in specific contexts, the use of bone as raw material is mentioned (Quintana and Mazzanti 2011; Escosteguy and Salemme 2012).

The vizcacha has traditionally been considered a complementary resource in the subsistence models proposed for the indigenous societies that inhabited the southern portion of the Pampas ecoregion (Quintana and Mazzanti 2001; Martínez and Gutiérrez 2004; Álvarez 2014). Although we agree with this idea, in this study, we wanted to evaluate in greater depth whether there were spatial and temporal differences in the economic importance of this resource. For this, through a Vizcacha Index, we compare the abundance of vizcacha remains with that of artiodactyls. Results show scarce evidence of the use of this resource in the Sandy and Depressed Pampas and the Ventania Hills. Although this could be a consequence of its lesser relevance as a resource, it cannot be ruled out that it may be due to poor bone preservation. In this sense, Morales (2022) noted these problems for the Ventania Hills. Contrarily, the importance of this resource in the Inter-hill Pampas and Tandilia is supported by the current data.

According to the VI, the exploitation of vizcacha was variable over time in the Inter-hill Pampas. Probably, this animal was exploited opportunistically when the *vizcacheras* were located within the daily foraging range of the residential camps, but the possibility of special procurement trips to known *vizcacheras*, especially in times of scarcity of the main resources, cannot be ruled out. In the case of Tandilia, the most marked trend line was observed, with a significant increase in the VI in the Late Holocene contexts. These high values correspond to a group of cave and rock shelter sites located at the eastern end of Tandilia, dated after 1000 years BP. Quintana and Mazzanti (2011, 2014), who have studied these sites, proposed that this rodent and other small vertebrates acquired greater importance in the final Late Holocene and were butchered with greater intensity. They also suggested that the annual availability and homogeneous distribution across the landscape allowed hunter-gatherers to exploit the vizcacha as a buffer resource when other small prey were scarce or difficult to obtain. According to these authors, this novel use of the faunal resources corresponds to a strategy of diversification and intensification of subsistence in which small game was predominant (Quintana and Mazzanti 2011). A recent study of the genetic structure of modern populations of vizcacha from Argentina showed a slight population size decline starting at approximately 1500 years BP, followed by its increase after 300 years BP (Gariboldi et al. 2019). The authors of this study suggest that the regional increase in human population density proposed by archaeologists for the Late Holocene and the intensive use of vizcacha caused a higher predation pressure that could have affected the population size of this species. However, the evidence of vizcacha exploitation at a regional level does not show such intensive use sufficient to generate hunting pressure on this resource.

Although indices such as VI are necessary heuristic tools for assessing changes in the relative representation of species over time, their results should be interpreted with caution, as factors other than the economic importance of a resource may affect their outcome. These include social, environmental, methodological, and taphonomic factors. Although the evaluation of the influence of each of these factors is beyond the scope of this chapter, it is essential to note that the highest VI values were obtained in cave sites (Cueva Tixi, Cueva El Abra, and Lobería I) or assemblages recovered from fluvial-lacustrine deposits with good bone preservation (Paso Otero 4 and Las Brusquillas 4). This condition shows how important it is to evaluate the role of differential preservation in the relative abundance of vizcacha in the archaeological record of this region in future work.

7.6 Conclusions

In this chapter, we reviewed the archaeological literature on the use of vizcacha by indigenous societies of the Pampas ecoregion and provided original data for some sites. This study showed that the vizcacha was an essential complementary resource for these societies from the Late Pleistocene to the Late Holocene, especially in the southern portion of the ecoregion. The evidence does not show a temporal change in the use of this resource at the regional level; only an increased use is observed in the eastern end of Tandilia, which researchers have linked to the regional process of intensification and diversification of subsistence. The species' ethological and ecological attributes made it an attractive economic resource for hunter-gatherer societies, among which we can mention the following: wide distribution in the landscape and year-round availability; predictable location; and ease of capture, either singly or in mass. This resource offered a moderate amount of meat and bone marrow, especially when caught in masse, as well as skin for manufacturing various handicrafts (e.g., bags, blankets) and bones occasionally used to elaborate instruments.

This study provided an up-to-date overview of the literature on this matter and contributed to new ideas and questions. It revealed the need for more taxon dates for this species. It highlighted research gaps in the knowledge of the past use of this resource, such as the causes of the differences observed in the areas of the ecoregion or the types of procurement methods employed by indigenous societies. This information is helpful not only for archaeologists interested in the subsistence strategies of past societies but also for paleontologists, paleoecologists, evolutionary geneticists, and conservationists seeking to assess the effects of long-term human predation on the population dynamics and evolutionary history of this species. Additionally, this study showed the need for methodological changes in zooarchaeological studies if we want to substantially improve our knowledge of the use of vizcacha by indigenous societies. These changes range from enhancing the recovery strategies and identifying the remains of this and other small vertebrates to developing new methodological tools, such as a fusion schedule for the species.

Finding and summarizing the information in the literature was difficult, primarily because it was produced over more than 40 years of research and, in some cases, dealt with sites excavated decades earlier. Excavation and recovery methods and zooarchaeological and taphonomical analyses have changed significantly over the years, conditioning the quality of the archaeological data, especially in the case of small fauna. In addition, there are differences in research efforts and traditions in each area of the ecoregion, even at present, causing profound inequalities in the amount and types of data generated. These differences should be considered, especially when interpreting the general trends that emerge when comparing and integrating the data. However, despite these challenges, the review study presented here constitutes a valuable state of the art about our current knowledge of indigenous societies' use of the vizcacha. Moreover, it provides helpful information that permits us to reflect on the methodological tools needed to achieve new research goals and answer further questions.

Acknowledgments The authors thank Luciano Rasia, Claudio Barbeito, and Francisco Acuña for their invitation to collaborate in this special volume. We are deeply grateful to those who contributed to obtaining some of the bibliography: Andrés Izeta, Pablo Messineo, and Mónica Salemme. Also, to Daniel Rafuse for his help in the initial stages of this work, to Luis Borrero and Juan B. Belardi for their reading and comments on a draft version of this chapter and to those who allowed us to use information that was difficult to access through publications: Gustavo Martínez, Romina Frontini, Nahuel Scheifler, and Mariano Bonomo. This research was granted by the Agencia Nacional de Promoción de la Investigación, el Desarrollo Tecnológico y la Innovación (PICT 2017-1969; PICT 2018-0686; PICT 2020-0789), and the Consejo Nacional de Investigaciones Científicas y Técnicas (CONICET-PUE N°22920160100079). We thank Instituto de Investigaciones Arqueológicas y Paleontológicas del Cuaternario Pampeano (UE INCUAPA-CONICET-UNCPBA) and Facultad de Ciencias Sociales (UNCPBA) for their support and collaboration.

References

Aldazabal V (2002) La Ocupación Humana en el Sector Centro Oriental de la Pampa Deprimida Provincia de Buenos Aires – Argentina. Unpublished doctoral thesis, Universidad de Buenos Aires, Buenos Aires

Aldazabal V, Eugenio EO (2013) La cerámica unguicular y corrugada en la pampa deprimida. Contextos y discusión. Cuad Inst Nac Antropol Pensam Latinoam 1(4):95–107

Aldazabal V, Eugenio E, Silveira M (2017) Nuevos datos sobre la subsistencia de los cazadores recolectores del sitio "Divisadero Monte 6" (partido de General Lavalle, provincia de Buenos Aires). Cuadernos de Antropología 18:105–120

Álvarez MC (2012) Análisis Zooarqueológicos en el Sudeste de la Región Pampeana. Patrones de Subsistencia durante el Holoceno Tardío. Unpublished doctoral thesis, Universidad Nacional del Centro de la Provincia de Buenos Aires, Olavarría

Álvarez MC (2014) Subsistence patterns during the Holocene in the Interserrana area (pampean region, Argentina): evaluating intensification in resource exploitation. J Anthropol Archaeol 34:54–65

Álvarez MC (2018) Análisis faunístico del sitio Laguna de los Pampas (partido de Lincoln, Región Pampeana). Aportes a los estudios de la subsistencia en el área Oeste de la Pampa Húmeda. Arqueología 24(1):127–146

Álvarez MC, Salemme MC (2015) Tendencias en la subsistencia en el Holoceno tardío en la región pampeana, Argentina: el caso del sitio La Toma (partido de Coronel Pringles, provincia de Buenos Aires). Relac Soc Argent Antropol 40(1):123–148

Álvarez MC, Kaufmann CA, Massigoge A, Gutiérrez MA, Rafuse DJ, Scheifler NA, González ME (2012) Bone modification and destruction patterns of leporid carcasses by Geoffroy'scat (Leopardus geoffroyi): an experimental study. Quat Int 278:71–80

Álvarez MC, Alcaraz AP, Gutiérrez MA, Martínez G (2013) Análisis zooarqueológico del sitio Paso Otero 4 (partido de Necochea). Aportes a la discusión de modelos de subsistencia de la región pampeana. Intersecc en Antropol 14:383–398

Andrade A, Boschín MT (2015) Explotación de roedores por las sociedades cazadoras-recolectoras de Patagonia durante el Holoceno tardío: de la evidencia arqueológica al registro histórico. Zephyrus 75:107–124

Apolinaire E, Castiñeira Latorre C (2017) Primeros estudios arqueológicos y sedimentológicos de un contexto estratigráfico en el interior entrerriano: el sitio Laguna del Negro 1. Comechingonia 21(2):1–10

Arenas P (2003) Etnografía y alimentación entre los Toba-Ñachilamole#ek y Wichí-Lhuku'tas del Chaco Central (Argentina). Pastor Arenas, Buenos Aires

Ávila JD (2011) Resultados de los fechados radiocarbónicos del sitio Laguna El Doce, departamento General López, provincia de Santa Fe. Relac Soc Argent Antropol 36:337–343

Barbarán FR (2003) Factibilidad de caza de subsistencia, comercial y deportiva en el Chaco semiárido de la provincia de Salta, Argentina. Fermentum Revista Venezolana de Sociología y Antropología 13(36):89–117

Barrientos G, Del Papa MC, Oliva FW (2002) Historia pre y postdepositacional del entierro secundario del sitio laguna Los Chilenos I (Provincia de Buenos Aires). Relac Soc Argent Antropol 27:303–325

Barros MP, Bellinzoni J, Kaufmann CA (2018a) Análisis de los instrumentos líticos de los sitios Laguna La Redonda y Laguna Muscar 2 (partido de Gral. Lamadrid, provincia de Buenos Aires). Intersecc en Antropol 19:111–122

Barros MP, Heider G, Álvarez MC, Kaufmann C, Bellinzoni J (2018b) First results of the hunter-gatherer weapon system studies in the middle basin of the Salado creek (Pampas Region, Argentina). J Lithic Stud 5(2):1–23

Bastourre L, Apolinaire E (2017) Estudios arqueofaunísticos en un contexto estratigráfico de las llanuras interiores de Entre Ríos: el sitio Laguna del Negro 1 (departamento Gualeguay, Argentina). Bol Mus Para Emilio Goeldi Ciencias Humanas 1(2):453–471

Bayón C, Politis GG (2014) The inter-tidal zone Site of La Olla: Early–Middle Holocene human adaptation on the Pampean coast of Argentina. In: Evans AM, Flatman JC, Flemming NC (eds) Prehistoric archaeology on the continental shelf: a global review. Springer, New York, pp 115–130

Bayón C, Pupio A, Frontini R, Vecchi R, Scabuzzo C (2010) Localidad arqueológica Paso Mayor: nuevos estudios 40 años después. Intersecc en Antropol 11(1):115–128

Bayón C, Frontini R, Vecchi R (2012) Middle Holocene settlements on coastal dunes, Southwest Buenos Aires province, Argentina. Quat Int 256:54–61

Behrensmeyer AK (1978) Taphonomic and ecologic information from bone weathering. Paleobiology 4:150–162

Behrensmeyer AK, Miller JH (2012) Building links between ecology and paleontology using taphonomic studies of recent vertebrate communities. In: Louys J (ed) Paleontology in ecology and conservation. Springer, Berlin, pp 69–91

Bellinzoni J (2018) Análisis zooarqueológico y tafonómico del sitio Laguna Muscar 2 (Pdo. de General Lamadrid, Pcia. de Buenos Aires). Unpublished bachelor thesis, Universidad Nacional del Centro de la Provincia de Buenos Aires, Olavarría

Bellinzoni J, Kaufmann CA, Álvarez MC (2021) El rol de la vizcacha (*Lagostomus maximus*) en los procesos de formación de los sitios arqueológicos de la llanura Interserrana Argentina. Revista del Museo de Antropología 14(1):47–64

Binford LR (1978) Nunamiut ethnoarchaeology. Academic, New York

Bonomo M (2005) Costeando las Llanuras. Arqueología del Litoral Marítimo Pampeano. Sociedad Argentina de Antropología, Buenos Aires

Bonomo M, Leon C (2010) Un contexto arqueológico en posición estratigráfica en los médanos litorales. El sitio Alfar (Pdo. Gral. Pueyrredón, Pcia. Bs. As.). In: Berón M, Luna L, Bonomo M, Montalvo C, Aranda C, Carrera Aizpitarte M (eds) Mamül Mapu: pasado y presente desde la arqueología pampeana, Tomo II. Libros del Espinillo, Ayacucho, pp 241–252

Bonomo M, Matarrese A (2012) Estado actual de las investigaciones arqueológicas en la localidad Nutria Mansa. Nuevos aportes de los artefactos picados y/o abradidos. Arqueología 18:153–180

Bonomo M, Leon DC, Turnes L, Apolinaire E (2008) Nuevas investigaciones sobre la ocupación prehispánica de la costa pampeana en el Holoceno tardío: el sitio arqueológico Claromecó 1 (partido de Tres Arroyos, provincia de Buenos Aires). Intersecc en Antropol 9:25–41

Bonomo M, Zucol AF, Gutiérrez Téllez B, Coradeghini A, Vigna MS (2009) Late Holocene palaeoenvironments of the Nutria Mansa 1 archaeological site, Argentina. J Paleolimnol 41:273–296

Bonomo M, Scabuzzo C, Leon DC (2013) Cronología y dieta en la costa atlántica pampeana. Argentina Intersecc en Antropol 14(1):123–136

Borella F, Frontini R, Bayón C (2017) Exploring the role of otariids in the subsistence of hunter gatherers from Pampean archaeological sites at the Late Holocene (Argentina). J Archaeol Sci Rep 12:142–150

Branch LC (1993) Social organization and mating system of the plains viscacha (*Lagostomus maximus*). J Zool 229(3):473–491

Broughton JM (1994) Declines in mammalian foraging efficiency during the late Holocene, San Francisco Bay, California. J Anthropol Archaeol 13:371–401

Brunazzo GA (1999) Investigaciones arqueológicas en el sitio La Norma (partido de Berisso, Provincia de Buenos Aires, Argentina). In: Actas del XII Congreso Nacional de Arqueología. Universidad Nacional de La Plata, La Plata, pp 101–106

Buc N, Loponte D (2016) Bone tools reflecting animal explotation. The case of *Lama guanicoe* in the lower Paraná Basin. Cuadernos del Instituto Nacional de Antropología y Pensamiento Latinoamericano-Series especiales 3(2):23–53

Bunn HT, Bartram LE Jr, Kroll EM (1988) Variability in bone assemblage formation from Hadza hunting, scavenging, and carcass processing. J Anthropol Archaeol 7(4):412–457

Cabrera AL, Willink AW (1980) Biogeografía de América Latina. In: Serie de Biología. Monografía Nro. 13. Secretaría General de la Organización de los Estados Americanos. Programa Regional de Desarrollo Científico y Tecnológico, Washington, DC

Cabrera A, Yepes J (1960) Zoogeografía de América del Sur. In: López HL, Ponte Gómez J (eds) Mamíferos Sud Americanos, Tomo I. Editorial Buenos Aires, Buenos Aires, pp 11–17

Catella L (2014) Movilidad y Utilización del Ambiente en Poblaciones Cazadoras-Recolectoras del Sur de la Región Pampeana: la Cuenca del Arroyo Chasicó como Caso de Estudio. Unpublished doctoral thesis, Universidad Nacional de La Plata, La Plata

Cohen BF, Kibii JM (2015) Experimental study of bone modification by captive caracal (*Caracal caracal*); a model for fossil assemblage analysis. J Archaeol Sci 62:92–104

Colombo M (2013) Los Cazadores Recolectores Pampeanos y sus Rocas. La Obtención de Materias Primas Líticas Vista desde las Canteras Arqueológicas del Centro de Tandilia. Unpublished doctoral thesis, Universidad Nacional de La Plata, La Plata

Cornaglia Fernández J (2013) Análisis taxonómicos e inferencias paleoambientales en el sudoeste santafecino. El sitio arqueológico Laguna El Doce. Rev Mus La Plata 13(87):109–124

Crivelli Montero E (1991) Laguna Del Trompa (Estancia La Herminia), Laprida, Pcia. De Bs. As. Excavaciones 1989–1991. Artefactos y Estructuras. Boletín del Centro de Registro Arqueológico y Paleontológico 3:18–29

Crivelli Montero E, Silveira M, Eugenio E, Escola P, Fernández M, Franco N (1987–88) El sitio Fortín Necochea (partido de General La Madrid, provincia de Buenos Aires). Estado actual de los trabajos Paleoetnológica 4:39–53

Crivelli Montero E, Eugenio E, Pardiñas U, Silveira M (1994) Archaeological investigations in the plains of the province of Buenos Aires (counties of General La Madrid and Laprida). Quaternary of South America y Antarctic Peninsula 10:167–209

Day Pilaria F (2018) Gestión de los Recursos Faunísticos en Sociedades Cazadoras, Recolectoras y Pescadoras. Análisis Arqueozoológico en Sitios del Litoral del Río de La Plata (partidos de Magdalena y Punta Indio, Provincia de Buenos Aires). Unpublished doctoral thesis, Universidad Nacional de La Plata, La Plata

Desmarest AG (1817) Gerboise première espèce: la grande gerboise *Dipus maximus* Blainv. Nouveau Dictionnaire d'Histoire Naturelle 13:117–119

Escosteguy P, Salemme MC (2012) Butchery evidence on rodent bones from archaeological sites in the Pampean Region. In: Lefévre C (ed) Proceedings of the general session of the 11th international Council for Archaeozoology conference. BAR International Series, Oxford, pp 227–236

Escosteguy P, Salemme MC (2017) Faunal subsistence resources in the Cañada Honda Locality (Northeastern Buenos Aires province, Argentina). In: Mondini M, Muñoz A, Fernández PM (eds) Zooarchaeology in the neotropics: environmental diversity and human-animal interactions. Springer, New York, pp 61–80

Escosteguy P, Salemme MC, González MI (2012) *Myocastor coypus* ("coipo", Rodentia, Mammalia) como recurso en los humedales de la Pampa bonaerense: patrones de explotación. Rev Mus Antropol 5:15–30

Escosteguy P, González MI, Frère MM (2015) New data on small fauna Salado River depression (province of Buenos Aires, Argentina): the case of San Ramon 7. Archaeofauna 24:295–313

Escosteguy P, González MR, Fiel MV, Vigna M (2017) A orillas de la laguna de Lobos: el sitio arqueológico Techo Colorado (Microrregión del Río Salado Bonaerense). Comechingonia 21:15–45

Escosteguy P, Salerno V, González MI, Frère MM (2023) Usos de la fauna en el río Salado bonaerense: saberes locales actuales. Mundo De Antes 17(1):109–140

Eugenio E (1991) Asentamientos arqueológicos en la Laguna del Trompa (Estancia La Herminia), Laprida, Pcia. de Bs. As. Boletín del Centro de Registro Arqueológico y Paleontológico 3:30–42

Eugenio E, Pardiñas U (1991) Análisis del material óseo del sitio Laguna de Sotelo, partido de Mar Chiquita. Boletín del Centro 3:43–52

Fernández FJ, Teta P, Pardiñas UFJ (2017) Evidencias arqueológicas de explotación antrópica de micromamíferos en el extremo austral de América del Sur. A Arqueol Etnol 72(1):9–32

Fisher JW (1995) Bone surface modifications in zooarchaeology. J Archaeol Method Theory 2:7–68

Flegenheimer N (1991) La Liebre, un sitio de cantera-taller. Boletín del Centro 2:58–64

Flegenheimer N, Bayón C, Scabuzzo C, Mazzia N, Vecchi R, Weitzel C, Frontini R, Colombo M (2010) Early Holocene human skeletal remains from the Argentinean Pampas. Curr Res Pleistocene 27:10–12

Flegenheimer N, Mazzia N, Weitzel C (2015) Landscape and rocks in the east-central portion of the Tandilia range (Buenos Aires Province, Argentina). PaleoAmerica 1(2):163–180

Frengüelli J (1946) Las grandes unidades físicas del territorio argentino. In: GAEA, Geografía de la República Argentina. Sociedad Argentina de Estudios Geográficos GAEA, Buenos Aires, p 3

Frère MM, González MI, Greco C (2016) Continuity in the use of shallow sites of the Salado river basin in the Pampean region. Argentina Radiocarbon 58(4):921–933

Frontini R (2012) El Aprovechamiento de Animales en Valles Fluviales y Lagunas del Sur Bonaerense durante el Holoceno. Unpublished doctoral thesis, Universidad de Buenos Aires, Buenos Aires

Frontini R (2013) Aprovechamiento faunístico en entornos acuáticos del sudoeste bonaerense durante el Holoceno (6900-700 años AP). Rel Soc Argent Antropol 38(2):493–519

Frontini R, Bayón C (2015) Consumo de recursos animales de porte menor durante el Holoceno tardío en el sudoeste de la provincia de Buenos Aires (Argentina). Archaeofauna 24:271–293

Frontini R, Bayón C (2017) Use of marine fauna and tool stones in the south of Buenos Aires province (Argentina) during the middle and late Holocene. In: Mondini M, Muñoz AS, Fernández PM (eds) Zooarchaeology in the neotropics: environmental diversity and human-animal interactions. Springer, New York, pp 25–46

Frontini R, Vecchi R (2014) Thermal alteration of small mammal from El Guanaco 2 site (Argentina): an experimental approach on armadillos bone remains (Cingulata, Dasypodidae). J Archaeol Sci 44:22–29

Frontini R, Tomassini RL, Bayón C (2018) Perinatal guanacos (*Lama guanicoe*) exploited by hunter-gatherers from the Holocene of Argentine pampas. Int J Osteoarchaeol 28(6):703–713

Frontini R, Fernández-Jalvo Y, Pesquero Fernández MD, Vecchi RJ, Bayón C (2019) Abrasion in archaeological fish bones from sand dunes. An experimental approach. Archaeol Anthropol Sci 11:4891–4907

Frontini R, Bayón C, Vecchi R (2021a) Fish capture strategies in Atlantic Littoral of Monte Hermoso district (Pampean Region Argentina) during middle Holocene. In: Belardi JB, Bozzuto DL, Fernández PM, Moreno EA, Neme GA (eds) Ancient Hunting Strategies in Southern South America. The Latin American Studies Book Series. Springer, Cham. https://doi.org/10.1007/978-3-030-61187-3_5

Frontini R, Vecchi R, Bayón C, Fiori S (2021b) Arqueología de la laguna Sauce Grande (partido de Monte Hermoso, provincia de Buenos Aires). Rel Soc Argent Antropol 46(1):121–140

Gariboldi MC, Felipe Inserra PI, Lucero S, Failla M, Perez SI, Vitullo AD (2019) Unexpected low genetic variation in the South American hystricognath rodent *Lagostomus maximus* (Rodentia: Chinchillidae). PLoS One 14(9):e0221559

Garrone MC, Cerda IA, Tomassini R (2021) Ontogenetic variability in the limb bones histology of plains vizcacha (*Lagostomus maximus*, Chinchillidae, Rodentia): implications for life history reconstruction of fossil representatives. Hist Biol 33(4):558–573

Gifford-González DP (2018) An introduction to zooarchaeology. Springer, New York

González MI (2005) Arqueología de alfareros, cazadores y pescadores pampeanos. Sociedad Argentina de Antropología, Buenos Aires

González de Bonaveri MI (1997) Potsherds, "Coypo" teeth, and fish bones: hunter-gatherer-fishers in the Río Salado (Pampa Region, Argentina). Quaternary of South America and Antarctic Peninsula 10:255–278

Grayson DK (1984) Quantitative zooarchaeology: topics in the analysis of archaeological faunas. Academic, Orlando

Gutiérrez MA, Martínez G, Luchsinger H, Grill S, Zucol AF, Hassan GS, Barros MP, Kaufmann CA, Álvarez MC (2011) Paleoenvironments in the Paso Otero locality during Late Pleistocene-Holocene (Pampean region, Argentina): an interdisciplinary approach. Quat Int 245(1):37–47

Gutiérrez MA, Kaufmann CA, González ME, Scheifler NA, Rafuse DJ, Massigoge A, Álvarez MC (2016) The role of small carnivores in the movement of bones: implications for the Pampas Archaeofaunal record. Argentina Archaeol Anthropol Sci 8(2):257–276
Hogg AG, Heaton TJ, Hua Q, Palmer JG, Turney CSM, Southon J, Bayliss A, Blackwell PG, Boswijk G, Bronk Ramsey C, Pearson C, Petchey F, Reimer P, Reimer R, Wacker L (2020) SHCal20 southern hemisphere calibration, 0-55,000 years Cal BP. Radiocarbon 62(4):759–778
Hoguin R, March RJ (2007–2008) Una primera aproximación al análisis tipo-tecnológico de los artefactos líticos del sitio La Represa (curso inferior del Quequén Salado, provincia de Buenos Aires). Arqueología 14:103–136
Jackson JE (1985) Peso de la canal y rendimiento en la vizcacha. RAPA 5:121–123
Jackson JE, Bucher EH, Chani JM (1996) Capture of blue-fronted amazons and hunting of vizcachas and tegu lizards in Argentina. In: Prescott-Allen R, Prescott-Allen C (eds) Assessing the sustainability of uses of wild species – case studies and initial assessment procedure. IUCN/Cambridge, Gland/Cambridge, pp 17–26
Johnson E (1985) Current developments in bone technology. In: Schiffer MB (ed) Advances in archaeological method and theory, 8th edn. Academic, New York, pp 157–235
Johnson E, Holliday VT, Martínez G, Gutiérrez MA, Politis GG (2012) Geochronology and landscape development along the middle Río Quequén Grande at the Paso Otero Locality, Pampa Interserrana. Argentina Geoarchaeol 27(4):300–323
Jones EL (2006) Prey choice, mass collecting, and the wild European rabbit (*Oryctolagus cuniculus*). J Anthropol Archaeol 25(3):275–289
Kaufmann CA, Álvarez MC (2007) La arqueofauna del sitio Calera (Sierras Bayas, región pampeana): un abordaje a los aspectos rituales del descarte de huesos de animales. In: Bayón C, González MI, Pupio A, Flegenheimer N, Frère MM (eds) Arqueología en las Pampas, vol 2. Sociedad Argentina de Antropología, Buenos Aires, pp 745–764
Kaufmann CA, Bellinzoni J, Álvarez MC, González ME, Favier-Dubois CM, Barros MP (2019) Aportes del estudio de los procesos de formación de sitio para el análisis de los restos humanos del sitio Laguna Muscar 2 (Lamadrid, Buenos Aires). Comechingonia 23(1):1–10
Kaufmann CA, Álvarez MC, Blasi AM, Barros MP (2023) Evidences of social interaction in the Pampas: the case study of Hangar site (province of Buenos Aires, Argentina). In: Bonnat F, Álvarez MC, Mazzanti DL, Barros MP, Bonomo M, Puente V (eds) Current research in archaeology of South American Pampas. Springer Nature, New York. (in press)
Lanzelotti SL, Bonaparte JF (2009) Contexto geoestratigráfico y procesos de formación del registro arqueológico en Cañada Honda: apuntes para su discusión y abordaje. In: Fucks EE, Deschamps C, Silva CG, Schnack EJ (eds) Trabajos del IV Congreso Argentino de Cuaternario y Geomorfología y II Reunión sobre el Cuaternario de América del Sur. Universidad Nacional de La Plata, La Plata, pp 247–258
Lanzelotti S, Politis GG, Carbonari JE, Huarte R, Bonaparte JF (2011) Aportes a la cronología del sitio 1 de Cañada Honda (partido de Baradero, provincia de Buenos Aires). Intersecc en Antropol 12:355–361
Leon DC (2014) Zooarqueología de Cazadores Recolectores del Litoral Pampeano: un Enfoque Multidimensional. Unpublished doctoral thesis, Universidad Nacional del Centro de la Provincia de Buenos Aires, Olavarría
Leon DC (2020) Análisis faunístico del sitio Quequén Salado 1. Subsistencia de cazadores recolectores del Holoceno tardío en el litoral marítimo pampeano. Rev Mus Antropol 13(2):7–22
Leon DC, Gutiérrez MA, Politis GG, Bayón C (2017) Análisis faunístico del sitio arqueológico La Olla (sectores 3 y 4), costa sudoeste del litoral atlántico bonaerense. Rel Soc Argent Antropol 42(1):1–10
Lloveras L, Moreno-García M, Nadal J (2008) Taphonomic analysis of leporid remains obtained from modern Iberian lynx (*Lynx pardinus*) scats. J Archaeol Sci 35(1):1–13
Loponte D, Acosta A, Tchilinguirián P (2010) Avances en la Arqueología de Pampa Ondulada: sitios Hunter y Meguay. In: Bárcena JR, Chiavazza H (eds) Arqueología Argentina en el Bicentenario de la Revolución de Mayo, Actas del XVII Congreso Nacional de Arqueología

Argentina, Tomo V. Facultad de Filosofía y Letras, Universidad Nacional de Cuyo- Instituto de Ciencias Humanas, Sociales y Ambientales, CONICET, Mendoza, pp 1811–1816
Lupo KD, Schmitt DN (2002) Upper Paleolithic net-hunting, small prey exploitation, and women's work effort: a view from the ethnographic and ethnoarchaeological record of The Congo Basin. J Archaeol Method Theory 9:147–179
Lyman RL (1994) Vertebrate taphonomy. Cambridge University Press, Cambridge
Lyman RL (2008) Quantitative paleozoology. Cambridge University Press, Cambridge
Madrid PEE, Barrientos G (2000) La estructura del registro arqueológico del sitio Laguna Tres Reyes I (provincia de Buenos Aires): nuevos datos para la interpretación del poblamiento humano en el sudeste de la región pampeana a inicios del Holoceno tardío. Relaciones de la Sociedad Argentina de Antropología 25:179–206
Madrid PEE, Politis GG, Bonomo M, March R (2002) Arqueología microrregional en el sudeste de la Región Pampeana Argentina: el curso del río Quequén Salado. Rel Soc Argent Antropol 27:327–355
Madsen DB, Schmitt D (1998) Mass collecting and the diet breadth model: a Great Basin example. J Archaeol Sci 25:445–455
March RJ, Rodríguez Loredo C, Hoguin RG, Foisset S, Joly D, Lucquin A (2011) Primeros resultados de la excavación del sitio de cazadores recolectores de La Represa: Oriente, partido de Coronel Dorrego, provincia de Bs. As. Argentina Arqueología Rosarina Hoy 3:9–74
Martínez G (1999) Tecnología, Subsistencia y Asentamiento en el Curso Medio del Río Quequén Grande: Un Enfoque Arqueológico. Unpublished doctoral thesis, Universidad Nacional de La Plata, La Plata
Martínez G (2001) 'Fish-tail'projectile points and megamammals: new evidence from Paso Otero 5 (Argentina). Antiquity 75(289):523–528
Martínez G (2006) Arqueología del curso medio del río Quequén Grande: estado actual y aportes a la arqueología de la región pampeana. Rel Soc Argent Antropol 31:249–275
Martínez G, Gutiérrez MA (2004) Tendencias en la explotación humana de la fauna durante el Pleistoceno final y Holoceno en la Región Pampeana (Argentina). In: Goñalons GM (ed) Zooarchaeology of South America. BAR International Series 1298, Oxford, pp 81–98
Martínez G, Gutiérrez MA (2011) Paso Otero 5: a summary of the interdisciplinary lines of evidence for reconstructing early human occupation and paleoenvironment in the Pampean region, Argentina. In: Vialou D (ed) Peuplements et Préhistoire de l'Amérique. Muséum National d'Histoire Naturelle, Département de Préhistoire, Paris, pp 271–286
Martínez G, Gutiérrez MA (2019) Early Holocene water well in the Pampas of Argentina: human responses to water shortage events. The Holocene 29(1):145–157
Martínez G, Messineo P, Piñeyro M, Kaufmann C, Barros M (2001) Análisis preliminar de la estructura faunística del sitio Paso Otero 3 (Área Interserrana, Región Pampeana, Argentina). In: Comisión Organizadora del IX Congreso Nacional de Arqueología (ed) Arqueología uruguaya hacia el fin del milenio, vol 1. Gráficos del Sur, Montevideo, pp 505–520
Martínez G, Mazzanti DL, Quintana C, Zucol AF, Colobig MM, Hassan GS, Brea M, Passeggi E (2013) Geoarchaeological and Paleoenvironmental context of the human settlement in the Eastern Tandilia Range, Argentina. Quat Int 299:23–37
Martínez G, Prates L, Flensborg G, Stoessel L, Alcaráz AP, Bayala P (2015) Radiocarbon trends in the Pampean region (Argentina). Biases and demographic patterns during the final Late Pleistocene and Holocene. Quat Int 356:89–110
Martínez G, Gutiérrez MA, Messineo PG, Kaufmann CA, Rafuse D (2016) Subsistence strategies in Argentina during the Late Pleistocene and Early Holocene. Quat Sci Rev 144:51–65. https://doi.org/10.1016/j.quascirev.2016.05.014
Maryañski JM, Nielsen AE (2015) Caza y extracción de pieles de Chinchillidae en la puna de Jujuy (Argentina): una aproximación arqueológica. Archaeofauna 24:205–218
Massigoge A (2007) Procesos de formación del registro arqueológico en el sitio Cortaderas (partido de San Cayetano, provincia de Buenos Aires). Intersecc en Antropol 8:197–214

Massigoge A (2009) Arqueología de los Cazadores-recolectores del Sudeste de la Región Pampeana: una Perspectiva Tafonómica. Unpublished doctoral thesis, Universidad Nacional de La Plata, La Plata

Massigoge A (2011) Nuevas evidencias arqueológicas del Holoceno Tardío en el área Interserrana: el sitio Las Brusquillas 2 (partido de San Cayetano, provincia de Buenos Aires, Argentina). Cazadores-recolectores del Cono Sur. Revista de Arqueología 5:179–195

Massigoge A (2012) Las Brusquillas 1 (partido de San Cayetano, provincia de Buenos Aires): un nuevo sitio del Holoceno tardío del área Interserrana de la región pampeana. Intersecc en Antropol 13(2):377–392

Massigoge A, Pal N (2011) Producción y uso de artefactos líticos en contextos cazadores-recolectores del Área Interserrana (Argentina): análisis integral de la diversidad tecno-morfológica y funcional. Revista Española de Antropología Americana 41(1):51–73

Massigoge A, Rodríguez MN, Rafuse D, Torino RF, Favier Dubois CM, Steffan PG (2018) Investigaciones arqueológicas en el sitio Las Brusquillas 3 (Holoceno tardío, región Pampeana, Argentina). Arqueología 24(1):147–171

Massigoge A, Rafuse D, Rodríguez JM, Torino RF, Favier Dubois CM (2021) Technology, subsistence, and mobility of Middle to Late Holocene hunter-gatherers in the southern Pampas: new archaeological data from las Toscas shallow lake (Argentina). Archaeol Anthropol Sci 13(4):69

Matteucci S (2012) Ecorregión Pampa. In: Morello J, Matteucci S, Rodríguez A, Silva M (eds) Ecorregiones y Complejos Ecosistémicos Argentinos. Orientación Gráfica, Buenos Aires, pp 391–446

Mazzanti DL (1997) Excavaciones arqueológicas en el sitio Cueva Tixi, Buenos Aires, Argentina. Lat Am Antiq 8:55–62. https://doi.org/10.2307/971592

Mazzanti DL (1999) Ocupaciones humanas tempranas en Sierras La Vigilancia y Laguna La Brava, Tandilia Oriental (Provincia de Buenos Aires). In: Actas del XII Congreso Nacional de Arqueología Argentina, vol 3. Universidad Nacional de La Plata, La Plata, pp 149–155

Mazzanti DL (2006) La constitución de territorios sociales durante el Holoceno tardío. El caso de las sierras de Tandilia, Argentina. Relaciones de la Sociedad Argentina de Antropología 31:277–300

Mazzanti D (2007) Arqueología de las Relaciones Interétnicas Posconquista en las Sierras de Tandilia. Unpublished doctoral thesis, Universidad de Buenos Aires, Buenos Aires

Mazzanti DL, Colobig MM, Zucol FA, Martínez GA, Porto López JM, Brea M, Passeggi E, Soria JL, Quintana C, Puente V (2010) Investigaciones arqueológicas en el sitio 1 de la localidad Lobería I. In: Berón M, Luna L, Bonomo M, Montalvo C, Aranda C, Carrera Aizpitarte M (eds) Mamül Mapu: Pasado y Presente desde la Arqueología Pampeana. Libros del Espinillo, Ayacucho, pp 215–230

Mazzanti D, Martínez GA, Quintana C (2012) Early settlements in eastern Tandilia, Buenos Aires Province, Argentina: archaeological contexts and site-formation processes. In: Miotti L, Salemme M, Flegenheimer N, Goebel T (eds) Southbound: late Pleistocene peopling of Latin America, current research in the Pleistocene. Texas A&M University Press, College Station, pp 99–103

Mazzanti DL, Martínez GA, Quintana C (2015) Asentamientos del Holoceno medio en Tandilia oriental. Relaciones de la Sociedad Argentina de Antropología 40(1):209–231

Mazzia N, Flegenheimer N (2021) Recursos orgánicos y objetos arqueológicos en la localidad Cerro La China: ilaciones a partir de las sustancias grasas. Arqueología 27(2):63–90

Messineo PG (2003) Análisis arqueofaunísticos en el sitio Laguna La Barrancosa 1 (Partido de Benito Juárez, provincia de Buenos Aires, Argentina). Archaeofauna 12:73–86

Messineo PG (2008) Investigaciones Arqueológicas en la Cuenca Superior del Arroyo Tapalqué (Partidos de Olavarría y Benito Juárez, Provincia de Buenos Aires). Unpublished doctoral thesis, Universidad Nacional de La Plata, La Plata

Messineo PG (2011) Investigaciones arqueológicas en la cuenca superior del Arroyo Tapalqué. Un modelo de ocupación humana para el centro de la subregión Pampa Húmeda durante el Holoceno tardío. Intersecc en Antropol 12(2):275–291

Messineo PG, Scheifler NA (2016) Investigaciones arqueológicas de cazadores-recolectores en el sitio Laguna Cabeza de Buey 2 (centro de los pastizales pampeanos, Buenos Aires): Cincuenta años después de las Industrias Culturales definidas por Bórmida. Intersecc en Antropol 17(2):213–226

Messineo PG, Álvarez MC, Favier Dubois C, Steffan PG, Colantonio MJ (2013) Estado de avance de las investigaciones arqueológicas en el sitio Empalme Querandíes 1 (centro de la subregión Pampa Húmeda, provincia de Buenos Aires). Comechingonia 17(1):123–148

Messineo PG, Kaufmann CA, Steffan PG, Favier Dubois C, Pal N (2014) Ocupaciones humanas en un valle intraserrano del sector noroccidental de Tandilia: sitio El Puente (partido de Olavarría, Buenos Aires). Rel Soc Arg Antrop 2:435–462

Messineo PG, Favier Dubois CMF, Politis GG, Vitale P (2021) Site formation process and megamammal bone radiocarbon dates in Campo Laborde (Pampas of Argentina): contribution towards a research methodology. Quat Int 586:53–65

Miotti L, Salemme MC (1999) Biodiversity, taxonomic richness and specialists-generalists during Late Pleistocene/Early Holocene times in Pampa and Patagonia (Argentina, Southern South America). Quat Int 53(54):53–68

Morales N (2015) Estudio de modificaciones de la superficie ósea en restos faunísticos del sector sur del área ecotonal húmedo-seca pampeana: El Sitio San Martín 1 como caso de estudio. Comechingonia 19(2):323–345

Morales N (2022) La Relación Humanos-Fauna en el Sur del Área Ecotonal Húmedo-Seca Pampeana desde una Perspectiva Zooarqueológica. Unpublished doctoral thesis, Universidad Nacional de La Plata, La Plata

Mucciolo L (2013) Cervids record from Río Luján 2 site (Campana, Buenos Aires province): results of the reanalysis. La Zaranda de Ideas 9:21–36

Oliva F, Moirano J, Saghessi M (1991) Estado de las investigaciones arqueológicas en el sitio Laguna de Puan 1. Boletín del Centro 2:127–138

Peres TM (2010) Methodological issues in zooarchaeology. In: Van Derwarker AM, Peres TM (eds) Integrating zooarchaeology and paleoethnobotany: a consideration of issues, methods, and cases. Springer, Berlin, pp 15–36

Pocheville A (2015) The ecological niche: history and recent controversies. In: Heams P, Huneman P, Lecointre G, Silberstein M (eds) Handbook of evolutionary thinking in the sciences. Springer, Sidney, pp 547–586

Politis GG (1984) Arqueología del Área Interserrana Bonaerense. Unpublished doctoral thesis, Universidad Nacional de La Plata, La Plata

Politis GG (2008) The pampas and Campos of South America. In: Silverman H, Isbell W (eds) Handbook of South American archaeology. Springer, New York, pp 235–260

Politis GG (2014) Discusión y consideraciones finales. In: Politis GG, Gutiérrez MA, Scabuzzo C (eds) Estado Actual de las Investigaciones en el Sitio Arqueológico Arroyo Seco 2. INCUAPA-CONICET-UNICEN, Tandil, pp 439–461

Politis GG, Beukens R (1991) Cronología radiocarbónica de la ocupación humana del área Interserrana Bonaerense (Argentina). Shincal 3(3):151–158

Politis GG, Gutiérrez MA (1998) Gliptodontes y cazadores-recolectores de la región pampeana (Argentina). Lat Am Antiq 9(2):111–134

Politis GG, Madrid PEE (2001) Arqueología Pampeana: Estado actual y perspectivas. In: Berberian E, Nielsen A (eds) Historia Argentina Prehispánica, II. Editorial Brujas, Córdoba, pp 737–814

Politis GG, Messineo PG (2008) The Campo Laborde site: new evidence for the Holocene survival of Pleistocene megafauna in the Argentine Pampas. Quat Int 191(1):98–114

Politis GG, Lozano P, Guzman LM (1994) Evidencias de la ocupación humana prehispánica del Litoral Bonaerense en el Sitio La Olla. In: Actas y Memorias del XI Congreso Nacional de Arqueología Argentina. Revista del Museo de Historia Natural de San Rafael XIV (Segunda Parte) 1(4):240

Politis GG, Martínez G, Bonomo M (2001) Alfarería temprana en sitios de cazadores-recolectores de la región pampeana (Argentina). Lat Am Antiq 12(2):167–181

Politis G, Martínez G, Bonomo M (2004) Revisión del sitio Zanjón Seco 2 (área Interserrana Bonaerense) en base a nuevos datos y análisis. In: Gradín CJ, Oliva FW (eds) La Región Pampeana -su pasado arqueológico-, Laborde Editor, Rosario, pp 73–88

Politis GG, Messineo PG, Kaufmann CA, Barros MP, Álvarez MC, Di Prado VS, Scalise R (2005) Persistencia ritual entre cazadores-recolectores de la llanura pampeana. Boletín de Arqueología PUCP 9:67–90

Politis GG, Prates L, Merino ML, Tognelli MF (2011) Distribution parameters of guanaco (*Lama guanicoe*), pampas deer (*Ozotoceros bezoarticus*) and marsh deer (*Blastocerus dichotomus*) in Central Argentina: archaeological and paleoenvironmental implications. J Archaeol Sci 38(7):1405–1416

Politis GG, Messineo PG, González ME, Álvarez MC, Dubois CF (2012) Primeros resultados de las investigaciones en el sitio Laguna de los Pampas (partido de Lincoln, provincia de Buenos Aires). Rel Soc Argent Antropol 37(2):463–472

Politis GG, Gutiérrez MA, Rafuse D, Blasi A (2016) The arrival of *Homo sapiens* into the southern cone at 14,000 years ago. PLoS One 11(9):e0162870

Politis GG, Messineo PG, Stafford TW Jr, Lindsey EL (2019a) Campo Laborde: a Late Pleistocene giant ground sloth kill and butchering site in the Pampas. Sci Adv 5(3):eaau4546. https://doi.org/10.1126/sciadv.aau4546

Politis GG, Leon DC, Mari F, Blasi A, Vitale P (2019b) Análisis y datación de los restos faunísticos de la colección Ameghino del sitio Cañada Rocha (Luján, provincia de Buenos Aires). Intersecc en Antropol 20(2):153–165

Pupio A (1995) Resultados preliminares del análisis del material faunísitico del sitio cantera-taller La Liebre. Jornadas Chivilcoyanas en Ciencias Sociales y Naturales 191–94. Centro de Estudios en Ciencias Sociales y Naturales, Chivilcoy

Quintana CA (2001) Composición y cambios en la secuencia faunística. In: Mazzanti DL, Quintana CA (eds) Cazadores y recolectores de las sierras de Tandilia Oriental. 1. Geología, Paleontología y Zooarqueología. Laboratorio de Arqueología. Universidad Nacional de Mar del Plata, Mar del Plata, pp 37–64

Quintana CA (2016) Microvertebrados del sitio arqueológico Cueva El Abra, Tandilia Oriental: tafonomía y paleoambiente. Comechingonia 20(1):203–229

Quintana CA, Mazzanti DL (2001) Selección y aprovechamiento de recursos faunísticos. In: Mazzanti DL, Quintana CA (eds) Cazadores y recolectores de las sierras de Tandilia Oriental. 1. Geología, Paleontología y Zooarqueología. Laboratorio de Arqueología, Universidad Nacional de Mar del Plata, Mar del Plata, pp 181–209

Quintana CA, Mazzanti DL (2011) Las vizcachas pampeanas (*Lagostomus maximus*, Rodentia) en la subsistencia indígena del Holoceno tardío de las Sierras de Tandilia Oriental (Argentina). Lat Am Antiq 22(2):253–270

Quintana CA, Mazzanti DL (2014) La emergencia de la diversificación de la caza en las sierras de Tandilia oriental durante el Holoceno tardío final. Comechingonia 18(2):41–64

Quintana CA, Valverde F, Mazzanti DL (2002) Roedores y lagartos como emergentes de la diversificación de la subsistencia durante el Holoceno tardío en sierras de la región Pampeana Argentina. Lat Am Antiq 13(4):455–473

Rafuse D (2013) Integridad del Registro Arqueofaunístico del Sitio Arroyo Seco 2 (Región Pampeana, Argentina) desde una Perspectiva Tafonómica. Unpublished doctoral thesis, Universidad Nacional del Centro de la Provincia de Buenos Aires, Olavarría

Rafuse DJ (2017) Early to Middle Holocene subsistence strategies in the Pampas region: evidence from the Arroyo Seco 2 site. J Archaeol Sci Rep 12:673–683

Rafuse DJ, González ME, Kaufmann CA, Álvarez MC, Gutiérrez MA, Massigoge A (2014) Análisis comparativo de los patrones de modificaciones óseas de dos carnívoros sudamericanos: el gato montés (*Leopardus geoffroyi*) y el zorro pampeano (*Lycalopex gymnocercus*). Aportes para la identificación de la acción de pequeños carnívoros en el registro arqueológico. Magallania 41(1):165–184

Rafuse D, Kaufmann CA, Gutiérrez MA, González M, Scheifler N, Álvarez MC, Massigoge A (2019) Taphonomy of modern communal burrow systems of the plains vizcacha (*Lagostomus maximus*, Chinchillidae) in the Pampas region of Argentina: implications for the fossil record. Hist Biol 31(5):517–534

Rasia LL (2016) Los Chinchillidae (Rodentia, Caviomorpha) Fósiles de la República Argentina: Sistemática, Historia Evolutiva y Biogeográfica, Significado Bioestratigráfico y Paleoambiental. Unpublished doctoral thesis, Universidad Nacional de La Plata, La Plata

Reitz EJ, Wing ES (2008) Zooarchaeology. Cambridge University Press, Cambridge

Ringuelet RA (1955) Panorama zoogeográfico de la provincia de Buenos Aires. Notas del Museo de La Plata. Sección Zoología 18(156):1–17

Ringuelet RA (1961) Rasgos fundamentales de la zoogeografía de la República Argentina. Physis 22:151–170

Rodríguez JM (2023) Preservación, Integridad y Resolución del Registro Arqueológico del Centro-este de Argentina durante el Holoceno Tardío. Unpublished doctoral thesis, Universidad Nacional del Centro de la Provincia de Buenos Aires, Olavarría

Salemme MC (2014) Zooarqueología y paleoambientes. In: Politis GG, Gutiérrez MA, Scabuzzo C (eds) Estado Actual de las Investigaciones en el Sitio Arqueológico Arroyo Seco 2. INCUAPA-CONICET-UNICEN, Tandil, pp 67–96

Salemme MC, Madrid PEE (2007) The archaeofaunas from Laguna Tres Reyes site: taxonomic richness and abundance during the beginning of the Late Holocene in the South-East Pampean region (Argentina). In: Gutiérrez M, Miotti L, Barrientos G, Mengoni-Goñalons G, Salemme M (eds) Taphonomy and zooarchaeology in Argentina. BAR, Oxford, pp 121–143. (International Series) 1601. Archaeopress

Salemme MC, Miotti LL (1987) Zooarchaeology and palaeoenvironments: some examples from the Patagonian and Pampean regions (Argentina). In: Rabassa J (ed) Quaternary of South America and Antarctic Peninsula. Routledge, London, pp 33–57

Salemme M, Tonni EP, Ceruti CN (1985) Los materiales faunísticos del sitio arqueológico "La Maza I" (partido de Berisso, prov. Buenos Aires): revisión crítica. Revista de la Asociación de Ciencias Naturales del Litoral 16(2):169–178

Salemme M, Escosteguy P, Frontini R (2012) La fauna de porte menor en sitios arqueológicos de la región pampeana, Argentina. Agente disturbador vs. recurso económico. Archaeofauna 21:163–185

Scheifler NA (2019) Ecología y subsistencia de los cazadores-recolectores en el campo de dunas del centro pampeano. Sociedad Argentina de Antropología, Buenos Aires

Scheifler NA (2020) Fogones, combustible óseo y procesamiento faunístico en el sitio de Huencú Nazar (Sistema Lagunar Hinojo-Las Tunas, región pampeana, Argentina): Implicancias para el uso del espacio por cazadores-recolectores. Lat Am Antiq 31(3):517–537

Scheifler NA, Messineo PG (2016) Exploitation of faunal resources by hunter-gatherers in the center of the Pampa grassland during the Holocene: the archaeofauna of the Laguna Cabeza de Buey 2 site (San Carlos de Bolivar, Buenos Aires, Argentina). Quat Int 391:61–73

Scheifler NA, Messineo PG, Pardiñas UFJ (2015) Implicancias tafonómicas y paleoambientales de los pequeños vertebrados del sitio arqueológico Campo Laborde (centro de los pastizales pampeanos, Buenos Aires, Argentina). Archaeofauna 24:187–208

Scheifler NA, Merino ML, Vitale P, Kaufmann CA, Messineo PG, Álvarez MC, Bocherens H (2023) Isotopic ecology in modern and Holocene populations of Pampas deer (*Ozotoceros bezoarticus*) from eastern Central Argentina. Implications for conservation biology and ecological models of hunter-gatherer subsistence. Environ Archaeol 28(2):45–61

Silveira MJ (1991) Análisis de los restos óseos de las excavaciones de los sectores H' y J' del sitio Laguna del Trompa, Estancia La Herminia – Laprida – Pcia. de Bs. As. Boletín del Centro de Registro Arqueológico y Paleontológico 3:105–115

Soriano A, León RJC, Sala OE, Lavado RS, Deregibus VA, Cauhépé MA, Scaglia OA, Velázquez CA, Lemcoff JH (1992) Río de la Plata Grasslands. In: Coupland RT (ed) Natural grasslands. Introduction and Western hemisphere. Elsevier, Amsterdam, pp 367–413

Speth JD, Spielmann KA (1983) Energy source protein metabolism and hunter-gatherer subsistence strategies. J Anthropol Archaeol 2:1–31
Steele DG, Baker BW (1993) Multiple predation: a definitive human hunting strategy. In: Hudson J (ed) From bones to behavior: Ethnoarchaeological and experimental contributions to the interpretation of faunal remains. Center for Archaeological Investigations, Carbondale, pp 9–37
Stuiver M, Reimer PJ (1993) Extended 14C data base and revised CALIB 3.0 14C Age calibration program. Radiocarbon 35(1):215–230
Szuter CR, Bayham FE (1989) Sedentism and prehistoric animal procurement among desert horticulturalists of the North American Southwest. In: Kent S (ed) Farmers as hunters: the implications of Sedentism. Cambridge University Press, Cambridge, pp 80–95
Tambussi C (1995) The fossil Rheiformes from Argentina. Cour Forschungsinst Senckenberg 181:121–129
Tchilinguirián P, Loponte D, Acosta A (2013–14) Geoarqueología de los sitios Hunter y Meguay, Pampa ondulada, provincia de Buenos Aires. Cuadernos del Instituto Nacional de Antropología y Pensamiento Latinoamericano 23(2):131–140
Toledo M (2011) El legado lujanense de Ameghino: revisión estratigráfica de los depósitos pleistocenos-holocenos del valle del río Luján en su sección tipo. Registro paleoclimático en la Pampa de los estadios OIS 4 al OIS 1. Rev Asoc Geol Argent 68(1):121–167
Tomassini RL, Montalvo CI, Beilinson E, Deschamps CM, Garrone MC, Gasparini GM, Zárate MA, Rabassa J, Ruella A, Tonni EP (2017) Microvertebrates preserved in mammal burrows from the Holocene of the Argentine Pampas: a taphonomic and paleoecological approach. Hist Biol 29:63–75
Tomassini RL, Montalvo CI, Fernández-Jalvo Y, Garrone MC, Kin MS (2019) Modern plains vizcacha (*Lagostomus maximus*, Chinchillidae, Rodentia) as a bone accumulating agent in the Argentine Pampas: application to the study of fossiliferous sites. J Arid Environ 161:11–24
Ugan A (2005) Does size matter? Body size, mass collecting, and their implications for understanding prehistoric foraging behavior. Am Antiq 70:75–90
Valverde F (2001) Huellas y marcas en restos óseos. In: Mazzanti DL, Quintana CA (eds) Cueva Tixi: Cazadores y recolectores de las Sierras de Tandilia Oriental. 1. Geología, Paleontología y Zooarqueología. Laboratorio de Arqueología. Universidad Nacional de Mar del Plata, Mar del Plata, pp 137–155
Vecchi R, Frontini R, Bayón MC (2013) Paso Vanoli: una instalación del Holoceno tardío en valles fluviales del sudoeste bonaerense. Revista del Museo de La Plata 13(87):77–93
Vecchi R, Stoessel L, Vaz EB (2023) Ocupaciones humanas en un ambiente lagunar del oeste del área Interserrana: Sitio Laguna del Indio. Intersecc Antropol 24(1):77–96
Vizcaíno S, Pardiñas UFJ, Bargo MS (1995) Distribución de los armadillos (Mammalia, Dasypodidae) en la región Pampeana (República Argentina) durante el Holoceno. Interpretación paleoambiental Mastozool Neotrop 2:149–165
Walker M, Head MJ, Berkelhammer M, Björck S, Cheng H, Cwynar L, Fisher D, Gkinis V, Long A, Lowe J, Newnham R, Rasmussen SO, Weiss H (2018) Formal ratification of the subdivision of the Holocene series/epoch (quaternary system/period): two new global boundary stratotype sections and points (GSSPs) and three new stages/subseries. Episodes 41:213–223
Yellen JE (1991) Small mammals: !Kung san utilization and the production of faunal assemblages. J Anthropol Archaeol 10(1):1–26

Chapter 8
From Pest to Vulnerable Species: Combining Ecological and Behavioural Knowledge for the Conservation and Management of *Lagostomus maximus*

Cecilia B. Contarde and Diego A. Guzmán

Abstract Despite being a native herbivore, in 1905, the plains vizcacha was declared a plague for agriculture by the Argentinean government. Since then, several eradication programmes have been conducted, leading to local extinction events and a considerable decrease in the vizcacha's abundance throughout its historical distribution range. Along with the waning of the vizcacha, its role as an ecosystem engineer also disappeared. Currently, efforts are being made to articulate the development of management and conservation projects for this gregarious burrowing mammal. In this chapter, we combine the known information about the species with our experimental results and field experience. Together, these elements have allowed us to develop a successful translocation and management protocol for the plains vizcacha, with the potential for adaptation for other fossorial mammals. The aim of this chapter is to provide a practical guide for managing vizcacha's populations based on their behavioural needs.

Keywords Ecosystem engineer · Native herbivore · Gregarious burrowing mammal · Conservation translocation protocol · Behavioural needs

8.1 Introduction

The plains vizcacha, *Lagostomus maximus*, is a caviomorph rodent exclusively found in southern South America. Detailed data on its actual geographical abundance and distribution are lacking; nevertheless, it is known that about 95% of its

C. B. Contarde · D. A. Guzmán (✉)
Área Biología del Comportamiento, Instituto de Investigaciones Biológicas y Tecnológicas, Universidad Nacional de Córdoba, CONICET, Córdoba, Argentina
e-mail: diego.guzman@unc.edu.ar

L. L. Rasia et al. (eds.), *Plains Vizcachas*, https://doi.org/10.1007/978-3-031-49487-1_8

geographical distribution is within Argentina, around 4% in Bolivia and only 1% in Paraguay (Roach 2016). Low levels of genetic diversity, the existence of a single genetically distinct population in Argentina (see Gariboldi et al. 2019) and changes in its effective size indicate that metapopulation processes and changes in human population dynamics during the late Holocene were important factors shaping the population genetic structure of *L. maximus*.

The vizcacha is the most sexually dimorphic and largest chinchillid (Weir 1974b). Adult females reach 65–70 cm in total length and weigh 2.5–4.5 kg, while adult males' total length is 65–90 cm, weighing up to 9 kg (Weir 1974a). Vizcachas have a large head with a blunt snout, a nose with folds that prevent dirt from entering (Jackson et al. 1996a) and bulging and black eyes. The front legs are short with four fingers on each hand (2nd to 5th), and the hind legs are robust and strong, with three fingers each (2nd to 4th) (Llanos and Crespo 1952; Weir 1970). The third toe of the hind legs is larger than the other two, and on the mid-ventral part of the second toe of the hind legs, they have specialized hairs for grooming (Hudson 1872; Llanos and Crespo 1952; Rieger and Walzthony 1987), located where they cannot be damaged while digging (Hudson 1892).

The fur is soft and abundant, in shades of grey mixed with cream on the back, and the belly is white. The head has a black or dark grey stripe that crosses the face horizontally from the nose, fading at the level of the ears (Fig. 8.1). Above and below this stripe, the colouration is clear, making it evident and characteristic with more emphasis in adult males. Above this stripe are the whiskers and thick black bristles, more marked in adult males. The forehead is dark grey, with a black sector at eye level. In general, females have lighter fur than males and offspring even more so, with a fairly whitish face (Llanos and Crespo 1952). Llanos and Crespo (1952) reported a seasonal change, with a lighter coat in the winter than in the summer, and Weir (1971) commented that the vizcachas she captured in Entre Ríos Province were greyer than those captured in Buenos Aires Province, suspecting that the colour difference would be related to the different types of soil. According to

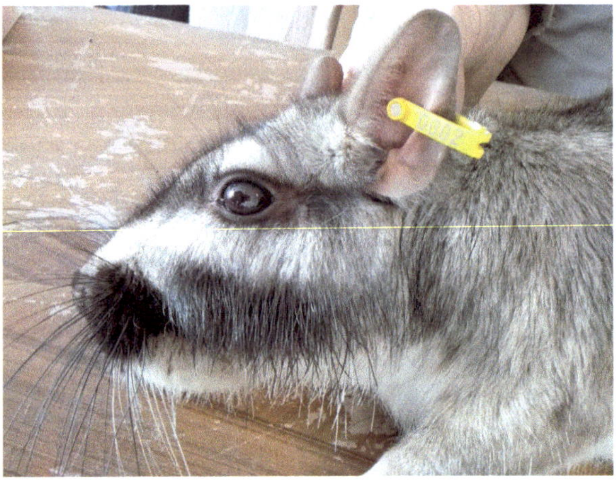

Fig. 8.1 Profile of an adult male vizcacha's face. It has a caravan on the left ear as it is a specimen that was part of a translocated social group

Jackson et al. (1996a), hair colour varies depending on the area since the substrate can dye it, being more light brown in sandy areas and dark grey in other areas.

The vizcacha is a gregarious rodent that lives in groups of 6–30 individuals on average, building underground burrows known as *vizcacheras* (Hudson 1892; Fernandez 1949; Llanos and Crespo 1952; Branch 1989, 1993a, c; Giulietti and Veneciano 2005). Vizcachas live 6–8 years (Llanos and Crespo 1952; Jackson 1989). Males reach sexual maturity at 12 months of age and females at 8 months (Jackson 1989; Branch 1993b); however, when environmental conditions are poor, males can remain in their natal *vizcachera* until 27 months of age (Jackson et al. 1996a). After reaching sexual maturity, males leave the natal burrow, while females usually remain in the burrow in which they were born throughout their lives. This difference results in a closely related kin group of females living together in the same burrow, along with newborns, juveniles, young males and adult males (Branch 1993a, c). Every year, there is competition between males to access burrow systems and the inhabiting females, which generally results in an annual rotation of breeding males (Branch 1993c). These families of vizcachas are clustered together forming colonies of many hectares, which are referred to as *vizcacherales* (Ambrosetti 1893; Branch 1993c).

Females have a principal oestrous cycle, and a secondary one occurs under some circumstances (Llanos and Crespo 1952), having offspring only once a year, with births occurring in spring (Weir 1970, 1971). If females lose their young at birth, they may come into oestrus approximately 17 days later, having a secondary cycle that same year, with births in late summer-early fall (Llanos and Crespo 1952; Weir 1970, 1971). Gestation lasts 154 days on average (Weir 1971).

There are some variations between regions in the dates of reproductive cycles. For example, in the main reproductive cycle in central Argentina, in Córdoba Province, copulations can be observed outside the burrows from June to August and births between October and December (Contarde and Guzmán, "unpublished data"). In Entre Ríos Province, oestrus has been reported to occur earlier, from March to April and births in July-August (Llanos and Crespo 1952). In La Pampa Province, Branch (1993b) recorded dates intermediate to the ones mentioned above. Variations between sites could be explained by climatic diversity and food availability. Also, a certain irregularity in the synchronization of the oestrus for the total number of females in the population may occur as females could be born as a result of a primary or a secondary cycle.

Female vizcachas have two offspring per litter, sometimes one and rarely three (Llanos and Crespo 1952). In general, the offspring are a female and a male or both females (Llanos and Crespo 1952; Weir 1970, 1971). The young are born quite developed, completely covered with hair, as most hystricomorph rodents (Llanos and Crespo 1952; Weir 1970, 1971), and remain inside the burrow for at least 2 months until they begin to go out to forage with the rest of the group (Branch 1993c). Females have two pectoral mammary glands and four nipples that emerge from the sides of the thorax. They produce a small amount of aqueous milk, which the young feed on repetitively (a situation observed only in captivity) (Llanos and Crespo 1952; Weir 1971).

The *vizcacheras* are formed by a series of interconnected tunnels and chambers with multiple exits (Fig. 8.2) (Branch et al. 1993). The number of active entrances to a burrow (entries with faeces, footprints and recent signs of excavation) can be related to the number of individuals living in it, but it is a measure that depends on the site since the burrow structure can vary according to the type of soil and the environment (Branch et al. 1994a). For example, in La Pampa Province, Branch et al. (1994a) recorded 0.57 *vizcacheras*/ha with an average of 33 active entrances per *vizcachera* and 1.4 individuals per entry (in a population that at the time of the study was abundant but in decline). In another population in Entre Ríos Province, they found a lower density of *vizcacheras* (0.32 *vizcacheras*/ha) with 3.6 average entries per *vizcachera* and approximately 2.5 vizcachas per entry. In Córdoba Province in central Argentina, Contarde (2019) studied a population with a density of 1.57 vizcacheras/ha with 11.97 average entries per *vizcachera* and an estimated 1.26 vizcachas per entry. Taking into account the great diversity of environments that its historical distribution range encompasses, these variations in the composition of the populations may be related to the great plasticity that the species has to adapt to the different conditions of availability and variety of food, climates and soil types. Furthermore, when surveyed, the colonies may have been in different phases of the population growth curve. In any case, for the same region, it is expected that,

Fig. 8.2 Natural *vizcachera* have multiple burrow entrances that can be seen as holes in the ground and the characteristic area of bare ground surrounding it

by sharing similar environmental conditions, the configuration of the populations will be conserved and allow estimations to be made.

The presence of curves, slopes of about 20–25° on average and lower and higher areas along the tunnels of the *vizcacheras* would protect vizcachas from floods, wind and predators (Hudson 1892; Llanos and Crespo 1952). Entering the *vizcachera*, between 1 and 2 m, a tunnel bifurcation or a large, relatively circular space (chambers) from which multiple tunnels diverge in different directions can be found (Hudson 1892). These chambers are where the females give birth, sleep (alone or in groups) during the day and where most of the underground behaviour takes place. Each *vizcachera* may have several chambers at different depths (Fernandez 1949).

All members of the social group participate in the construction and maintenance of the burrows (Branch 1993c). Vizcachas dig mainly with their front legs, but when the ground is too hard, they can help themselves with their incisors (Fernandez 1949). The earth is then pulled back by the action of the hind legs with a strong and rapid movement. This action forms a ditch, and when too much soil accumulates, they change direction, creating another ditch that together forms a typical Y in front of the entrance to the burrow (Hudson 1892). These ditches become increasingly deeper, wider and longer as the construction of the burrow progresses and eventually merges, forming a fan-shaped area of bare ground. Due to the movement of the earth carried out when excavating the tunnels, which are piled on top of the objects they accumulate, in front of the entrances, a C-shaped rise of the earth is formed that also functions as a protection against floods (Ambrosetti 1893).

Near the burrows, about 10–20 m away, there are individual burrows (Llanos and Crespo 1952) known as satellite *vizcacheras* (Branch 1993a) that consist of an entrance, tunnel and chamber. These burrows are built and used by migrant males when they try to access a group of females (Branch 1993c) or as shelter when they are foraging near them and some danger is perceived.

The stimulus that causes a vizcacha to form a new group is unknown. The social group is close-knit, but at some point, one, or a small group, separates and forms a new, independent burrow (Hudson 1892). It has also been observed that when resources are scarce or when the original group is greatly reduced, the remaining individuals move to a nearby burrow, merging with the group that lives there (Branch et al. 1993). As vizcachas actively defend their burrow against intruders, migrating in groups would allow them to evade the defence and be able to settle in the occupied burrow, while belonging to a larger group would function as an anti-predatory strategy (Branch et al. 1993). Entrances to the burrow and its surrounding area are defended by all members of the social group (Branch 1989, 1993a). The defence is so strong that even if a vizcacha is in danger near a burrow it does not belong to, it will make the greatest effort to run towards its burrow or to any satellite *vizcachera* even though it is further away and the risk is greater. It has been reported that if, in the face of an imminent threat and the absence of other options, vizcachas enter the burrow of another group to take refuge, members of that group will aggressively repel it, which poses an additional risk, as the intruding animal is forced to leave the refuge before the threat has ceased (Hudson 1892; Ambrosetti 1893). In situations of imminent danger faced far from the burrow to which the animal belongs, the main source of emergency shelter is the satellite *vizcacheras* scattered on the perimeters of the main *vizcacheras*. In any case, visits between neighbouring *vizcacheras* are common every night and actually marked paths are formed between burrows (Hudson 1892; Llanos and Crespo 1952). The presence in the area is tolerated, but the entrance is where they draw the limit. Furthermore, only a partial overlap is observed between the territories or home ranges of neighbouring social groups (Branch 1993a).

They are herbivores with a preference for grasses (Llanos and Crespo 1952; Giulietti and Jackson 1986; Branch and Sosa 1994). They have occasionally been seen drinking water, after a drought or in captivity, but usually obtain water from the vegetation they consume (Hudson 1892; Llanos and Crespo 1952; Weir 1970). They defecate anywhere, both inside and outside burrows (Llanos and Crespo 1952). Their faeces are cylindrical with rounded edges, 1 cm long in the youngest individuals and up to 3 or 3.5 cm in adults. The thickness varies from 0.5 to 0.8 cm, and the colour is dark green or brown, shiny when fresh.

Vizcachas remain active throughout the year, may show activity inside their burrows during the day and are generally observed outside the burrows from dusk to dawn (Llanos and Crespo 1952; Branch 1993a, b). Females and juveniles emerge together at dusk and feed in one or more groups near burrows (Branch 1989, 1993b). In winter, adult males forage with females and juveniles. During this season, activity patterns are more similar between age classes and sexes. From spring to autumn, males tend to feed alone, spending much of their time on the *vizcachera,* while the rest of the group feeds (Branch 1993b). Due to their habit of feeding close to the burrow, they create patches of low vegetation and bare ground, creating a mosaic in the landscape that is characteristic of the areas occupied by the species (Branch 1989; Branch et al. 1999). Depending on food availability, they can travel up to 400 m from the burrow to forage (Llanos and Crespo 1952) and even further on nights with low moonlight.

At night, adult males carry branches and objects to burrow entrances and use them for scent marking (Branch 1989). This marking is done by rubbing the sides of their faces or by sitting and urinating on them (Branch 1993a). Branch (1993a) recorded the onset of marking in spring with the arrival of migrating males and, to a greater extent, in autumn, coinciding with female oestrus.

Vizcachas have a wide range of vocalizations that they use to communicate with their social group and with other groups in the same colony. Some of these vocalizations can be heard for miles around and are repeated when groups hear them (Branch 1993c).

The vizcacha is considered an ecosystem engineer (Villarreal et al. 2008), creating local and regional differences in both species diversity and structural complexity of the landscape. Through grazing and burrowing, vizcachas produce spatial patterns in the soil composition, biomass and amount of nutrients available for the vegetation (Villarreal et al. 2008). These changes affect processes such as productivity and nutrient cycling and influence the distribution and dynamics of other animal species (Machicote et al. 2004).

The activity of native herbivores such as the vizcacha would allow the maintenance of low and open physiognomies dominated by grasses, such as lawns and pastures (Bontti et al. 1999; Villarreal et al. 2008). This allows for the establishment of a greater diversity of plant species than in the absence of the herbivore (Branch et al. 1994b, 1999). This is because the feeding and mowing of all vegetation prevent invasive species from dominating the area. Arias et al. (2005) found that plant species diversity was much higher at the edges of vizcacha foraging areas than in more distant areas or areas without vizcacha.

The presence of a greater variety of plant species also allows other animal species to establish themselves, both trophically associated species and those that use different types of vegetation for nesting, for example.

The *vizcacheras* also serve as a refuge for a large number of species that coexist with the vizcacha. A very common host is the burrowing owl (*Athene cunicularia*), which uses the tunnels to make its own burrows. There are also toads, frogs, insects, arachnids and other smaller rodents such as southern mountain cavy (*Microcavia australis*), lizards (*Salvator merianae*), boa constrictor (*Boa constrictor*), other snakes and even foxes (*Lycalopex griseus*). The vizcacha also plays an important role in mitigating the conflict between the puma (*Puma concolor*, one of its main predators) and livestock producers. Pessino et al. (2001) recorded a marked preference of the puma for vizcacha even near livestock and guanaco (*Lama guanicoe*) populations, so the presence of vizcacha would divert the puma's attention and reduce attacks on livestock.

When vizcachas build their burrows, they move soil from the lower layers to the surface and vice versa, and they feed on vegetation and make their depositions both outside and inside the burrow, thus exchanging and concentrating nutrients and minerals in the area of the *vizcachera*, increasing its availability in the environment (Arias et al. 2003; Arias et al. 2005; Villarreal et al. 2008). For this reason, Jackson et al. (1996b) consider them to be efficient converters of low-quality pasture, as they found that the vegetation near the burrow had a better nutritional composition than that further away (mainly higher amounts of nitrogen and phosphorus; Arias et al. 2005; Villarreal et al. 2008). In addition, the accumulation of nitrogen inside the burrows would be a useful reserve in ecosystems with high fire frequencies (Villarreal et al. 2008). The construction of burrows also increases water infiltration into the soil (Arias et al. 2005). Thus, burrows have an effect on the landscape at various scales in geomorphology, hydrology and soil dynamics (Arias et al. 2005).

The species come into conflict with humans over the use of space and food (Pereira and Quintana 2009). Although the real impact of the vizcacha on productive activities is unknown, it was declared a national pest in Argentina in 1905 under National Law No. 4863 (Llanos and Crespo 1952; Rendel 1990), a status that was endorsed for the species in several provincial legislations. Due to unrestricted hunting permits (as their meat is highly valued among local hunters) and eradication campaigns promoted by national and regional governmental authorities, vizcachas have experienced local extinctions in numerous regions of Argentina (Rendel 1990; Barberis et al. 2015).

When a significant decrease in populations was noted and, as a result, its role as a native ecosystem engineer was lost, measures were taken by the competent Argentinian authorities to manage the populations and ensure their repopulation. Nationally, Law N°4863 was repealed in 2018 by Decree 27/2018 and replaced by National Law 27444/2018. Currently, the vizcacha is extinct in large areas within its historical distribution and is abundantly conserved in others. In Córdoba Province, for example, in central Argentina, it is extinct in almost all of the south and east of the province and has therefore been classified as a 'vulnerable species' for the region (Abba et al. 2018). This situation places the vizcacha in an ambiguous state: on the one hand, its important ecosystemic role is being protected with new legislation and the repeal of its pest species status. On the other hand, it is in conflict with rural and peri-urban human populations due to the use of space, the construction of burrows and its potential competition with domestic livestock. Despite the change in the legal status of the species, certain sections of society still perceive the vizcacha as a pest to be controlled or eradicated.

8.2 Translocation and Management Protocol for the Plains Vizcacha

Burrowing mammals face a common threat: the widespread persecution due to farmers' perception of them as competitors to livestock (Branch et al. 2002; Hoogland 2006; Pereira and Quintana 2009). Governments spend millions annually to 'control' burrowing mammal populations for the benefit of the livestock industry. However, multiple research studies have clearly demonstrated that these campaigns are not cost-effective and result in indiscriminate poisoning of other wildlife species (Hoogland 2006; Wesche et al. 2007; Delibes-Mateos et al. 2011).

In Córdoba Province, the vizcacha has been classified as a pest since 1885 (Provincial Law 1005, Article 247) and has retained this status in subsequent legislation (Decree 6373/79, 1979). Nevertheless, the initial investigations into the harm inflicted by this species in central Argentina began in 1990 (Rendel 1990; Navarro et al. 1996; Bontti et al. 1999; Javier et al. 2003). Rendel (1990) reported that the vizcacha was problematic in 54% (13 out of 24) of the districts in Cordoba Province. However, only six districts had a notable economic impact (25% of the total departments) despite several years of population control and eradication efforts by the provincial government and private entities. In the same study, the 'damage' caused by vizcacha was also mentioned as a limiting factor in production, but in no department, it was considered to be the main factor. In other species (e.g. tucura *Tropidacris collaris*, parrots *Myiopsitta monachus* and hare *Lepus europaeus*), meteorological problems such as drought and hail and even the lack of adequate tools for the work or the fact that the areas were not suitable for crops were mentioned as more important problems. In fact, burrowing mammals, particularly at low-to-moderate densities, can have a positive impact on forage quality and productivity (Whicker and

Detling 1988; Smith and Foggin 1999; Noble et al. 2007). If burrowing mammal colonies cover less than 30% of the landscape, livestock weight gains are minimally affected (Derner et al. 2006), and it has been observed that livestock mass declines when burrowing mammals are poisoned over the same period (Smith and Foggin 1999). Although burrowing mammals are frequently held responsible for grassland degradation, they have been coexisting with free-ranging, native megaherbivores for millions of years. The interaction between livestock and burrowing mammals may be mutually beneficial, and together, they can enhance the grassland ecosystem through synergistic effects (Davidson et al. 2010). Burrowing animals generally have more localized effects due to their sedentary colonial tendencies, whereas megaherbivores have a broader impact at a larger spatial scale (Davidson et al. 2010; Yoshihara et al. 2010).

As Rendel (1990) points out, unplanned use of the environment causes imbalances, in most cases with negative consequences not only for the environment but also for humans themselves. So, looking to the future, it is necessary to seek solutions that do not end up in the indiscriminate application of the labels 'harmful' or 'pest' and that promote coexistence with the rest of the species.

Two key situations could be highlighted where human intervention and the development and application of management plans for the vizcachas are necessary. The first situation arises in places where the vizcacha was historically present but was exterminated, resulting in a noticeable loss of its ecosystemic function, and the aim is to recover it. A tool widely used in these cases is the reintroduction of species, in this context defined by Zlatanova (2016) as conservation translocations, which consist of the voluntary human movement of living organisms from one place to another within their historical distribution, with the main objective of achieving a conservation benefit (improving the conservation status of the species locally or globally, and/or restoring the functions or processes of a natural ecosystem). The second situation arises where the vizcacha is still present, its activity being the main source of conflict with humans. In these situations, it is necessary to provide communities with tools to manage the population and thus facilitate the coexistence of humans and vizcachas. It is necessary to find ways to keep vizcachas away from the resources that the local people are trying to protect and prevent them from recolonizing the area.

One example of the first situation is what happened in the Quebrada del Condorito National Park (PNQC) in Córdoba Province. This National Park was created in 1996 with the aim of transforming agro-pastoral systems into protected areas. In the PNQC, the last records of vizcachas date back to the 1950s. Vizcachas' absence, together with the prohibition of burning and the exclusion of livestock as management measures after the creation of the National Park, would have favoured the proliferation of tall grasslands (a risk factor for the spread of fires). As vizcachas, along with guanacos (*Lama guanicoe*), form part of the native fauna that inhabited this region, the reintroduction of these species is a main objective of the Park's management plan in order to recover their ecosystemic role and increase the heterogeneity and complexity of the landscape.

Knowledge of the biological, ecological and behavioural characteristics of the species is essential for effectively managing their populations and facilitating their reintroduction, making it more feasible to devise solutions that, while prioritizing animal welfare, can be effective and easy to apply to manage wild populations and minimize conflicts between the species and humans.

To ensure the success of a wildlife management plan, one of the main factors to be taken into account is the effect of each action performed by humans that could increase the stress response of the animals involved. Distress (when stress affects biological functions; Teixeira et al. 2007) causes negative effects such as reproductive failure, stunted growth, increased susceptibility to disease and, in the worst case, death (Teixeira et al. 2007). Some of the commonly known sources of stress in a translocation project are disturbances in the environment of the donor population, capture, captivity, health control, transport, release, adaptation to the new site, human presence, etc.

8.2.1 Study and Selection of Potential Donor Populations

Understanding the biological, ecological, demographic and behavioural aspects of wild populations of vizcachas would be an important first step for any translocation in order to analyse the status and health of the population, determine whether a particular population could be a potential donor of specimens for translocation and/or determine whether the removal of specimens would compromise the subsistence of the target population.

In addition, to select source populations, it is necessary to consider government regulations. The capture, handling and treatment of native species, the transport of individuals and the release on the new site can have different regulations according to the province where it will be held.

Before capturing for translocation, it is important to analyse the ability to donate individuals from potential target populations. The social structure of the groups to be translocated is considered a key factor for the future adaptation of the animals to the new habitat (Griffith et al. 1989; Fischer and Lindenmayer 2000; Shier 2006; Germano et al. 2015). Considering that these animals live in family groups with task differentiation (guarding, care of young, etc.) and a defined social structure (Branch 1993c), the most parsimonious option to be adopted is the capture and translocation of entire social groups (individuals sharing the same burrow or *vizcachera* will be housed in the same adaptation pen). The number of animals in a social group can be calculated using the Branch et al. (1994a) method (see above). To ensure that the capture of specimens for translocation does not reduce the population sufficiently to negatively affect it, care should be taken to ensure that the number of captured individuals does not exceed 10% of the total population. This percentage depends on the characteristics of the species; for the vizcacha, reproductive productivity is considered to be 1.9 offspring per year for each female, with a 55% survival rate of the offspring in the first year of life, and they live 6–8 years (Navarro et al. 1996; Bodmer and Robinson 2004).

In addition, the study of wild populations allows us to obtain basal parameters (e.g. time budget and faeces cortisol concentration), which should be taken into account for comparison with translocated groups and, in this way, to define the degree of adaptation to the new site, for example.

8.2.2 Host Site Preparation

The initiation of any translocation process is the selection and preparation of the site where the animals will be reintroduced. In the case of vizcachas, plains or valleys with abundant biomass and clear visibility are ideal. It is common to see *vizcacheras* built under medium-to-large trees, in most cases with the base of the trunk free of branches and associated shrubs. In those cases where the physical environment is adequate but the vegetation is not (e.g. high pastures with a high density of shrubs due to the removal of herbivores), the preparation of the chosen site should be initiated, ideally at least 1 year before beginning the process of capturing animals. Mechanical methods, such as mowing and/or the temporary introduction of domestic animals, can be utilized to create short and compact grasses. The objective is to achieve an area of low grassland (which facilitates the detection of predators) with sufficient biomass and space to accommodate all the individuals to be translocated.

Good habitat quality for a fossorial mammal such as the vizcacha is determined by the amount of available food, the environmental carrying capacity, climate, number of potential predators and soil type, especially the presence of underground refuges. In addition to providing protection from predators and the environment, the burrow is the place where important social interactions occur. There are informal reports collected as a result of our activities that there is a greater chance of success in the translocation of species when animals are released in abandoned *vizcacheras* than when they are released in rustic or improvised shelters. In reintroductions of other species, such as the prairie dog (*Cynomys ludovicianus*), if there were no remaining burrow sites dug by extinct populations, artificial burrows were provided (Koprowski et al. 2013). For this reason, where abandoned *vizcacheras* do not exist, it is imperative to include the construction of artificial burrows in the process of preparing the receptor site, ideally using information collected from wild *vizcacherales* so that the artificial burrows have the greatest possible similarity to natural burrows. Therefore, to plan the construction of artificial burrows, the physical and environmental parameters measured in natural burrows should be taken into account, i.e. the number of entrances to the burrow in which a social group of the target size lives, the shape and size of tunnels and chambers, the depth at which the first chambers are located and the characteristics of the terrain (slope, vegetation cover, vegetation type and soil type). Including these characteristics would favour the expression of a natural behavioural repertoire and encourage the adaptation of the vizcachas to the new site. The artificial burrows must provide the vizcachas with sufficient thermal insulation (temperature inside natural burrows at a depth of 1 m is around 20–21 °C, with an annual amplitude of no more than 10 °C; Llanos and Crespo 1952; Giulietti and Veneciano 2005), and they must allow the vizcachas to modify, extend them and must be feasible to build. The material from which they are constructed is an important detail to cover these characteristics. The thermal insulation of natural burrows is determined by the depth at which they are located, the low-temperature transmission of the compacted soil, the length and shape of the tunnels and the number of burrow entrances (which define the type of air currents generated inside). When compacted soil cannot be used, adobe is an ideal material for this work, as it is similar and easy to shape, can be sourced from the site, has good thermal insulation and can be easily modified by the vizcachas.

When setting up artificial burrows, another factor to consider is the social dynamics of the group. For a gregarious species such as the vizcacha, maintaining the integrity of the social group is very important to ensure adaptation and settlement on the new site. However, it can happen that not the entire group is captured and/or that neighbouring individuals have also been captured, adding the restructure of the hierarchies to the stress of the translocation process. Furthermore, Contarde et al. (2021b) observed that the different age classes of the social group use the burrow entrances differently, with young and juveniles using only some of the entrances. This pattern could reflect the influence of age and/or hierarchy on their use, be part of specific anti-predatory strategies, or it could mean that not all burrow entrances and the tunnel system are connected or available to all members of the group. It is therefore important to provide the artificial burrow with more than one chamber and entrance, so they can enter/exit the burrow from different places and modulate their social interaction according to their behavioural needs. The underground artificial chambers should be constructed at least 1 m deep so that the temperature inside them can reach the 20 °C of the natural burrows and give the vizcachas translocated a certain thermal comfort until they modify the artificial burrows enough. Connecting the chambers to each other and outside through subterranean tunnels makes the artificial burrows more complex and similar to the natural burrows (see above). In our translocation project, we used an innovative method for the construction of artificial burrows, which was the assembly of underground tunnels by arranging adobe tiles in a row from the chambers to the outside. In this way, we could design the tunnels in a similar shape to that observed in the natural *vizcacheras*, incorporating curves and bifurcations (Contarde and Guzmán, "unpublished data"). It is best to incorporate as many of the features measured in natural burrows as possible (height, length, slope of the tunnels, etc.), as they will satisfy the requirements mentioned above, while increasing the likelihood of satisfying those needs that may not have been met due to differences between their perception of their environment and ours.

To encourage translocated individuals to adapt and remain on site, a soft release is recommended. Soft releases refer to any measures that promote a gradual transition to the new site, including protection from predators, food supplementation and provision of shelter (De Milliano et al. 2016). These measures are applied for a certain time and then withdrawn. One of these measures is adaptation pens, which provide translocated animals with protection against predation, allow them to get to know the environment with little or no danger, prevent them from dispersing and encourage adaptation to the new site as they remain in place. Adaptation pens should be sized according to the number of individuals and have one or more artificial burrows inside. A social group will be placed in each of the adaptation pens, so they should be constructed with the necessary artificial burrow(s) for each social group to be translocated. The perimeter of the adaptation pen should be high enough so that predators cannot enter, the mesh should have small holes (less than 5 cm) so that the youngest individuals cannot pass through it, and it has to have buried veneers or even buried mesh to avoid escapes by tunnels. Another technique to prevent escapes by tunnel construction underneath the perimeter fence consists of placing wire mesh on the floor of the perimeter of the pens, covering at least 50 cm on each side of the vertical perimeter. Ideally, the edges of this mesh, placed horizontally on the ground, should be buried 10 cm and tensioned with a wire and stakes to the ground, as we observed that the vizcachas focus on the perimeter of the pen to start making underpasses to escape. Therefore, covering the ground in this area would prevent escapes by digging and additionally reduce the effort required to build and dismantle the adaptation pens (the dismantling process for the release of the translocated animals will be described in the corresponding section). However, it is crucial to ensure that it is tightly secured and regularly checked, as even a small lift could result in vizcachas digging their way out.

The separation between pens should be similar to the separation between natural *vizcacheras*, as vizcachas are expected to modify the artificial burrows and remain in them throughout their lives (especially females). The spatial distribution patterns of natural burrows may be determined by the spatial or temporal distribution of resources, anti-predatory strategies and/or intraspecific (inter-group) competition, so it is important to conserve them. For example, in a *vizcacheral* in central Argentina, the mean separation was 40.08 ± 18.03 m (Contarde et al. 2021a).

As vizcachas will be confined, it is mandatory to provide them with water ad libitum in shallow water containers. The diet has to be periodically supplemented with soaked maize, alfalfa and fresh vegetables (Weir 1970; Maik-Siembida 1974). Ideally, food should be provided ad libitum as well; nevertheless, the calculated amount of food vizcachas need is around 100 grams per kilogram of animal weight, with half of it being dry matter like alfalfa (Maik-Siembida 1974; Hagen et al. 2015). Fresh vegetables can then constitute 50–70% of the daily food portion, while dry food can be provided in an unrestricted quantity, allowing any excess to be consumed on subsequent days.

The monitoring of the translocated groups should be performed periodically during the adaptation period, and once the liberation of the animals is done, the monitoring frequency can be reduced gradually. One way of monitoring is by using night vision cameras or camera traps placed inside and outside (for predator detection) of the adaptation pens. The idea is to record indicators of injuries and pathologies, the social interaction between the members of each social group and the frequency and duration of other behaviours (e.g. vigilance, modification of the artificial burrow, feeding, fighting, presence of predators and reaction of the vizcachas to predators).

The selection of types of cameras and places to install them could be done before vizcachas arrive at the adaptation pens, and in this way, the installation and design of the monitoring system are incorporated into the construction of the pens.

An additional measure for site preparation entails building satellite burrows in the vicinity of the adaptation pens (Renison et al. 2023). This will provide vizcachas with secure shelters when they venture away from the primary artificial burrow while foraging and could help to encourage the construction of new burrows.

8.2.3 Capture

The capture method to be used must allow capturing a complete social group or a representative group of it (in relation to the proportion of age classes and sexes) (Letty et al. 2007). There are several methods for live capture of vizcachas: three for nocturnal capture and one for diurnal capture (Llanos and Crespo 1952; Weir 1970; Branch et al. 1993, 1994a). The diurnal method (flooding of burrows) is not recommended due to ethical issues and the high stress it causes the individuals. Nocturnal methods consist of baited Tomahawk traps, gravity-operated tube traps and baited corral traps. The first two methods are for individual trapping, and the third is for group trapping. Bait choices include fresh green leafy vegetables, carrots, alfalfa, corn and oat (listed in order of preference based on our experience).

For any procedure involving the capture of vizcacha specimens and their handling, one of the features that has to be taken into account is the local reproductive cycle of the species. Although a main and secondary reproductive cycle is recognized, the dates on which they occur vary between regions. For this reason, the study of wild populations that are potential donors of specimens is important to estimate these dates and, from these, the optimal capture window. For reasons detailed above, capturing individuals between September and January could result in the capture of females who are at a very advanced stage of pregnancy that can be negatively affected by the stress of translocation or females that have recently given birth and are still caring for their offspring underground. In the latter case, if the mothers are captured, the young would remain inside the burrows with no chance of survival. In addition, the young of a secondary cycle are born approximately between the end of March and May, so trapping between March and June could lead to the same situation. Thus, there would be two possible trapping windows in the year: (1) July to August and (2) late January to February. The first capture window has an extra advantage in that it coincides with the winter in the Southern Hemisphere, so the amount of food available is scarce, which would increase the willingness of the vizcachas to enter the traps if they are baited. During these two time windows, especially during the first, there is a high chance of capturing females in the early stages of pregnancy. The translocation of females in this stage of gestation can be advantageous for the settlement of the individuals in the new place because of two factors linked to their behaviour, the care of the offspring and their social group formation dynamics. The offspring remain inside the burrow for the first 2 months without going outside, which means that the female refrains from venturing far, as the young would not be ready to follow her yet.

Another important characteristic is that the female offspring remain in their natal burrow, and as the offspring are born before the final release of the animals on the new site, there is a greater probability that they will remain even in the face of a migration of the adults that were not born on the site. These capture windows are general, and ideally, precise confirmation of the dates of the breeding cycle should be made for the population where the individuals will be removed.

A translocation project is highly costly, time-consuming and labour-intensive. When choosing the trapping method, a balance is sought between the resources allocated, the effectiveness of the trapping method and whether it promotes the welfare of the individuals. The efficiency of trapping methods will be determined by the number of individuals caught per time and the degree to which the captured individuals are representative of the age and sex structure of the social groups trapped.

The individual traps, as their name indicates, capture one vizcacha at a time (with exceptions in which up to two or three vizcachas have been captured simultaneously). The original Tomahawk traps are square or rectangular wire mesh cages. Their mechanism is based on attracting the animal into the cage by luring it with bait. Once inside the cage, while seeking the bait at the end, the animal pulls the trigger (pedal or stick), which releases the door, leaving it trapped. This method is widely used to capture small mammals (Weir 1970). As vizcachas forage close to the burrow in which they live and defend their territory against intruders, to enhance the capture of vizcachas from the same social group these traps should be placed in the close vicinity of the target *vizcachera* (Fig. 8.3).

In relation to their original design, our experience suggests that some modifications can be made, by reducing some of the aversive components of these traps for the vizcachas, which would increase their capture rate. One of the first modifications would be about the material. The use of wooden panels instead of metal grids has been successfully adopted in the construction of this type of trap for the capture of rabbits. Using wood for the construction of rabbit traps was found to be less aversive, possibly because it is a more natural material, with a texture similar to those found in the wild. In addition, the floor of the trap can be covered with soil obtained from the target burrow. This soil serves several functions: it covers the odours of the trap, it reduces the potential aversion that vizcachas may experience walking on a wooden floor and it also acts as a track canvas to allow the detection of vizcachas entering the trap. Another modification that has proven to be effective in catching vizcachas is to add an opening at the end of the wooden trap, covered with a 2 cm square grid-welded mesh. The presence of the opening allows air currents to pass through the trap, carrying the scent of the bait and increasing its attractiveness; this modification also gives the trap the sensation of being a tunnel (rather than an enclosed space), so that the vizcachas would be more prone to enter. The size of the mesh used is important because the vizcacha, especially once inside the trap, will try to get out by gnawing on the bars, and if it gets past the nose through the mesh, the friction can injure it.

Fig. 8.3 Wooden tomahawk traps positioned for capture near a *vizcachera*

Attempting to free trapped individuals is another peculiar behaviour of the vizcacha. This behaviour has been reported in burrows deliberately blocked by humans (with earth, branches and stones) with the intention of eradicating the population; in the following days, vizcachas have been observed digging from outside the burrow. Similar situations are observed when vizcachas are caught inside a trap and vocalize for help. Attention should be paid to this 'help from the outside', and it should be considered that the trap door must be heavy enough so that it cannot be pushed from the outside by another vizcacha. Another solution would be to use a different trap closing system, where the door closes with a guillotine action rather than a swing action, preventing it from being pushed in from the outside.

One way to increase the effectiveness of the traps is through habituation. This can be done by placing the traps, but leaving them open with bait inside but unarmed, for three or more nights before capture. Through this procedure, the vizcachas get used to enter the traps and consume the bait, thus reducing the capture effort when the traps are active.

The advantages of these Tomahawk traps include protection from predators while the animals are in the trap, the possibility of using the trap to transport the captured animals and the fact that these traps are easy to set and remove. The disadvantage is that the number of individuals captured each night is limited because, once a vizcacha is caught, its attempts to escape and vocalizations alert the rest of the group and discourage them from exploring the remaining traps still actively.

Tube traps (Llanos and Crespo 1952) are not baited individual traps and consist of a wire mesh tube, which is placed directly at the entrance of the burrow (Fig. 8.4),

Fig. 8.4 Three tube traps are positioned for capture in a *vizcachera*. Each trap is installed in a burrow entrance and covered with cloth, earth and branches

taking advantage of the slope of the entrance to use gravity as the mechanism of action. At the end attached to the burrow entrance, the trap has a swinging door that only opens towards the inside of the trap. The vizcacha must push the door to get out of the burrow (entering the trap) and gets trapped when the door closes behind it by the action of gravity. These traps should be placed at each entrance to the target burrow, as the tunnels can be connected. The first to leave the burrow each night are the adult males, who watch from the entrance to spot dangers and then lead the rest of the group out; therefore, by using tube traps adult males will most likely be the first to be caught. Tube traps should be left in place for several consecutive days to capture a complete social group, with the precaution of not exceeding 5 days in a row. Food deprivation for longer periods of time has been reported to be detrimental to the species (Clauss et al. 2007; Hagen et al. 2015). If during these 5 days the number of individuals captured is not the intended number, or the group captured is not representative of the age and/or sex ratio of the population, the traps should be removed for a few days and then placed again for further 5 consecutive days. One option for removing the traps is to open the end of the trap furthest from the burrow and lock the swinging door (leaving the trap open) so that individuals can exit and enter the burrow, through the trap. This would generate a habituation to the trap that would increase the chances of capture in the following days.

The same caution should be taken with tube traps as with Tomahawk traps, in terms of the grid size of the wire mesh. Also, it is necessary to periodically check the traps as vizcachas on the outside will try to help by digging around the trap.

We recommend the implementation of a variation we made to the original design (Llanos and Crespo 1952), which is to cover the upper part of the trap with cloth and then place soil and branches on top. This complement would have several functions: by functioning as an extension of the entrance tunnel, it also favours the continuity of this environment, which is characterized by a minor thermal amplitude compared to the one recorded in the open air, providing thermal comfort to the captured vizcachas. This complement would also protect the captured animals from dew and predator harassment, and it would give a tunnel-like sensation that would probably encourage the vizcachas to enter.

The disadvantages of this type of trap are that it is necessary to have several traps equal to the number of entrances to the burrow where the target group is located, they catch one animal at a time, and it is necessary to capture and collect animals on several consecutive days to obtain a balanced group in terms of age classes and sexes. The advantages are that they do not require adaptation time or baiting, they are easy to transport and relatively easy to place and it is possible to use the traps to transport the captured animals. Furthermore, it would be the appropriate method to use when wanting to capture only adult males to supplement a population, for example.

The third type of trap is a baited corral trap (Branch et al. 1993, 1994a). Unlike the traps mentioned above, it allows several vizcachas to be captured at the same time; thus, as we already mentioned, by placing the trap close to a particular *vizcachera*, the chances of capturing several individuals of the same social group at the same time are increased. Another advantage of this method is that trapping and keeping the animals together may reduce the stress response of the vizcachas while inside the trap. The corral trap enclosures are circular, between 5 and 7 m in diameter and made of strong wire mesh with small holes (Fig. 8.5). The vizcacha is capable of jumping up to 2 m and bouncing over a vertical surface, such as mesh, to achieve greater heights. We have observed vizcachas even climbing the wire mesh. Therefore, it is crucial to adjust the pen's top height and shape to avoid escapes. The height can be adapted in two ways: the first is to make it as high as possible, about 2 m at least; the second way is to make the mesh 1–1.5 m high and add another mesh (or the same mesh folded) angled inwards at the top.

Being a burrowing animal, vizcachas will also attempt to escape by digging a tunnel under the perimeter. The circular shape of the pen has the benefit of not having corners, which would be the first places where the vizcachas would concentrate their efforts to escape. In addition, the same structure against digging as the one mentioned for the adaptation pens could be applied: a sturdy wire mesh of at least 1 m wide, 50 cm on each side of the vertical perimeter mesh, placed on the ground. The vizcacha will first try to dig at the edge of the enclosure, where it will come up against the mesh placed on the floor, which will discourage the behaviour. The mesh must be strong enough so that they cannot cut it with their teeth and must be tightly stretched and secured to the ground so that they cannot get under it.

Fig. 8.5 Circular trap pen installed for capture near a *vizcachera*. It has three entrances with a squared-tube shape. On top of the perimeter mesh, there is a mesh projected inwards to avoid escapes. There is a cage inside the pen that works as a refuge for the capture vizcachas and to facilitate their collection. On the left, there is a door made so the operators can enter the trap pen

One or more gates should be placed around the perimeter of the pen. To activate the trap pen, a guillotine gate activated manually with a rope can be used as was done by Ferreyra et al. (2007), and this mechanism requires the operators to be in the immediate area, which could persuade the vizcachas to enter the pen. At the same time, it has the advantage that the operators will remove the animals immediately, so they will not have time to escape by digging, for example.

The vizcachas are still perceived as a hunting target, even if it is not legally allowed (poaching). Considering that staying near *vizcacherales* overnight may represent a certain risk to one's physical integrity, and the possibility of human presence deterring vizcachas from entering the trap, the optimal trap activating method would be the one in which the operators do not have to be present. An alternative method could be to train the vizcachas to use a swinging door, which they have to push to enter or leave the trap. When the decision to start trapping is made, the door can be locked to prevent it from opening outwards. To achieve this habituation, the pen should be baited several days before capture and a progression should be made from a simple perimeter opening, through a swing gate of half the total length, to a full swing gate. The size of the gate should be large enough for one vizcacha to pass through but no more than that, as when one vizcacha pushes the gate to get in, a conspecific can take advantage of the space to get out. This method has an adaptation period of about 10–12 days (3 nights per gate type) before being activated.

The advantage of the corral trap is its capacity to capture a large number of animals keeping them together. The disadvantages are that it requires more labour and materials to set up than other types of traps and that for 12–15 days it is necessary to keep it baited so that the vizcachas can habituate to enter.

Based on the results obtained from several trapping campaigns, we concluded that the corral trap had the higher trap rate, capturing an average of 11 vizcachas per night, with a balanced sex distribution of 48% females and 52% males, and a breakdown of 12% young, 18% juveniles and 70% adults. In contrast, the wooden tomahawk traps had a moderate capture rate, averaging 0.79 vizcachas per night, skewed towards females (68% females and 32% males), while capturing 15% young and 85% adults. The tube traps recorded the lowest capture rate of 0.62 vizcachas per night, predominantly capturing adult males (95%), with minimal female presence (5%) and no young individuals captured.

The corral trap seems to be the best option for social group capture intended as part of a translocation project. Wooden Tomahawk traps would be a good option if, for example, it is not possible to construct a corral trap. In addition, tube traps would be the option of choice for capturing individuals to supplement populations. As they capture mostly adult males, they could be used to simulate the dynamics of male migration in populations that are not connected.

8.2.4 Handling and Transportation

Once caught, it is necessary to examine the vizcachas for injuries, diseases, parasites and overall body condition. This examination should be performed in the presence of a veterinarian, who will be able to anesthetize and treat them, if necessary. Vizcachas that become injured or sick during the translocation process and cannot be expediently treated should be kept in captivity or euthanized (according to the veterinarian's criteria). The euthanized animal and animals dying for unknown reasons should be submitted to a veterinary pathologist for necropsy (Truett et al. 2001).

To prevent the introduction of diseases to the host site, a comprehensive examination of the individuals and the administration of broad-spectrum antiparasitics to the vizcachas are essential before transporting them to the new location.

Animals must be transported in accordance with stipulated guidelines such as CITES (IUCN/SSC 2013). Vizcachas can be transported together (always keeping the social groups of origin) in a sufficiently large transport cage or individually in smaller transport cages. In both cases, cages must be well-ventilated, sheltered from the sun and weather and large enough to allow for postural adjustment. The cages should have a padded floor (made of straw, alfalfa, local vegetation or cardboard) to absorb urine and keep the animals dry and comfortable. Fresh food such as green leafy vegetables or carrots can be placed inside. Cardboard or wood should be placed between the transport cages and the vehicle to prevent loud banging noises during transport.

Once in the new location, the vizcachas must be released gently into the adaptation pens, always keeping the members of the original social group in the same enclosure.

8.2.5 Adaptation to the New Location

The period immediately following the release of semi-captive animals is critical for successful translocation (Truett et al. 2001). If the translocated vizcachas have not been able to modify their burrows sufficiently to feel safe during their time in the adaptation pens, or if the original social group has not been captured or if they have not formed a strong group, the vizcachas are likely to disperse or suffer from social disorientation upon release. This would result in the vizcachas dispersing outside the host site or attempting to build new burrows, with the consequent risk of death from predation or low temperatures.

Monitoring while the vizcachas are in the adaptation pens and during the first few weeks after release is crucial. The monitoring should be intense during the first few days of the vizcacha's stay in the new location and can be gradually reduced as signs of adaptation, and the absence of pathologies or undesirable behaviours are noted. In this way, decisions can be made regarding the level of stress experienced by the *vizcachas* during the process; excessive levels of stress could lead to the acquisition of myopathy, diseases and aversive reactions to conspecifics, among other things. Behavioural observations can also be used to determine whether or not only members of the same social group have been captured and grouped together. This is defined, for example, by the observation of fights, aggression and segregation of individuals. When the vizcachas are released from the adaptation pens, it is advisable to carry out another intensive monitoring, looking for dispersal and/or abundance of individuals in the site, indicators of predator presence, new entrances and tunnels and modifications to the artificial burrows provided. The amount of soil excavated from the inside is a good indicator of the animals' adaptation to the new site, as this would imply greater depth and complexity of the tunnels provided (with the consequent protection from predators and weather) and the adaptation of the artificial burrow to the group's needs (improved group cohesion).

In our experience, 3 months is a reasonable period of time for vizcachas to remain in the adaptation pens. During this time, the vizcachas are expected to modify and build new entrances to the artificial burrows, make and respond to wild-type alarm calls and have their first offspring.

The release of the translocated groups can be done by removing the buried metal sheets or wire mesh from the ground and lifting the perimeter mesh. The lifting is done by folding up the bottom edge of the mesh, leaving a space of approximately 30 cm from the ground. This provides some protection against predators by allowing free passage for the vizcachas but adds some difficulty for the predator and domestic livestock to pass through.

In the months following release, the diet can be supplemented with alfalfa and occasionally fresh vegetables, reducing the amount gradually to encourage foraging and feeding on local flora.

After release, one way to monitor translocated groups is to conduct a monthly faecal pellet count, which can be used to estimate population density per area (e.g. faeces/ha), a method widely used to estimate rabbit densities (Palomares 2001; Cabezas and Moreno 2007; Fernandez-de-Simon et al. 2011; Rouco et al. 2016). Basically, this technique involves placing circular plots on the ground, randomly distributed throughout the population's range, and counting the amount of faecal pellets gathered in each plot on a monthly basis.

Since vizcachas defecate similarly to rabbits (in the form of easily recognizable pellets scattered throughout the area) and their populations are also stationary, this technique has the potential to estimate vizcacha population density in an area. In addition, this method would allow us to determine the use of space and the size of the action area during the different seasons. Other advantages are that it can be carried out throughout the year during daylight hours, it requires very few elements, it is a low-cost technique and the same setup can be used over several years to estimate population growth.

In a translocation project, with the aim of reintroducing a species, it is necessary to consider the genetic dynamics associated with the establishment of the population. Genetic variability is an important factor as it increases the chances of adaptation to the new environment.

A management measure that could be implemented once the groups have been translocated is the supplementation of the nucleus population with reproductive males at the end of the summer, around February and March (before the initiation of the oestrus). To significantly enhance a population genetically, it is necessary to conduct supplementation at rates of at least 5% of the recipient population over a period of two to six generations. Only in cases where the genetic distance between the source and recipient populations is relatively large, short supplementation programmes (<2 generations) should be considered (Pacioni et al. 2019). Nevertheless, the calculations regarding the number of individuals required to translocate as a supplemental measure are based on the assumption that a source population of significant size and genetic health is readily available. If this is not the case, specific models must be utilized to maximize the potential for long-term conservation and recovery of rare and threatened species. As mentioned at the end of Sect. 8.2.3 (capture), metal tube traps would be an optimal trap choice to capture adult males to supplement populations.

References

Abba AM, Barquez RM, Castilla MC, Coda JA, Damino V, Díaz MM, Periago ME, Pia M, Priotto J, Superina M (2018) Mamíferos de Córdoba y su estado de conservación. Editorial de la UNC, Córdoba

Ambrosetti J (1893) Notas Biológicas: Contribución al estudio de la Biología Argentina. Revista del Jardín Zoológico de Buenos Ayres I:39–52

Arias SM, Madanes N, Quintana RD (2003) Estructura y composición de la vegetacion de vizcacheras activas e inactivas en el delta del Paraná. Mastozool Neotrop 10(1):9–20

Arias SM, Quintana RD, Cagnoni M (2005) Vizcacha's influence on vegetation and soil in a wetland of Argentina. Rangel Ecol Manag 58(1):51–57

Barberis IM, Romano M, Montani E, Cordini C, Derlindati EJ (2015) Registro de vizcachas (*Lagostomus maximus*) en Pampa de las Lagunas, sur de Santa Fe, Argentina. Nót Faun (Segunda Serie) 182:1–6

Bodmer RE, Robinson JG (2004) Evaluating the sustainability of hunting in the Neotropics. In: People in nature: wildlife conservation in South and Central America. Columbia University Press, New York, pp 299–323

Bontti EE, Boo RM, Lindstrom LI, Elia OR (1999) Botanical composition of cattle and vizcacha diets in central Argentina. J Range Manag 52(4):370–377

Branch LC (1989) Demography and social organization of the plains viscacha. University of California, Berkeley Dec, p 1989

Branch LC (1993a) Intergroup and intragroup spacing in the plains vizcacha, *Lagostomus maximus*. J Mammal 74(4):890–900

Branch LC (1993b) Seasonal patterns of activity and body mass in the plains vizcacha, *Lagostomus maximus* (family Chinchillidae). Can J Zool 71(5):1041–1045

Branch LC (1993c) Social organization and mating system of the plains viscacha (*Lagostomus maximus*). J Zool 229(3):473–491

Branch LC, Villarreal D, Machicote M (2002) Conservation challenges of ecosystem engineers: case studies from grasslands and shrublands of North and South America. The Open Country 4:37–48

Branch L, Sosa R (1994) Foraging behavior of the plains vizcacha, *Lagostomus maximus* (Rodentia: Chinchillidae), in semi-arid scrub of central Argentina. Vida Silv Neotrop 3(2):96–99

Branch LC, Villarreal D, Fowler GS (1993) Recruitment, dispersal, and group fusion in a declining population of the plains vizcacha (*Lagostomus maximus*; Chinchillidae). J Mammal 74(1):9–20

Branch LC, Villarreal D, Sosa A, Pessino M, Machicote M, Lerner P, Borraz P, Urioste M, Hierro JL (1994a) Estructura de las colonias de vizcacha y problemas asociados con la estimación de la densidad poblacional en base a la actividad de las vizcacheras. Mastozool Neotrop 1(2):135–142

Branch LC, Villarreal D, Sbriller AP, Sosa RA (1994b) Diet selection of the plains vizcacha (*Lagostomus maximus*, family Chinchillidae) in relation to resource abundance in semi-arid scrub. Can J Zool 72(12):2210–2216

Branch LC, Hierro JL, Villarreal D (1999) Patterns of plant species diversity following local extinction of the plains vizcacha in semi-arid scrub. J Arid Environ 41(2):173–182

Cabezas S, Moreno S (2007) An experimental study of translocation success and habitat improvement in wild rabbits. Anim Conserv 10(3):340–348

Clauss M, Besselmann D, Schwarm A, Ortmann S, Hatt JM (2007) Demonstrating coprophagy with passage markers? The example of the plains viscacha (*Lagostomus maximus*). Comp Biochem Physiol A Mol Integr Physiol 147(2):453–459

Contarde C (2019) Caracterización espacial y patrones de actividad en una colonia de vizcachas (*Lagostomus maximus*) de la zona serrana del centro de Córdoba. Universidad Nacional de Córdoba

Contarde C, De Luca A, Guzman D (2021a) Patrón de distribución de las madrigueras en una colonia de vizcachas (*L. maximus*). In: XXIX Reunión Argentina de Ecología. Tucumán, Argentina, p 337
Contarde C, Kembro J, Guzman D (2021b) Differential use of the burrow system entrances by a gregarious Caviomorph (*Lagostomus maximus*). In: 58th annual conference of the animal behaviour society. Estados Unidos
Davidson AD, Ponce E, Lightfoot DC, Fredrickson EL, Brown JH, Cruzado J, Brantley SL, Sierra-Corona R, List R, Toledo D (2010) Rapid response of a grassland ecosystem to an experimental manipulation of a keystone rodent and domestic livestock. Ecology 91(11):3189–3200
De Milliano J, Di Stefano J, Courtney P, Temple-Smith P, Coulson G (2016) Soft-release versus hard-release for reintroduction of an endangered species: an experimental comparison using eastern barred bandicoots (*Perameles gunnii*). Wildl Res 43(1):1–12
Delibes-Mateos M, Smith AT, Slobodchikoff CN, Swenson JE (2011) The paradox of keystone species persecuted as pests: a call for the conservation of abundant small mammals in their native range. Biol Conserv 144(5):1335–1346
Derner JD, Detling JK, Antolin MF (2006) Are livestock weight gains affected by black-tailed prairie dogs? Front Ecol Environ 4(9):459–464
Fernandez M (1949) Sobre la vizcacha (*Lagostomus trichodactylus* Brookes) sus vivienda y su protección. Bol Acad Nac Ci Córdoba 38:347–378
Fernandez-de-Simon J, Díaz-Ruiz F, Villafuerte R, Delibes-Mateos M, Ferreras P (2011) Assessing predictors of pellet persistence in European rabbits *Oryctolagus cuniculus*: towards reliable population estimates from pellet counts. Wildl Biol 17(3):317–325
Ferreyra H, Uhart MM, Romano MC, Beldoménico PM, Samartino L, Paolicchi F, Lauricella M, Jorge MC, Schettino A, Guida N, Martín A (2007) Inmovilización química y evaluación de salud de vizcachas salvajes (*Lagostomus maximus*) en el chaco arido argentino. Arq Ciênc Vet Zool UNIPAR 10(2):91–99
Fischer J, Lindenmayer DB (2000) An assessment of the published results of animal relocations. Biol Conserv 96(1):1–11
Gariboldi MC, Inserra PIF, Lucero S, Failla M, Perez SI, Vitullo AD (2019) Unexpected low genetic variation in the South American hystricognath rodent *Lagostomus maximus* (Rodentia: Chinchillidae). PLoS One 14(9):e0221559
Germano JM, Field KJ, Griffiths RA, Clulow S, Foster J, Harding G, Swaisgood RR (2015) Mitigation-driven translocations: are we moving wildlife in the right direction? Front Ecol Environ 13(2):100–105
Giulietti JD, Jackson JE (1986) Composición anual de la dieta de vizcacha (*Lagostomus maximus*) en pastizales naturales en la provincia de San Luis, Argentina. Rev Arg Prod Anim 6(3–4):229–237
Giulietti J, Veneciano J (2005) La vizcacha. Informativo Rural, EEA INTA San Luis 2(7):8–10
Griffith B, Scott JM, Carpenter JW, Reed C (1989) Translocation as a species conservation tool: status and strategy. Science 245(4917):477–480
Hagen KB, Besselmann D, Cyrus-Eulenberger U, Vendl C, Ortmann S, Zingg R, Kienzle E, Kreuzer M, Hatt JM, Clauss M (2015) Digestive physiology of the plains viscacha (*Lagostomus maximus*): a large herbivorous hystricomorph rodent. Zoo Biol 34(4):345–359
Hoogland J (2006) Conservation of the black-tailed prairie dog: saving North America's western grasslands, 1st edn. Island Press. The Canadian Field-Naturalist, Washington, DC
Hudson WH (1872) On the habits of the vizcacha (*Lagostomus trichodactylus*). Proc Zool Soc Lond 1872:822–833
Hudson WH (1892) The naturalist in La Plata. JM Dent & Company, London
Jackson J (1989) Reproductive parameters of the plains vizcacha (*Lagostomus maximus*) in San Luis Province, Argentina. Vida Silv Neotrop 2(1):57–62
Jackson JE, Branch LC, Villarreal D (1996a) Lagostomus maximus. Mamm Species 543:1–6

Jackson JE, Bucher EH, Chani JM (1996b) Capture of blue-fronted amazons and hunting of vizcachas and tegu lizards in Argentina. Assessing the sustainability of uses of wild species: case studies initial assessment procedure. IUCN, Gland, pp 17–25

Javier AP, Quintana RD, Susana M (2003) Diets of plains vizcacha, greater rhea and cattle in Argentina. J Range Manag 56(1):13–20

Koprowski J, Hale S, Hicks H (2013) A review of black-tailed prairie dog reintroduction strategies and site selection: Arizona reintroduction. Environment 44(1):1–6

Letty J, Marchandeau S, Aubineau J (2007) Problems encountered by individuals in animal translocations: lessons from field studies. Ecoscience 14(4):420–431

Llanos AC, Crespo JA (1952) Ecología de lo vizcacha (*Lagostomus maximus maximus* Blainv) en el Nordoeste de la Provincia de Entre Ríos. Rev Investig Agríc 6:289–378

Machicote M, Branch LC, Villarreal D (2004) Burrowing owls and burrowing mammals: are ecosystem engineers interchangeable as facilitators? Oikos 106(3):527–535

Maik-Siembida J (1974) Breeding viscachas: *Lagostomus maximus*: at Lodz Zoo. Int Zoo Yearb 14:116–117

Navarro JL, Rosati VR, Fraire EC (1996) Incidencia de vizcachas (*Lagostomus maximus*) en un cultivo de soja. Mastozool Neotrop 4(2):137–144

Noble JC, Mueller WJ, Detling JK, Pfitzner GH (2007) Landscape ecology of the burrowing bettong: warren distribution and patch dynamics in semiarid eastern Australia. Austral Ecol 32(3):326–337

Pacioni C, Wayne AF, Page M (2019) Guidelines for genetic management in mammal translocation programs. Biol Conserv 237:105–113

Palomares F (2001) Comparison of 3 methods to estimate rabbit abundance in a Mediterranean environment. Wildl Soc Bull 29:578–585

Pereira JA, Quintana RD (2009) Trophic interactions among plains vizcacha (*Lagostomus maximus*), greater rhea (*Rhea americana*), and cattle in a wetland of the Paraná River Delta Region, Argentina. Stud Neotropical Fauna Environ 44(1):1–6

Pessino ME, Sarasola JH, Wander C, Besoky N (2001) Long-term response of puma (*Puma concolor*) to a population decline of the plains vizcacha (*Lagostomus maximus*) in the Monte desert, Argentina. Ecol Austral 11(02):061–067

Rendel C (1990) Estimación de daños causados por la vizcacha (*Lagostomus maximus*) en la provincia de Córdoba. Rev Arg Prod Anim 10(1):63–79

Renison D, Cingolani AM, Contarde C, Guzmán D (2023) Asistiendo a la reintroducción de vizcachas (*Lagostomus maximus*): ¿Cómo aumentar el área de pastoreo seguro? Ecol Austral 33(1):020–029

Rieger I, Walzthony D (1987) Hair brush organs in Viscachas, *Lagostomus maximus*. Z Säugetierkd 52(2):125–127

Roach N (2016) *Lagostomus maximus*. The IUCN red list of threatened species 2016:e. T11170A78320596. Retrieved 20 Sept 2023

Rouco C, Santoro S, Delibes-Mateos M, Villafuerte R (2016) Optimization and accuracy of faecal pellet count estimates of population size: the case of European rabbits in extensive breeding nuclei. Ecol Indic 64:212–216

Shier DM (2006) Effect of family support on the success of translocated black-tailed prairie dogs. Conserv Biol 20(6):1780–1790

Smith AT, Foggin JM (1999) The plateau pika (*Ochotona curzoniae*) is a keystone species for biodiversity on the Tibetan plateau. Animal Conservation Forum, Cambridge University Press

Teixeira CP, De Azevedo CS, Mendl M, Cipreste CF, Young RJ (2007) Revisiting translocation and reintroduction programmes: the importance of considering stress. Anim Behav 73(1):1–13

Truett JC, Jo Ann LDD, Matchett MR, Owens E, Seery D (2001) Translocating prairie dogs: a review. Wildl Soc Bull (1973–2006) 29(3):863–872

Villarreal D, Clark KL, Branch LC, Hierro JL, Machicote M (2008) Alteration of ecosystem structure by a burrowing herbivore, the plains vizcacha (*Lagostomus maximus*). J Mammal 89(3):700–711

Weir BJ (1970) The management and breeding of some more hystricomorph rodents. Lab Anim 4(1):83–97
Weir BJ (1971) The reproductive physiology of the plains viscacha, *Lagostomus maximus*. J Reprod Fertil 25(3):355–363
Weir BJ (1974a) The tuco-tuco and plains viscacha. Symp Zool Soc Lond 34:113–130
Weir BJ (1974b) Reproductive characteristics of hystricomorph rodents. Symp Zool Soc Lond 34:265–299
Wesche K, Nadrowski K, Retzer V (2007) Habitat engineering under dry conditions: the impact of pikas (*Ochotona pallasi*) on vegetation and site conditions in southern Mongolian steppes. J Veg Sci 18(5):665–674
Whicker AD, Detling JK (1988) Ecological consequences of prairie dog disturbances. Bioscience 38(11):778–785
Yoshihara Y, Okuro T, Buuveibaatar B, Undarmaa J, Takeuchi K (2010) Complementary effects of disturbance by livestock and marmots on the spatial heterogeneity of vegetation and soil in a Mongolian steppe ecosystem. Agric Ecosyst Environ 135(1–2):155–159
Zlatanova D (2016) IUCN guidelines for reintroductions and conservation translocations of species – problems and solutions. In: First national conference of reintroduction of conservation-reliant species. University Press, Sofia, pp 18–29

Chapter 9
Plains Vizcacha Endoparasites: An Exploration of Diversity, Health and Ecology

Victoria Canova, María del Rosario Robles, Agustín M. Abba, and Graciela T. Navone

Abstract Parasites are an integral part of all ecosystems and have gained wide acceptance as biological tags since they can provide important information about their hosts, including biology, diet, feeding behaviour, population structure, movements, geographical ranges and phylogenies. The plains vizcacha is a good model for studies on parasites since it has a particular behaviour (gregarious, coprophagic and allogrooming habits), a wide geographical distribution occurring in diverse ecoregions, economic importance for causing damage on the fields and for having meat for consumption and a sanitary interest as a potential host for zoonotic or enzootic parasites. Different populations of this rodent have been studied from economic, biological and ecological aspects in Argentina; however, studies on the diversity of parasites or their implications for human and animal health are very scarce. This chapter aims to explore the available information on plains vizcacha endoparasites considering their diversity, sanitary aspects, ecological data and distribution from host and environment. In this chapter, we present the following five sections: (1)—Diversity of Parasitic Species, Geographical Records and Ecological Data; (2)—Animal and Human Health Considerations; (3)—Structure of Parasitic Populations and Communities; (4)—Potential Use of the Parasitic Species as Biotags; and (5)—Conclusions.

Keywords *Lagostomus maximus* · Protozoa · Helminths · Sanitary importance · Biotags

V. Canova (✉) · M. d. R. Robles · A. M. Abba · G. T. Navone
Centro de Estudios Parasitológicos y de Vectores CEPAVE (CCT La Plata CONICET-UNLP), La Plata, Buenos Aires, Argentina
e-mail: victoriac@cepave.edu.ar; rosario@cepave.edu.ar; abbaam@cepave.edu.ar; gnavone@cepave.edu.ar

L. L. Rasia et al. (eds.), *Plains Vizcachas*, https://doi.org/10.1007/978-3-031-49487-1_9

9.1 Introduction

Parasitology is currently trapped between apparently insurmountable data limitations and the urgent need to understand how parasites respond to global change. Parasitism is arguably the most species-rich mode of animal life on the earth (Rohde 1982; Dobson et al. 2008; Larsen et al. 2017), and parasites probably comprise a majority of the undescribed or undiscovered species left to modern science (Larsen et al. 2017; Okamura et al. 2018). In recent years, the global diversity and distribution of parasite richness have become a topic of particular concern (Dobson et al. 2008; Carlson et al. 2017; Dallas et al. 2018), both in light of the accelerating rate of disease emergence in wildlife, livestock and humans (Han et al. 2016) and growing recognition of the ecological significance of many parasites (Dougherty et al. 2016). Parasitic taxa are expected to face disproportionately high extinction rates in the coming century, causing a cascade of unknown but possibly massive ecological repercussions (Carlson et al. 2017; Cizauskas et al. 2017). Understanding the impacts of global change relies on baseline knowledge about the richness and biogeography of parasite diversity, but some groups are better studied than others. Emerging and potentially zoonotic viruses dominate this field (Han et al. 2015; Carlson et al. 2019); macroparasites receive comparatively less attention (Carlson et al. 2020).

In this context, parasites are an integral part of all ecosystems and have gained wide acceptance as biological tags since they can provide important information about their hosts, including biology, diet, feeding behaviour, population structure, movements, geographical ranges and phylogenies (Poulin and Morand 2004; Luque and Poulin 2007). Different aspects of populations and communities of helminths have been used as tags in different host groups (e.g. prevalence, abundance, mean abundance, specific richness, diversity, presence of particular parasitic species and gradients in the distribution of parasite assemblages) (Roberts 1991; Timi 2007; Alarcos et al. 2016); however, few studies have been recorded in mammals (Aznar et al. 1995; Balbuena et al. 1995) and even fewer in terrestrial systems (Roberts 1991; Holmstad et al. 2004).

The plains vizcacha, *Lagostomus maximus* (Desmarest, 1817) (Rodentia: Chinchillidae), is a good model for studies on parasites since it has a particular behaviour (gregarious, coprophagic and allogrooming habits), a wide geographical distribution occurring in diverse ecoregions, economic importance for causing damage on the fields and for having meat for consumption and a sanitary interest as a potential host for zoonotic or enzootic parasites (Jackson et al. 1996; Spotorno and Patton 2015; Cirignoli and Lartigau 2019; Canova et al. 2023; Canova, 2023, "unpublished data").

Different populations of this rodent have been studied from economic, biological and ecological aspects in Argentina (e.g. Llanos and Crespo 1952; Branch 1993a, b, c; Branch et al. 1993, 1994a, b, c; Jackson et al. 1996; Kufner and Monge 1998; Puig et al. 1998; Arias et al. 2003, 2005; Dorfman et al. 2016; Gariboldi et al. 2019; Acuña et al. 2021, 2022; Renison et al. 2023). However, studies on the diversity of parasites or their implications for human and animal health are very scarce (e.g. Rossanigo et al. 1986; Martínez 1988; Foster et al. 2002; Ferreyra et al. 2007; Canova et al. 2023). There are still fewer proposals to explore ecological aspects of parasites, both from their host distribution, in terms of indices such as prevalence, intensity and abundance, as well as other variables that contribute to understanding the structure of populations and communities (richness, diversity, evenness, among others) in this host species (Rossanigo et al. 1986; Foster et al. 2002).

In *L. maximus*, parasites could be used as biological tags. The ecological aspects can be explored since both the parasite community structure and species compositions are influenced by different natural and artificial variables. Several endoparasite species have been found on the plains vizcacha (Railliet and Henry 1909; Schuurmans Stekhoven 1951; Rossanigo et al. 1986; Sutton and Durette-Desset 1987, 1995; Martínez 1988; Couch et al. 2001; Foster et al. 2002; Ferreyra et al. 2007; Toso et al. 2011; Sangenis et al. 2016; Canova et al. 2021; Cwirenbaum et al. 2021; Canova et al. 2023; Canova, 2023, "unpublished data"), but there are practically no studies that exceed the records in the host and different areas in Argentina. Likewise, some taxonomic diagnoses should be reviewed.

This study aims to explore the available information on plains vizcacha endoparasites considering their diversity, sanitary aspects, ecological data and distribution from host and environment.

This chapter consists of five sections: (1)—Diversity of Parasitic Species, Geographical Records and Ecological Data; (2)—Animal and Human Health Considerations; (3)—Structure of Parasitic Populations and Communities; (4)—Potential Use of the Parasitic Species as Biotags; and (5)—Conclusions.

9.2 Diversity of Parasitic Species, Geographical Records and Ecological Data

Currently, a total of 21 endoparasite species, seven protozoa and 14 helminths, have been found along the geographical distribution of *L. maximus* in Argentina (Tables 9.1 and 9.2, Figs. 9.1, 9.2, and 9.3).

Protozoa parasites from this rodent have different types of cycles (e.g. Couch et al. 2001; Toso et al. 2011; Sangenis et al. 2016; Cwirenbaum et al. 2021; Canova et al. 2023; Canova, 2023, "unpublished data"). Some of them have a direct or monoxenous life cycle, with a single host where asexual and sexual reproduction in the small intestine occurs with the formation of oocysts that come out with the faeces, contaminating the soil, water and food that will later be ingested by the same specimen or others becoming infected (e.g. *Eimeria*) (Fig. 9.4). The development of the infectious stage (oocyst) depends on environmental conditions such as temperature and humidity (Robben and Sibley 2004; Poulin and Kamiya 2015; Cizauskas et al. 2017; Berkhout et al. 2020; Martorelli Di Genova and Knoll 2020). Other ones present an indirect or heteroxenous life cycle with two types of hosts, an

Table 9.1 Protozoa species from *Lagostomus maximus*

ID species	Species	Authors	Type host	Other hosts	Geographical data[a]	Literature
1	*Eimeria lagostomi*	Couch, Foster, Machicote and Branch, 2001	*Lagostomus maximus*		La Pampa, La Rioja, Entre Ríos, Santiago del Estero (Argentina)	Couch et al. (2001), Ferreyra et al. (2007) and Canova, 2023, "unpublished data"
2	*Eimeria vizcacha*	Couch, Foster, Machicote and Branch, 2001	*Lagostomus maximus*		La Pampa, Entre Ríos, Buenos Aires (Argentina)	Couch et al. (2001) and Canova, 2023, "unpublished data"
3	*Eimeria* cf. *chinchillae*	De Vos and Van der Westhuizen, 1968	*Chinchilla lanigera*	*Lagostomus maximus*	Argentina	Couch et al. (2001)
4	Cysts of Sarcocystidae		*Lagostomus maximus*		Buenos Aires (Argentina)	Cwirenbaum et al. (2021)
5	*Sarcocystis* spp.		*Lagostomus maximus*		Buenos Aires, Santiago del Estero (Argentina)	Canova et al. (2023)
6	*Toxoplasma gondii*	Nicolle and Manceaux, 1908	*Ctenodactylus gundi*	*Lagostomus maximus*	Buenos Aires, Entre Ríos, Santiago del Estero (Argentina)	Canova, 2023, "unpublished data"
7	*Trypanosoma cruzi*	Chagas, 1909		*Lagostomus maximus*[b], *Dasyprocta* sp.[b]	Argentina	Toso et al. (2011) and Sangenis et al. (2016)

[a]Information for *Lagostomus maximus* as host
[b]The record of *T. cruzi* in these hosts should be verified considering new evidence

intermediate host (IH), the prey, in which the asexual reproduction occurs with the production of tissue cysts that are ingested by a definitive host (DH), the predator, in which the sexual reproduction occurs with the formation of oocysts that come out with the faeces, and later, these are ingested by the IH through contaminated soil, food or water (e.g. *Besnoitia and Sarcocystis*) (Robben and Sibley 2004; Dubey et al. 2016; Cwirenbaum et al. 2021; Rosenthal 2021; Canova et al. 2023) (Fig. 9.5). Moreover, other protozoa species have life cycles with one vertebrate host and one vector (e.g. *Trypanosoma cruzi*) (Herrera 2010; Toso et al. 2011; Filigheddu et al.

2017; Busselman and Hamer 2022; Chacón et al. 2022) (Fig. 9.6). Some protozoan species with these life cycles were recorded in plains vizcacha (Couch et al. 2001; Ferreyra et al. 2007; Toso et al. 2011; Sangenis et al. 2016; Cwirenbaum et al. 2021; Canova et al. 2023; Canova, 2023, "unpublished data") (Table 9.1).

Among the helminths, the nematodes registered in plains vizcacha have a direct life cycle (Morand et al. 2006; Beveridge et al. 2014). Although in these cases, only the presence of a single host is necessary, the development of the infectious stage (egg) depends on environmental conditions such as temperature and humidity (Poulin and Kamiya 2015; Cizauskas et al. 2017; Berkhout et al. 2020) (Table 9.2, Fig. 9.7). On the other hand, the cestodes have an indirect life cycle, with an arthropod as an intermediate host and *L. maximus* as a definitive host (Beveridge 1994; Georgiev et al. 2006; Mariaux et al. 2017) (Table 9.2, Fig. 9.8).

The parasites registered in *L. maximus* are listed in Tables 9.1 (Protozoa) and 9.2 (Helminths), showing data on the host type and other hosts and geographical distribution. Moreover, ecological data available as values of prevalence (*P* in %), mean intensity (MI) and mean abundance (MA) are provided in the text, following Bush et al. (1997). *P* is the number of hosts infected with a particular parasite species divided by the number of hosts examined; MI is the total number of individuals of a particular parasite species divided by the number of hosts infected with that parasite species; and MA is the total number of individuals of a particular parasite species divided by the total number of hosts examined (including both infected and uninfected hosts).

Three species of *Eimeria* (Apicomplexa, Eimeriidae) were recorded in plains vizcachas from Argentina. Although the site of infection is reported as unknown, these protozoa inhabit the small intestine, and oocysts were recovered from faeces. *Eimeria lagostomi* Couch, Foster, Machicote and Branch, 2001, was found in plains vizcachas of Caleu Caleu Department from La Pampa Province [*P* = 10% (17/163)] (Couch et al. 2001), General Belgrano Department from La Rioja Province (Ferreyra et al. 2007), Federal Department from Entre Ríos Province and Aguirre Department from Santiago del Estero Province (Canova, 2023, "unpublished data") (Fig. 9.1a). *Eimeria vizcacha* Couch, Foster, Machicote and Branch, 2001, was found in plains vizcachas from Caleu Caleu Department from La Pampa Province [*P* = 1% (2/163)] (Couch et al. 2001), Federal Department from Entre Ríos Province and Tornquist Department from Buenos Aires Province (Canova, 2023, "unpublished data") (Fig. 9.1b). *Eimeria* cf. *chinchillae* De Vos and Van der Westhuizen, 1968, was found in one plains vizcacha from Argentina [*P* = 1% (1/163)] (Couch et al. 2001). *Eimeria chinchillae* was originally described from *Chinchilla lanigera* Bennett, 1829 [named *Chinchilla laniger* (Molina, 1782) by Couch et al. (2001)], from a hatchery in Pretoria District (South Africa) in 1968. Thirty years later, this species was recorded in one plains vizcacha from Argentina as *E.* cf. *chinchillae*. The specific identity was not reached since some morphological discrepancies were observed (Couch et al. 2001).

Table 9.2 Helminth species from *Lagostomus maximus*

ID species	Species	Authors	Type host	Other hosts	Geographical data[a]	Literature
8	*Graphidioides affinis*	Mégnin, 1895	*Dolichotis patagonum*	*Lagostomus maximus*	La Pampa (Argentina)	Foster et al. (2002)
9	*Graphidioides rudicaudatus*	Railliet and Henry, 1909	*Lagostomus maximus*	*Lagidium viscacia*	Santiago del Estero, Tucumán, San Luis, Entre Ríos, Buenos Aires, La Pampa (Argentina)	Schuurmans Stekhoven (1951), Rossanigo et al. (1986), Matínez (1988), Sutton and Durette-Desset (1995), Foster et al. (2002)
10	*Graphidioides* spp.		*Lagostomus maximus*		Santiago del Estero, Entre Ríos, Buenos Aires, Salta, Córdoba (Argentina)	Canova, 2023, "unpublished data"
11	*Lagostonema ecasiense*	Sutton and Durette-Desset, 1987	*Lagostomus maximus*		Mendoza, Buenos Aires, La Pampa, Santiago del Estero, Entre Ríos, Córdoba (Argentina)	Sutton and Durette-Desset (1987), Foster et al. (2002), and Canova, 2023, "unpublished data"
12	*Viannella viscaciae*	Goodey, 1925	*Lagidium viscacia*	*Lagostomus maximus*	San Luis, La Pampa, Santiago del Estero, Entre Ríos, Buenos Aires, Córdoba (Argentina)	Rossanigo et al. (1986), Martínez (1988), Foster et al. (2002), and Canova, 2023, "unpublished data"
13	*Ascaris lumbricoides*	Linnaeus, 1758	*Homo sapiens*	*Lagostomus maximus*	San Luis (Argentina)	Rossanigo et al. (1986)
14	*Oxyuris* sp.		*Lagostomus maximus*		Entre Ríos (Argentina)	Martínez (1988)
15	*Wellcomia hugoti*	Canova and Robles, 2021	*Lagostomus maximus*		Entre Ríos (Argentina)	Canova et al. (2021)
16	*Heteroxynema (Cavioxyura) viscaciae*	Hugot and Sutton, 1989	*Lagidium viscacia boxi*	*Lagostomus maximus*	La Pampa, La Rioja (Argentina)	Foster et al. (2002) and Ferreyra et al. (2007)
17	*Heteroxynema (Cavioxyura)* n. sp.		*Lagostomus maximus*		Buenos Aires, Santiago del Estero, Entre Ríos, Salta, Córdoba (Argentina)	Canova, 2023, "unpublished data"

18	*Trichuris* sp.		*Lagostomus maximus*		La Pampa, Santiago del Estero, Entre Ríos, Buenos Aires (Argentina)	Martínez (1988), Foster et al. (2002), and Canova, 2023, "unpublished data"
19	*Viscachataenia quadrata*	von Linstow, 1904	*Lagidium peruanum*	*Lagidium viscacia, Lagostomus maximus*	San Luis (Argentina)	Rossanigo et al. (1986) and Denegri et al. (2003)
20	*Monoecocestus* sp.		*Lagostomus maximus*		La Pampa, Santiago del Estero, Entre Ríos, Buenos Aires (Argentina)	Foster et al. (2002), and Canova, 2023, "unpublished data"
21	*Cyclophyllidea* gen. sp.		*Lagostomus maximus*		Santiago del Estero, Buenos Aires (Argentina)	Canova, 2023, "unpublished data"

[a]Information for *Lagostomus maximus* as host

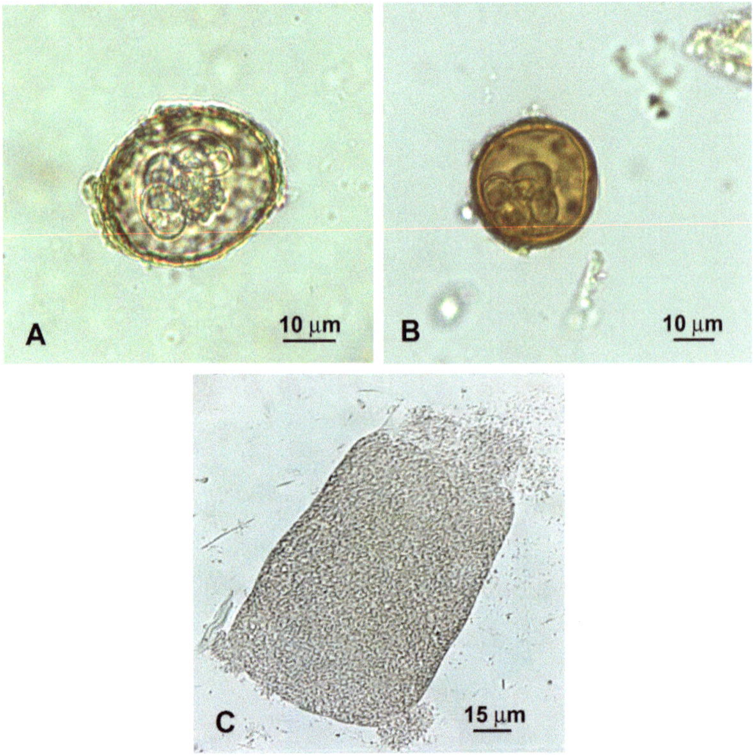

Fig. 9.1 Protozoan from plains vizcacha. (**a**) *Eimeria lagostomi*. (b) *Eimeria vizcacha*. (**c**) Cyst of *Sarcocystis* sp. (Canova et al. 2023)

In addition, cysts of Sarcocystidae (Apicomplexa) were reported from 15 plains vizcachas from Berazategui Department, Buenos Aires Province (Cwirenbaum et al. 2021). The sites of infection reported were ovaries, mammary glands, uterus, brain, spleen, adrenal, superficial fascia, fluid from the amniotic sac and embryonic pituitary gland. The size and morphology of the cysts and their location predominantly in non-neural tissues allowed Cwirenbaum et al. (2021) to attribute the observed cysts to the genus *Besnoitia*. The distribution among the hosts was evaluated, indicating a $P = 12\%$ (15/128) (Cwirenbaum et al. 2021).

Also, *Sarcocystis* spp. (Apicomplexa, Sarcocystidae) from *L. maximus* were found in Buenos Aires (Punta Indio Department) and Santiago del Estero (Aguirre Department) provinces (Canova et al. 2023). The sites of infection reported were different muscles (tongue, diaphragm, masseter, heart and hind limbs). Canova et al. (2023), through morphological and molecular studies with two genetic fragments (*18S rRNA* and *coxI*), reported the presence of at least two *Sarcocystis* species in *L. maximus*. The analysis of genetic distances with the *18S rRNA* and *coxI* sequences showed that all *Sarcocystis* spp. of *L. maximus* were grouped with a large support, being in a sister clade with the species *S. nesbitti* Mandour, 1969, *S. canis* Dubey and Speer, 1991, and *Sarcocystis* sp., parasites of small mammals and snakes.

Fig. 9.2 Helminths from plains vizcacha. (**a**) Male of *Lagostonema ecasiense*. (**b**) Female of *Wellcomia hugoti*. (**c**) Female of *Trichuris* sp. (**d**) *Monoecocestus* sp.

Moreover, the distribution among the hosts was evaluated, indicating a *P* = 3.6% (1/28) in Buenos Aires Province and *P* = 46.1% (6/13) in Santiago del Estero Province (Canova et al. 2023) (Fig. 9.1c).

Toxoplasma gondii (Nicolle and Manceaux, 1908) (Apicomplexa, Sarcocystidae) was found in eight plains vizcachas from Berazategui Department, Buenos Aires Province [*P* = 12.8% (6/47)]; Federal Department, Entre Ríos Province [*P* = 2.1% (1/47)]; and Aguirre Department, Santiago del Estero Province [*P* = 2.1% (1/47)]. This parasite was found by three different diagnostic methods that included serological, molecular and histopathological studies (Canova, 2023, "unpublished data").

Finally, surveys on *Trypanosoma cruzi* Chagas, 1909 (Trypanosomatida, Trypanosomatidae), from plains vizcachas showed negative results with different methods (Wisnivesky-Colli et al. 1992; Ceballos et al. 2006; Ferreyra et al. 2007; Wehrendt et al. 2019). However, Toso et al. (2011) and Sangenis et al. (2016) reported to Argentina one case of infection positive for a child who, during a hunting excursion, fed exclusively on plains vizcachas (*L. maximus*) and agoutis (*Dasyprocta* sp.). Subsequently, the child suffered myocarditis that caused his death 20 days later.

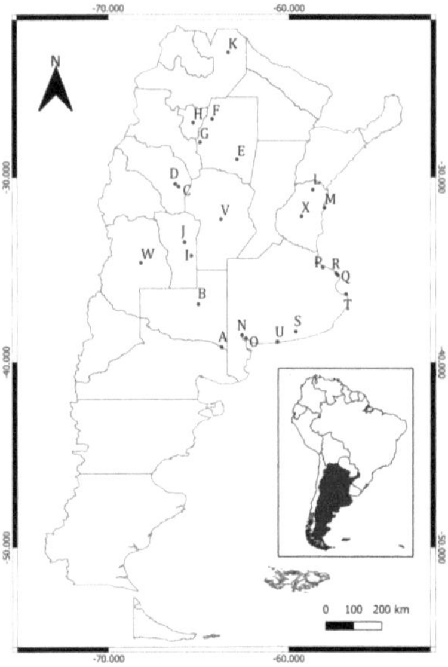

Reference map	Department	Province	ID Species
A	Caleu Caleu	La Pampa	1, 2, 8, 9, 11, 12, 16, 18, 20
B	Toay	La Pampa	8, 9, 11, 12, 18, 20
C	General Belgrano	La Rioja	1
D	Chamical	La Rioja	16
E	Aguirre	Santiago del Estero	1, 5, 6, 10, 11, 12, 17, 18, 20, 21
F	Jiménez	Santiago del Estero	9
G	Guasayán	Santiago del Estero	9
H	Leales	Tucumán	9
I	General Pedernera	San Luis	9, 12, 19
J	Coronel Pringles	San Luis	9, 12, 13, 19
K	General José de San Martín	Salta	10, 17
L	Federal	Entre Ríos	1, 2, 6, 10, 11, 12, 15, 17, 18, 20
M	Concordia	Entre Ríos	10, 11, 12, 15, 18
N	Tornquist	Buenos Aires	2, 10, 11, 12, 17, 18, 20
O	Bahía Blanca	Buenos Aires	10, 11, 12, 17, 21
P	Berazategui	Buenos Aires	4, 6, 9, 10, 11, 12, 18
Q	Punta Indio	Buenos Aires	5, 10, 11, 12, 17, 18
R	Magdalena	Buenos Aires	10, 11, 12, 17
S	San Cayetano	Buenos Aires	10, 11, 12
T	La Costa	Buenos Aires	10, 11, 12, 17, 18
U	Tres Arroyos	Buenos Aires	11, 12, 17
V*	-	Córdoba	10, 11, 12, 17
W*	-	Mendoza	11
X*	-	Entre Ríos	9, 12, 14, 18

* Only the province is known.

Fig. 9.3 Parasite species recorded in plains vizcacha from Argentina. The table shows references to letters on the map corresponding to ID species mentioned in Tables 9.1 and 9.2

Among Helminths, different species of *Graphidioides* (Nematoda, Trichostrongylidae) located in the stomach were recorded in plains vizcacha. *Graphidioides affinis* (Mégnin, 1895) was originally recorded for *Dolichotis patagonum* (Zimmermann, 1780) in Argentina (Mégnin 1895). Subsequently, this species was recorded in plains vizcachas occurring in Caleu Caleu [*P* = 59% (16/28) and MI = 6] and Toay [*P* = 10% (1/10) and MI = 2] departments from La Pampa Province (Foster et al. 2002). *Graphidioides rudicaudatus* (Railliet and Henry, 1909) was reported in plains vizcachas from provinces of Santiago del Estero (Jiménez and Guasayán departments), Tucumán (Leales Department) (Schuurmans Stekhoven 1951), Entre Ríos (without georeferences) (Martínez 1988) and Buenos Aires (Berazategui Department) (Sutton and Durette-Desset 1995). Also, this nematode species was recorded in Coronel Pringles [*P* = 97.4% (77/79)] and General Pedernera [*P* = 100% (22/22)] departments with a MA = 75 to both sites from San Luis Province (Rossanigo et al. 1986) and Caleu Caleu [*P* = 68% (19/28) and MI = 27] and Toay [*P* = 100% (10/10) and MI = 18] departments from La Pampa Province (Foster et al. 2002). Moreover, *Graphidioides* spp. were found in plains vizcachas in Aguirre Department from Santiago del Estero Province [*P* = 69.2% (9/13), MI = 10.4 and MA = 7.2]; Bahía Blanca and Tornquist [*P* = 75% (9/12), MI = 27.9 and MA = 20.9], Berazategui [*P* = 100% (12/12), MI = 142.4 and MA = 142.4], San Cayetano [*P* = 33.3% (2/6), MI = 23.5 and MA = 7.8], Magdalena, Punta Indio and La Costa departments from Buenos Aires Province; Córdoba Province [*P* = 50% (2/4), MI = 98 and MA = 49]; Concordia [*P* = 80% (4/5), MI = 53 and MA = 42.4] and Federal [*P* = 91.7% (11/12), MI = 22 and MA = 20.2]

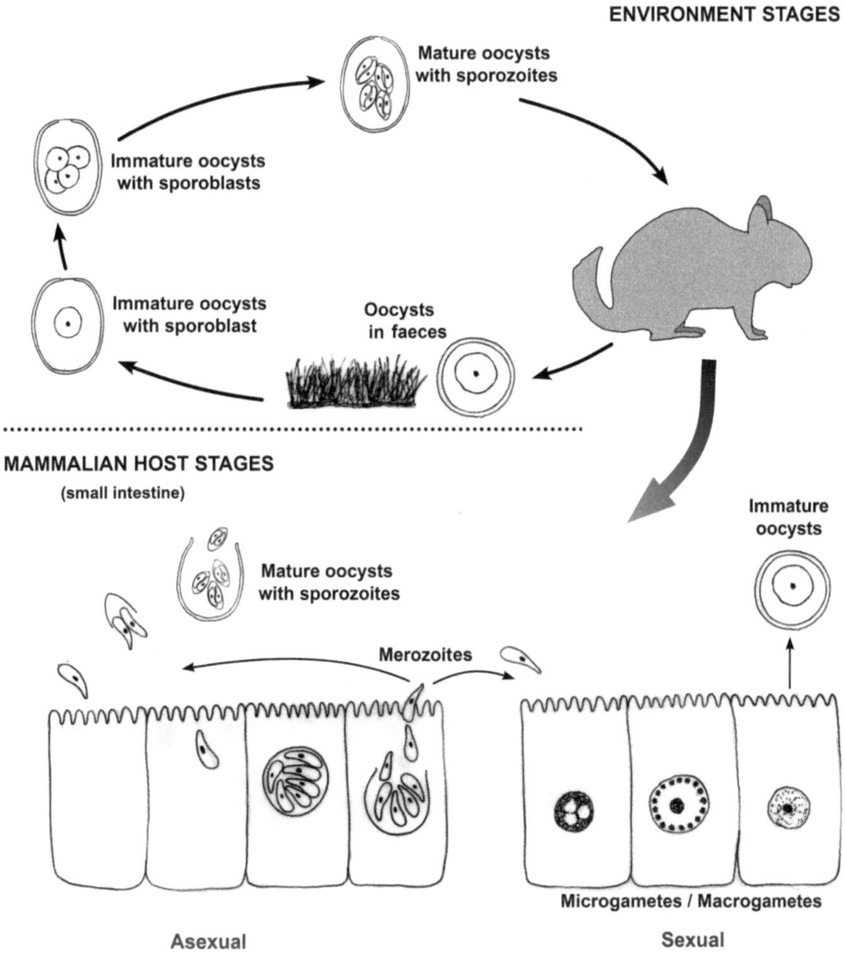

Fig. 9.4 Life cycle of *Eimeria* sp.: plains vizcacha (host) becomes infected with ingestion of mature oocysts (sporulated). The excystation occurs inside the small intestine, and sporozoites are released. Sporozoites penetrate into intestinal cells, and asexual (schizogony/merogony with the formation of merozoites) and sexual (gamogony with the formation of micro- and macrogametes) reproduction occurs. The fertilization between micro- and macrogametes results in the formation of immature oocysts that are released into the environment with the host's faeces. In the environment, oocysts mature, under optimal environmental conditions, into infective oocysts (sporogony)

departments from Entre Ríos Province; and General José de San Martín Department from Salta Province (Canova, 2023, "unpublished data").

Lagostonema ecasiense Sutton and Durette-Desset, 1987 (Nematoda, Molineidae), was originally recorded in the stomach and small intestine of *L. maximus* in Mendoza and Buenos Aires provinces (Sutton and Durette-Desset 1987). Posteriorly, this species was recorded in Caleu Caleu [*P* = 97% (27/28) and MI = 636] and Toay [*P* = 80% (8/10) and MI = 6] departments from La Pampa

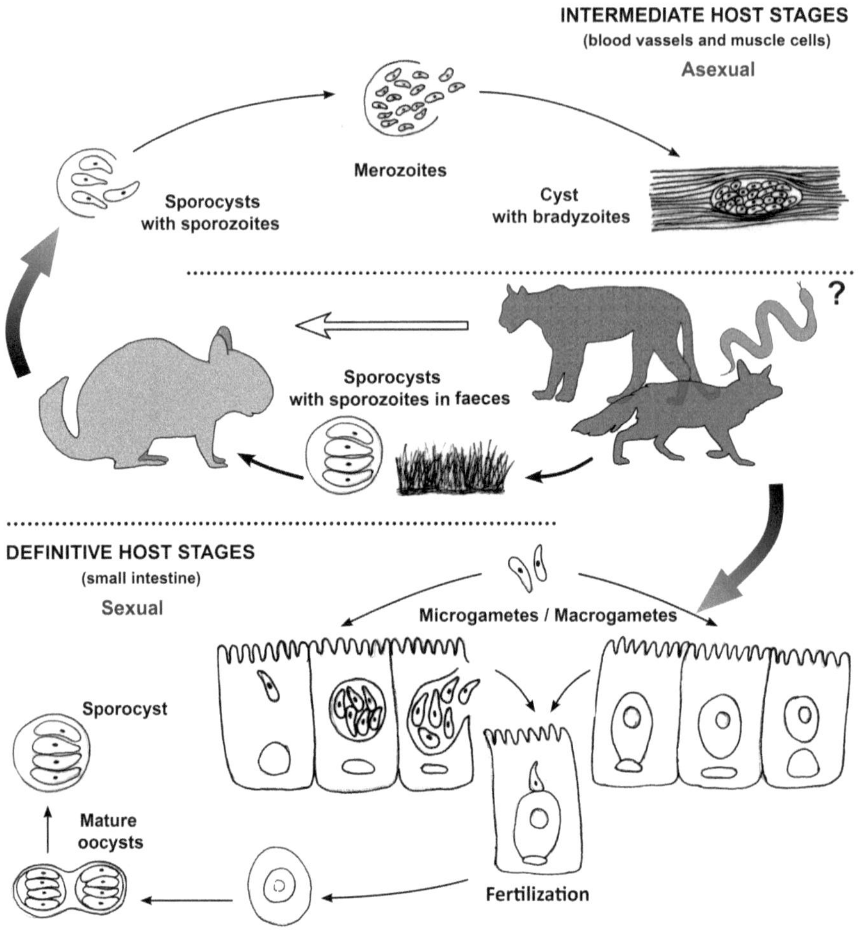

Fig. 9.5 Life cycle of *Sarcocystis* sp.: plains vizcacha (IH), the prey, becomes infected with the ingestion of sporocysts. Sporocyst rupture occurs by releasing sporozoites that enter in endothelial cells of blood vessels, giving rise to asexual reproduction (schizogony/merogony). The merozoites invade muscle cells and develop into sarcocysts containing bradyzoites. Carnivores/omnivores (DH), the predator, becomes infected with the ingestion of IH's tissues with sarcocysts. Excystation occurs inside the small intestine, and bradyzoites are released invading intestinal cells. Bradyzoites differentiate into macro- and microgametes (gamogony) and sexual reproduction occurs. The fusion of macro- and microgametes results in the formation of oocysts/sporocysts that are released into the environment with the host's faeces

Province (Foster et al. 2002); Aguirre Department from Santiago del Estero Province [*P* = 100% (13/13), MI = 313.8 and MA = 313.8]; Concordia [*P* = 80% (4/5), MI = 22.75 and MA = 18.2] and Federal [*P* = 100% (12/12), MI = 173.3 and MA = 173.3] departments from Entre Ríos Province; Bahía Blanca and Tornquist [*P* = 91.7% (11/12), MI = 55.7 and MA = 51.1], Berazategui [*P* = 100% (12/12), MI = 53.9 and MA = 53.9], San Cayetano [*P* = 16.7% (1/6), MI = 9 and MA = 1.5],

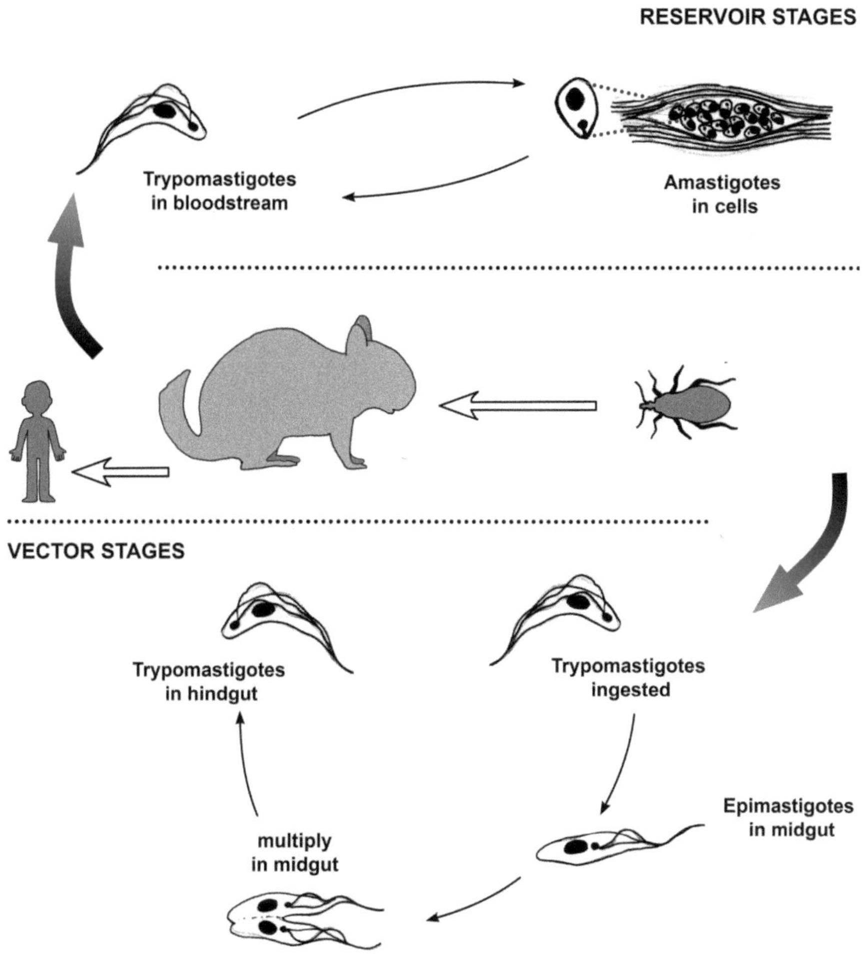

Fig. 9.6 Life cycle of *Trypanosoma cruzi*: two possible infection routes of plains vizcachas (reservoir): (**a**) triatomine insect vector takes a blood meal and releases trypomastigotes in its faeces near the site of the bite wound and (**b**) ingestion of infected triatomine insect vectors or food contaminated with their faeces contaminated with trypomastigotes. Inside the plains vizcacha, the trypomastigotes invade cells (e.g. muscle and digestive) and differentiate into intracellular amastigotes. The amastigotes multiply and differentiate into trypomastigotes that are released into the circulation. Trypomastigotes can enter another cell or be ingested by another vector. Triatomine insect vector becomes infected by feeding reservoir blood that contains circulating trypomastigotes. The ingested trypomastigotes transform into epimastigotes in the vector's midgut. Epimastigotes differentiate into infective metacyclic trypomastigotes in the hindgut. In the case reported by Toso et al. (2011) and Sangenis et al. (2016), the child could have contracted *T. cruzi* from eating raw or undercooked meat and ingesting infective trypomastigotes

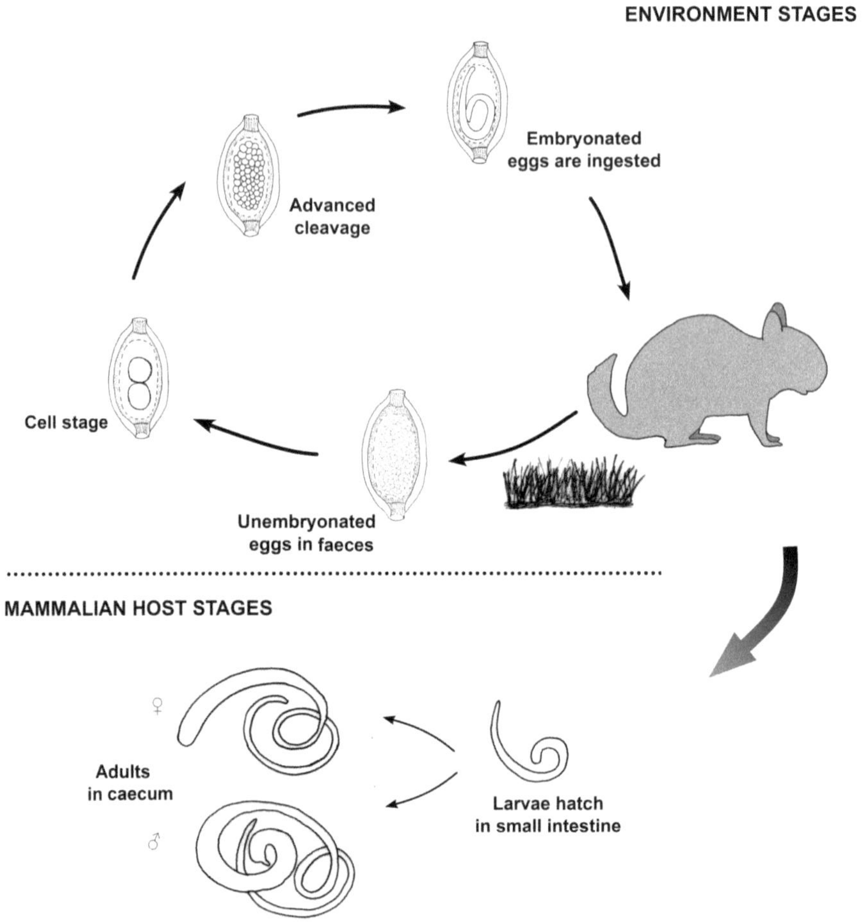

Fig. 9.7 Life cycle of *Trichuris* sp.: plains vizcacha becomes infected with the ingestion of embryonated eggs. Eggs hatch and release larvae inside the small intestine, which mature and establish as adults in the caecum. The adult worms copulate, and unembryonated eggs are released in the environment with the faeces' host. In the environment, eggs mature, under optimal environmental conditions, into the infective stage

Magdalena, Punta Indio, Tres Arroyos and La Costa departments from Buenos Aires Province; and Córdoba Province [*P* = 25% (1/4), MI = 16 and MA = 4] (Canova, 2023, "unpublished data") (Fig. 9.2a).

Viannella viscaciae Goodey, 1925 (Nematoda, Viannaiidae), was originally recorded in the small intestine of *Lagidium viscacia* (Molina, 1782) [named *Viscacia viscacia* by Goodey (1925)] in South America (Goodey 1925). Posteriorly, this species was recorded in *L. maximus* from Coronel Pringles [*P* = 83.5% (66/79)] and General Pedernera [*P* = 77.2% (17/22)] departments with a MA = 2461 to both sites from San Luis Province (Rossanigo et al. 1986); Caleu Caleu [*P* = 97% (27/28) and MI = 883] and Toay [*P* = 100% (10/10) and MI = 500] departments from La Pampa

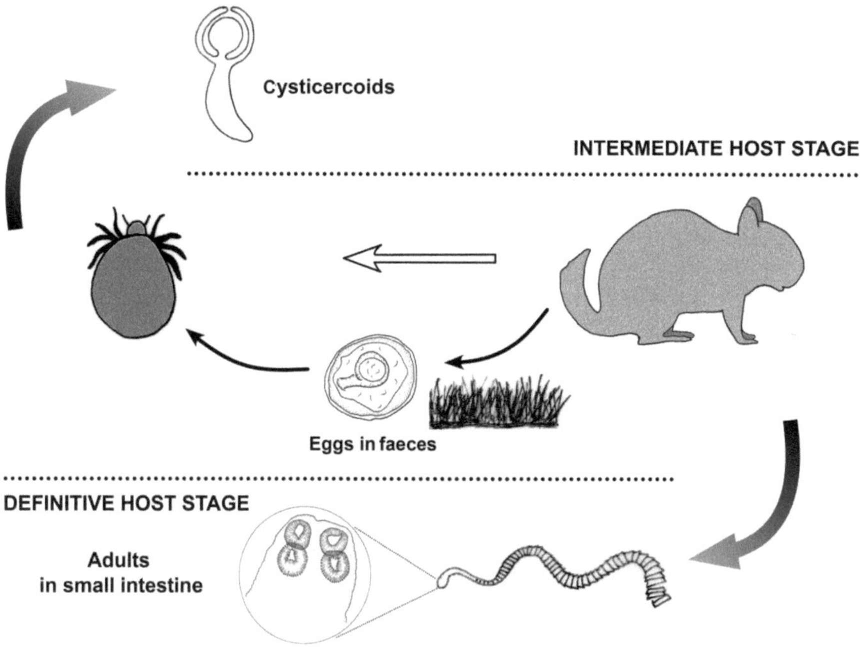

Fig. 9.8 Life cycle of *Monoecocestus* sp.: eggs of *Monoecocestus* sp. are released into the environment with the faeces' plains vizcacha (DH). The mature eggs are ingested by arthropods (IH). Inside IH, oncospheres are released from the eggs and penetrate the intestinal wall, which develops into cysticercoid larvae. Plains vizcacha (DH) becomes infected after ingestion of IH carrying the cysticercoid larvae. After ingestion, the arthropod's tissues are digested and the cysticercoid larvae are released in the stomach and small intestine. Maturation of the parasites into adult worms occurs. Eggs are released in the environment with the faeces' host

Province (Foster et al. 2002); Aguirre Department from Santiago del Estero Province [*P* = 100% (13/13), MI = 55.2 and MA = 55.2]; Bahía Blanca and Tornquist [*P* = 100% (12/12), MI = 480 and MA = 480], Berazategui [*P* = 91.7% (11/12), MI = 313.3 and MA = 287.2], San Cayetano [*P* = 16.7% (1/6), MI = 4 and MA = 0.7], Magdalena, Punta Indio, Tres Arroyos and La Costa departments from Buenos Aires Province; and Córdoba Province [*P* = 25% (1/4), MI = 13 and MA = 3.25] (Canova, 2023, "unpublished data"). Also, this nematode species was recorded in Concordia [*P* = 80% (4/5), MI = 318.5 and MA = 254.8] and Federal [*P* = 100% (12/12), MI = 875.2 and MA = 875.2] departments from Entre Ríos Province (Canova, 2023, "unpublished data"). Previously, this species was found in the same province by Martínez (1988).

Ascaris lumbricoides Linnaeus, 1758 (Nematoda, Ascarididae), was recorded in the small intestine of plains vizcachas in Coronel Pringles Department from San Luis Province [*P* = 1.2% (1/79) and MA = 0.09] (Rossanigo et al. 1986). In this case, this nematode species is considered an accidental parasite of plains vizcacha (Rossanigo et al. 1986).

Oxyuris sp. (Nematoda, Oxyuridae) was recorded in plains vizcachas from Entre Ríos Province (Martínez 1988).

Wellcomia hugoti Canova and Robles, 2021 (Nematoda, Oxyuridae), was originally recorded in the caecum and ascending colon of *L. maximus* in Concordia and Federal [*P* = 28.6% (2/7), MI = 959 and MA = 274] departments from Entre Ríos Province (Canova et al. 2021) (Fig. 9.2b).

Heteroxynema (*Cavioxyura*) *viscaciae* Hugot and Sutton, 1989 (Nematoda, Heteroxynematidae), was originally recorded in *Lagidium viscacia boxi* Thomas, 1921 (Hugot and Sutton 1989), from Languiñeo Department, Chubut Province. Also, it was recorded in plains vizcachas in Caleu Caleu Department from La Pampa Province [*P* = 75% (21/28) and MI = 14] (Foster et al. 2002) and Chamical Department from La Rioja Province [*P* = 40% (2/5)] (Ferreyra et al. 2007).

A new species of *Heteroxynema* (*Cavioxyura*) (Nematoda, Heteroxynematidae) was recorded in the caecum and ascending colon of plains vizcachas in Aguirre Department from Santiago del Estero Province [*P* = 76.9% (10/13), MI = 32.4 and MA = 24.9]; Federal Department from Entre Ríos Province [*P* = 75% (9/12), MI = 3.4 and MA = 2.6]; Bahía Blanca and Tornquist [*P* = 83.3% (10/12), MI = 24 and MA = 20], Magdalena, Punta Indio, La Costa and Tres Arroyos departments from Buenos Aires Province; General José de San Martín Department from Salta Province; and Córdoba Province [*P* = 25% (1/4), MI = 2 and MA = 0.5] (Canova, 2023, "unpublished data").

Trichuris sp. (Nematoda, Trichuridae) was recorded in the caecum of plains vizcachas in Caleu Caleu and Toay departments from La Pampa Province with *P* = 7% (2/28) and 30% (3/10), respectively, and a MI = 1 in both sites (Foster et al. 2002); Aguirre Department from Santiago del Estero Province [*P* = 7.7% (1/13), MI = 1 and MA = 0.1]; and Tornquist [*P* = 58.3% (7/12), MI = 1.4 and MA = 0.8], Berazategui [*P* = 8.3% (1/12), MI = 2 and MA = 0.2], Punta Indio, and La Costa departments from Buenos Aires Province (Canova, 2023, "unpublished data"). Also, this nematode species was found in Concordia [*P* = 20% (1/5), MI = 1 and MA = 0.2] and Federal [*P* = 33.3% (4/12), MI = 3.5 and MA = 1.2] departments from Entre Ríos Province (Canova, 2023, "unpublished data") (Fig. 9.2c). Previously, this taxon was found in the same province by Martínez (1988).

Viscachataenia quadrata (von Linstow, 1904) (Eucestoda, Anoplocephalidae) was recorded in the small intestine of plains vizcachas as *Cittotaenia pectinata* (Goeze, 1782; Rossanigo et al. 1986). Later, Denegri et al. (2003) reviewed different species and suggested synonymizing the names *Cittotaenia pectinata* (Goeze, 1782); *Cittotaenia quadrata* von Linstow, 1904; *Mosgovoyia viscaciae* Spasskii, 1951; *C. viscaciae* (Spasskii, 1951) Beveridge, 1978; and *Bertiella finlayi* Mazza, Parodi and Fiora, 1932 (in part), with *Viscachataenia quadrata* (von Linstow, 1904). Rossanigo et al. (1986) found this species in the small intestine of plains vizcachas in Coronel Pringles [*P* = 5% (4/79)] and General Pedernera [*P* = 13.6% (3/22)] departments with a MA = 1 to both sites from San Luis Province.

Monoecocestus sp. (Eucestoda, Anoplocephalidae) was recorded in the small intestine of plains vizcachas in Caleu Caleu [*P* = 29% (8/28) and MI = 5] and Toay [*P* = 10% (1/10) and MI = 2] departments from La Pampa Province (Foster et al.

2002), Aguirre Department from Santiago del Estero Province, Federal Department from Entre Ríos Province and Tornquist Department from Buenos Aires Province (Canova, 2023, "unpublished data") (Fig. 9.2d).

Cyclophyllidea gen. sp. (Eucestoda) was recorded in the small intestine of plains vizcachas in Aguirre Department from Santiago del Estero Province and Bahía Blanca Department from Buenos Aires Province (Canova, 2023, "unpublished data").

9.3 Animal and Human Health Considerations

The finding of some protozoa of health importance reveals the role of plains vizcacha in the transmission of zoonotic (species transmitted from animals to humans) and/or enzootic pathogens (species transmitted between animals) such as the cases of *T. cruzi* and *Sarcocystis* spp. (Herrera 2010; Noya and González 2015; Rosenthal 2021).

Trypanosoma cruzi, the causal agent of Chagas disease or American trypanosomiasis, is a flagellated protozoan that infects mammals and is transmitted by several species of hematophagous triatomine insects (Hemiptera, Reduviidae, Triatominae) that act as vectors, being *Triatoma*, *Panstrongylus* and *Rhodnius* the genera of greatest epidemiological importance (Herrera 2010; Toso et al. 2011; Filigheddu et al. 2017; Busselman and Hamer 2022; Chacón et al. 2022). Among these, *Triatoma infestans* (Klug, 1834) presents high adaptability to the domestic environment and is considered the main vector of Chagas disease in Argentina, Bolivia, Brazil, Chile, Paraguay, Peru and Uruguay (Chacón et al. 2022).

When the triatomine vector feeds on the blood of a parasitized mammalian host, it ingests the trypomastigote stage. Once in the midgut of the vector, the trypomastigotes differentiate into epimastigotes (the extracellular and replicative stage in the triatomine), which migrate to the hindgut of the vector where they differentiate into trypomastigotes. These forms are excreted in the vector's faeces and are infective to mammalian hosts. When the triatomine feeds, it ingests blood and simultaneously defecates, eliminating the infective trypomastigotes that enter the mammal through the wound caused by the bite or through mucous membranes. Once in the mammal's blood, the trypomastigotes invade cells from various tissues (e.g. muscle and digestive) and differentiate into intracellular amastigotes (the replicative stage in vertebrates). After several replicative cycles, amastigotes differentiate into trypomastigotes that are released into the circulation and can infect other cells or a triatomine vector (Toso et al. 2011; Murillo-Godínez 2018; Busselman and Hamer 2022) (Fig. 9.6).

The form of vectorial transmission of *T. cruzi* through hematophagous triatomines (previously described) is considered the main one; however, the parasite can be transmitted through other routes, such as transfusion, transplacental, orally, through infected organ transplants or laboratory accidents. In these ways, there has been an increase in oral transmission, which can occur by ingesting drinks or food contaminated with crushed parasitized vectors, their droppings or anal secretions of infected mammals (e.g. marsupials), by ingesting blood or meat (raw or partially cooked) from infected mammalian reservoirs such as armadillos (*Dasypus* spp.) or through milk during lactation (Toso et al. 2011; Rueda et al. 2014; Filigheddu et al. 2017; Murillo-Godínez 2018; Busselman and Hamer 2022). Two cases of oral transmission of *T. cruzi* by consumption of raw or undercooked meat or blood involving wild mammals *Dasypus* sp., *Dasyprocta* sp. and *L. maximus* have been reported in children from Argentina (Toso et al. 2011; Sangenis et al. 2016). Among the reservoirs, more than 150 species of wild, domestic and synanthropic mammals, belonging to the orders Didelphimorphia, Lagomorpha, Chiroptera, Rodentia, Pilosa, Cingulata, Carnivora, Primates and Perissodactyla, are naturally infected with *T. cruzi*. Rodents include both wild [e.g. *Akodon* sp., *Calomys expulsus* (Lund, 1841), *Cavia* sp., *Dasyprocta leporina* (Linnaeus, 1758), *Galea spixii* (Wagler, 1831), *Nectomys squamipes* (Brants, 1827), *Oligoryzomys stramineus* Bonvicino and Weksler, 1998, and *Thrichomys apereoides* (Lund, 1839)] and synanthropic species [e.g. *Rattus rattus* (Linnaeus, 1758) *and Rattus norvegicus* (Berkenhout, 1769)] (Herrera 2010; Toso et al. 2011; Noya and González 2015; Filigheddu et al. 2017; Murillo-Godínez 2018; Chacón et al. 2022; Morales et al. 2023). More studies are necessary to evaluate the role of *L. maximus* as a reservoir and potential transmitter of *T. cruzi* to other animals and humans, as well as the role that vizcacheras may play in the maintenance of *T. infestans*.

Sarcocystis spp. are worldwide-distributed protozoan parasites closely related to other apicomplexans such as *Besnoitia* spp., *Neospora caninum* Dubey, Carpenter, Speer, Topper and Uggla, 1988, and *Toxoplasma gondii* (Dubey et al. 2016). Most species have a predator-prey life cycle, with one intermediate host (IH) or prey, where tissue cysts (sarcocysts) are produced in the muscles, and carnivorous or omnivorous as definitive hosts (DH) where oocysts/sporocysts are produced in the intestinal mucosa (Dubey et al. 2016; Rosenthal 2021). The IH becomes infected by the ingestion of food or water contaminated with the faecal matter of the DH infected with oocysts. Once ingested by IH, the parasite will reproduce asexually with the formation of tissue cysts (sarcocysts) in the muscles. The DH becomes infected by feeding on IH with sarcocysts in its tissues. In DH, the cycle takes place in the small intestine, where the sexual reproduction of the parasite occurs with the formation of oocysts that will be released into the environment with faecal matter (Dubey et al. 2016, Rosenthal 2021) (Fig. 9.5). Around 200 species of this genus infecting mammals, birds and reptiles around the world have been reported (Robben and Sibley 2004; Dubey et al. 2016). On the other hand, rodents from different families (e.g. Muridae, Cricetidae and Caviidae) act as natural IH of at least 46 *Sarcocystis* species, which have carnivorous mammals, raptors and snakes as DH (Dubey et al. 2016; Prakas et al. 2019; Rudaitytė-Lukošienė et al. 2022). From a zoonotic point of view, humans can act as DH for *S. hominis* (Railliet and Lucet, 1891) and *S. heydorni* Dubey, Wilpe, Calero-Bernal, Verma and Fayer, 2015 (cattle as IH), and *S. suihominis* (Tadros and Laarman, 1976) (swine as IH). In addition, humans can act as IH of *S. nesbitti* with snakes as DH (Fayer et al. 2015; Dubey et al. 2016; Rosenthal 2021). However, contact with other animals could expose humans to other yet undiscovered *Sarcocystis* species (Rosenthal 2021). Some *Sarcocystis* species are clinically important for humans and domestic and wild animals due to their pathogenic effects, such as abortion, weight loss, encephalitis and myositis (Kirillova et al. 2018).

The presence of tissue cysts places *L. maximus* as an intermediate host of the *Sarcocystis* species, which is consistent with its herbivorous feeding mode. More studies are necessary to confirm the number of *Sarcocystis* species present in *L. maximus* and their taxonomic identity, the DH and the zoonotic/enzootic potential. On the other hand, the molecular similarities observed between *Sarcocystis* spp. of *L. maximus* and those of other hosts could suggest that they have some interaction in the life cycle (Canova et al. 2023).

Regarding helminths, they present a great diversity of life cycles (with one or more hosts) and modes of infection and ecological characteristics (Georgiev et al. 2006; Morand et al. 2006) (Figs. 9.7 and 9.8). Some helminths have medical-veterinary implications [e.g. *Trichinella spiralis* (Owen, 1835), *Toxocara canis* (Werner, 1782), *Trichuris trichiura* (Linnaeus, 1771), *Necator americanus* (Stiles, 1902), *Enterobius vermicularis* (Linnaeus, 1758), *Fasciola hepatica* Linnaeus, 1758, *Taenia solium* Linnaeus, 1758, *and Echinococcus granulosus* (Batsch, 1786)] and can generate health risks and economic losses (Karesh et al. 2012; Apt 2014; Vidal-Anzardo et al. 2020; Lima et al. 2021; Sánchez et al. 2022). Likewise, rodents can act as intermediate or definitive hosts of numerous species of helminths, including some of epidemiological importance [e.g. *Angiostrongylus cantonensis* (Chen, 1935), *Rodentolepis nana* (von Siebold, 1852), *Hymenolepis diminuta* (Rudolphi, 1819), *Calodium hepaticum* (Bancroft, 1893), *Echinococcus multilocularis* (Leuckart, 1863), *Hydatigera taeniaeformis* (Batsch, 1786), *and Toxocara* spp.] (Stojcevic et al. 2004; Easterbrook et al. 2007; Reperant et al. 2009; Hancke et al. 2011; Lima et al. 2021).

About the helminths recorded in *L. maximus*, none had been considered of health importance. However, it is known that the interactions between parasites and hosts can be altered because of environmental modifications (Budria and Candolin 2014). Thus, some parasites that are rarely pathogenic can become important population regulators when hosts are stressed (Eberhardt et al. 2013). In this sense, prolonged stress situations decrease the immune function and the general physiological condition of the hosts, making them more susceptible to infections, which further weakens and predisposes them to infections that can be more serious (Beldomenico and Begon 2009, 2015; Eberhardt et al. 2013). Moreover, this situation can generate a modification of the parasite assemblages, which can become made up of new parasite species, a scenario that suggests that these hosts can act as potential reservoirs of pathogenic species that can reach humans and become diseases of zoonotic importance (Beldomenico et al. 2008; Pedersen and Greives 2008; Beldomenico and Begon 2009, 2015; Eberhardt et al. 2013; Glidden et al. 2021). Likewise, given the synergistic effect of stress and parasitic infections, it is expected that stressed individuals and, therefore, those with poorer physiological condition will have more intense infections since the parasites will find less opposition to their survival and proliferation (Beldomenico and Begon 2009, 2015).

On the other hand, parasites can respond directly (stressful effects on parasite stages) or indirectly (stressful effects on hosts) to environmental stressors. For example, in a scenario of environmental degradation, the number, composition and abundance of species in a community will change with respect to safe environment. Due to the complexity of the interaction with stress factors or environmental conditions in general, parasites can reflect those states or conditions, and therefore, many of them can be used as bioindicators to assess environmental conditions (Sures et al. 2023).

To advance in the analysis of these possibilities, we will develop the following points of this chapter, mainly addressing studies on helminths.

9.4 Structure of Parasitic Populations and Communities

The presence of a parasitic species in a host is determined by numerous environmental factors (e.g. soil, temperature and humidity) and host characteristics (e.g. taxonomy, age, sex, geographic distribution, phylogenetic relationships and immunological, physiological and hormonal conditions) (Arneberg et al. 1998; Esch and Fernández 1993; Poulin and Morand 2000, 2004; Arneberg 2002; Stojcevic et al. 2004; Cattadori et al. 2006).

The description of parasite diversity can be presented at different scales of study. All parasites of different species within the same host form an infracommunity whose composition can be determined by the interactions between the species included (Holmes and Price 1986). The set of infracommunities of a host population is known as the component community (Holmes and Price 1986). While infracommunities are considered short-lived since they remain until the host dies, the component community is considered long-lived since it can last for several generations. However, the composition of component community can change with the extinction of some species and the migration of others from different populations of the same or different host. In this way, it can be observed that parasites can be studied at different levels that allow to give a notion at different spatial and temporal scales of the associations between the different parasite species and with their hosts, adding to the study assembled over the entire phylogenetic history by evolutionary processes such as host switching and extinction (Poulin 1997).

In addition to the study at the infracommunity and component community, parasites can be studied at the population level where individuals of a single species that parasitize all the individuals of the host population in a given place and time are considered. The aggregation of parasites within infrapopulations, that are, groups of conspecific parasites infecting single host individuals, has attracted extensive attention due to its hypothesized consequences for the ecology and evolution of host-parasite interactions (Anderson and May 1978; Jaenike 1996; Poulin 2006; Morrill et al. 2017). In this sense, most potential hosts are uninfected or harbour a few parasites and a small number of hosts carry most of the parasite population (Shaw and Dobson 1995; Poulin 2013).

Despite being a parasitology law, parasite aggregation still shows considerable variation from moderate to extreme cases (Poulin 2007, 2013; Johnson and Hoverman 2014), likely due to variation in the type or strength of its underlying causes (Poulin 2007). Factors promoting aggregation such as reproduction and recruitment of parasites on hosts (Grear and Hudson 2011), the attraction of infective stages of parasites to one another and environmental aggregation (Morrill and Forbes 2016) and host condition-linked immunity (Morrill and Forbes 2012) must be considered alongside factors reducing aggregation such as density-dependent mortality of parasites (Anderson and May 1978; Luong et al. 2011) and parasite-induced host mortality (Anderson and Gordon 1982).

In this sense, the structure of parasitic populations and communities and the composition of parasite species in each host population can be influenced and modified by variations in biotic and abiotic factors, such as human activities, translocation of host species, contaminants and numerous variables related to different ecological aspects (e.g. habitat, ecology and environmental conditions) (Sapp and Esch 1994; Thieltges et al. 2008; Ibrahim 2012; Muñoz and Castro 2012).

On the other hand, parasite communities can be largely shaped by host assemblages and environment in two different ways. First, the environment shapes the host assemblage and this shapes the parasite community. Second, environmental variables could directly shape parasite assemblages (Berkhout et al. 2020). In this second case, environmental variables such as precipitation, soil type and temperature can directly influence the survival of a parasite; for example, excessively high temperatures can cause stress and desiccation of eggs and larvae (Poulin and Kamiya 2015; Cizauskas et al. 2017; Berkhout et al. 2020).

In both aquatic and terrestrial systems, the biotic characteristics of a locality and the composition of its biota are generally very similar to those of nearby localities within a radius of a few kilometres, but very different from those located hundreds or thousands of kilometres away (Poulin and Kamiya 2015). In this sense, between host populations parasite abundance or community similarity can vary widely, especially if host populations are distant from each other (Poulin and Kamiya 2015; Berkhout et al. 2020).

Regarding the structure of parasite populations and communities of *L. maximus*, Rossanigo et al. (1986) found five species of helminths (four nematodes and one cestode) and differences in parasite prevalence in plains vizcacha populations from two sites in San Luis Province, Argentina. On the other hand, Foster et al. (2002) found seven species of helminths (six nematodes and one cestode), showing differences in prevalence and parasite intensities in plains vizcacha populations from two sites in La Pampa Province, Argentina. In addition, for nematode *Heteroxynema (Cavioxyura) viscaciae* differences in prevalence and intensity in two seasons of the year were found (Foster et al. 2002).

Exploratory studies showed differences in ecological parameters when comparing the parasite fauna of *L. maximus* from sites with contrasting environmental characteristics, which correspond to different ecoregions of Argentina. Among the ecological parameters, differences in the values of abundance, mean abundance, mean intensity, specific richness, diversity, evenness and dominance of the parasitic communities between the sampling sites were observed. On the other hand, these exploratory studies showed that there is no strong association between parasite loads and host characteristics (weight, body length and digestive tract length), but a certain trend dependent on environmental characteristics was observed (Canova, 2023, "unpublished data"). Some of these exploratory results obtained are detailed in greater depth in the following section.

9.5 Potential Use of the Parasitic Species as Biotags

Several works have used parasites as biological tags for the identification and discrimination of different host populations through the results of the richness, composition and abundance of parasite species. In this context, a parasite can be present at one site and absent at another, or parasites can be present at different sites, but more prevalent or abundant at one site than another one (Zischke et al. 2013; Catalano et al. 2014; Bray and Cribb 2015; Cantatore and Timi 2015; Poulin and Kamiya 2015; Alarcos et al. 2016; Irigoitia et al. 2017).

Exploratory studies on the parasite fauna of *L. maximus* showed differences when comparing sites with a significant anthropic impact (semi-captivity state) in relation to sites with less anthropic impact. The most anthropized site showed the lowest species richness and the highest evenness, showing a more equitable distribution among the parasitic species present, than the less anthropized sites. On the other hand, the most anthropized site showed central helminth species with high values of P, MI and MA, while the less anthropized sites, in addition to presenting central species, had satellite species with lower values of P, MI and MA (Canova, 2023, "unpublished data") (Fig. 9.9). This scenario agrees partially with results observed from other rodent species (e.g. capybaras) (Eberhardt et al. 2013) and with what is proposed by Beldomenico and Begon (2009, 2015) and Sures et al. (2023) in a stressful environment (like cases of plains vizcachas in semi-captivity) where the hosts show a lower specific richness and a higher parasite intensity.

On the other hand, different abundances of three taxa of nematodes present in all the populations of *L. maximus* analysed were observed. Likewise, the presence of some parasite species in some sites and absence in others were also observed. In this way, the plains vizcacha parasites allowed us to differentiate populations of this rodent with different environmental characteristics, being able to act as biological tags (Canova, 2023, "unpublished data").

In this way, the parasitological information not only allows us to differentiate populations of *L. maximus* that occur in contrasting environment but also could indicate possible stress scenarios.

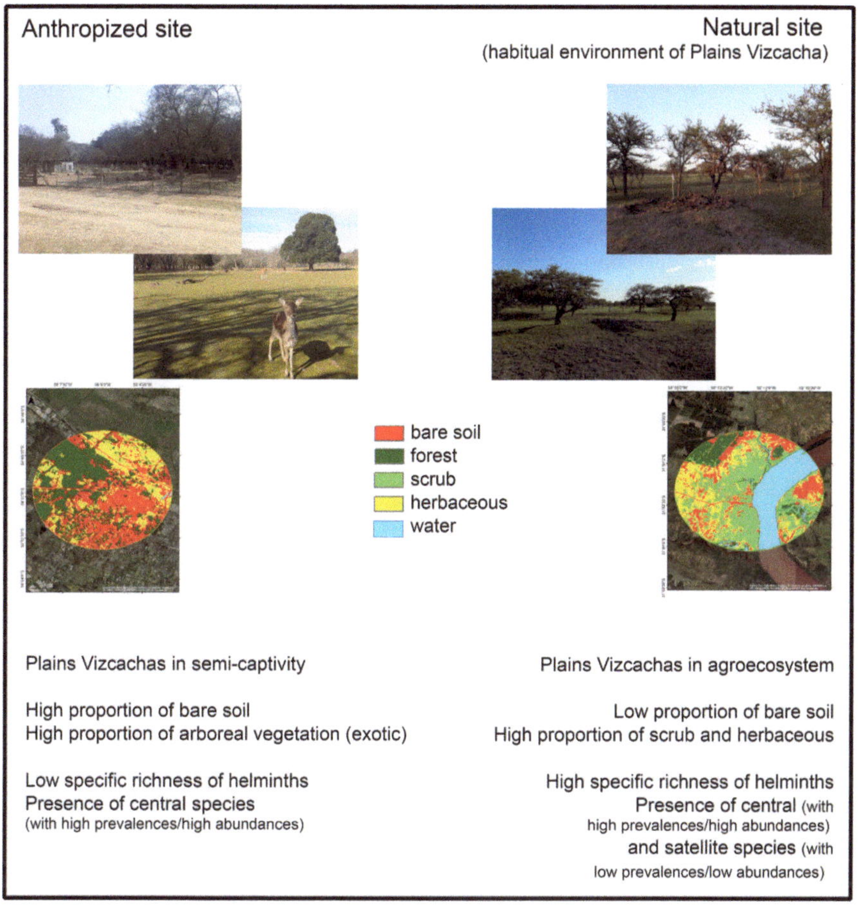

Fig. 9.9 Characteristics of anthropized and natural sites observed in exploratory studies on the parasite fauna of plains vizcacha (Canova, 2023, "unpublished data")

9.6 Conclusions

Wildlife, throughout history, has been an important source of infectious diseases transmitted to humans, giving rise to zoonoses (Ferri 2021; González-Barrio 2022). Currently, it is estimated that 61.6% of the pathogens that affect humans are zoonotic, a context in which wild animals seem to be involved, acting as reservoirs for the transmission of infectious agents not only for humans but also for domestic animals (González-Barrio 2022; Sánchez et al. 2022). Among the factors associated with the emergence of zoonoses are anthropic changes in land use (with the replacement of areas of native forest and grasslands by plantations of exotic trees, deforestation, urbanization, industrial development, management of water channels, increase in agricultural and livestock production), the consequences derived from climate change that directly affect the ecology of vectors and reservoirs, animal trade, translocation of species, hunting and the trade in meat of wild animals (Manzoli et al. 2011; Glidden et al. 2021; Rupasinghe et al. 2022; Sánchez et al. 2022). These factors can influence the transmission and life cycle of parasites and modify parasite-host interactions through a variety of mechanisms that involve changes in the abundance, behaviour and immune response of the hosts, the composition and structure of the host community and the encounter rate and/or parasite/host compatibility (Budria and Candolin 2014; Glidden et al. 2021; Riquelme et al. 2021). Therefore, anthropogenic changes can be an important factor not only for the loss of biodiversity but also for the disturbance of the balance that could expose wildlife, domestic fauna and humans to new parasites and pathogens, interpreting it as a worrisome epidemiological scenario (Beldomenico and Begon 2009; Manzoli et al. 2011; Glidden et al. 2021; Riquelme et al. 2021; Rupasinghe et al. 2022; Calero-Bernal and García-Bocanegra 2023).

Rodents play an important role in the maintenance and spread of zoonoses, being able to act as reservoirs and/or hosts for various microorganisms that can affect humans and other animals (Jittapalapong et al. 2009; Lima et al. 2021; Riquelme et al. 2021). The presence of some parasites of zoonotic importance (e.g. *Trypanosoma cruzi*, *Toxoplasma gondii*, *Cryptosporidium* sp., *Sarcocystis* sp. and *Trichinella* sp.) in some groups of synanthropic and wild rodents allows warning of the probability of infection in other less explored groups distributed in different habitats (Herrera 2010; Ribicich et al. 2010; Palmezano-Díaz et al. 2015; Dubey et al. 2016; Riquelme et al. 2021; Xu et al. 2022).

In this context, *L. maximus* is involved as a wild rodent that can be found in close contact with other wild animals, as well as with domestic animals and humans, being able to act as a host and/or reservoir for a wide variety of parasites and pathogens. Likewise, it highlights not only the importance of the study of the parasite fauna of this rodent from the point of view of biodiversity or the importance in medical-veterinary health but also from the value that parasites have in providing us with information on different populations of this mammal found in sites with contrasting environmental characteristics, including those that can generate stressful situations. For all this, the importance of deepening the knowledge of the parasitosis of different wild hosts is highlighted, both from the perspective of biodiversity and from aspects of sanitary importance, for which a better understanding of the distribution of the species is necessary, of the epidemiology, the mechanisms of transmission, the diagnosis, the prevention and the control of the same (OPS 2003; Poulin and Morand 2004).

The analysis carried out throughout this chapter allows us to observe the close relationship that exists between wildlife, the environment and humans in the transmission of different parasites and pathogens. In relation to this, the concept of One Health arises, which recognizes the interconnectivity between human, animal and environmental health (He et al. 2022; Ogunseitan 2022; Tull et al. 2022). In this regard, it is essential to adopt a multidisciplinary approach to explore the social dimensions and human behaviours associated with disease transmission and understand the circumstances in which zoonotic diseases emerge and spread (He et al. 2022).

References

Acuña F, Barbeito CG, Miglino MA et al (2021) Estudio de la muerte embrionaria en mamíferos a través de un modelo no convencional: *Lagostomus maximus*, vizcacha de llanura (Rodentia, Chinchillidae). Anal Acad Nac Agron Veterin LXXIV 74:1–9

Acuña F, Barbeito CG, Portiansky EL et al (2022) Spontaneous embryonic death in plains viscacha (*Lagostomus maximus* – Rodentia), a species with unique reproductive characteristics. Theriogenology 185:88–96. https://doi.org/10.1016/j.theriogenology.2022.03.024

Alarcos AJ, Pereira AN, Taborda NL et al (2016) Parasitological evidence of stocks of *Paralichthys isosceles* (Pleuronectiformes: Paralichthyidae) at small and large geographical scales in South American Atlantic coasts. Fish Res 173:221–228. https://doi.org/10.1016/j.fishres.2015.07.018

Anderson RM, Gordon DM (1982) Processes influencing the distribution of parasite numbers within host populations with special emphasis on parasite-induced host mortalities. Parasitology 85(2):373–398. https://doi.org/10.1017/S0031182000055347

Anderson RM, May RM (1978) Regulation and stability of host-parasite population interactions: I. Regulatory processes. J Anim Ecol 47(1):219–247. https://doi.org/10.2307/3933

Apt BW (2014) Infecciones por parásitos más frecuentes y su manejo. Rev Méd Clín Las Condes 25(3):485–528. https://doi.org/10.1016/S0716-8640(14)70065-3

Arias SM, Madanes N, Quintana RD (2003) Estructura y composición de la vegetación en vizcacheras activas e inactivas en el delta del Paraná. Mastozool Neotrop 10(1):9–20

Arias SM, Quintana RD, Cagnoni M (2005) Vizcacha's influence on vegetation and soil in a wetland of Argentina. Rangel Ecol Manag 58(1):51–57. https://doi.org/10.2111/1551-5028(2005)58<51:VIOVAS>2.0.CO;2

Arneberg P (2002) Host population density and body mass as determinants of species richness in parasite communities: comparative analyses of directly transmitted nematodes of mammals. Ecography 25(1):88–94. https://doi.org/10.1034/j.1600-0587.2002.250110.x

Arneberg P, Skorping A, Grenfell B, Read AF (1998) Host densities as determinants of abundance in parasite communities. Proc R Soc B Biol Sci 265(1403):1283–1289. https://doi.org/10.1098/rspb.1998.0431

Aznar FJ, Raga JA, Corcuera J, Monzón F (1995) Helminths as biological tags for franciscana (*Pontoporia blainvillei*) (Cetacea, Pontoporiidae) in Argentinian and Uruguayan waters. Mammalia 59(3):427–435. https://doi.org/10.1515/mamm.1995.59.3.427

Balbuena JA, Aznar FJ, Fernández M, Raga JA (1995) Parasites as indicators of social structure and stock identity of marine mammals. Dev Mar Biol 4:133–139. https://doi.org/10.1016/S0163-6995(06)80017-X

Beldomenico PM, Begon M (2009) Disease spread, susceptibility and infection intensity: vicious circles? Trends Ecol Evol 25(1):21–27. https://doi.org/10.1016/j.tree.2009.06.015

Beldomenico PM, Begon M (2015) Interacciones entre el estrés, el parásito y el hospedador: ¿Un triángulo vicioso? FAVE Cs. Veterinarias 14:42–56. https://doi.org/10.14409/favecv.v14i1/3.5160

Beldomenico PM, Telfer S, Gebert S et al (2008) Poor condition and infection: a vicious circle in natural populations. Proc R Soc B Biol Sci 275:1753–1759. https://doi.org/10.1098/rspb.2008.0147

Berkhout BW, Borregaard MK, Brandl R et al (2020) Host assemblage and environment shape β-diversity of freshwater parasites across diverse taxa at a continental scale. Glob Ecol Biogeogr 29:38–49. https://doi.org/10.1111/geb.13005

Beveridge I (1994) Family Anoplocephalidae Cholodkovsky, 1902. In: Khalil LF, Jones A, Bray RA (eds) Keys to the cestode parasites of vertebrates. CAB International, Wallingford, pp 315–366

Beveridge I, Spratt DM, Durette-Desset MC (2014) Order Strongylida (Railliet & Henry, 1913). In: Schmidt-Rhaesa A (ed) Handbook of Zoology. Gastrotricha, Cycloneuralia and Gnathifera. Volume 2, Nematoda. De Gruyter, Berlin, pp 557–612

Branch LC (1993a) Social organization and mating system of the plains viscacha (*Lagostomus maximus*). J Zool 229:473–491. https://doi.org/10.1111/j.1469-7998.1993.tb02649.x

Branch LC (1993b) Seasonal patterns of activity and body mass in the plains vizcacha, *Lagostomus maximus* (family Chinchillidae). Can J Zool 71:1041–1045. https://doi.org/10.1139/z93-137

Branch LC (1993c) Intergroup and intragroup spacing in the plains vizcacha, *Lagostomus maximus*. J Mammal 74(4):890–900. https://doi.org/10.2307/1382427

Branch LC, Villarreal D, Fowler GS (1993) Recruitment, dispersal, and group fusion in a declining population of the plains vizcacha (*Lagostomus maximus*; Chinchillidae). J Mammal 74(1):9–20. https://doi.org/10.2307/1381901

Branch LC, Villarreal D, Fowler GS (1994a) Factors influencing population dynamics of the plains viscacha (*Lagostomus maximus*, Mammalia, Chinchillidae) in scrub habitat of central Argentina. J Zool 232(3):383–395. https://doi.org/10.1111/j.1469-7998.1994.tb01580.x

Branch LC, Villarreal D, Sbriller AP, Sosa RA (1994b) Diet selection of the plains vizcacha (*Lagostomus maximus*, family Chinchillidae) in relation to resource abundance in semi-arid scrub. Can J Zool 72:2210–2216. https://doi.org/10.1139/z94-296

Branch LC, Villarreal D, Sosa A et al (1994c) Estructura de las colonias de vizcacha y problemas asociados con la estimación de la densidad poblacional en base a la actividad de las vizcacheras. Mastozool Neotrop 1(2):135–142

Bray RA, Cribb TH (2015) Are cryptic species a problem for parasitological biological tagging for stock identification of aquatic organisms? Parasitology 142(1):125–133. https://doi.org/10.1017/S0031182014000092

Budria A, Candolin U (2014) How does human-induced environmental change influence host-parasite interactions? Parasitology 141(4):462–474. https://doi.org/10.1017/S0031182013001881

Bush AO, Lafferty KD, Lotz JM, Shostak AW (1997) Parasitology meets ecology on its own terms: Margolis et al. revisited. J Parasitol 83:575–583. https://doi.org/10.2307/3284227

Busselman RE, Hamer SA (2022) Chagas disease ecology in the United States: recent advances in understanding *Trypanosoma cruzi* transmission among triatomines, wildlife, and domestic animals and quantitative synthesis of vector-host interactions. Annu Rev Anim Biosci 10:325–348. https://doi.org/10.1146/annurev-animal-013120-043949

Calero-Bernal R, García-Bocanegra I (2023) Parasites and wildlife. Animals 13(628):1–4

Canova V, Robles MR, Abba AM (2021) A new species of *Wellcomia* (Nematoda: Oxyuridae) in the plains viscacha (Rodentia: Chinchillidae) from Argentina, an emended diagnosis and an update of the genus *Wellcomia*. Parasitol Res 120(3):929–940. https://doi.org/10.1007/s00436-020-06980-1

Canova V, Helman E, Robles MR et al (2023) First report of *Sarcocystis* spp. (Apicomplexa, Sarcocystidae) in *Lagostomus maximus* (Desmarest, 1917) (Rodentia, Chinchillidae) in Argentina. Int J Parasitol Parasites Wildl 20:180–186. https://doi.org/10.1016/j.ijppaw.2023.03.001

Cantatore DMP, Timi JT (2015) Marine parasites as biological tags in South American Atlantic waters, current status and perspectives. Parasitology 142:5–24. https://doi.org/10.1017/S0031182013002138

Carlson CJ, Burgio KR, Dougherty ER et al (2017) Parasite biodiversity faces extinction and redistribution in a changing climate. Sci Adv 3:e1602422. https://doi.org/10.1126/sciadv.1602422

Carlson CJ, Zipfel CM, Garnier R, Bansal S (2019) Global estimates of mammalian viral diversity accounting for host sharing. Nat Ecol Evol 3:1070–1075. https://doi.org/10.1038/s41559-019-0910-6

Carlson CJ, Dallas TA, Alexander LW et al (2020) What would it take to describe the global diversity of parasites? Proc R Soc B 287:20201841. https://doi.org/10.1098/rspb.2020.1841

Catalano SR, Whittington ID, Donnellan SC, Gillanders BM (2014) Parasites as biological tags to assess host population structure: guidelines, recent genetic advances and comments on a holistic approach. Int J Parasitol Parasites Wildl 3:220–226. https://doi.org/10.1016/j.ijppaw.2013.11.001

Cattadori IM, Haukisalmi V, Henttonen H, Hudson PJ (2006) Transmission ecology and the structure of parasite communities in small mammals. In: Morand S, Krasnov BR, Poulin R (eds) Micromammals and macroparasites: from evolutionary ecology to management. Springer, Tokyo, pp 349–369

Ceballos LA, Cardinal MV, Vazquez-Prokopec GM et al (2006) Long-term reduction of *Trypanosoma cruzi* infection in sylvatic mammals following deforestation and sustained vector surveillance in northwestern Argentina. Acta Trop 98:286–296. https://doi.org/10.1016/j.actatropica.2006.06.003

Chacón F, Bacigalupo A, Álvarez-Duhart B et al (2022) The parasite load of *Trypanosoma cruzi* modulates feeding and defecation patterns of the Chagas disease vector *Triatoma infestans*. Microorganisms 10:1003. https://doi.org/10.3390/microorganisms10051003

Cirignoli S, Lartigau B (2019) *Lagostomus maximus*. In: SAyDS–SAREM (eds) Categorización 2019 de los mamíferos de Argentina según su riesgo de extinción. Lista Roja de los mamíferos de Argentina. http://cma.sarem.org.ar/es/especie-nativa/lagostomus-maximus. Accessed 15 June 2023

Cizauskas CA, Carlson CJ, Burgio KR et al (2017) Parasite vulnerability to climate change: an evidence-based functional trait approach. R Soc Open Sci 4:160535. https://doi.org/10.1098/rsos.160535

Couch L, Foster GW, Machicote M, Branch LC (2001) Descriptions of two new species of *Eimeria* (Apicomplexa: Eimeriidae) and of *Eimeria Chinchillae*-like oocysts from the plains vizcacha *Lagostomus maximus* (Desmarest, 1817) (Rodentia: Chinchillidae) from Argentina. J Parasitol 87(1):144–147. https://doi.org/10.1645/0022-3395(2001)087[0144:DOTNSO]2.0.CO;2

Cwirenbaum R, Schmidt AR, Cortasa SA et al (2021) First record of an infection by tissue cyst-forming coccidia in wild vizcachas (*Lagosotmus maximus*, Rodentia) of Argentina. Int J Parasitol Parasites Wildl 16:52–58. https://doi.org/10.1016/j.ijppaw.2021.08.002

Dallas TA, Aguirre AA, Budischak S et al (2018) Gauging support for macroecological patterns in helminth parasites. Glob Ecol Biogeogr 27:1437–1447. https://doi.org/10.1111/geb.12819

Denegri G, Dopchiz MC, Elissondo MC, Beveridge I (2003) *Viscachataenia* n. g., with the redescription of *V. quadrata* (von Linstow, 1904) n. comb. (Cestoda: Anoplocephalidae) in *Lagidium viscacia* (Rodentia: Chinchillidae) from Argentina. Syst Parasitol 54:81–88. https://doi.org/10.1023/A:1022589725500

Dobson A, Lafferty KD, Kuris AM et al (2008) Homage to Linnaeus: how many parasites? How many hosts? Proc Natl Acad Sci 105(1):11482–11489. https://doi.org/10.1073/pnas.0803232105

Dorfman VB, Inserra PIF, Leopardo NP et al (2016) The South American Plains Vizcacha, *Lagostomus maximus*, as a valuable animal model for reproductive studies. JSM Anat Physiol 1(1):1004

Dougherty ER, Carlson CJ, Bueno VM et al (2016) Paradigms for parasite conservation. Conserv Biol 30:724–733. https://doi.org/10.1111/cobi.12634

Dubey JP, Calero-Bernal R, Rosenthal BM, Speer CA, Fayer R (2016) Sarcocystosis of animals and humans, 2nd edn. CRC Press, Boca Raton

Easterbrook JD, Kaplan JB, Vanasco NB et al (2007) A survey of zoonotic pathogens carried by Norway rats in Baltimore, Maryland, USA. Epidemiol Infect 135(7):1192–1199. https://doi.org/10.1017/S0950268806007746

Eberhardt AT, Costa SA, Marini MR et al (2013) Parasitism and physiological trade-offs in stressed capybaras. PLoS One 8(7):e70382. https://doi.org/10.1371/journal.pone.0070382

Esch GW, Fernández JC (1993) A functional biology of parasitism. Ecology and evolutionary implications. Chapman and Hall, University California Press, USA

Fayer R, Esposito DH, Dubey JP (2015) Human infections with Sarcocystis species. Clin Microbiol Rev 28(2):295–311. https://doi.org/10.1128/CMR.00113-14

Ferreyra H, Uhart MM, Romano MC et al (2007) Inmovilización química y evaluación de salud de vizcachas salvajes (*Lagostomus maximus*) en el chaco árido argentino. Arq Ciênc Vet Zool Unipar 10(2):91–99

Ferri EFR (2021) Una salud y zoonosis (y viceversa). An R Acad Doc Esp 6(2):319–340
Filigheddu MT, Górgolas M, Ramos JM (2017) Enfermedad de Chagas de transmisión oral. Med Clin (Barc) 148(3):125–131. https://doi.org/10.1016/j.medcli.2016.10.038
Foster GW, Branch LC, Machicote M et al (2002) Gastrointestinal helminths of the plains vizcacha (*Lagostomus maximus*) from Argentina, with observations on interspecific interactions between nematodes and cestodes. Comp Parasitol 69(1):26–32. https://doi.org/10.1654/1525-2647(2002)069[0026:GHOTPV]2.0.CO;2
Gariboldi MC, Inserra PIF, Lucero S et al (2019) Unexpected low genetic variation in the South American hystricognath rodent *Lagostomus maximus* (Rodentia: Chinchillidae). PLoS One 14(9):e0221559. https://doi.org/10.1371/journal.pone.0221559
Georgiev BB, Bray RA, Littlewood DTJ (2006) Cestodes of small mammals: taxonomy and life cycles. In: Morand S, Krasnov BR, Poulin R (eds) Micromammals and macroparasites: from evolutionary ecology to management. Springer, Tokyo, pp 29–62
Glidden CK, Nova N, Kain MP et al (2021) Human-mediated impacts on biodiversity and the consequences for zoonotic disease spillover. Curr Biol 31:R1342–R1361. https://doi.org/10.1016/j.cub.2021.08.070
González-Barrio D (2022) Zoonoses and wildlife: One Health approach. Animals 12(480):1–7. https://doi.org/10.3390/ani12040480
Goodey T (1925) *Viannella viscaciae* n. sp., a nematode parasite of the South American rodent *Viascacia viscacia*. J Helminthol 3(3–4):157–162. https://doi.org/10.1017/S0022149X00029473
Grear DA, Hudson P (2011) The dynamics of macroparasite host-self-infection: a study of the patterns and processes of pinworm (Oxyuridae) aggregation. Parasitology 138(5):619–627. https://doi.org/10.1017/S0031182011000096
Han BA, Schmidt JP, Bowden SE, Drake JM (2015) Rodent reservoirs of future zoonotic diseases. Proc Natl Acad Sci U S A 112(22):7039–7044. https://doi.org/10.1073/pnas.1501598112
Han BA, Kramer AM, Drake JM (2016) Global patterns of zoonotic disease in mammals. Trends Parasitol 32:565–577. https://doi.org/10.1016/j.pt.2016.04.007
Hancke D, Navone GT, Suárez OV (2011) Endoparasite community of *Rattus norvegicus* captured in a shantytown of Buenos Aires City, Argentina. Helminthologia 48(3):167–173. https://doi.org/10.2478/s11687-011-0025-3
He J, Guo Z, Yang P et al (2022) Social insights on the implementation of One Health in zoonosis prevention and control: a scoping review. Infect Dis Poverty 11(48). https://doi.org/10.1186/s40249-022-00976-y
Herrera L (2010) Una revisión sobre reservorios de *Trypanosoma* (*Schizotrypanum*) *cruzi* (Chagas, 1909), agente etiológico de la enfermedad de Chagas. Bol Mal Salud Amb 50(1):3–15
Holmes JC, Price PW (1986) Communities of parasites. In: Kikkawa J, Anderson DJ (eds) Community ecology: patterns and processes. Blackwell, Oxford, pp 187–213
Holmstad PR, Holstad Ø, Karbøl G et al (2004) Parasite tags in ecological studies of terrestrial hosts: a study on ptarmigan (*Lagopus* spp.) dispersal. Ornis Fenn 81:128–136
Hugot JP, Sutton CA (1989) Contribution à la connaissance de la faune parasitaire d' Argentina, XV. Etude morphologique de *Heteroxynema* (*Cavioxyura*) *viscaciae* n. sp. (Nematoda, Heteroxynematidae) parasite de *Lagidium viscacia boxi* (Mammalia, Rodentia). Syst Parasitol 13:111–120. https://doi.org/10.1007/BF00015219
Ibrahim MM (2012) Variation in parasite infracommunies of *Tilapia zillii* in relation to some biotic and abiotic factors. Int J Zool Res 8(2):59–70
Irigoitia MM, Incorvaia IS, Timi JT (2017) Evaluating the usefulness of natural tags for host population structure in chondrichthyans: parasite assemblages of *Sympterygia bonapartii* (Rajiformes: Arhynchobatidae) in the Southwestern Atlantic. Fish Res 195:80–90. https://doi.org/10.1016/j.fishres.2017.07.006
Jackson JE, Branch LC, Villarreal D (1996) Lagostomus maximus. Mamm Species 543:1–6
Jaenike J (1996) Population-level consequences of parasite aggregation. Oikos 76:155–160. https://doi.org/10.2307/3545757

Jittapalapong S, Herbreteau V, Hugot JP et al (2009) Relationship of parasites en pathogens diversity to rodents in Thailand. Agric Nat Resour 43(1):106–117

Johnson PT, Hoverman JT (2014) Heterogeneous hosts: how variation in host size, behaviour and immunity affects parasite aggregation. J Anim Ecol 83:1103–1112. https://doi.org/10.1111/1365-2656.12215

Karesh WB, Dobson A, Lloyd-Smith JO et al (2012) Ecology of zoonoses: natural and unnatural histories. Lancet 380(9857):1936–1945. https://doi.org/10.1016/S0140-6736(12)61678-X

Kirillova V, Prakas P, Calero-Bernal R et al (2018) Identification and genetic characterization of *Sarcocystis arctica* and *Sarcocystis lutrae* in red foxes (*Vulpes vulpes*) from Baltic States and Spain. Parasit Vectors 11(173). https://doi.org/10.1186/s13071-018-2694-y

Kufner MB, Monge S (1998) Dieta de *Lagostomus maximus* (Rodentia, Chinchillidae) en áreas sometidas a intervención humana en el desierto de monte, Argentina. Iheringia Ser Zool 84:175–184

Larsen BB, Miller EC, Rhodes MK, Wiens JJ (2017) Inordinate fondness multiplied and redistributed: the number of species on earth and the new pie of life. Q Rev Biol 92:229–265. https://doi.org/10.1086/693564

Lima VFS, Ramos RAN, Giannelli A et al (2021) Occurrence of zoonotic gastrointestinal parasites of rodents and the risk of human infection in different biomes of Brazil. Braz J Vet Med 43(1):e113820. https://doi.org/10.29374/2527-2179.bjvm113820

Llanos AC, Crespo JA (1952) Ecología de la vizcacha (*Lagostomus maximus maximus* Blainv.) en el nordeste de la provincia de Entre Ríos. Rev Invest Agr 6(3–4):289–378

Luong LT, Vigliotti BA, Hudson PJ (2011) Strong density-dependent competition and acquired immunity constrain parasite establishment: implications for parasite aggregation. Int J Parasitol 41:505–511. https://doi.org/10.1016/j.ijpara.2010.12.003

Luque JL, Poulin R (2007) Metazoan parasite species richness in Neotropical fishes: hotspots and the geography of biodiversity. Parasitology 134:865–878. https://doi.org/10.1017/S0031182007002272

Manzoli DE, Antoniazzi LR, Beldomenico PM (2011) Cambio ambiental global, parásitos y la salud de sus hospedadores: las moscas parásitas del género *Philornis* en pichones de aves. Hornero 26(1):45–53

Mariaux J, Tkach VV, Vasileva GP et al (2017) Cyclophyllidea van Beneden in Braun, 1900. In: Caira J, Jensen K (eds) Planetary biodiversity inventory (2008–2017): tapeworms from vertebrate bowels of the earth. University of Kansas, Natural History Museum, Lawrence, pp 77–148

Martínez FA (1988) Helmintofauna del *Myocastor coypus* y *Lagostomus maximus*, Nematodes. Vet Argent 5(41):33–37

Martorelli Di Genova B, Knoll LJ (2020) Comparisons of the sexual cycles for the coccidian parasites *Eimeria* and *Toxoplasma*. Front Cell Infect Microbiol 10:604897. https://doi.org/10.3389/fcimb.2020.604897

Mégnin P (1895) Note sur un nématode nouveau parasite du Mara. Bull Soc Zool Fr 20:173–176

Morales ME, Campo Verde Arbocco F, Muñoz-San Martín C et al (2023) High *Trypanosoma cruzi* prevalence in armadillo (*Zaedyus pichiy*; Xenarthra: Chlamyphoridae) populations from Mendoza, Argentina. Parasitol Res 122:1593–1604. https://doi.org/10.1007/s00436-023-07861-z

Morand S, Bouamer S, Hugot JP (2006) Nematodes. In: Morand S, Krasnov BR, Poulin R (eds) Micromammals and macroparasites: from evolutionary ecology to management. Springer, Tokyo, pp 63–79

Morrill A, Forbes MR (2012) Random parasite encounters coupled with condition-linked immunity of hosts generate parasite aggregation. Int J Parasitol 42(7):701–706. https://doi.org/10.1016/j.ijpara.2012.05.002

Morrill A, Forbes MR (2016) Aggregation of infective stages of parasites as an adaptation and its implications for the study of parasite-host interactions. Am Nat 187(2):225–235. https://doi.org/10.1086/684508

Morrill A, Dargent F, Forbes MR (2017) Explaining parasite aggregation: more than one parasite species at a time. Int J Parasitol 47:185–188. https://doi.org/10.1016/j.ijpara.2016.11.005

Muñoz G, Castro R (2012) Comunidades de parásitos eumetazoos de peces labrisómidos de Chile central. Rev Biol Mar Oceanogr 47(3):565–571. https://doi.org/10.4067/S0718-19572012000300018

Murillo-Godínez G (2018) Enfermedad de Chagas (tripanosomiasis americana). Med Int Méx 34(6):959–970. https://doi.org/10.24245/mim.v34i6.2217

Noya BA, González ON (2015) An ecological overview on the factors that drives to Trypanosoma cruzi oral transmission. Acta Trop 151:94–102. https://doi.org/10.1016/j.actatropica.2015.06.004

Ogunseitan OA (2022) One Health and the environment: from conceptual framework to implementation science. Environ Sci Policy Sustain Develop 64(2):11–21. https://doi.org/10.1080/00139157.2022.2021792

Okamura B, Hartigan A, Naldoni J (2018) Extensive uncharted biodiversity: the parasite dimension. Integr Comp Biol 58:1132–1145. https://doi.org/10.1093/icb/icy039

Organización Panamericana de la Salud (OPS) (2003) Zoonosis y enfermedades transmisibles comunes al hombre y a los animales, Parasitosis, vol III, 3rd edn, Washington, DC

Palmezano-Díaz JM, Plazas-Rey LK, Rojas-Carvajal D (2015) Infección por *Toxoplasma*: panorama actual. Spei Domus 11(22):47–56. https://doi.org/10.16925/sp.v11i22.1154

Pedersen AB, Greives TJ (2008) The interaction of parasites and resources cause crashes in a wild mouse population. J Anim Ecol 77:370–377. https://doi.org/10.1111/j.1365-2656.2007.01321.x

Poulin R (1997) Species richness of parasite assemblages: evolution and patterns. Annu Rev Ecol Syst 28:341–358. https://doi.org/10.1146/annurev.ecolsys.28.1.341

Poulin R (2006) Parasite aggregation: causes and consequences. In: Poulin R (ed) Evolutionary ecology of parasites. Princeton University Press, Princeton, pp 134–159

Poulin R (2007) Evolutionary ecology of parasites, 2nd edn. Princeton University Press, Princeton

Poulin R (2013) Explaining variability in parasite aggregation levels among host samples. Parasitology 140:541–546. https://doi.org/10.1017/S0031182012002053

Poulin R, Kamiya T (2015) Parasites as biological tags of fish stocks: a meta-analysis of their discriminatory power. Parasitology 142:145–155. https://doi.org/10.1017/S0031182013001534

Poulin R, Morand S (2000) The diversity of parasites. Q Rev Biol 75(3):277–293

Poulin R, Morand S (2004) Parasite biodiversity. Smithsonian Books, Washington

Prakas P, Kirillova V, Gavarāne I et al (2019) Morphological and molecular description of *Sarcocystis ratti* n. sp. from the black rat (*Rattus rattus*) in Latvia. Parasitol Res 118:2689–2694. https://doi.org/10.1007/s00436-019-06393-9

Puig S, Videla F, Cona M et al (1998) Diet of the vizcacha *Lagostomus maximus* (Rodentia, Chinchillidae), habitat preferences and food availability in Northern Patagonia, Argentina. Mammalia 62(2):191–204. https://doi.org/10.1515/mamm.1998.62.2.191

Railliet A, Henry A (1909) Sur la classification des Strongylidae: I. Metastrongylinae. CR Soc Biol 66:85–88

Renison D, Cingolani AM, Contarde C, Guzmán D (2023) Asistiendo a la reintroducción de vizcachas (*Lagostomus maximus*): ¿Cómo aumentar el área de pastoreo seguro? Ecol Austral 33:020–029. https://doi.org/10.25260/EA.23.33.1.0.1961

Reperant LA, Hegglin D, Tanner I et al (2009) Rodents as shared indicators for zoonotic parasites of carnivores in urban environments. Parasitology 136(3):329–337. https://doi.org/10.1017/S0031182008005428

Ribicich M, Gamble HR, Bolpe J et al (2010) *Trichinella* infection in wild animals from endemic regions of Argentina. Parasitol Res 107:377–380. https://doi.org/10.1007/s00436-010-1873-3

Riquelme M, Salgado R, Simonetti JA et al (2021) Intestinal helminths in wild rodents from native forest and exotic pine plantations (*Pinus radiata*) in central Chile. Animals 11(384):1–10. https://doi.org/10.3390/ani11020384

Robben PM, Sibley LD (2004) Food- and waterborne pathogens: you are (infected by) what you eat! Microbes Infect 6:406–413. https://doi.org/10.1016/j.micinf.2003.12.016

Roberts M (1991) The parasites of the Polynesian rat within and beyond New Zealand. Int J Parasitol 21(7):777–783. https://doi.org/10.1016/0020-7519(91)90145-W

Rohde K (1982) Ecology of marine parasites. Univ. Queensland Press, St. Lucia

Rosenthal BM (2021) Zoonotic Sarcocystis. Res Vet Sci 136:151–157. https://doi.org/10.1016/j.rvsc.2021.02.008

Rossanigo CE, Jackson JE, Lukovich R (1986) Endoparásitos de la vizcacha (*Lagostomus maximus*). Primera descripción en la República Argentina. Rev Med Vet 67(4):219–223

Rudaitytė-Lukošienė E, Jasiulionis M, Balčiauskas L et al (2022) Morphological and molecular description of Sarcocystis myodes n. sp. from the bank vole (Clethrionomys glareolus) in Lithuania. Biology 11(4):512. https://doi.org/10.3390/biology11040512

Rueda K, Trujillo JE, Carranza JC, Vallejo GA (2014) Transmisión oral de *Trypanosoma cruzi*: una nueva situación epidemiológica de la enfermedad de Chagas en Colombia y otros países suramericanos. Biomedica 34:631–641. https://doi.org/10.7705/biomedica.v34i4.2204

Rupasinghe R, Chomel BB, Matínez-López B (2022) Climate change and zoonoses: a review of the current status, knowledge gaps, and future trends. Acta Trop 226:106225. https://doi.org/10.1016/j.actatropica.2021.106225

Sánchez A, Contreras A, Corrales JC, Fe C (2022) En el principio fue la zoonosis: One Health para combatir esta y futuras pandemias. Informe SESPAS 2022. Gac Sanit 36(S1):S61–S67. https://doi.org/10.1016/j.gaceta.2022.01.012

Sangenis LHC, Nielebock MAP, Santos CS et al (2016) Chagas disease transmission by consumption of game meat: systematic review. Rev Bras Epidemiol 19(4):803–811. https://doi.org/10.1590/1980-5497201600040010

Sapp KK, Esch GW (1994) The effects of spatial and temporal heterogeneity as structuring forces for parasite communities in Helisoma anceps and Physa gyrina. Am Midl Nat 132(1):91–103. https://doi.org/10.2307/2426204

Schuurmans Stekhoven JH (1951) Nematodos parasitarios de anfibios, pájaros y mamíferos de la República Argentina. Acta Zool Lillo 10:315–400

Shaw DJ, Dobson AP (1995) Patterns of macroparasites abundance and aggregation in wildlife populations: a quantitative review. Parasitology 111(S1):S111–S133. https://doi.org/10.1017/S0031182000075855

Spotorno AE, Patton JL (2015) Superfamily chinchilloidea Bennett, 1833. In: Patton JL, Pardiñas UFJ, D'Elia G (eds) Mammals of South America Volume 2: Rodents. University of Chicago Press, Chicago, pp 762–786

Stojcevic D, Mihaljevic Z, Marinculic A (2004) Parasitological survey of rats in rural regions of Croatia. Vet Med 49(3):70–74

Sures B, Nachev M, Schwelm J et al (2023) Environmental parasitology: stressor effects on aquatic parasites. Trends Parasitol 39(6):461–474. https://doi.org/10.1016/j.pt.2023.03.005

Sutton CA, Durette-Desset MC (1987) Contribution à la connaissance de la faune parasitologique Argentine XVIII. *Lagostonema ecasiense* n. gen., n. sp. (Trichostrongyloidea, Nematoda), parasite de *Lagostomus maximus* (Chinchillidae, Caviomorpha). Bull Mus Natl Hist Nat 4A 9(1):127–131

Sutton CA, Durette-Desset MC (1995) A description of *Graphidioides kravetzi* n. sp. and the revision of *Graphidioides* Cameron, 1923 (Nematoda Trichostrongyloidea), parasites of Neotropical rodents. Syst Parasitol 31:133–145. https://doi.org/10.1007/BF02185545

Thieltges DW, Jensen KT, Poulin R (2008) The role of biotic factors in the transmission of free-living endohelminth stages. Parasitology 135(4):407–426. https://doi.org/10.1017/S0031182007000248

Timi JT (2007) Parasites as biological tags for stock discrimination in marine fish from South American Atlantic waters. J Helminthol 81:107–111. https://doi.org/10.1017/S0022149X07726561

Toso MA, Vial UF, Galanti N (2011) Transmisión de la enfermedad de Chagas por vía oral. Rev Med Chile 139:258–266. https://doi.org/10.4067/S0034-98872011000200017

Tull A, Valdmann H, Tammeleht E et al (2022) High overlap of zoonotic helminths between wild mammalian predators and rural dogs – an emerging One Health concern? Parasitology:1–10. https://doi.org/10.1017/S0031182022001032

Vidal-Anzardo M, Moscoso MY, Fabian MB (2020) Parasitosis intestinal: Helmintos. Prevalencia y análisis de la tendencia de los años 2010 a 2017 en el Perú. An Fac Med 81(1):26–32. https://doi.org/10.15381/anales.v81i1.17784

Wehrendt DP, Gómez-Bravo A, Ramirez JC et al (2019) Development and evaluation of a duplex TaqMan qPCR assay for detection and quantification of *Trypanosoma cruzi* infection in domestic and sylvatic reservoir hosts. Parasit Vectors 12:567. https://doi.org/10.1186/s13071-019-3817-9

Wisnivesky-Colli C, Schweigmann NJ, Alberti A et al (1992) Sylvatic American trypanosomiasis in Argentina, Trypanosoma cruzi infection in mammals from the Chaco forest in Santiago del Estero. Trans R Soc Trop Med Hyg 86:38–41. https://doi.org/10.1016/0035-9203(92)90433-D

Xu J, Liu H, Jiang Y et al (2022) Genotyping and subtyping of *Cryptosporidium* spp. and *Giardia duodenalis* isolates from two wild rodent species in Gansu Province, China. Sci Rep 12:12178. https://doi.org/10.1038/s41598-022-16196-1

Zischke MT, Griffiths SP, Tibbetts IR, Lester RJ (2013) Stock identification of wahoo (*Acanthocybium solandri*) in the Pacific and Indian Oceans using morphometrics and parasites. ICES J Mar Sci 70(1):164–172. https://doi.org/10.1093/icesjms/fss164

Part II
Morphology, Development and Physiology

Chapter 10
Skull Ontogeny and Sexual Dimorphism

Luciano Luis Rasia

Abstract The skull ontogeny and sexual dimorphism of the plains vizcacha, *Lagostomus maximus*, is here described from quantitative approaches. The quantitative study of allometry was performed based on 19 skull linear measurements taken from 78 specimens, using two methods: bivariate regression and multivariate allometry. Seven variables show positive allometry for both sexes in the bivariate and multivariate analyses: length of the nasals, length of the zygoma, length of the upper diastema, length of the upper cheek teeth series, height of the mandibular condyle, length of the mandible, and length of the lower cheek teeth series. Nine variables show sexual dimorphism in allometric terms: breadth of zygoma, breadth of anterior palate, breadth of posterior palate (for the bivariate analysis), height of the rostrum, length of the zygoma, length of the upper teeth series, length of the lower teeth series, length anterior to the masseteric notch, and length of the mandible (for the multivariate analysis). The results show that the braincase develops in prenatal stages, and growth of the skull after birth is mainly given by the lengthening of the rostrum, as was observed in other mammals, including caviomorph rodents. Additionally, sexual dimorphism of the skull can be related to differences in mastication.

Keywords Sexual dimorphism · Ontogeny · Skull morphology · Morphometry · Mammals

10.1 Introduction

The plains vizcacha, *Lagostomus maximus*, is one of the most sexually dimorphic living rodents (e.g., Llanos and Crespo 1952; Branch 1993a; Jackson et al. 1996; Schulte-Hostedde 2007), being males much larger than females (male-biased sexual

L. L. Rasia (✉)
División Paleontología Vertebrados, Museo de La Plata, La Plata, Argentina

CONICET, La Plata, Argentina
e-mail: lucianorasia@fcnym.unlp.edu.ar

L. L. Rasia et al. (eds.), *Plains Vizcachas*, https://doi.org/10.1007/978-3-031-49487-1_10

size dimorphism). Besides the size, males of this species have a massive head and a more pronounced facial mask color (Weir 1974; Jackson et al. 1996).

There have been several ontogenetic studies of the skull of caviomorph rodents, focusing on skull allometry (Nelson and Shump 1978; Monteiro et al. 1999; Vasallo 2000; Verzi et al. 2010; Aeschbach et al. 2016; Nasif and Abdala 2015; Tavares et al. 2016), skull developmental patterns (Wilson and Sánchez-Villagra 2009, 2010; Wilson 2010), ontogeny of sexual dimorphism (Lammers et al. 2001; Farmer and German 2004), brain development (Ferreira et al. 2022), and ontogeny of extinct taxa (Vassallo and Mora 2007; Rinderknecht et al. 2018).

Studies of the ontogeny of the plains vizcacha have been focused on different subjects such as reproductive morphology (e.g., Acuña et al. 2020, 2021), reproductive physiology and breeding (Weir 1970, 1971), bone histology (Garrone et al. 2021), brain development (Núñez and Bee-De-Speroni 1997), and growth rates (Llanos and Crespo 1952; Jackson 1990). Only Rasia et al. (in preparation) study the skull ontogeny from a qualitative approach.

With respect to sexual dimorphism during ontogeny, many rodents exhibit heterochronic patterns such as males growing faster and for longer periods than females (e.g., Lima et al. 1997; Yoshinaga et al. 1997; Begall and Burda 1998; Jackson and Van Aarde 2003). Rodents show sexual size dimorphism generally after weaning (Boutin and Larsen 1993; Kam and Degen 1994; Derrickson 1988; Jackson and Van Aarde 2003), which takes place at 2nd–3rd month in *Lagostomus maximus* (Llanos and Crespo 1952), but sexual size dimorphism in this species becomes evident after the 1st month (Jackson 1990). Growth rates of both sexes are fast until the 16th–18th month; then, the growth rate decreases in males and stabilizes in females (Jackson 1990).

Male-biased sexual size dimorphism occurs in many fossorial rodents (e.g., Bathyergidae, Geomyidae), related to strong male-male competition and polygyny (Schulte-Hostedde 2007), and is observed in other mammals (Weckerley 1998). In the plains vizcacha, male-biased sexual size dimorphism co-occur with these reproductive behaviors (male-male competition and polygyny; see Branch 1993a, b). The marked sexual dimorphism of *Lagostomus maximus* may be due to differential use of resources by each sex or sexual selection in males (Branch 1993a). There is a significant size variability among males of the same age that may be due to social status (Jackson 1990), given that each social group is dominated by one male (e.g., Cabrera and Yepes 1960).

In this chapter, the postnatal ontogeny of the skull in a growth sample of males and females of *Lagostomus maximus* is quantified, evaluating the statistical differences in the rate of growth between sexes in different skull dimensions.

10.2 Material and Methods

The results of this chapter are part of the unpublished PhD Dissertation of the author (Rasia 2016).

An ontogenetic series of 78 specimens (33 females and 45 males) deposited in the mammalogical collections of the Museo Argentino de Ciencias Naturales (MACN) in Buenos Aires (Argentina) and the Museo de La Plata (MLP) in La Plata (Argentina) is analyzed. All individuals were collected in localities within Entre Ríos Province (Argentina) to avoid intraspecific variations due to ecogeographic factors, given that the distribution of *Lagostomus maximus* subspecies is uncertain (Jackson et al. 1996; see Chap. 4). Most of the studied material was collected by J.A. Crespo and A.C. Llanos and studied in their classic work about the ecology of vizcachas (Llanos and Crespo 1952).

A total of 19 linear measurements are used (Fig. 10.1), representing the length, width, and height of skull structures, in order to estimate the allometric growth of skull components. These measurements were adapted from studies performed in several mammal groups such as caviomorph rodents (Nasif and Abdala 2015), marsupials (Abdala et al. 2001; Flores et al. 2006; Giannini et al. 2004), carnivorans (Giannini et al. 2010; Tarnawski et al. 2014a, b, 2015), and primates (Flores and Casinos 2011) and are abbreviated as follow: BAP, anterior breadth of palate; BB, breadth of braincase; BPP, posterior breadth of palate; BZ, zygomatic breadth; HC, height of mandibular condyle; HM, height of mandibular body; HO, height of occipital plate; HR, height of rostrum; LB, length of basicranium; LM, length of the mandible; LLD, length of lower diastema; LLS, length of lower cheek teeth series;

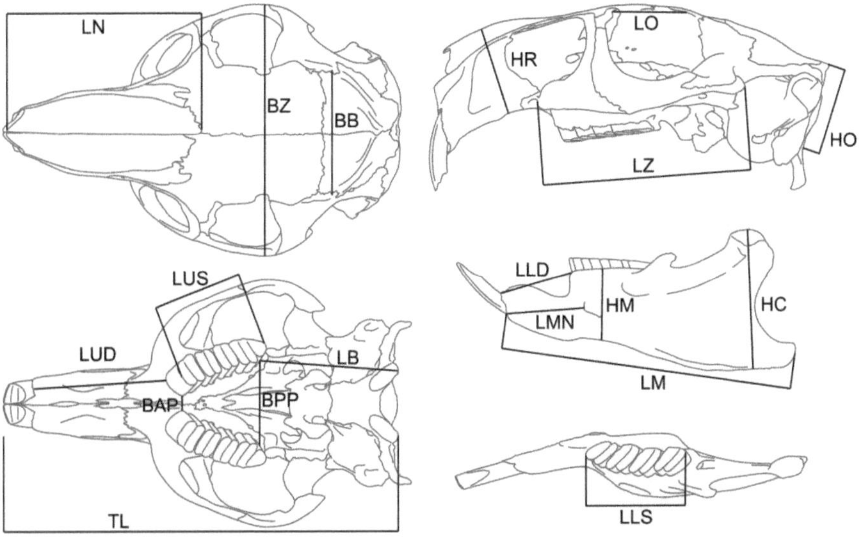

Fig. 10.1 Skull measurements of *Lagostomus maximus* used in this study. *BAP* anterior breadth of palate, *BB* breadth of braincase, *BPP* posterior breadth of palate, *BZ* zygomatic breadth, *HC* height of mandibular condyle, *HM* height of mandibular body, *HO* height of occipital plate, *HR* height of rostrum, *LB* length of basicranium, *LM* length of the mandible, *LLD* length of lower diastema, *LLS* length of lower cheek teeth series, *LMN* length of mandible anterior to masseteric notch, *LN* length of the nasals, *LO* length of the orbit, *LUD* length of upper diastema, *LUS* length of upper cheek teeth series, *LZ* zygomatic length, *TL* total length

LMN, length of mandible anterior to masseteric notch; LN, length of the nasals; LO, length of the orbit; LUD, length of upper diastema; LUS, length of upper cheek teeth series; LZ, zygomatic length; TL, total length.

Two approaches were taken: bivariate regression and multivariate allometry. The statistic allometric analyses were performed independently in each sex, and growth trend differences were described in order to detect dimorphism in allometric terms.

10.2.1 Bivariate Regression

For the bivariate analysis, the skull geometric mean was taken as a proxy of the total size (e.g., Tarnawski et al. 2014b, 2015). The geometric mean is a size variable derived from the Nth root of the product of N measurements, and it is a good predictor of the size of an individual.

To estimate the change in each skull variable with respect to the total size and following studies in other groups of mammals (e.g., Abdala et al. 2001; Nasif and Abdala 2015), we used the logarithmic transformation of the allometry equation.

$$\log y = \log b_0 + \log b_1\, x + \log e$$

where y is the dependent variable, log b_0 is the y intercept (constant of normalization), x is the geometric mean (independent variable), b_1 is the slope or coefficient of allometry, and e is the error term. Deviations from isometry were assessed by testing the significance of the coefficients of allometry by a two-tailed t-test, with the null coefficient set at 1.0 and the Type I error at a significance level of $p = 0.01$. Additionally, coefficients were considered marginally significant given $0.05 > P > 0.01$. Isometry is the condition in which an allometry coefficient is statistically indistinguishable from the unit. Deviation from unit represents negative allometry if $b_1 < 1.0$ and positive if $b_1 > 1.0$. To calculate b_1, we used Reduced Major Axis (RMA, e.g., Niklas 1994; Radinsky 1981a, b), where both variables involved in the bivariate relation ($y1$ and $y2$) are interchangeable. That means that the size dependence is not implicit and the residues are oblique components representing variations in $y1$ and $y2$. RMA determines the slope of the largest axis of the ellipse formed by the bivariate regression.

10.2.2 Multivariate Allometry

The multivariate allometry analysis was based on the generalization of the allometry equation (see Jolicoeur 1963a, b), following studies in other mammal groups (e.g., Giannini et al. 2004; Flores et al. 2006). Size is regarded as a latent variable affecting all measured variables simultaneously. The first eigenvector of a principal component analysis (PCA) expresses various allometric relationships of all variables with the latent size, provided that it is extracted from a variance-covariance matrix of log-transformed variables and scaled to unity (see Jolicoeur 1963a). Under isometry, all variables respond equally to growth, and the isometric autovector is equal to $1/p^{0.5}$, where p is the number of variables (i.e., 0.229 for 19 variables). The value of the first autovector of the PC1 represents the observed multivariate allometry coefficient of the corresponding variable. The comparison of each empirical element of the autovector of the first unit with the isometric autovector permits to detect negative (<0.229) and positive (>0.229) deviations from isometry in each variable.

Statistical departures from isometry were estimated using a jackknife (Quenouille 1956; Tukey 1958; Manly 1997; Giannini et al. 2004; Flores et al. 2006). This resampling technic, where pseudovalues are generated with one specimen removed at a time, produces confidence intervals (extracted from the standard deviation of the pseudovalue sample) for the empirically obtained autovector elements. The confidence interval may include the null value 0.229, therefore being indistinguishable from isometry, or may exclude the null value and therefore can be considered significantly allometric, either positive if the element is >0.229 or negative if it is <0.229. With each resampling, a pseudovalue is calculated for each variable using the traditional formulation for the first-order jackknife:

$$\hat{e}^*_j = n\hat{e} - (n-1)\hat{e}_{-j}$$

where one pseudovalue $\hat{e}^*_j$ corresponds to the removal of specimen j from the sample of size n, $\hat{e}$ is the observed element of the unit eigenvector that corresponds to the multivariate coefficient of allometry of the skull variable x, and $\hat{e}_{-j}$ is the value of the coefficient obtained with specimen j removed. From the collection of 1 to n ($n = 33$ for females and $n = 45$ for males) pseudovalues per variable, a mean is calculated for each variable, representing the jackknife estimate of the multivariate allometry coefficient for that variable. The difference between that estimate and the actual value from the complete sample is a measure of the bias. The standard deviation and the corresponding 99% confidence interval (for n–1 degrees of freedom) are calculated for each allometry coefficient.

Finally, to avoid the effect of extreme values, confidence intervals were calculated with the complete sample of pseudovalues (untrimmed) and eliminating upper and lower extreme values (trimmed). Data that showed lower mean absolute value from bias were considered.

10.3 Results

10.3.1 Bivariate Regression

Through the R^2 coefficient (determination coefficient adjusted by $df = N2$), it was evaluated that the correlation of the examined variables varies between 1.9% and 98.6% in females and between 62% and 99.6% in males (Tables 10.1 and 10.2).

Table 10.1 Bivariate allometry analysis for females of *Lagostomus maximus* ($N = 33$)

	Regression		Intercept			Slope			
Variable	R^2	F	$\log(b_0)$	t	p_{b0}	$b1$	F_{iso}	p_{b1}	Trend
LN	0.949	579.266	−0.453	−5.328	8.37764E-06	1.446	86.957	1.65E+04	+
BZ	0.984	1869.549	0.284	8.17	3.15401E-09	1.041	3.099	0.088	Iso
BB	**0.754**	94.905	0.913	16.653	5.14101E-17	0.423	118.368	4.12093E-12	−
HR	0.974	1141.497	−0.035	−0.863	0.394	0.962	1.728	0.198	Iso
LO	0.917	343.454	−0.472	−5.194	1.23172E-05	1.21	13.768	0.0008	+
LZ	0.985	2095.053	0.075	2.119	0.042	1.12	27.246	0.00001	+
HO	0.889	248.976	0.615	12.284	1.90413E-13	0.577	93.742	6.89E-11	−
LUD	0.979	1456.613	−0.258	−5.719	2.73241E-06	1.196	48.262	8.56936E-08	+
LUS	0.969	981.777	−0.724	−11.11	2.44619E-12	1.427	133.833	8.84848E-13	+
BAP	**0.019**	0.615	−0.683	−3.012	0.005	0.877	0.547	0.465	Iso
BPP	0.919	353.054	−0.209	−2.674	0.012	1.052	1.001	0.325	Iso
LB	0.888	245.625	0.326	4.437	0.0001	0.841	8.36	0.007	−
TL	0.985	1995.823	0.372	10.501	9.85246E-12	1.099	18.003	0.0002	+
LLD	0.951	596.067	−0.323	−5.056	0.00002	1.101	5.848	0.022	+
LMN	0.939	482.837	−0.277	−3.856	0.0005	1.119	6.546	0.016	+
HM	0.971	1038.866	−0.223	−5.008	2.09761E-05	1.002	0.003	0.957	Iso
HC	0.968	929.114	−0.396	−6.523	2.79448E-07	1.294	65.163	4.08E-09	+
LM	0.986	2207.571	0.24	7.063	6.21298E-08	1.107	23.404	0.00003	+
LLS	0.962	795.013	−0.556	−8.249	2.56129E-09	1.332	69.709	1.97474E-09	+

R^2 correlation coefficient (low values in bold), F F-test for regression, $log(b_0)$ intercept from RMA, t Student t-test for intercept coefficients, p_{b0} p value for $b_0 = 0$, $b1$ slope from RMA, F_{iso} F isometry test, p_{b1} p value for $b_1 = 1$

Anterior breadth of palate (BAP) is the least correlated variable with the skull size (1.9% in females and 62% in males). Breadth of the braincase (BB) is correlated with skull size a 75% in females and 81% in males. In the remaining variables, the correlation with skull size is close to 90% or higher.

Ten variables show positive allometric growth in both sexes (LN, LZ, LUD, LUS, TL, LLD, LMN, HC, LM, LLS). One variable shows isometry in both sexes (HR). Three variables show negative allometry in both sexes (BB, HO, LB). Sexual dimorphism was observed in the growth trends of four of the analyzed variables (BZ, LO, BAP, BPP).

Table 10.2 Bivariate allometry analysis for males of *Lagostomus maximus* ($N = 45$)

	Regression		Intercept			Slope			
Variable	R^2	F	$\log(b_0)$	t	p_{b0}	$b1$	F_{iso}	p_{b1}	Trend
LN	0.992	5303.357	−0.307	−11.237	2.23E-14	1.343	479.366	0	+
BZ	0.996	9949.245	0.301	19.697	3.83E-23	1.029	7.985	0.007	+
BB	**0.806**	178.812	1.050	32.335	8.29E-32	0.326	417.554	0	−
HR	0.985	2765.979	−0.126	−4.367	7.79E-05	1.027	1.961	0.169	Iso
LO	0.960	1037.112	−0.063	−1.526	0.134	0.915	8.568	0.005	−
LZ	0.996	11689.546	0.126	8.526	8.63E-11	1.079	68.386	1.97E-10	+
HO	0.974	1641.052	0.365	13.308	7.59E-17	0.759	131.691	1.13E-14	−
LUD	0.993	5972.556	−0.328	−13.763	2.34E-17	1.245	293.050	0	+
LUS	0.969	1350.270	−0.395	−8.394	1.32E-10	1.182	39.415	1.44E-07	+
BAP	**0.620**	70.248	−0.390	−3.965	0.0003	0.704	14.548	0.0004	−
BPP	0.980	2055.113	−0.055	−1.795	0.079	0.945	6.794	0.013	−
LB	0.972	1503.363	0.272	8.062	3.88E-10	0.892	20.152	5.29E-05	−
TL	0.995	7993.807	0.444	25.500	1.37E-27	1.052	20.500	4.67E-05	+
LLD	0.980	2094.719	−0.353	−9.752	1.84E-12	1.128	31.222	1.45E-06	+
LMN	0.986	2980.834	−0.348	−11.052	3.80E-14	1.167	72.644	8.72E-11	+
HM	0.988	3418.882	−0.322	−12.005	2.54E-15	1.062	12.627	0.001	+
HC	0.987	3179.327	−0.372	−11.149	2.87E-14	1.277	196.203	0	+
LM	0.993	6292.419	0.235	11.363	1.55E-14	1.109	68.226	2.04E-10	+
LLS	0.971	1424.617	−0.226	−5.351	3.19E-06	1.091	11.170	0.002	+

R^2 correlation coefficient (low values in bold), F F-test for regression, $log(b_0)$ intercept, t Student t-test for the intercept coefficient, P_{b0} P value for $b_0 = 0$, $b1$ slope, F_{iso} F isometry test, P_{b1} p value for $b_1 = 1$

When comparing the trajectories of females and males (Table 10.3), 14 of the 19 variables show similar slopes in both sexes. In four of the variables with different slopes in both sexes, the slope is higher in females than in males (BB, LO, LUS, LLS), and only in HO, the slope is higher in males than in females.

Ten variables show positive allometric trends in both sexes (LN, BZ, LZ, LUD, TL, LLD, LMN, HM, HC, LM). Two variables show a negative allometric trend in both sexes (ABP, LB). Two variables show isometric trend in both sexes (HR, BPP).

10.3.2 *Multivariate Allometry*

Taking into account the observed allometry coefficients (Table 10.4), seven variables show positive allometry for both sexes (LN, LZ, LUD, LUS, HC, LM, and LLS), four variables show negative allometry for both sexes (BB, HO, BAP, and LB), two variables show isometry for both sexes (BZ and TL), and six variables show sexual dimorphism in allometric terms (HR, LO, BPP, LLD, LMN, and HM).

The mean of the absolute bias in females for untrimmed values (0.0011) is 2.18 times lower than for trimmed values (0.0024). Similarly, in males, the mean of the

Table 10.3 Test for common slope, common intercept, and shift for females ($n = 33$) and males ($n = 45$) of *Lagostomus maximus*

Variable	Common slope				Common intercept			Shift	
	Lr_{b1}	p_{b1}	$b1$ com	Trend	W_{b0}	p_{b0}	$b0$ com	W_{shift}	p_{shift}
LN	2.838	0.092	1.353	+	0.729	0.393	−0.317	2.248	0.134
BZ	0.229	0.632	1.031	+	0.264	0.608	0.300	2.389	0.122
BB	**5.234**	**0.022**	**Female > male**						
HR	3.319	0.068	1.008	Iso	2.087	0.149	−0.106	2.788	0.095
LO	**17.815**	**2.43512E-05**	**Female > male**						
LZ	2.32	0.128	1.085	+	19.659	9.26E-06	Female > male		
HO	**14.831**	**0.0001**	**Male > female**						
LUD	1.824	0.177	1.235	+	0.065	0.799	−0.315	2.511	0.113
LUS	**17.84**	**2.40243E-05**	**Female > male**						
BAP	1.172	0.279	0.739	–	10.775	0.001	Male > female		
BPP	3.571	0.059	0.960	Iso	0.717	0.397	−0.072	2.186	0.139
LB	0.799	0.371	0.884	–	30.041	4.23E-08	Male > female		
TL	2.937	0.087	1.061	+	3.639	0.056	0.424	2.740	0.098
LLD	0.277	0.598	1.122	+	6.780	0.009	Male > female		
LMN	0.747	0.387	1.160	+	0.357	0.550	−0.334	2.331	0.127
HM	2.7	0.100	1.048	+	16.980	3.78E-05	Female > male		
HC	0.129	0.719	1.281	+	0.213	0.644	−0.375	2.369	0.124
LM	0.004	0.951	1.109	+	1.986	0.159	0.242	2.258	0.133
LLS	**17.728**	**2.54848E-05**	**Female > male**						

Lr_{b1} likelihood ratio test of common slope, p_{b1} p value for Lr parameter, b_1 com common slope for RMA, W_{b0} Wald statistic for intercept, p-W p value for W_{b0}, $b0$ com common intercept for RMA, W_{shift} Wald statistic for shift, p-W_{shift} p value for W_{shift}. Bold indicates variables with different slope for both sexes absolute bias for untrimmed values (0.0018) is 2.11 times lower than for trimmed values (0.0038). Therefore, we took into account the untrimmed values for multivariate allometry interpretations.

Thirteen variables show the same allometric trend in females and males. Three variables show positive allometry for females and males (LN, LUD, and HC). Six variables show isometry for both sexes (BZ, LO, BPP, TL, LLD, and HM), with BZ being the most isometric variable. Four variables show negative allometry for both sexes (BB, HO, BAP, andLB).

Six variables show sexual dimorphism in allometric terms (HR, LZ, LUS, LMN, LM, and LLS).

10.4 Discussion and Conclusions

The allometric trends observed in *Lagostomus maximus* are similar to those reported for other caviomorph rodents (Nelson and Shump 1978; Nasif and Abdala 2015) and also for other mammals such as primates (Flores and Casinos 2011), carnivorans (Giannini et al. 2010), and marsupials (Giannini et al. 2004; Flores et al. 2006). The breadth of the braincase, height of the occipital plate, and length of the basicranium show negative allometric trends, and the length of the orbit shows isometry in both sexes, so the neurocranium grows slower than the splanchnocranium during postnatal ontogeny, indicating a higher growth rate of the neurocranium in prenatal stages, as observed in other mammals (e.g., Emerson and Bramble 1993). Length of the nasals and length of the upper diastema, both with a positive allometric trend in both sexes, indicate a rapid development of the rostrum compared with the rest of the skull.

Five variables showed sexual dimorphism in allometric terms: zygomatic breadth (for the bivariate analysis), height of the rostrum, zygomatic length, length of the mandible anterior to the masseteric notch, and length of the mandible (for the multivariate analyses). These variables are closely related to the degree of development of temporal and masseteric muscles, a feature that was previously reported for this species (Rasia et al. 2011, in preparation).

Four variables showed allometric differences between sexes that could be related to variation in masticatory habits (anterior and posterior breadth of palate in the bivariate analysis and length of the upper and lower cheek teeth in the multivariate analysis).

The results of bivariate and multivariate analyses are in agreement with qualitative differences observed by Rasia et al. (2011, in preparation) and indicate that sexual dimorphism in the skull of *Lagostomus maximus* is related to differences in mastication/chewing. Variation between sexes in the masticatory musculature has been reported for murids (Daniels et al. 2008) without stating functional implications. Sexual size dimorphism in plains vizcachas may be a response to environmental pressure, as has been observed in other vertebrates (see Badyaev 2002).

Table 10.4 Multivariate allometry analysis of *Lagostomus maximus*

				Untrimmed				Trimmed			
Variable	Sex	Observed CA	Observed departure	Resampled CA	Bias	%99 confidence interval	Growth trend	Resampled CA	Bias	%99 confidence interval	Growth trend
LN	Female	0.309	0.08	0.306	0.002	0.267–0.345	+	0.305	0.002	0.276–0.334	+
	Male	0.298	0.069	0.295	0.001	0.276–0.314	+	0.305	−0.003	0.296–0.314	+
BZ	Female	0.225	−0.004	0.225	−0.0003	0.219–0.232	=	0.226	−0.0005	0.219–0.232	=
	Male	0.229	0	0.228	0.0003	0.220–0.236	=	0.229	−0.0003	0.224–0.234	=
BB	Female	0.079	−0.15	0.081	−0.0008	0.067–0.099	–	0.078	0.001	0.066–0.091	–
	Male	0.065	−0.164	0.066	−0.0003	0.054–0.077	–	0.066	−0.0004	0.057–0.075	–
HR	Female	0.207	−0.022	0.207	0.0004	0.185–0.228	–	0.204	0.002	0.191–0.216	–
	Male	0.227	−0.002	0.222	0.002	0.194–0.250	=	0.239	−0.006	0.232–0.247	+
LO	Female	0.252	0.023	0.259	−0.003	0.195–0.322	=	0.223	0.014	0.201–0.246	=
	Male	0.199	−0.03	0.206	−0.003	0.171–0.241	=	0.189	0.005	0.177–0.201	–
LZ	Female	0.243	0.014	0.242	0.0003	0.224–0.259	=	0.241	0.001	0.229–0.252	+
	Male	0.24	0.011	0.242	−0.001	0.232–0.252	+	0.237	0.001	0.233–0.242	+
HO	Female	0.118	−0.111	0.116	0.001	0.094–0.138	–	0.119	−0.001	0.109–0.130	–
	Male	0.167	−0.062	0.167	0.0001	0.155–0.179	–	0.168	−0.0005	0.161–0.175	–
LUD	Female	0.258	0.029	0.258	0.0002	0.229–0.286	+	0.258	0.0002	0.241–0.274	+
	Male	0.276	0.047	0.273	0.001	0.254–0.293	+	0.284	−0.004	0.276–0.291	+
LUS	Female	0.307	0.078	0.309	−0.001	0.289–0.331	+	0.311	−0.002	0.295–0.327	+
	Male	0.259	0.03	0.271	−0.006	0214–0328	=	0.235	0.012	0.226–0.244	=
BAP	Female	0.018	−0.211	0.025	−0.004	- 0.051- 0.102	–	0.007	0.006	- 0.055- 0.068	–
	Male	0.122	−0.107	0.122	−0.00005	0.09–0.155	–	0.125	−0.001	0.097–0.152	–

BPP	Female	0.219	−0.01	0.219	0.0004	0.192–0.246	=	0.214	0.003	0.194–0.234	=
	Male	0.208	−0.021	0.214	−0.003	0.183–0.245	=	0.194	0.007	0.186–0.202	−
LB	Female	0.173	−0.056	0.169	0.001	0.152–0.189	−	0.173	−0.000006	0.156–0.189	−
	Male	0.196	−0.033	0.192	0.002	0.167–0.217	−	0.203	−0.004	0.194–0.212	−
TL	Female	0.238	0.009	0.238	0.0003	0.223–0.253	=	0.24	−0.001	0.230–0.249	+
	Male	0.234	0.005	0.231	0.001	0.214–0.248	=	0.242	−0.004	0.237–0.246	+
LLD	Female	0.233	0.004	0.233	0.0002	0.199–0.267	=	0.243	−0.005	0.224–0.262	=
	Male	0.249	0.02	0.241	0.004	0.199–0.284	=	0.268	−0.009	0.259–0.276	+
LMN	Female	0.236	0.007	0.238	−0.0008	0.213–0.262	=	0.237	−0.001	0.219–0.255	=
	Male	0.258	0.029	0.257	0.0004	0.248–0.268	+	0.259	−0.001	0.252–0.267	+
HM	Female	0.215	−0.014	0.217	−0.0009	0.178–0.255	=	0.226	−0.005	0.211–0.241	=
	Male	0.235	0.006	0.239	−0.002	0.22–0.259	=	0.229	0.003	0,222-0,237	=
HC	Female	0.277	0.048	0.279	−0.0008	0.242–0.316	+	0.279	−0.001	0.258–0.300	+
	Male	0.283	0.054	0.283	−0.0001	0.271–0.296	+	0.283	−0.0002	0.272–0.295	+
LM	Female	0.239	0.01	0.24	−0.00008	0.222–0.259	=	0.242	−0.001	0.229–0.254	+
	Male	0.246	0.017	0.245	0.0008	0.232–0.257	+	0.25	−0.002	0.245–0.256	+
LLS	Female	0.285	0.056	0.29	−0.002	0.248–0.333	+	0.285	0.0002	0.262–0.308	+
	Male	0.239	0.01	0.248	−0.004	0.208–0.287	=	0.224	0.008	0.216–0.233	=

AC allometry coefficient; isometry value = 0.229

In summary, the neurocranuim of *Lagostomus maximus* grows slower than the splanchnocranium during postnatal stages, with the fast development of the rostrum. In addition, there is sexual dimorphism in the growth rates of some portions of the skull related to the origin/attachment of temporal and masseteric muscles (involved in chewing) and other portions of the skull also closely related to mastication.

Acknowledgments I want to thank the curators P. Teta (MACN), S. Lucero (MACN), D. Verzi (MLP), and I. Olivares (MLP) for allowing access to the collections under their care. I am also grateful to my PhD advisors, Adriana Candela and David Flores, for their support, especially David for his help with the analyses, and to Valentina Segura for the discussion that improved this chapter.

References

Abdala F, Flores DA, Giannini NP (2001) Postweaning ontogeny of the skull of *Didelphis albiventris*. J Mammal 87:190–200

Acuña F, Barbeito CG, Portiansky EL, Ranea G, Nishida F, Miglino MA, Flamini MA (2020) Early and natural embryonic death in *Lagostomus maximus*: association with the uterine glands, vasculature, and musculature. J Morphol 281(7):710–724

Acuña F, Barbeito CG, Portiansky EL, Miglino MA, Flamini MA (2021) Prenatal development in *Lagostomus maximus* (Rodentia, Chinchillidae): a unique case among eutherian mammals of physiological embryonic death. J Morphol 282(5):720–732

Aeschbach M, Carrillo JD, Sánchez-Villagra MR (2016) On the growth of the largest living rodent: postnatal skull and dental shape changes in capybara species (*Hydrochoerus* spp.). Mamm Biol 81:558–570

Badyaev AV (2002) Growing apart: an ontogenetic perspective on the evolution of sexual size dimorphism. TREE 17(8):369–378

Begall S, Burda H (1998) Reproductive biology (behaviour, breeding and post natal development) in subterranean mole-rats, *Cryptomys hottentotus* (Bathyergidae). Z Saugetierkd 54:360–376

Boutin S, Larsen KW (1993) Does food availability affect growth and survival of males and females differently in a promiscuous small mammal, *Tamiasciurus hudsonicus*? J Anim Ecol 62:364–370

Branch LC (1993a) Social organization and mating system of the plains viscacha (*Lagostomus maximus*). J Zool 229:473–491

Branch LC (1993b) Seasonal patterns of activity and body mass in the plains vizcacha, *Lagostomus maximus* (family Chinchillidae). Can J Zool 71:1041–1045

Cabrera A, Yepes J (1960) Mamíferos sudamericanos, vol 2. Ediar, Buenos Aires, p 160

Daniels DW, Tian Z, Barton ER (2008) Sexual dimorphism of Murine masticatory muscle function. Arch Oral Biol 53(2):187–192

Derrickson EM (1988) Patterns of postnatal growth in a laboratory colony of *Peromyscus leucopus*. J Mammal 69(1):57–66

Emerson SB, Bramble DM (1993) Scaling, allometry, and skull design. In: Hanken J, Hall BK (eds) The skull. Volume 3: functional and evolutionary mechanisms. Chicago University Press, Chicago

Farmer MA, German RZ (2004) Sexual dimorphism in the craniofacial growth of the guinea pig (*Cavia porcellus*). J Morphol 259:172–181

Ferreira JD, Dozo MT, Bubadué J, Kerber L (2022) Morphology and postnatal ontogeny of the cranial endocast and paranasal sinuses of capybara (*Hydrochoerus hydrochaeris*), the largest living rodent. J Morphol 283:66–90
Flores DA, Casinos A (2011) Cranial ontogeny and sexual dimorphism in two New World monkeys: *Alouatta caraya* (Atelidae) and *Cebus apella* (Cebidae). J Morphol 272(6):744–757
Flores DA, Giannini N, Abdala F (2006) Comparative postnatal ontogeny of the skull in the australidelphian metatherian *Dasyurus albopunctatus* (Marsupialia: Dasyuromorpha: Dasyuridae). J Morphol 267:426–440
Garrone MC, Cerda IA, Tomassini RL (2021) Ontogenetic variability in the limb bones histology of plains vizcacha (*Lagostomus maximus*, Chinchillidae, Rodentia): implications for life history reconstruction of fossil representatives. Hist Biol 33(4):558–573
Giannini NP, Abdala F, Flores DA (2004) Comparative postnatal ontogeny of the skull in *Dromiciops gliroides* (Marsupialia: Microbiotheriidae). Am Mus Novit 3460:1–17
Giannini NP, Segura V, Giannini MI, Flores D (2010) A quantitative approach to the cranial ontogeny of the puma. Mamm Biol 75:547–554
Jackson JE (1990) Growth rates in vizcacha (*Lagostomus maximus*) in San Luis, Argentina. Vida Silv Neotrop 2(2):52–55
Jackson TP, Van Aarde RJ (2003) Sex- and species-specific growth patterns in cryptic African rodents, *Mastomys natalensis* and *M. coucha*. J Mammal 84(3):851–860
Jackson JE, Branch LC, Villareal D (1996) *Lagostomus maximus*. Mamm Species 543:1–6
Jolicoeur P (1963a) The multivariate generalization of the allometry equation. Biometrics 19:497–499
Jolicoeur P (1963b) The degree of generality of robustness in *Martes americana*. Growth 27:1–27
Kam M, Degen AA (1994) Body mass and growth rate of fat sand rat (*Psammomys obesus*) pups: effect of litter size and water content of *Atriplex halimus* consumed by pregnant and lactating females. Funct Ecol 8:351–357
Lammers AR, Dziech HA, German RZ (2001) Ontogeny of sexual dimorphism in *Chinchilla lanigera* (Rodentia: Chinchillidae). J Morphol 82(1):179–189
Lima M, Bozinovic F, Jaksic FM (1997) Body mass dynamics and growth patterns of leaf-eared mice *Phyllotis darwini* in a semi-arid region of the Neotropics. Acta Theriol 42:15–24
Llanos AC, Crespo JA (1952) Ecología de la vizcacha (*Lagostomus maximus maximus* Blainv.) en el nordeste de la Provincia de Entre Ríos. Rev Invest Agrícol 6:289–378
Manly BFJ (1997) Randomization, bootstrap, and Monte Carlo methods in biology, 2nd edn. Chapman & Hall, New York
Monteiro LR, Lessa LG, Abe AS (1999) Ontogenetic variation in skull shape of *Thrichomys apereoides* (Rodentia: Echimyidae). J Mammal 80(1):102–111
Nasif NL, Abdala F (2015) Craniodental ontogeny of the pacarana *Dinomys branickii* Peters 1873 (Rodentia, Hystricognathi, Caviomorpha, Dinomyidae). J Mammal 96(6):1224–1244
Nelson TW, Shump KA Jr (1978) Cranial variation and size allometry in *Agouti paca* from Ecuador. J Mammal 59(2):387–394
Niklas KJ (1994) Plant allometry. In: The scaling of form and process. University of Chicago Press, Chicago
Núñez HA, Bee-De-Speroni N (1997) Análisis cuantitativo del desarrollo encefálico prenatal en la vizcacha de las pampas *Lagostomus maximus maximus* (Rodentia: Chinchillidae), Argentina. Rev Chil Hist Nat 70:59–65
Quenouille MH (1956) Notes on bias in estimation. Biometrika 43:353–360
Radinsky LB (1981a) Evolution of the skull shape in carnivores. 1. Representative modern carnivores. Biol J Linn Soc 15:369–388
Radinsky LB (1981b) Evolution of the skull shape in carnivores. 2. Additional modern carnivores. Biol J Linn Soc 16:337–355
Rasia LL (2016) Los Chinchillidae (Rodentia, Caviomorpha) fósiles de la República Argentina: sistemática, historia evolutiva y biogeográfica, significado bioestratigráfico y paleoambiental. PhD dissertation, Universidad Nacional de La Plata, La Plata, Argentina

Rasia LL, Candela AM, Flores DA (2011) Sexual dimorphism in living and fossil plains viscachas (Rodentia, Caviomorpha, Chinchillidae) of Argentina: a qualitative approach. Ameghiniana 48(4):R196–R197

Rasia LL, Segura V, Candela AM, Flores DA (in preparation) Skull ontogeny and sexual dimorphism in the caviomorph rodent *Lagostomus maximus* (Chinchillidae): a qualitative approach

Rinderknecht A, Bostelmann E, Ubilla M (2018) Making a giant rodent: cranial anatomy and ontogenetic development in the genus *Isostylomys* (Mammalia, Hystricognathi, Dinomyidae). J Syst Palaeontol 16(3):245–261

Schulte-Hostedde AI (2007) Chapter 10: sexual size dimorphism in rodents. In: Wolff JO, Sherman PW (eds) Rodent societies: an ecological & evolutionary perspective. The Chicago University Press, Chicago/London

Tarnawski BA, Cassini GH, Flores DA (2014a) Allometry of the postnatal cranial ontogeny and sexual dimorphism in *Otaria byronia* (Otariidae). Acta Theriol 59(1):81–97

Tarnawski BA, Cassini GH, Flores DA (2014b) Skull allometry and sexual dimorphism in the ontogeny of the southern elephant seal (*Mirounga leonina*). Can J Zool 92(1):19–31

Tarnawski BA, Flores DA, Cassini GH, Cappozzo LH (2015) A comparative analysis on cranial ontogeny of South American fur seals (Otariidae: *Arctocephalus*). Zool J Linnean Soc 173:249–269

Tavares WC, Pessôa LM, Seuánez HN (2016) Phylogenetic and size constrains on cranial ontogenetic allometry of spiny rats (Echimyidae, Rodentia). J Evol Biol 29(9):1752–1765

Tukey JW (1958) Bias and confidence in not quite large samples. Ann Math Stat 29:614

Vasallo AI (2000) Alometría e isometría en varias especies de roedores caviomorfos, con comentarios sobre la estructura del aparato masticatorio del orden Rodentia. Mastozool Neotrop 7(1):37–46

Vassallo AI, Mora MS (2007) Intraspecific scaling and ontogenetic growth patterns of the skull in living and fossil ctenomyid and octodontid rodents (Caviomorpha: Octodontoidea). In: Kelt D, Kaspin D (eds) The quintessential naturalist: honoring the life and legacy of Oliver P. Pearson. Special Publication of the University of California Press

Verzi DH, Álvarez A, Olivares AI, Morgan CC, Vassallo AI (2010) Ontogenetic trajectories of key morphofunctional cranial traits in South American subterranean ctenomyid rodents. J Mammal 91(6):1508–1516

Weckerley FW (1998) Sexual-size dimorphism: influence of mass and mating systems in the most dimorphic mammals. J Mammal 79(1):33–52

Weir BJ (1970) The management and breeding of some more hystricomorph rodents. Lab Anim 4:83–97

Weir BJ (1971) The reproductive physiology of the plains viscacha, *Lagostomus maximus*. J Reprod Fertil 25(3):355–363

Weir BJ (1974) The tuco-tuco and plains viscacha. In: Rowlands IW, Weir BJ (eds) The biology of hystricomorph rodents, vol 34. Symposia of the Zoological Society of London, pp 113–130

Wilson LAB (2010) The evolution of morphological diversity in rodents: patterns of cranial ontogeny. University of Zurich

Wilson LAB, Sánchez-Villagra MR (2009) Heterochrony and patterns of cranial suture closure in hystricognath rodents. J Anat 214:339–354

Wilson LAB, Sánchez-Villagra MR (2010) Diversity trends and their ontogenetic basis: an exploration of allometric disparity in rodents. Proc R Soc B 277:1227–1234

Yoshinaga Y, Okayama T, Ohno W, Shiraishi S (1997) Growth, development, and reproductive patterns in the Japanese field vole, *Microtus montebelli*. J Mammal 78:830–838

Chapter 11
Morphophysiology of the Female Reproductive System

Francisco Acuña, Mirta Alicia Flamini, Enrique Leo Portiansky, and Claudio Gustavo Barbeito

Abstract The organs of the vizcacha's reproductive system are like those of other eutherian mammals. However, some organs exhibit species-specific morphological characteristics. The ovaries of the vizcacha are flat and have multiple folds, which increase their surface area. This unique feature enables the polyovulation of 200–800 oocytes per estrus, the highest ovulatory rate observed among mammals. The morphology of the oviducts and uterine horns is comparable to that of other mammals. However, the caudal segments of the uterine horns have significantly thicker walls, tunics, and glandular and vascular areas as compared to the cranial and middle segments. These anatomical differences have not been reported in other mammals. Studies on female fetuses have shown that the uterine horns already exhibit differences along the craniocaudal direction during prenatal life, which have been associated with spontaneous embryonic death in the cranial and middle implantation sites. Additionally, the vizcacha has a non-muscular cervix and a well-developed paraurethral gland (female prostate) found in all analyzed specimens. The expression of glycoconjugates and glycosidic residues in the organs of the female reproductive system varies according to the reproductive physiological state.

Keywords Vizcacha · Ovary · Uterine tubes · Uterus · Vagina · Female prostate

F. Acuña (✉) · C. G. Barbeito (✉)
Laboratorio de Histología y Embriología Descriptiva, Experimental y Comparada, Facultad de Ciencias Veterinarias, Universidad Nacional de La Plata, La Plata, Argentina

Consejo Nacional de Investigaciones Científicas y Técnicas (CONICET), La Plata, Argentina
e-mail: facuna@fcv.unlp.edu.ar; barbeito@fcv.unlp.edu.ar

M. A. Flamini
Laboratorio de Histología y Embriología Descriptiva, Experimental y Comparada, Facultad de Ciencias Veterinarias, Universidad Nacional de La Plata, La Plata, Argentina

E. L. Portiansky
Consejo Nacional de Investigaciones Científicas y Técnicas (CONICET), La Plata, Argentina

Laboratorio de Análisis de Imágenes, Facultad de Ciencias Veterinarias, Universidad Nacional de La Plata, La Plata, Argentina

L. L. Rasia et al. (eds.), *Plains Vizcachas*, https://doi.org/10.1007/978-3-031-49487-1_11

11.1 Introduction

In the 1970s, Dr. Barbara Weir, a researcher at the Institute of Comparative Physiology of the Zoological Society of London, described morphological and physiological characteristics of the reproductive system of the female *Lagostomus maximus*, the plains vizcacha, based on studies carried out in a colony made up of specimens of both sexes that were transported from Argentina to England (Carter et al. 2023). Although the organs of the reproductive system of the female vizcacha do not differ from that of other eutherian mammals, certain particularities are exclusive to the species (Weir 1971b, c) (Fig. 11.1).

The estrous cycle of the vizcacha lasts 45 days. During estrus, the opening of the vaginal closing membrane, swelling of the vulva, and the secretion of yellowish mucus are observed. The vagina remains open for 7 days, and mating usually occurs towards the end of this period. Intercourse can be identified by the presence of blood at the vaginal opening and confirmed by the detection of copulatory plugs or sperm in a vaginal smear (Weir 1971b). In the vizcacha, poly-implantation occurs since each uterine horn has between 5 and 6 implantation sites (IS). At 35 days after coitus, the cranial and medial IS are reabsorbed. In contrast, caudal IS continue growing and developing. After 155 days of gestation, the female delivers two precocial pups (Weir 1971a, b; Roberts and Weir 1973; Jackson 1989; Flamini et al. 2019).

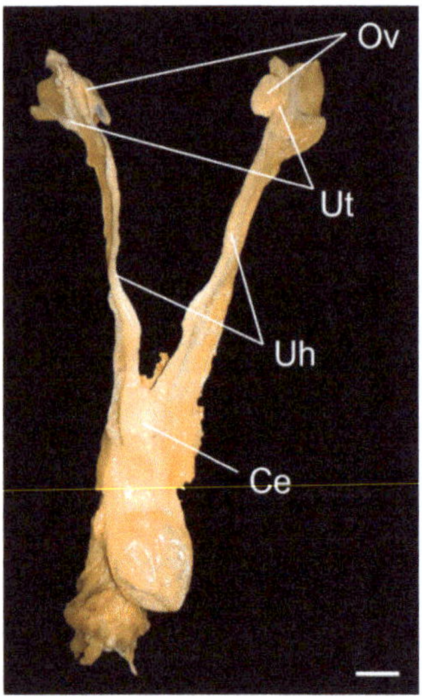

Fig. 11.1 Reproductive organs of a *L. maximus* female. Abbreviations: Ce, cervix; Ov, ovaries; Ut, uterine tubes; Uh, uterine horns. Scale bar = 1 cm. (License Number 5619441104760)

11.2 Ovaries

The shape of the ovaries of *L. maximus* is atypical as compared to other hystricomorphs and eutherian mammals. They are flat and irregular due to multiple folds which increase the surface of the organ. This is related to its polyovulatory capacity of 200–800 oocytes per estrus (Weir 1971c) (Fig. 11.2a). There are examples of variations in ovarian morphology between mammals of different orders and even within species of the same order. In *Xenarthra*, the ovaries of *Chaetophractus vellerosus* (piche llorón) and *Zaedyus pichiy* (pichi ciego) are cylindrical with rounded ends. They have two different zones, a ventral parenchymal zone called "cortex" and a vascular dorsal zone known as "medulla," with a typical structure. In contrast,

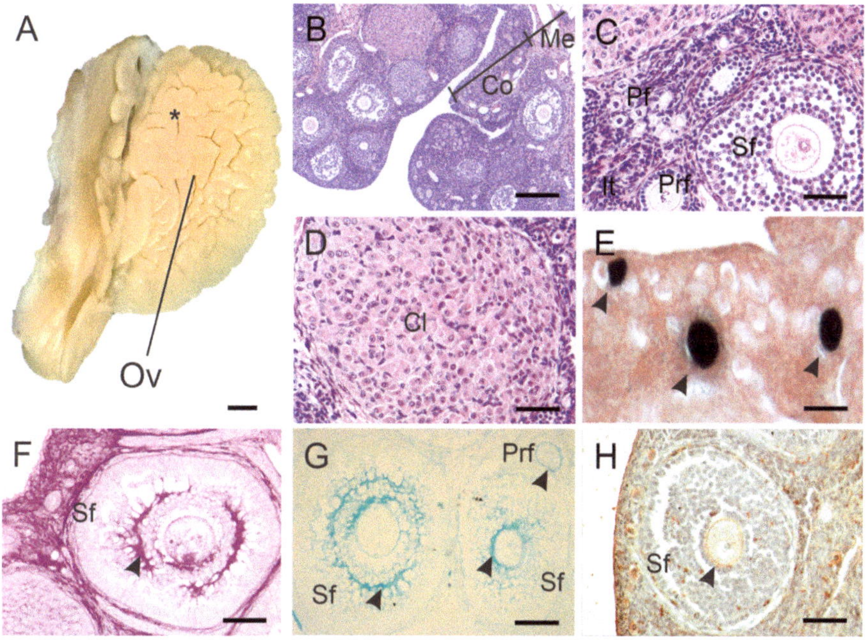

Fig. 11.2 Ovary and oviduct of a *L. maximus* female. (**a**) Macroscopic morphology. The asterisk indicates an ovarian fold. (**b**) Microscopic morphology of the ovarian folds of females with multiple gestations. Ovarian cortex and medulla. (**c**) Structures in the ovary cortex of a female in follicular phase: interstitial tissue, primordial follicles, primary follicles, and secondary follicles. (**d**) Primary corpus luteum in the ovary cortex of a full-term pregnant female. (**e**) Follicles with dark and homogeneous deposits (arrowheads). von Kossa staining technique (License Number 5623250265228). (**f**) Secondary follicle with positive secretion to the periodic acid-Schiff reagent histochemical technique (arrowhead). (**g**) Primary and secondary follicles with positive to the Alcian blue pH 2.5 histochemical technique of secretory and pellucid zones. (**h**) Secondary follicle with a positive reaction for the lectin *Ricinus communis agglutinin*-I (RCA-I) of the zona pellucid (License number 5623641038011). Abbreviations: Co, cortex; It, interstitial tissue; Me, medulla; Pf, primary follicles; Prf, primary follicles; Sf, secondary follicles. (**b–d**) hematoxylin-eosin staining. Scale bar = 0.2 cm (**a**), 500 μm (**b**), 200 μm (**c, d**), 100 μm (**f–h**) and 50 μm (**e**)

in *Dasypus hybridus* (mulita), the ovaries are reniform, where the concavity corresponds to the parenchymal zone and the convexity to the vascular zone (Codón and Casanave 2000). In the ovaries of the perissodactyl *Equus ferus caballus* (horse), the region where the corpora lutea and ovarian follicles are located is limited to an area of the ovarian surface called the ovulation fossa. This fossa is the only site where the follicles approach the ovarian surface and ovulation occurs (Alves et al. 2018). However, these variations are not related to the ovulatory rate as occurs in the vizcacha. The insectivore *Elephantulus myurus* follows the vizcacha in its polyovulatory rate (50–120 oocytes per estrus), although its ovarian morphology does not include structures such as folds that increase the surface area and facilitate ovulation (van der Horst and Gillman 1941; Tripp 1971). Therefore, the relationship between the ovarian morphology represented by the folds and the ovulatory rate of the vizcacha is unique among eutherians.

The microscopic morphology of each ovarian fold in the vizcacha includes a cortex and a medulla (Flamini et al. 2009) (Fig. 11.2b). This differentiation is already observed in prenatal life, as observed in fetuses at 90 days of gestation (Flamini et al. 2020). In the ovaries of adult females, there are variations in the ratio of cortex and medulla, depending on the physiological state studied (anestrus, follicular phase, early pregnancy, and term pregnancy). During anestrus, the ovarian cortex is proportionally smaller than the medulla, and growing ovarian follicles, few corpora lutea, and atretic follicles are seen. In contrast, in females in the follicular phase, the ovarian folds present a predominance of cortex with a greater number of developing ovarian follicles, especially tertiary ovarian follicles, as compared to anestrous females (Fig. 11.2c). In addition, primary and secondary corpora lutea are evident. In early pregnant females, the ovarian cortex is proportionally larger than the medulla, and there is a predominance of primary corpora lutea and few growing ovarian follicles. In females pregnant at term, the predominance of the cortex is maintained, and the corpora lutea are the most predominant structures (Flamini et al. 2009) (Fig. 11.2d).

Secondary corpora lutea (SCL) are those that contain a retained oocyte (Weir and Rowlands 1974) with a degenerated zona pellucida (Felipe and Masson 2005). In vizcacha, SCL are abundant during mid-gestation and originate through pseudo-ovulation, which occurs due to reactivation of the hypothalamic-pituitary-ovarian axis caused by a decrease in progesterone serum levels in mid-gestation pregnancy (Dorfman et al. 2013; Fraunhoffer et al. 2017). In other hystricomorphs, such as chinchillas and African porcupines, SCL originate from infertile cycle (Weir and Rowlands 1974). In the nutria *Myocastor coypus*, SCL were also observed, although their origin has not been determined (Felipe and Masson 2005). In other eutherian taxa, such as the horse, SCL are also formed, but in this case, they are a consequence of the activity of equine chorionic gonadotropin secreted by the endometrial cups (Stewart and Allen 1981; Klein 2015).

The ovarian structures observed in the four physiological states of the vizcacha present partial similarities with the follicular and luteal phases of the estrous cycle of other mammals (Goldman et al. 2007). However, these structures are smaller in size as compared to other species, possibly due to their greater number in the vizcacha, especially of the tertiary follicles and corpora lutea (Flamini et al. 2009). The presence of numerous primordial follicles could be related to a follicular renewal mechanism through neo-oogenesis in adult life (Inserra et al. 2013). This topic will be discussed in greater depth in Chap. 16. It is important to highlight that the ovaries in the four stages studied by Flamini et al. (2009) usually present calcified follicles (Fig. 11.2e). Although the presence of calcium deposits in the ovaries is uncommon in mammals, it has also been observed in the ovarian cortex of baboons (Ishmael 1975) and the ovarian stroma of human females (Clement and Cooney 1992).

In the ovarian cortex of the vizcacha, there are interstitial cells that form the interstitial tissue (Gil et al. 2007). These cells surround the follicles and the corpora lutea in pregnant females (Flamini et al. 2009) (Fig. 11.2c). The ovarian interstitial cells of the vizcacha are eosinophilic and have hyperchromatic nuclei (Flamini et al. 2009). The presence of eosinophilia was also observed in ovarian interstitial cells during pregnancy in the hystricomorph rodent *Octodon degus* (Octodontidae) (Weir 1974). The distribution and morphology of ovarian interstitial tissue vary among different mammalian species (Guraya and Greenwad 1964). In other hystricomorphs such as *Cavia porcellus* (Caviidae) and *Myoprocta acouchy* (Dasyproctidae), interstitial tissue is scant (Rood and Weir 1970; Weir 1971a; Weir and Rowlands 1974). However, in *Dasyprocta leporina* (Dasyproctidae, Weir 1971a) and *M. coypus* (Myocastoridae, Hillermann et al. 1958; Felipe et al. 1999), interstitial tissue is abundant. These interstitial tissue cells are generated from thecal cell differentiation of atretic follicles (Guraya and Greenwad 1964; Weir and Rowlands 1974). The histological, histochemical, and ultrastructural characteristics of the ovarian interstitial tissue of the vizcacha are related to the storage of precursors for steroidogenesis (Gil et al. 2007).

Beyond the morphological differences between the different physiological states mentioned, there are biochemical variations between them. Thus, differences were observed in the zona pellucida of the ovarian follicles, noting that the types of glycoconjugates and specific sugar residues vary between non-pregnant and pregnant females. Using histochemical and lectin-histochemical techniques, it was determined that the zona pellucida of the ovarian follicles of non-pregnant and pregnant females expresses several glycoconjugates and glycoside residues. For example, the zona pellucida of non-pregnant females expresses β-galactose, while this sugar is lacking in pregnant females. On the other hand, residues such as β-D-mannose and α-D-glucose are expressed in the zona pellucida of the ovarian follicles of pregnant females, but not in non-pregnant ones (Flamini et al. 2009; Acuña et al. 2018) (Fig. 11.2f–h). These differences could be a consequence of the variations in the serum levels of ovarian hormones, such as estrogens and progesterone (Dorfman et al. 2013), and be related to the fertilization of only 10% of the ovulated oocytes (Roberts and Weir 1973).

11.3 Uterine Tubes

The uterine tubes of the vizcacha have a very tortuous disposition and are connected to the ovary through the mesovarium (Weir 1971c) (Fig. 11.3a–c). Flamini et al. (2009) also described the microscopic morphology of the uterine tubes (fimbriae, infundibulum, ampulla, isthmus) in different reproductive states (Fig. 11.3d, e). As in other mammals, the vizcacha uterine tube epithelium is composed of ciliated cells and mucous-secreting cells. In anestrus females, there are abundant mucous secretory cells in the luminal epithelium, while in follicular phase females ciliated cells predominate (Fig. 11.3e). In pregnant females, both early and term gestations, ciliated and secretory cells are observed in variable proportions (Flamini et al. 2014). These ciliated and secretory cells perform functions related to the production and movement of oviductal fluid, as well as the transport of oocytes and blastocysts,

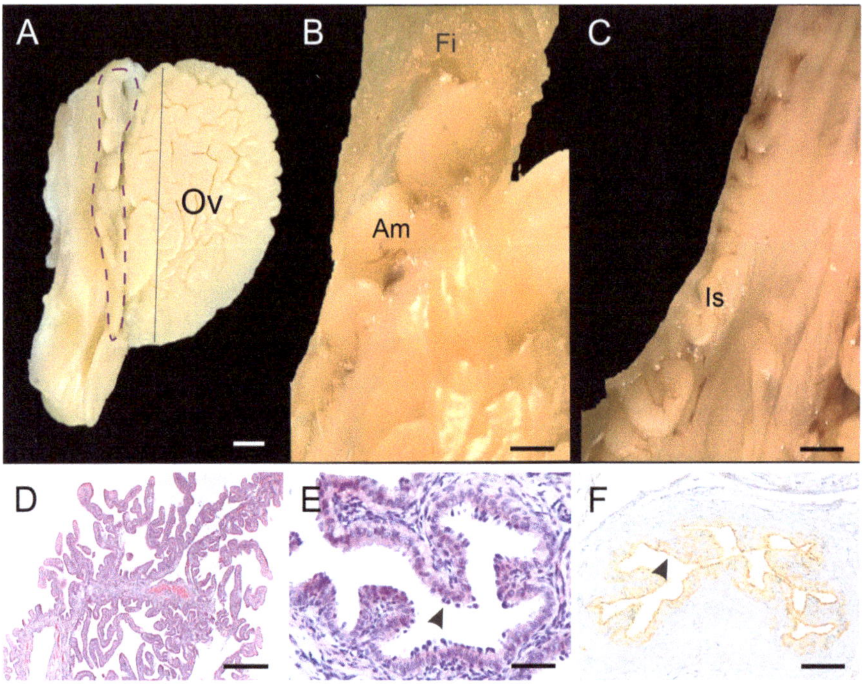

Fig. 11.3 Uterine tubes of a *L. maximus*. (**a**) Macroscopic morphology of the uterine tubes, enclosed with a broken line, in closed relation to the ovary. (**b**) Macroscopic view of the fimbriae and ampulla (License Number 5623710018591). (**c**) Macroscopic view of the isthmus. (**d**) Microscopic morphology of the digitiform projections that form the fimbriae. (**e**) Folds of the mucosa of the ampulla. The arrowhead indicates the secretory cells. (**f**) Glycocalyx of the luminal epithelium of the isthmus containing α-D-N-acetylgalactosamine glycosidic residues, according to labeling with the lectin *Dolichos biflorus agglutinin* (DBA). (**d, e**), hematoxylin-eosin stain. Abbreviations: Am, ampoule; Fi, fimbriae Is, isthmus; Ov, ovary. Scale bar = 0.2 cm (**a**), 1 cm (**b, c**), 500 μm (**d**), 200 μm (**f**), 100 μm (**e**)

functions like those found in other eutherians. In addition, the characteristics of the glycoconjugates present in mucosal cells and in oviductal secretions also vary between pregnant and non-pregnant vizcachas. During the follicular phase, the oviductal glycocalyx shows a higher intensity of expression of glycosidic residues, such as N-acetylgalactosamine, L-fucose, and N-acetylneuraminic acid, as compared to anestrus females and those with early and intermediate gestations (Fig. 11.3f). These changes are related to hormonal variations (Flamini et al. 2014), since ovarian hormones regulate the activity of glycosyltransferases in the uterine tubes and uterus (Tulsiani et al. 1997).

11.4 Uterus

The uterus of the vizcacha is bicornuate, and its histological structure is like that of other eutherians. Like the ovaries and uterine tubes, the uterus presents variations related to reproductive status (Fig. 11.1a). It is composed of three tunics endometrium, myometrium, and perimetrium (Fig. 11.4a, b). The endometrium contains epithelial tissue like that of the uterine tubes. In anestrous females, the luminal epithelium is composed mainly of secretory cells and has few ciliated cells, whereas, in the other stages, ciliated cells are more abundant. The uterine glands are scarce during anestrus and abundant during the follicular phase. In early gestating females,

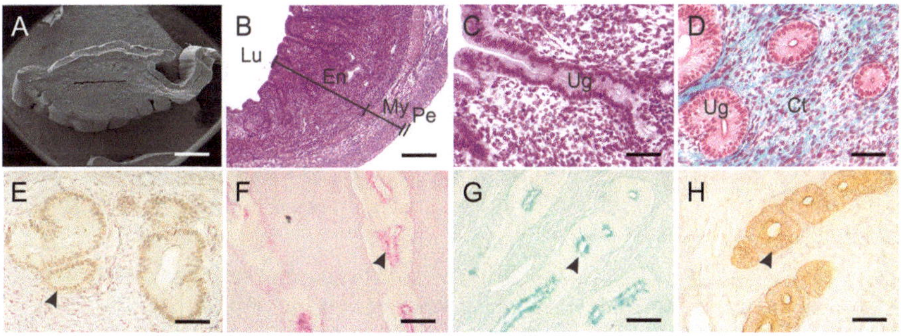

Fig. 11.4 Uterine horns of *L. maximus*. (**a**) Cross section of a uterine horn observed under a scanning electron microscope. (**b**) Microscopic morphology. Organization of the uterine wall into endometrium, myometrium, and perimetrium. (**c, d**) Uterine glands of the uterine horns of non-pregnant females. (**e**) Nuclear expression of the progesterone receptor in cells from the uterine glands of non-pregnant females (arrowhead). (**f**) Glycocalyx with neutral glycoconjugates, periodic acid-Schiff reagent histochemical technique (arrowhead). (**g**) Identification of acid glycoconjugates of the glandular glycocalyx (arrowheads) using the Alcian blue pH 2.5 histochemical technique. (**h**) Uterine horn of a non-pregnant female with positive reaction for the lectin *Ulex europaeus agglutinin* (UEA-I) of the endometrial glands (arrowhead). Abbreviations: En, endometrium; Ct, connective tissue; Lu, lumen; My, myometrium; Pe, perimetrium; Ug, uterine gland. (**b, c**) hematoxylin-eosin staining; (**d**) Gomori trichrome stain. Scale bar = 1 mm (**a**), 200 μm (**b**), 100 μm (**d–h**). (**e**) License Number 5623840850494

the glands are deep and tortuous, decreasing in number in females pregnant at term (Flamini et al. 2019) (Fig. 11.4c). The uterine glands express progesterone receptor (RE). Variations in ER expressions have been related to histotroph secretion and its relationship with the embryonic death and resorption (Fig. 11.4d).

The endometrial stroma, also known as the lamina propria, varies depending on the physiologic state (Fig. 11.4d). During anestrus, there are abundant fibroblasts and migrating cells, mainly lymphocytes. Collagen fibers are thin and sparse. In pregnant females, the endometrial connective tissue is less cellular and looser. The blood vessels are small, congested, and close to the surface epithelium (Flamini et al. 2019).

Histochemical, lectin-histochemical, and morphometric studies carried out in the uterus of the vizcacha revealed a heterogeneous pattern of glycoconjugates in the luminal and glandular epithelia, depending on the physiological state (Fig. 11.4f–h) (Flamini et al. 2019).

A particularly interesting aspect of the vizcacha uterine horns is its heterogeneity. Morphometric studies revealed that the thickness of the uterine wall and its tunics, as well as the areas occupied by the uterine glands and blood vessels in the endometrium and myometrium, increase significantly in the craniocaudal direction in the organ of non-pregnant females. This heterogeneity has been associated with spontaneous embryonic death in this species (Acuña et al. 2020). These aspects are analyzed in greater depth in Chap. 12.

In vizcacha females with early pregnancy, it was observed that both the glandular area and the vascularization remain in the same direction, but with significantly higher values as compared to non-pregnant females. This suggests a possible role for both uterine components during early prenatal development (Acuña et al. 2021). On the other hand, in females with advanced gestation, it was observed that the thickness of the myometrium decreases as compared to non-pregnant females. This is related to the dilation that the uterus experiences during pregnancy, which results in a thinner muscle layer (Flamini et al. 2019).

Anatomical studies carried out on vizcacha fetuses and adult females also showed differences in uterine vascularity along the uterine horns (Flamini et al. 2020). This was confirmed by angiography (Giacchino et al. 2020). It should be noted that uterine heterogeneity is also observed in other eutherians. For example, in the South American camelids *Vicugna pacos* and *Llama glama*, the left uterine horn, where embryonic/fetal and placental development occurs, has a greater diameter and vascularity than the right uterine horn, where implantation does not occur (Mendoza et al. 2014). In the left uterine horns of pregnant *V. pacos*, a decrease in glycocalyx staining was observed when histochemical techniques such as periodic acid-Schiff and Alcian blue were used, as well as in the relative abundance of mucin transcripts 1 (MUC 1) and 16 (MUC16) (Barraza et al. 2018). In addition, an increase in the expression of transcripts of extracellular matrix (ECM) molecules such as collagen type 1 alpha 2 (COL1A2) and collagen type 3 alpha 1 (COL3A1) was observed, as well as enzymes that degrade the ECM, such as metalloproteases (3, −7, −9, and −14) (Barraza et al. 2020). In certain bats, such as *Megaderma lyra* and *Molossus ater*, it was observed that the cranial uterine segments have greater vascularity, being the right uterine horn in *Megaderma lyra* and the left uterine horn in *Molossus ater* (Gopalakrishna and Khaparde 1978; Rasweiler 1990). Similarly, the insectivore *E. myurus* has a greater expansion of the uterine glands in the caudal region of both uterine horns (Tripp 1971). These examples show that uterine heterogeneity is not exclusive to the vizcacha, since in these species uterine development is concentrated in the embryonic IS. In contrast, in the vizcacha, implantation occurs along both uterine horns, although the IS are subsequently reabsorbed in the cranial and middle regions, surviving only the embryos implanted in the thickest areas of the uterine wall and its tunics, with greater vascularity and glandular area (Weir 1971a; Roberts and Weir 1973; Flamini et al. 2019; Acuña et al. 2022).

The uterus of the vizcacha lacks a body. It is composed of the uterine horns and the neck or cervix. The cervix, which is not seen externally, is double due to the presence of an intermediate septum that extends from the bifurcation of the uterine horns to the middle third of the vagina (Flamini et al. 2019) (Fig. 11.5a, b). This

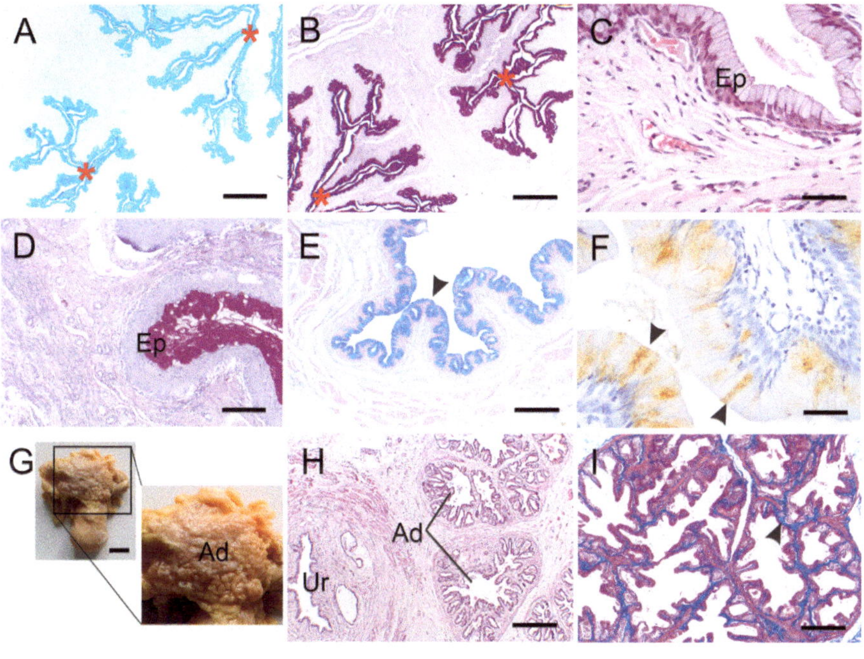

Fig. 11.5 Cervix, vagina, and paraurethral gland of *L. maximus*. (**a**) Glycocalyx and secretion from both uterine endocervix with positive reaction to the histochemical technique of Alcian blue pH 2.5 (AB pH 2.5) (asterisks). (**b**) Same as A, but with the periodic acid-Schiff reagent (PAS) histochemical technique (asterisks). (**c**) Vaginal epithelium of a female in anestrus, with a predominance of mucous cells. (**d**) Stratified cornified epithelium of the vagina of a female in follicular phase. PAS histochemical technique. (**e**) Vagina of pregnant females. PAS-ABpH 2.5 histochemical techniques (arrowhead). (**f**) β-D-Mannose and/or α-D-Glucose glycosidic residues in the cytoplasm of mucosal epithelial cells. Lectin histochemistry using the lectin *Concanavalia ensiformis agglutinin* (CON-A) (arrowheads). (**g**) Macroscopic view of the paraurethral gland. (**h**) Microscopic morphology of the alveolar adenomers of the paraurethral gland. (**i**) Connective tissue of the adenomer wall. Gomori trichrome stain. Abbreviations: Ad, adenomers; Ep, epithelium; Ur, urethra. Scale bar = 0.5 mm (**g**), 500 μm (**a**, **b**, **h**), 200 μm (**d**, **e**, **i**), 100 μm (**f**)

characteristic is also observed in other hystricomorphs such as the chinchilla, the coypu (Hafez 1970), and the black agouti (Mayor et al. 2011). However, in other hystricomorphs such as *C. porcellus* (Breazile and Brown 1976; dos Santos et al. 2016), *Galea spixii* (dos Santos et al. 2014), and *Atherurus africanus* (Mayor et al. 2003), the cervix is unique. This type of double cervix is also found in myomorphic rodents such as rats and mice (da Costa Ferreira Jr 1995; Graham 1966). Therefore, the type of cervix does not appear to have taxonomic value among rodents (Flamini et al. 2019).

11.5 Cervix

As occurs in other hystricomorphs (Felipe et al. 1998), the cervix is divided into an endocervix and an exocervix according to the characteristics of its mucosa. The mucosa of the endocervix is like that of the uterine horn, but its beginning is recognized by the absence of glands (Flamini et al. 2019), as is also observed in the nutria (*M. coypus*) (Felipe et al. 2002), and the black agouti (*Dasyprocta fuliginosa*) (Mayor et al. 2011).

At some moments of the reproductive cycle of the plain vizcacha, the endocervix shows a bistratified epithelium with mucous cells in the superficial layer. No species has been found in the consulted bibliography that presents this type of epithelial tissue in the endocervix. In contrast, the exocervix presents a stratified epithelium. The transition between both types of epithelial tissue is like that found in rats, mice (da Costa Ferreira Jr 1995; Graham 1966), and nutria (Felipe et al. 2001a). The exocervix lacks glands, unlike what occurs in another hystricomorph, *G. spixii*, where they persist (dos Santos et al. 2014). The connective tissue of the cervix is highly cellular, like that observed in the murid rodent *Notomys mitchellii*, whereas in other murid, like the Australian plains mouse, *Pseudomys australis*, collagen fibers are abundant (Breed 1985). No significant differences were found in terms of the abundance and distribution of elastic fibers between the exocervix and the endocervix of the vizcacha, unlike what was observed in the rat (Battlehner et al. 2003). These fibers influence the dilation of the birth canal (Flamini et al. 2019), which should be greater in females who give birth to proportionally larger pups, such as in hystricomorphs.

In the preliminary description of the cervix of the vizcacha, Weir (1971a) argued that this organ does not possess muscle tissue. However, Flamini et al. (2019) observed the presence of bundles of muscle fibers, separated by abundant connective tissue, which does not form a continuous muscle layer as observed in other species of hystricomorphs, including the porcupine (Mayor et al. 2003) and the black agouti (Mayor et al. 2011).

11.6 Vagina

The vagina of the vizcacha exhibits a remarkable length, a characteristic also observed in other hystricomorph rodents such as the chinchilla (Weir 1967), the agouti (Weir 1971c), the guinea pig (Breazile and Brown 1976), and the nutria (Felipe et al. 2000). However, the vagina of the vizcacha presents a distinctive feature not described in previous species: a median raphe that extends from the cervix to the middle of the vagina, resulting in the presence of two cranial cavities and a single caudal cavity in this organ (Weir 1971c; Flamini et al. 2019). It is important to highlight that this median raphe is not exclusive to hystricomorphs, since it is also found in Australian rodents of the genera *Notomys* and *Pseudomys*, although in the latter its extension is considerably less and ends in a wedge shape inside the vaginal lumen (Breed 1985). In addition, the vaginal lumen of the vizcacha exhibits a remarkable irregularity and roughness due to the presence of folds, a common feature in most mammals studied (Weir 1971a, b). The epithelium of the vaginal mucosa of the vizcacha shows notable variations in terms of the number of layers and the characteristics of the superficial cells, depending on its physiological state (Fig. 11.5c, d). These variations in the vaginal epithelium are common in several mammalian species but have not been observed in the nutria (Felipe et al. 2000). In addition, vaginal mucosal epithelial cells present differential expression patterns of glycoconjugates and glycosidic residues, which have been determined by histochemical and lectin-histochemical techniques (Fig. 11.5e, f). These differences seem to be related to the variations in serum hormone levels during the estrous and reproductive cycle of the species (Flamini et al. 2014).

The vaginal mucosa of the vizcacha presents a high concentration of leukocytes, especially during anestrus (Flamini et al. 2014). The presence of leukocytes in the vagina, along with secretions, is a common feature in other hystricomorphs, such as the agouti (Weir 1971a, b, c). Colpocytograms of the vizcacha showed the presence of leukocytes in all physiological states, although they were more abundant during anestrus and pregnancy. This observation coincides with what was described for the nutria, where leukocytes were found not only during the luteal stage but also during the follicular phase (Silva et al. 1997; Felipe et al. 2001b).

In most rodents, the vaginal opening is partially closed until puberty, when it opens, and remains so throughout reproductive life. After puberty, the presence of vaginal occlusions varies according to the species. Unlike what is observed in murine animals, in which the vagina of the non-pregnant female always remains open, in hystricomorphs, the presence of a cellular vaginal closure membrane is observed that disappears during estrus and before delivery. This membrane has been described in several species, including the *C. porcellus*, *Ch. laniger*, *Lagidium viscacia*, and *O. degus*, and it has also been found in other species, although less frequently (Weir 1971b; Weir and Rowlands 1974). In the coypu, belonging to the same suborder, the vaginal closure membrane was not found (Felipe et al. 2001a, b). However, this structure was observed in both pregnant and anestrous vizcachas. The membrane completely seals the vaginal opening, often making it difficult to identify the location of the vaginal opening. In two pregnant females with fetuses similar in size to a newborn calf, the vaginas were observed without the presence of this membrane, which suggests that its disappearance occurs shortly before parturition (Flamini et al. 2012).

11.7 Paraurethral Gland

The female vizcacha has a prostate gland, also known as the paraurethral gland or Skene gland. This structure is constant in several species of mammals, such as the gerbil and the rat (Biancardi et al. 2017), although it is not so in other hystricomorphs. The female prostate is homologous to the male prostate. Its microscopic morphology was described by Flamini et al. (2002) (Fig. 11.5g–i). Recent ultrastructural, histochemical, and lectin-histochemical studies, carried out in females with and without pregnancy, showed the absence of differences between both physiological stages. This would seem to indicate that the organ is not modified by variations in sex hormones (Flamini et al. 2021). The presence of the prostate in female mammals is not constant. In rodents, it has been found in species such as *Rattus rattus* (Shehata 1972), *Arvicanthis niloticus* and *Meriones libycus* (Shehata 1975), *Rattus norvegicus* (Shehata 1980), *Praomys natalensis* (Gross and Didio 1987), and *Meriones unguiculatus* (Inomata et al. 1992; de Jesus et al. 2021). It has also been observed in females, as well as in some species of lagomorphs, carnivores, and bats. Although its function is still unknown, future embryological, functional, and evolutionary studies are required (Biancardi et al. 2017) to determine its role during the reproductive process. In plains vizcacha, histological, histochemical, and ultrastructural studies have revealed characteristics of active secretory cells in glandular cells like those of the male prostate. Morphological studies on this gland should be complemented by future physiological studies to infer its possible functions (Flamini et al. 2002, 2021).

11.8 Mammary Gland

The mammary gland is another organ that has a functional, but not anatomical, relationship with the female reproductive system (König and Liebich 2004). In hystricomorphs, the position of this gland determines the position of the pups during lactation, which is important for predator detection and evasion (Kleiman 1974). In the plains vizcacha, there are two pairs of breasts located on the thorax and laterally (Weir 1971a), although only the first pair is involved in lactation (Goode et al. 1981). Milk, the secretion product of the mammary glands, is of vital importance for the pups. Its composition can vary according to the species (Skibiel et al. 2013). Milk provides nutrients, growth factors, hormones, and immunoglobulins. In plains vizcacha, the chemical composition of the milk includes Na^+, Cl^-, K^+, nitrogen, citrate, lactose, proteins, and fats (Goode et al. 1981). During the final stages of gestation, the mammary glands undergo significant growth and development of the mammary alveoli, as occurs in female mammals in general (Biswas et al. 2022). Lactation in precocial viscacha pups lasts approximately 55 days (Weir 1971a).

11.9 Conclusion

The morphophysiology of the plains vizcacha reproductive system presents similarities with that of other placental mammals, but it also exhibits unique characteristics in terms of ovarian and uterine morphology, as well as its relationship with polyovulation, polyimplantation, and embryonic death. To broaden our understanding of these relationships, multiple studies are needed that encompass molecular, morphological, comparative, energetic, and evolutionary aspects. These investigations will contribute to a better understanding of the reproductive physiology of this species and its place in the context of mammalian reproduction.

Acknowledgments We would like to thank ECAS (Estación de Cría de Animales Silvestres) from the Ministry of Agroindustry of the Province of Buenos Aires (Argentina) for capturing the animals and the Universidad Nacional de La Plata and CONICET for their support.

References

Acuña F, Tano de la Hoz MF, Díaz AO, Portiansky EL, Barbeito CG, Flamini MA (2018) Histochemistry of the zona pellucida of the ovary of a species with natural polyovulation: *Lagostomus maximus* (Rodentia, Hystricomorpha, Chinchillidae). Reprod Domest Anim 54(2):207–215

Acuña F, Barbeito CG, Portiansky EL, Ranea G, Nishida F, Miglino MA, Flamini MA (2020) Early and natural embryonic death in *Lagostomus maximus*: association with the uterine glands, vasculature, and musculature. J Morphol 281(7):710–724

Acuña F, Barbeito CG, Portiansky EL, Miglino MA, Flamini MA (2021) Prenatal development in *Lagostomus maximus* (Rodentia, Chinchillidae): a unique case among eutherian mammals of physiological embryonic death. J Morphol 282(5):720–732

Acuña F, Barbeito CG, Portiansky EL, Ranea G, Miglino MA, Flamini MA (2022) Spontaneous embryonic death in plains viscacha (*Lagostomus maximus* – Rodentia), a species with unique reproductive characteristics. Theriogenology 185:88–96

Alves BG, Alves KA, Gastal GDA, Gastal MO, Figueiredo JR, Gastal EL (2018) Spatial distribution of preantral follicles in the equine ovary. PLoS One 13(6):e0198108

Battlehner CN, Caldini EG, Pereira JCR, Luque EH, Montes GS (2003) How to measure the increase in elastic system fibres in the lamina propria of the uterine cervix of pregnant rats. J Anat 203:405–418

Barraza DE, Zampini R, Apichela SA, Pacheco JI, Argañaraz ME (2018) Changes in mucins and matrix metalloproteases in the endometrium of early pregnant alpacas (*Vicugna pacos*). Acta Histochem 120(5):438–445

Barraza DE, Zampini R, Apichela SA, Pacheco JI, Argañaraz ME (2020) Modifications of extracellular matrix features in the left and right uterine horns during the embryo pre-implantation period in *Vicugna pacos*. Theriogenology 157:440–448

Biancardi MF, Dos Santos FCA, de Carvalho HF, Sanches BDA, Taboga SR (2017) Female prostate: historical, developmental, and morphological perspectives. Cell Biol Int 41(11):1174–1183
Biswas SK, Banerjee S, Baker GW, Kuo CY, Chowdhury I (2022) The mammary gland: basic structure and molecular signaling during development. Int J Mol Sci 23(7):3883
Breazile JE, Brown EM (1976) Anatomy. In: Wagner JE, Manning PJ (eds) The biology of the guinea pig. Cap 6. Academic Press, New York, pp 53–62
Breed WG (1985) Morphological variation in the female reproductive tract of Australian rodents in the genera *Pseudomys* and *Notomys*. J Reprod Fertil 73:379–384
Carter AM, Acuña F, Barbeito CG (2023) I. W. Rowlands, Barbara J. Weir and the biology of the hystricomorph rodents. Reproduction 166(1):R1–R13
Clement PB, Cooney TP (1992) Idiopathic multifocal calcification of the ovarian stroma. Arch Pathol Lab Med 116:204–205
Codón SM, Casanave EB (2000) Comparative morphology of the ovaries of three species of Dasypodidae (Mammalia Xenarthra). Rev Chil Anat 18(2):251–257
da Costa Ferreira Jr JM (1995) Estudo morfológico da modulação da fibras do sistema elástico induzida por esteróides ováricos na lâmina própia da cuello uterino da rata. (Tesis). Instituto de Biociências, Universidade de São Pablo, Brazil
de Jesus Gomes L, Rodrigues GA, Medeiros BCM, Manso LA, Ramos JG, de Azevedo Brito PV, Taboga SR, de Carvalho HF, Dos Santos FCA, Biancardi MF (2021) Correction to: the influence of pregnancy on female prostate morphophysiology in gerbils (*Meriones unguiculatus*). Reprod Sci 28(9):2480
Dorfman VB, Saucedo L, Di Giorgio NP, Inserra PI, Fraunhoffer N, Leopardo NP, Halperín J, Lux-Lantos V, Vitullo AD (2013) Variation in progesterone receptors and GnRH expression in the hypothalamus of the pregnant South American plains viscacha, *Lagostomus maximus* (Mammalia, Rodentia). Biol Reprod 89(5):115
dos Santos AC, Oliveira GB, Viana DC, Oliveira FD, Silva Rdos S, Rici RE, de Oliveira MF, de Assis-Neto AC (2014) Development and morphological changes in the vaginal closure membrane throughout gestation in *Galea spixii* (Rodentia: Caviidae). Microsc Res Tech 79(5):359–364
dos Santos AC, Oliveira FP, Carvalho VD, Olivotti FA, Oliveira M, Menezes Oliveira SF, Alcântara D, Gomes Vasconcelos B, Grassi Rici RE, Chaves de Assis-Neto A, Miglino MA (2016) Intrauterine development of female genital organs in *Cavia porcellus* (Rodentia: Caviidae). Pak J Zool 46:389–397
Felipe AE, Masson PG (2005) Un modelo descriptivo del sistema reproductor hembra del coipo. Rev Electrón Vet 6(12):1–19
Felipe AE, Callejas SS, Cabodevila JA (1998) Anatomicohistological characteristics of female genital tubular organs of the South American nutria (*Myocastor coypus*). Anat Histol Embryol 27:245–250
Felipe AE, Cabodevila JA, Callejas SS (1999) Anatomicohistological characteristics of the ovary of the coypu (*Myocastor coypus*). Anat Histol Embryol 28:89–95
Felipe AE, Castro A, Callejas SS, Cabodevila JA (2000) Morphological study of the vagina of the *Myocastor coypus* (Coypo). Rev Chil Anat 18:231–235
Felipe AE, Callejas SS, Cabodevila JA (2001a) Anatomical and histological description of the uterine neck of the *Myocastor coypus* (coypo). Revista FAVE 15:15–22
Felipe AE, Cabodevila JA, Callejas SS (2001b) Characterization of the estrous cycle of the *Myocastor coypus* (coypo) by means of exfoliative colpocitology. J Neotrop Mammal 8:129–137
Felipe AE, Cabodevila JA, Callejas SS (2002) Morphology of the utero-tubal junction in the *Myocastor coypus* (coypo). Rev Chil Anat 20:139–144
Flamini MA, Barbeito CG, Gimeno EJ, Portiansky EL (2002) Morphological characterization of the female prostate (Skene's gland or paraurethral gland) of *Lagostomus ma*ximus maximus. Ann Anat 184(4):341–345
Flamini MA, Barbeito CG, Gimeno EJ, Portiansky EL (2009) Histology, histochemistry and morphometry of the ovary of the adult plains viscacha (*Lagostomus maximus*) in different reproductive stages. Acta Zool 90(4):390–400

Flamini MA, Díaz AO, Barbeito CG, Portiansky EL (2012) Morphology, morphometry, histochemistry and lectin histochemistry of the vagina of the plains viscacha (*Lagostomus maximus*). Biotech Histochem 87(2):81–94

Flamini MA, Barbeito CG, Portiansky EL (2014) A morphological, morphometric and histochemical study of the oviduct in pregnant and non-pregnant females of the plains viscacha (*Lagostomus maximus*). Acta Zool 95(2):186–195

Flamini MA, Barbeito CG, Portiansky EL (2019) Morphological characteristics of the uterus and uterine cervix of the plains viscacha (*Lagostomus maximus*). Acta Zool 101(4):353–365

Flamini MA, Barreto RSN, Matias GSS, Birbrair A, de Castro H, Sasahara T, Barbeito CG, Miglino MA (2020) Key characteristics of the ovary and uterus for reproduction with particular reference to poly ovulation in the plains viscacha (*Lagostomus maximus*, Chinchillidae). Theriogenology 142:184–195

Flamini MA, Barbeito CG, Díaz AO, Portiansky EL (2021) Comparison of the structural and ultrastructural characteristics of the female prostate between pregnant and non-pregnant plains viscacha (*Lagostomus maximus*). Tissue Cell 68:101458

Fraunhoffer NA, Jensen F, Leopardo N, Inserra PIF, Abuelafia AM, Espinosa MB, Charif SE, Dorfman VB, Vitullo AD (2017) Hormonal behavior correlates with follicular recruitment at mid-gestation in the South American plains vizcacha, *Lagostomus maximus* (Rodentia, Caviomorpha). Gen Comp Endocrinol 250:162–174

Giacchino M, Claver JA, Inserra PI, Lange FD, Gariboldi MC, Ferraris SR, Vitullo AD (2020) Nutritional deficiency and placenta calcification underlie constitutive, selective embryo loss in pregnant South American plains viscacha, *Lagostomus maximus* (Rodentia, Caviomorpha). Theriogenology 155:77–87

Gil E, Forneris M, Domínguez S, Penissi A, Fogal T, Piezzi RS, Scardapane L (2007) Morphological and endocrine study of the ovarian interstitial tissue of viscacha (*Lagostomus maximus maximus*). Anat Rec (Hoboken) 290(7):788–794

Goldman JM, Murr AS, Cooper RL (2007) The rodent estrous cycle: characterization of vaginal cytology and its utility in toxicological studies. Birth Defects Res B Dev Reprod Toxicol 80(2):84–97

Goode JA, Peaker M, Weir BJ (1981) Milk composition in the plain viscacha (*Lagostomus maximus*). J Reprod Fertil 62:563–566

Gopalakrishna A, Khaparde MS (1978) Early development, implantation and amniogenesis in the Indian vampire bat, *Megaderma lyra lyra* (Geoffroy). Proc Indian Acad Sci 87:91–104

Graham CH (1966) Cyclic changes in the squamo-columnar junction of the mouse cuello uteri. Anat Rec 155:251–260

Gross SA, Didio IJA (1987) Comparative morphology of the prostate in adult male and female *Praomys* (*Mastomys*) *natalensis* studied with electron microscopy. J Submicrosc Cytol 19:77–84

Guraya SS, Greenwad GS (1964) A comparative histochemical study of interstitial tissue and follicular atresia in the mammalian ovary. Anat Rec 149:411–434

Hafez ESE (1970) Female reproductive organs. In: Hafez ESE (ed) Reproduction and breeding techniques for laboratory animals. Lea and Febiger, Philadelphia, pp 74–106

Hillermann HH, Gaynor AI, Stanley HP (1958) The genital systems of nutria (*Myocastor coypus*). Anat Rec 130:513–531

Inomata T, Ninomiya H, Somiya H, Saito H, Mochizuki K (1992) Histological observation on the female prostate in Mongolian gerbils (*Meriones unguiculatus*). Exp Anim 41(3):383–386

Inserra PI, Leopardo NP, Willis MA, Freysselinard AL, Vitullo AD (2013) Quantification of healthy and atretic germ cells and follicles in the developing and post-natal ovary of the South American plains viscacha, *Lagostomus maximus*: evidence of continuous rise of the germinal reserve. Reproduction 147(2):199–209

Ishmael J (1975) Occurrence of calcium deposits in the ovaries of the baboon. Lab Anim 9:383–386

Jackson J (1989) Reproductive parameters of the plains viscacha(*Lagostomus maximus*) in San Luis province, Argentina. Vida Silv Neotrop 2:57–62

Kleiman DG (1974) Patterns of behaviour in hystricomorph rodents. Symp Zool Soc Lond 34:171–208

Klein C (2015) Pregnancy recognition and implantation of the conceptus in the mare. Adv Anat Embryol Cell Biol 216:165–188

König HE, Liebich HG (2004) Veterinary anatomy of domestic mammals. Textbook and colour atlas. Schattauer, Stuttgart

Mayor P, Lopez Bejar M, Jori F, Fenech M, Lopez-Gatius F (2003) Reproductive functional anatomy and oestrous cycle pattern of the female brush-tailed porcupine (*Atherurus africanus*, Gray 1842) from Gabon. Anim Reprod Sci 77:247–259

Mayor P, Bodmer RE, Lopez-Bejar M (2011) Functional anatomy of the female genital organs of the wild black agouti (*Dasyprocta fuliginosa*) female in the Peruvian Amazon. Anim Reprod Sci 123(3–4):249–257

Mendoza G, Echevarría L, Llerena C, Castro A, Domínguez M, Gómez S, Ghezzi MD, Barbeito CG (2014) Comparación morfológica entre el útero fetal y el útero adulto de la alpaca (*Vicugna pacos*) y la llama (*Lama glama*). Salud Tecnol Vet 1(1):1

Rasweiler JJ (1990) Implantation, development of the fetal membranes, and placentation in the captive black mastiff bat, *Molossus ater*. Am J Anat 187:109e36

Roberts CM, Weir BJ (1973) Implantation in the plains viscacha, *Lagostomus maximus*. J Reprod Fertil 33(2):299–307

Rood JP, Weir BJ (1970) Reproduction in female wild guinea pigs. J Reprod Fertil 23:393–409

Shehata R (1972) Female prostate in the house rat *Rattus rattus*. Acta Anat 83:426–434

Shehata R (1975) Female prostate in *Arvicanthis niloticus* and *Meriones libycus*. Acta Anat 92:513–523

Shehata R (1980) Female prostate and urethral glands in the home rat, *Rattus norvegicus*. Acta Anat 107:286–288

Silva LB, Alonso CR, Idiart JR (1997) Colpocitología exfoliativa en el coipo (*Myocastor coypus*). Vet Argent XIV:18–24

Skibiel AL, Downing LM, Orr TJ, Hood WR (2013) The evolution of the nutrient composition of mammalian milks. J Anim Ecol 82(6):1254–1264

Stewart F, Allen WR (1981) Biological functions and receptor binding activities of equine chorionic gonadotrophins. J Reprod Fertil 62(2):527–536

Tripp HR (1971) Reproduction in elephant-shrews (Macroselididae) with special reference to ovulation and implantation. J Reprod Fertil 26:149–159

Tulsiani DR, Yoshida-Komiya H, Araki Y (1997) Mammalian fertilization: a carbohydrate-mediated event. Biol Reprod 57(3):487–494

van der Horst CJ, Gillman J (1941) The numbers of eggs and survivingembryos in *Elephantulus*. Anat Rec 80:443–452

Weir BJ (1967) Aspects of reproduction in some hystricomorph rodents. Ph.D. Thesis, University of Cambridge.

Weir BJ (1971a) Some observation on reproduction in the female agouti, *Dasyprocta agouti*. J Reprod Fertil 24:203–211

Weir BJ (1971b) The reproductive physiology of the plains viscacha, *Lagostomus maximus*. J Reprod Fertil 25:355–363

Weir BJ (1971c) The reproductive organs of the female plains viscacha, *Lagostomus maximus*. J Reprod Fertil 25(3):365–373

Weir BJ (1974) Reproductive characteristics of hystricomorph rodents. Symp Zool Soc Lond 34:265–301

Weir BJ, Rowlands I (1974) The functional anatomy of the Hystricomorph ovary. Symp Zool Soc Lond 34:303–332

Chapter 12
Poly-implantation and Embryonic Death

Francisco Acuña, Enrique Leo Portiansky, María Angélica Miglino, Mirta Alicia Flamini, and Claudio Gustavo Barbeito

Abstract Embryo implantation in the vizcacha begins at 18 days post-coitum (dpc). It occurs in the caudocranial direction, that is, from the caudal to the cranial uterine segments. For each uterine horn, between 5 and 6 implantation sites (IS) are found. In each IS, an interstitial and antimesometrial implantation is observed, with an elongated blastocyst due to the growth of the visceral endoderm. Between 26 and 70 dpc, the death and resorption of the embryos begins in the cranial and middle IS; on the other hand, the caudal IS keep their growth and development rate, giving rise to two precocial pups after 5 months of gestation. Former studies concluded that death was not due to infectious agents, superfetation, or macro and microscopic variations in the uterine horns. However, later studies revealed that the uterine horns are heterogeneous in their anatomy and physiology along their craniocaudal extension. It is proposed that the uterine glands and blood vessels would have a differential role between the IS that are resorbed and those that will survive. It was observed that embryonic death in this species follows a morphological pattern like that registered in murine rats and mice, considered classic physiological and experimental models. That pattern is characterized by the presence of hemorrhage, neutrophilic

F. Acuña (✉) · C. G. Barbeito (✉)
Laboratorio de Histología y Embriología Descriptiva, Experimental y Comparada, Facultad de Ciencias Veterinarias, Universidad Nacional de La Plata, La Plata, Argentina

Consejo Nacional de Investigaciones Científicas y Técnicas (CONICET), La Plata, Argentina
e-mail: facuna@fcv.unlp.edu.ar; barbeito@fcv.unlp.edu.ar

E. L. Portiansky
Consejo Nacional de Investigaciones Científicas y Técnicas (CONICET), La Plata, Argentina

Laboratorio de Análisis de Imágenes, Facultad de Ciencias Veterinarias, Universidad Nacional de La Plata, La Plata, Argentina

M. A. Miglino
Departamento de Cirurgia, Faculdade de Medicina Veterinária y Zootecnia, Universidad de São Pablo, San Pablo, San Pablo, Brazil

M. A. Flamini
Laboratorio de Histología y Embriología Descriptiva, Experimental y Comparada, Facultad de Ciencias Veterinarias, Universidad Nacional de La Plata, La Plata, Argentina

L. L. Rasia et al. (eds.), *Plains Vizcachas*, https://doi.org/10.1007/978-3-031-49487-1_12

infiltration, necrosis, and fibrosis. Morphological and physiological similarities observed between related and phylogenetically distant species allow us to propose the vizcacha as an unconventional model for studying spontaneous embryonic death.

Keywords Prenatal development · *Conceptus* · Embryo loss · Uterus

12.1 Introduction

The initial information regarding early gestation in *Lagostomus maximus* (vizcacha) was derived from research conducted by Roberts and Weir (1973). Several of these findings align with those observed in other hystricomorph rodents (Roberts and Perry 1974).

In the vizcacha, fertilization occurs within the uterine tubes, where approximately 10% of the 200–800 ovulated oocytes are successfully fertilized. The rate of mitosis during cleavage appears to be slow, as morulae do not enter the uterus until the sixth-day post coitum (dpc) with only four cells. Up to 10 dpc, the uterus exhibits signs of post-estrous hyperemia and does not display macroscopic evidence of embryo implantation. It is not until dpc 13 that a small number of normal-appearing morulae are observed floating in the uterine lumen (Roberts and Weir 1973). Blastocysts in all the hystricomorphs studied share similar morphology. They feature a prominent cavity known as the blastocoel, along with a narrow perivitelline space located between the trophoblast and the zona pellucida. Additionally, they possess an inner cell mass enveloped by a thin trophoblast. In contrast, the abembryonic trophoblast, situated at the pole opposite to the inner cell mass, is thickened and measures approximately 95–100 μm in diameter.

The timing of embryo implantation during gestation varies among hystricomorph rodents. In *Chinchilla lanigera* (chinchilla), it occurs at 5.5 dpc, while in *Proechimys guairae* (casiragua) occurs between 6 and 6.5 dpc. In contrast, *Octodon degus* (degu) and *Cavia porcellus* (guinea pig) experience implantation between 6 and 7 dpc. However, in the case of the vizcacha, this process takes place notably later, at 18 dpc.

The interval between ovulation and implantation in the vizcacha is the longest reported among rodents. There has been a hypothesis that this could be an implantation diapause, as suggested by Roberts and Weir in 1973, although no studies have been conducted to confirm this hypothesis to date. Additionally, the observed period appears to be relatively short as compared to the diapause duration observed in other mammals (Fenelon and Renfree 2018). During the preimplantation period, the blastocyst is situated at the antimesometrial end of the uterine lumen, and the abembryonic trophoblast is oriented toward the uterine epithelium (Roberts and Perry 1974). In the guinea pig, the zona pellucida remains intact until pre-implantation. The loss of this zone and the attachment of the blastocyst to the uterine epithelium involve pseudopodial processes that develop in the trophoblast at the abembryonic pole of the blastocyst (Roberts and Perry 1974). These findings confirm the results of

Blandau (1949), who showed that as the pseudopodial processes become more numerous, they penetrate the zona pellucida, which gradually disappears in that region, and form a "junctional cone" that extends into the uterine epithelium. Although it is uncertain whether this process occurs in the vizcacha, given the similar developmental pattern in hystricomorphs, it is plausible (Roberts and Perry 1974).

Observations during implantation in casiragua, *Myocastor coypus* (coypu), degu, chinchilla, and vizcacha indicate that embryo spacing along the uterine horns is not random. In the vizcacha, most blastocysts were located at the ovarian ends of the uterine horns, suggesting that blastocyst spacing occurs near the time of implantation. It has been suggested that this spacing could result from muscular activity (Roberts and Weir 1973). In some species, like *Rattus norvegicus* (Kaulenas et al. 1991) and the European hare (Drews et al. 2013), the contractile activity of the myometrium is responsible for blastocyst distribution; however, this observation was not confirmed in hystricomorphs. In a study conducted on non-pregnant vizcacha uterine horns, we observed a significant increase in myometrial thickness in the cranial-caudal direction. We hypothesize that this may be related to the spacing of blastocysts along each uterine horn (Acuña et al. 2020). Implantation in the vizcacha is interstitially and antimesometrially consistent with all the hystricomorphs studied (Roberts and Weir 1973). In the vizcacha, blastocysts located near the neck of each uterine horn are the first to implant at 18 dpc. These implantation sites (IS) exhibit features such as apposition, adhesion, endometrial invasion, and substantial lymphocytic infiltration (Roberts and Weir 1973). In uteri with a gestation period of 18.5 dpc, blastocysts from both uterine horns were found at three different spatiotemporal moments. The caudal ones were located immediately inside the endometrial lamina propria, those in the middle region penetrated the luminal epithelium, while the cranial ones were floating in the lumen of the organ. In almost all pregnancies during this period, 5–6 IS were observed in each uterine horn (Roberts and Weir 1973) (Fig. 12.1a). Similar results were reported by Flamini et al. (2019) in early pregnancies of vizcachas from the Pampas region.

In species that develop an invasive placenta, such as the vizcacha (Flamini et al. 2011; Barbeito et al. 2021), multiple molecules participate in the invasion of the blastocyst into the uterine endometrium, many of which are glycoconjugates (Wooding and Burton 2008). In a study conducted during early gestation in the vizcacha, it was observed that the pattern of glycosidic residues did not differ between the IS with 26 dpc. This could indicate that although implantation in the vizcacha occurs at different times (Roberts and Weir 1973), there would be no regional differences in the molecules that participate in this process (Acuña et al. 2021).

The post-implantation development of hystricomorph rodents, including the vizcacha, is not well understood as in mice. Moreover, many descriptions of this phase of development are outdated, dating back to the 1970s (Roberts and Perry 1974). However, there is more recent information available on the post-implantation development of the vizcacha, which was described by Leopardo and Vitullo (2017) and is discussed in Chap. 15.

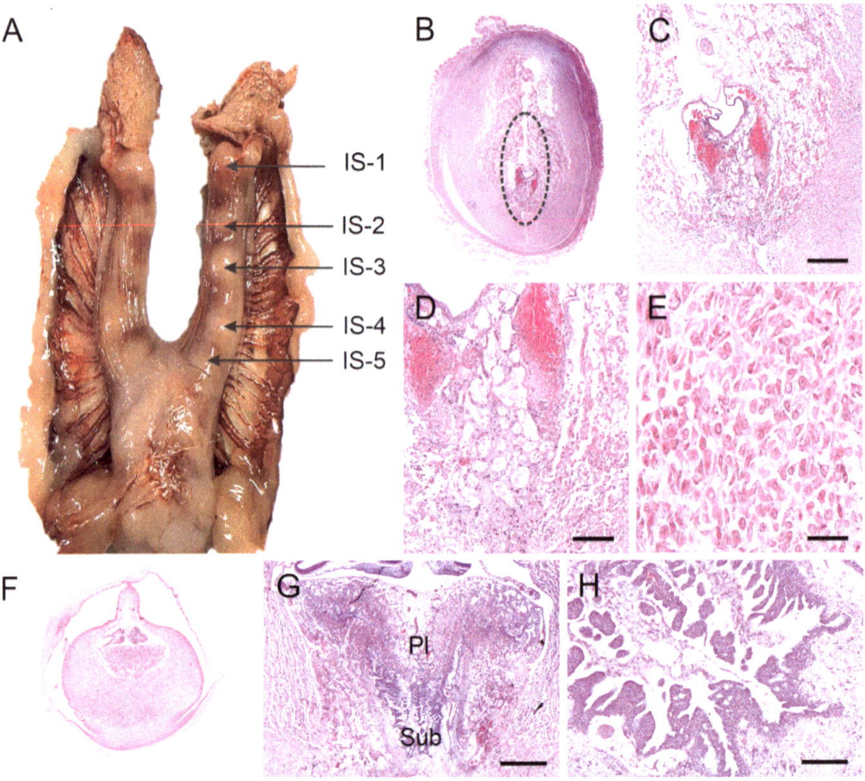

Fig. 12.1 Post-implantation development in *L. maximus*. (**a–e**) Early gestation, 26 days postcoitus. (**a**) 5 IS are observed per uterine horn. (**b**) Interstitial and antimesometrial *conceptus*. The dotted oval delimits the embryo. (**c**) Detail of the *conceptus* in Fig. 12.1b. (**d**) Trophoblastic branches extending towards the decidua. (**e**) Decidua. (**f–h**) IS of a 46 days postcoitus gestation. (**h**) *Conceptus*. (**g**) Placenta and subplacenta. (**h**) Detail of the subplacenta. Abbreviations: IS, implantation site; Pl, placenta; Sub, subplacenta. (**b–d**, **g**, **h**) hematoxyly-eosin staining; (**e**, **f**) periodic acid-Schiff reagent histochemical technique. Scale bar: 500 μm (**b**, **f**), 200 μm (**e**, **g**, **h**), 100 μm (**c**, **d**). License Number 5630341065605 (**a–e**), License Number 5630350137084 (**f–h**)

While there is ample information available regarding the exomorphology (external morphology) of embryos and fetuses of hystricomorph rodents (Fortes et al. 2013; de Oliveira et al. 2017; El Bizri et al. 2017), there is limited knowledge about their internal morphology. Some insights into the internal morphology of embryos and fetuses of the *Octodon degus* were provided by Rojas et al. (1982). However, recent data on the vizcacha, specifically from females with a gestational age of 45–53 dpc, have been added, incorporating information about various organ systems during this stage (Acuña et al. 2023). From a comparative morphology perspective, the characteristics reported in both species, *Octodon degus* and the vizcacha, do not differ significantly from those described for other eutherian mammals (Franciolli et al. 2011; Marcela et al. 2012; Alberto et al. 2013; Pieri et al. 2015). Additionally, Acuña et al. (2023) concur with an observation made by

Roberts and Perry (1974) that the rate of embryo development is slower compared to that of the placenta in the vizcacha. These authors suggest that this fact may be linked to the lengthy gestation period of the vizcacha (5 months) and the development of precocial offspring. However, it is noted that further research is needed to confirm this observation (Fig. 12.1b–h).

Early embryonic development in the vizcacha includes a unique aspect of its reproductive biology: embryonic death and resorption. This process begins between 26–35 dpc and primarily affects the cranial and middle IS, culminating in 70 dpc. In contrast, the caudal IS continue to develop, eventually giving rise to two precocial pups (Weir 1971a, b; Roberts and Weir 1973; Flamini et al. 2019) (Fig. 12.2a). While embryonic death is relatively common in the vizcacha, it occurs less frequently in other hystricomorph species like chinchillas (Weir 1967) and nutrias (Newson 1966) and is only occasional in casiraguas (Roberts 1973). This embryonic death is observed in various eutherian mammals, including murine rodents (Flores et al. 2014; Fonseca et al. 2014), species of Chiroptera (bats) (Bouchard et al. 2001), Carnivora (carnivores) (Ortega-Pacheco et al. 2007), and Lagomorpha (hares and rabbits) (Schroeder et al. 2013), among others. Consequently, embryonic

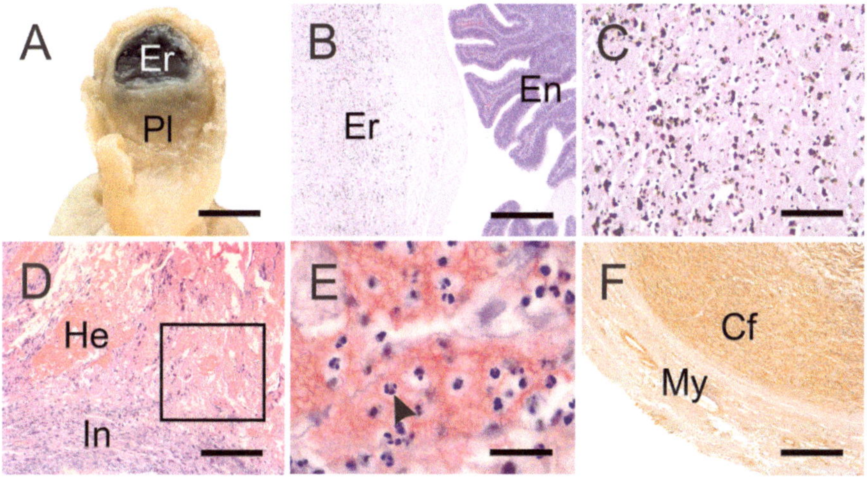

Fig. 12.2 Death and embryonic resorption in *L. maximus* at 46 days postcoitus. (**a**) Implantation site in resorption. Two zones are differentiated, embryonic (Er), and placental (Pl). (**b**) Relationship between the endometrium (En) and the embryonic zone in resorption (Er). (**c**) Embryonic zone in which embryonic structures are not recognized. The dark material corresponds to cellular debris. (**d**) Microscopic morphology of the placental area. Hemorrhage and inflammatory infiltration are observed. (**e**) Higher magnification of Fig. 12.1d. Some of the cells of the inflammatory infiltrate, such as neutrophils (arrowheads). (**f**) Layer of collagen fibers and fibroblasts (Cf) underlying the myometrium (My) and delimiting the embryonic and placental zones. Positive immunostaining with the primary antibody vimentin for fibroblasts. (**b–e**), hematoxylin-eosin stain. Scale bar: 0.5 cm (**a**), 200 µm (**b, d**), 100 µm (**c, e**). License number: 5630360198277 (All figures)

death is a reproductive process that is conserved among eutherian mammals (Drews et al. 2020), although in no other case is such extensive or localized loss observed as in *L. maximus*.

The initial studies on vizcacha reproduction aimed to elucidate the causes of embryonic death, but no histological or vascularization differences were found between the cranial and caudal regions of the uterus that could explain this phenomenon. Furthermore, no etiological agents were identified as potential factors. Consequently, this embryonic death was considered to be spontaneous (Roberts and Weir 1973). In later research, Flamini et al. (2019) noted variations in color, size, and consistency between resorbing IS (cranial and median) and viable ones in early gestation. The resorbing IS appeared brownish-black, smaller, and soft, while the viable ones were yellowish-white, larger, and firm. Additionally, in a study involving 26 dpc vizcachas, it was observed that all IS contained an embryo, a developing placenta, and deciduas, although with some particularities. In this study, cranial and medial IS displayed immune cell infiltration and the presence of collagenous fibers encircling the embryo and part of the extraembryonic trophoblast. In contrast, caudal IS lacked these characteristics. Notably, different hypotheses linking the survival of IS to potential variations in the expression of progesterone and prolactin hormone receptors, as well as to molecules associated with cell renewal and adhesion (essential processes for successful implantation), were all ruled out. The study found that the expression patterns of these molecules were similar in all analyzed IS, regardless of whether the implanted embryos survived (Acuña et al. 2021). In subsequent research conducted at more advanced gestational stages (46–53 dpc), alterations in histological organization were observed in the cranial and medial IS. These alterations included remnants of the yolk sac, infiltration of inflammatory cells, foci of fibrinoid necrosis (Flamini et al. 2020; Giacchino et al. 2020), hemorrhages, and the presence of a network of collagenous fibers and fibroblasts surrounding the entire resorption area (Acuña et al. 2022) (Fig. 12.2b–h).

The studies conducted on females with early and intermediate gestations have enabled the identification of a sequential pattern of events during embryonic death and resorption in the vizcacha, specifically affecting the cranial and medial IS undergoing resorption. This pattern comprises the following stages:

1. Delayed growth and embryonic development
2. Cardiac dysfunction, as assessed by echo Doppler
3. Embryo death, disintegration of the extraembryonic adnexa, and loss of placental integrity
4. Hemorrhage, necrosis, and fibrosis (Acuña et al. 2021, 2022)

It is noteworthy that these characteristics observed in the vizcacha are consistent with those found in other species with lower rates of embryonic death (Flores et al. 2014; Fonseca et al. 2014).

Sensitivity to oxygen deprivation appears to differ between the embryo (higher sensitivity) and the placenta (lower sensitivity) in humans (Burton et al. 2021). Based on our observations in the vizcacha, we propose that a similar dynamic could exist in this species. Consequently, the lower blood supplies in the cranial and medial IS could lead to the death of the embryo (Acuña et al. 2021). Some authors suggest that the uteroplacental compartment functions as a sensor for danger and for "foreign" inflammatory stressors (Nadeau-Vallée et al. 2016), which can originate, for example, from molecules resulting from tissue necrosis due to hypoxia caused by poor vascularization (Nakashima et al. 2020). The destruction of cells by necrosis leads to the release of intracellular molecules into the extracellular space. These molecules, which are not common in that space, are considered damage-associated molecular patterns (DAMPs). DAMPs trigger a non-infectious and sterile inflammatory response by binding to specific pattern receptors (RRPs). The binding of DAMPs to RRPs, as in pathogen-induced inflammation, activates both non-immune cells and cells of innate immunity. This activation leads to the production of several cytokines (Chen and Nuñez 2010; Brien et al. 2019) that, in turn, recruit cells and activate adaptive immune responses (Gong et al. 2020). We propose that tissue destruction and release of cellular contents resulting from the necrosis of embryos in the cranial and middle IS could indeed involve different DAMPs (Acuña et al. 2022). In resorbing IS in the vizcacha, there is an abundant infiltration of neutrophils and other typical inflammatory cell types, such as macrophages and foreign body giant cells. These cells are likely related to the presence of cellular debris generated from the necrosis of embryos in the resorbing IS. In conclusion, tissue disorganization and cell disintegration, first of the embryo and then of the placenta in the cranial and median resorbing IS of the vizcacha, act as DAMPs and trigger a sterile inflammatory response (Acuña et al. 2022). It is noteworthy that this inflammation coexists with the normal growth and development of the caudal IS.

12.2 Conclusions

The features of poly-implantation and embryonic death in the vizcacha, along with polyovulation, represent unique events in the reproductive biology of eutherian mammals. To gain a deeper understanding of the mechanisms underlying these processes, further molecular and morphological studies are required to investigate and test different hypotheses.

Acknowledgments We would like to thank ECAS (Estación de Cría de Animales Silvestres) from the Ministry of Agroindustry of the Province of Buenos Aires, (Argentina) for capturing the animals, and the Universidad Nacional de La Plata and CONICET for their support.

References

Acuña F, Barbeito CG, Portiansky EL, Ranea G, Nishida F, Miglino M, Flamini MA (2020) Early and natural embryonic death in *Lagostomus maximus*: association with the uterine glands, vasculature, and musculature. J Morphol 281(7):710–724

Acuña F, Barbeito CG, Portiansky EL, Miglino MA, Flamini MA (2021) Prenatal development in *Lagostomus maximus* (Rodentia, Chinchillidae): a unique case among eutherian mammals of physiological embryonic death. J Morphol 282(5):720–732

Acuña F, Barbeito CG, Portiansky EL, Ranea G, Miglino MA, Flamini MA (2022) Spontaneous embryonic death in plains viscacha (*Lagostomus maximus* – Rodentia), a species with unique reproductive characteristics. Theriogenology 185:88–96

Acuña F, Portiansky EL, Miglino MA, Flamini MA, Barbeito CG (2023) Embryonic-placental relationship in *Lagostomus maximus* as compared to other hystricognath rodents and eutherian mammals. Zoology 158:126082

Alberto MLV, Meirelles FV, Perecin F, Ambrósio CE, Favaron PO, Franciolli ALR, Mess AM, Dos Santos JM, Rici REG, Bertolini M, Miglino MA (2013) Development of bovine embryos derived from reproductive techniques. Reprod Fertil Dev 25:907–917

Barbeito CG, Acuña F, Miglino MA, Portiansky EL, Flamini MA (2021) Placentation and embryo death in the plains viscacha (*Lagostomus maximus*). Placenta 108:97–102

Blandau RJ (1949) Embryo-endometrial interrelationship in the rat and guinea pig. Anat Rec 104(3):331–359

Bouchard S, Zigouris J, Fenton MB (2001) Automn mating and likely resorption of an embryo by a Hoary Bat, *Lasiurus cinereus* (Chiroptera: Vespertilionidae). Am Midl Nat 145:210–212

Brien M-E, Baker B, Duval C, Gaudreault V, Jones RL, Girard S (2019) Alarmins at the maternal-fetal interface: involvement of inflammation in placental dysfunction and pregnancy complications 1. Can J Physiol Pharmacol 97(3):206–212

Burton GJ, Cindrova-Davies T, Yung HW, Jauniaux E (2021) Hypoxia and reproductive health: oxygen and development of the human placenta. Reproduction 161(1):F53–F65

Chen GY, Nuñez G (2010) Sterile inflammation: sensing and reacting to damage. Nat Rev Immunol 10(12):826–837

de Oliveira GB, de Araújo Júnior HN, da Silva Costa H, Silva AR, de Moura CEB, de Oliveira RHA, Miglino MA, de Oliveira MF (2017) Post-implantation development of red-rumped agouti (*Dasyprocta leporina* Linnaeus, 1758). Anim Reprod Sci 182:35–47

Drews B, Ringleb J, Waurich R, Hildebrandt TB, Schröder K, Roellig K (2013) Free blastocyst and implantation stages in the European brown hare: correlation between ultrasound and histological data. Reprod Fertil Dev 25(6):866–878

Drews B, Landaverde LF, Kühl A, Drews U (2020) Spontaneous embryo resorption in the mouse is triggered by embryonic apoptosis followed by rapid removal via maternal sterile purulent inflammation. BMC Dev Biol 20(1):1

El Bizri HR, Barros Monteiro FO, Andrade RDS, Valsecchi J, de Araújo Guimaraes DA, Mayor P (2017) Corrigendum to 'Embryonic and fetal morphology in the lowland paca (*Cuniculus paca*): a precocial hystricomorph rodent'. Theriogenology 128:225

Fenelon JC, Renfree MB (2018) The history of the discovery of embryonic diapause in mammals. Biol Reprod 99(1):242–251

Flamini MA, Portiansky EL, Favaron PO, Martins DS, Ambrósio CE, Mess AM, Miglino MA, Barbeito CG (2011) Chorioallantoic and yolk sac placentation in the plains viscacha (*Lagostomus maximus*) – a caviomorph rodent with natural polyovulation. Placenta 32(12):963–968

Flamini MA, Barbeito CG, Portiansky EL (2019) Morphological characteristics of the uterus and uterine cervix of the plains viscacha (*Lagostomus maximus*). Acta Zool 101(4):353–365

Flamini MA, Barreto RSN, Matias GSS, Birbrair A, Harumi de Castro Sasahara T, Barbeito CG, Miglino MA (2020) Key characteristics of the ovary and uterus for reproduction with particular reference to poly ovulation in the plains viscacha (Lagostomus maximus, Chinchillidae). Theriogenology 142:184–195

Flores LE, Hildebrandt TB, Kühl AA, Drews B (2014) Early detection and staging of spontaneous embryo resorption by ultrasound biomicroscopy in murine pregnancy. Reprod Biol Endocrinol 12(1):38

Fonseca BM, Almada M, Costa MA, Teixeira NA, Correia-da-Silva G (2014) Rat spontaneous foetal resorption: altered α2-macroglobulin levels and uNK cell number. Histochem Cell Biol 142(6):693–701

Fortes EA, Ferraz MS, Bezerra DO, Júnior AMC, Cabral RM, Sousa F, Almeida HM, Pessoa GT, Menezes DJ, Guerra SP, Sampaio IB, Assis Neto AC, Carvalho MA (2013) Prenatal development of the agouti (*Dasyprocta prymnolopha* Wagler, 1831): external features and growth curves. Anim Reprod Sci 140(3–4):195–205

Franciolli AL, Cordeiro BM, da Fonseca ET, Rodrigues MN, Sarmento CA, Ambrosio CE, de Carvalho AF, Miglino MA, Silva LA (2011) Characteristics of the equine embryo and fetus from days 15 to 107 of pregnancy. Theriogenology 76(5):819–832

Giacchino M, Claver JA, Inserra PI, Lange FD, Gariboldi MC, Ferraris SR, Vitullo AD (2020) Nutritional deficiency and placenta calcification underlie constitutive, selective embryo loss in pregnant South American plains viscacha, Lagostomus maximus (Rodentia, Caviomorpha). Theriogenology 155:77–87

Gong T, Liu L, Jiang W, Zhou R (2020) DAMP-sensing receptors in sterile inflammation and inflammatory diseases. Nat Rev Immunol 20(2):95–112

Kaulenas A, Parkington HC, Coleman HA (1991) Response of the rat myometrium to phenylephrine in early pregnancy and the effects of 6-hydroxydopamine. Br J Pharmacol 103(2):1429–1434

Leopardo NP, Vitullo AD (2017) Early embryonic development and spatiotemporal localization of mammalian primordial germ cell-associated proteins in the basal rodent *Lagostomus maximus*. Sci Rep 7(1):594

Marcela SG, Cristina RMM, Angel PGM, Manuel AM, Sofía DC, Patricia DLRS, Bladimir RR, Concepción SG (2012) Chronological and morphological study of heart development in the rat. Anat Rec 295:1267–1290

Nadeau-Vallée M, Obari D, Palacios J, Brien MÈ, Duval C, Chemtob S, Girard S (2016) Sterile inflammation and pregnancy complications: a review. Reproduction 152(6):R277–R292

Nakashima A, Shima T, Tsuda S, Aoki A, Kawaguchi M, Yoneda S, Yamaki-Ushijima A, Cheng SB, Sharma S, Saito S (2020) Disruption of placental homeostasis leads to preeclampsia. Int J Mol Sci 21(9):3298

Newson RM (1966) Reproduction in the Ferral coypu (*Myocastor coypu*). Symp Zool Soc Lond 15:323–334

Ortega-Pacheco A, Segura-Correa JC, Jimenez-Coello M, Linde Forsberg C (2007) Reproductive patterns and reproductive pathologies of stray bitches in the tropics. Theriogenology 67(2):382–390

Pieri NCG, Souza AF, Casals JB, Roballo KCS, Ambrósio CE, Martins DS (2015) Comparative development of embryonic age by organogenesis in domestic dogs and cats. Reprod Domest Anim 50:625–631

Roberts CM (1973) The embryology of certain hystricomorph rodents. Ph.D. Thesis, University of London

Roberts CM, Perry JS (1974) Hystricomorph embryology. In: Rowlands IW, Weir BJ (eds) The biology of hystricomorph rodents. Academic Press, New York, pp 333–360

Roberts CM, Weir BJ (1973) Implantation in the plains viscacha, *Lagostomus maximus*. J Reprod Fertil 33(2):299–307

Rojas MA, Montenegro MA, Morales B (1982) Embryonic development of the degu, Octodon degus. J Reprod Fertil 66(1):31–38

Schroeder K, Drews B, Roellig K, Goeritz F, Hildebrandt TB (2013) Embryonic resorption in context to intragestational corpus luteum regression: a longitudinal ultrasonographic study in the European brown hare (*Lepus europaeus* PALLAS, 1778). Theriogenology 80(5):479–486

Weir BJ (1967) The care and management of laboratory hystricomorph rodents. Lab Anim 1(2):95–104

Weir BJ (1971a) The reproductive physiology of the plains viscacha, *Lagostomus maximus*. J Reprod Fertil 25(3):355–363

Weir BJ (1971b) The reproductive organs of the female plains viscacha, *Lagostomus maximus*. J Reprod Fertil 25(3):365–373

Wooding P, Burton G (2008) Comparative Placentation: Structures, Functions and Evolution. 8th ed. Springer, Germany.

Chapter 13
Placentation and Placenta

Claudio Gustavo Barbeito, Francisco Acuña, Enrique Leo Portiansky, María Angélica Miglino, and Mirta Alicia Flamini

Abstract Gestation in the vizcacha lasts approximately 153 days. At 26–32 days post-coitus (dpc), extensions of the trophoblast invading the decidua are observed at caudal implantation sites (IS), marking the beginning of placentation. During later gestational stages (46–53 dpc), both the main placenta and a subplacenta are observed. The formed placenta is hemochorial and chorioallantoic. Its shape, size, and organization have been related to the development and delivery of precocial offspring. The subplacenta, unique to hystricomorphs, varies in size during gestation and serves as a source of trophoblasts. In advanced stages, the subplacenta exhibits necrotic areas with dystrophic calcification. Based on the positive response to the PAS histochemical technique and comparisons with findings in other hystricomorphs, it is hypothesized that this organ may have an endocrine function. The main placenta is covered by an avascular parietal sac. As occurs in other hystricomorphs, the visceral yolk sac forms an inverted vitelline placenta that persists throughout gestation. In early and intermediate stages of vizcacha gestation, two cell types are observed: giant trophoblastic cells (GTC) located at the trophospongium edges and infiltrating the decidua, and uterine natural killer cells (*uNK*), found

C. G. Barbeito (✉) · F. Acuña (✉)
Laboratorio de Histología y Embriología Descriptiva, Experimental y Comparada, Facultad de Ciencias Veterinarias, Universidad Nacional de La Plata, La Plata, Argentina

Consejo Nacional de Investigaciones Científicas y Técnicas (CONICET), La Plata, Argentina
e-mail: barbeito@fcv.unlp.edu.ar; facuna@fcv.unlp.edu.ar

E. L. Portiansky
Laboratorio de Análisis de Imágenes, Facultad de Ciencias Veterinarias, Universidad Nacional de La Plata, La Plata, Argentina

Consejo Nacional de Investigaciones Científicas y Técnicas (CONICET), La Plata, Argentina

M. A. Miglino
Departamento de Cirurgia, Faculdade de Medicina Veterinária y Zootecnia, Universidad de São Pablo, San Pablo, San Pablo, Brazil

M. A. Flamini
Laboratorio de Histología y Embriología Descriptiva, Experimental y Comparada, Facultad de Ciencias Veterinarias, Universidad Nacional de La Plata, La Plata, Argentina

L. L. Rasia et al. (eds.), *Plains Vizcachas*, https://doi.org/10.1007/978-3-031-49487-1_13

near the uterine arteries. These *uNK* cells may be associated with the invasion and remodeling of uterine arteries and, consequently, to placentation. However, both cells show inconsistencies among different hystricomorph species. The division in placenta and subplacenta and the fibrovascular ring of the visceral yolk sac are characters common to all hystricomophs studied and found in *L. maximus* too, these characters reforze the hypothesis of the monophyletic origin of the group.

Keywords Placenta · Subplacenta · Mesoplacenta · Precocial young

13.1 Introduction

The formation of the placenta (placentation) represents one of the evolutionary milestones of the oviparity-viviparity transition. This event involved morphophysiological changes in the female reproductive system (Van Dyke et al. 2014; Blackburn 2015), as well as the loss of genes related to the provision of yolk to oocytes (Brawand et al. 2008). The placenta is a transitory organ that develops exclusively during pregnancy. It has a mixed character since it results from the interaction between maternal and embryonic/fetal tissues (Mossman 1937, 1987, 1991; Chavatte-Palmer and Tarrade 2016; Roberts et al. 2016). Morphological studies at the macro and microscopic levels demonstrated great variability in the structure and characteristics of the placenta among the different groups of eutherians (Jones et al. 2007; Kanashiro et al. 2009; Mess et al. 2012; Rezende et al. 2012; Carter and Mess 2017; Wilsher et al. 2019; Diessler et al. 2023). In fact, the placenta is considered one of the most variable organs in this group of animals (Wagner 2018). Different criteria are used to classify this organ, among them, the distribution of the chorionic villi, the shape of the interdigitation between the chorionic villi and the endometrium, and the components of the maternal-embryonic/fetal barrier (Chavatte-Palmer and Tarrade 2016). The description and classification of the placenta not only allow us to expand the reproductive knowledge of the species but also help us to understand aspects of their evolutionary history and the reproductive adaptations that have developed over time (Roberts et al. 2016; Griffith 2021).

13.1.1 Characteristics of Hystricomorphs Placenta

Hystricomorphs, like most rodents, have an inverted yolk placenta and a chorioallantoic placenta. The chorioallantoic placenta is discoidal, with a distribution of chorionic villi, and labyrinthine due to interdigitation between the embryonic/fetal and maternal components. Furthermore, this placenta is hemochorial due to the structure of the placental barrier (Wooding and Burton 2008). In contrast to the placenta of other studied rodents, the chorioallantoic placenta of hystricomorphs is divided into a main placenta and a subplacenta (Barbeito et al. 2021) (Fig. 13.1a, b).

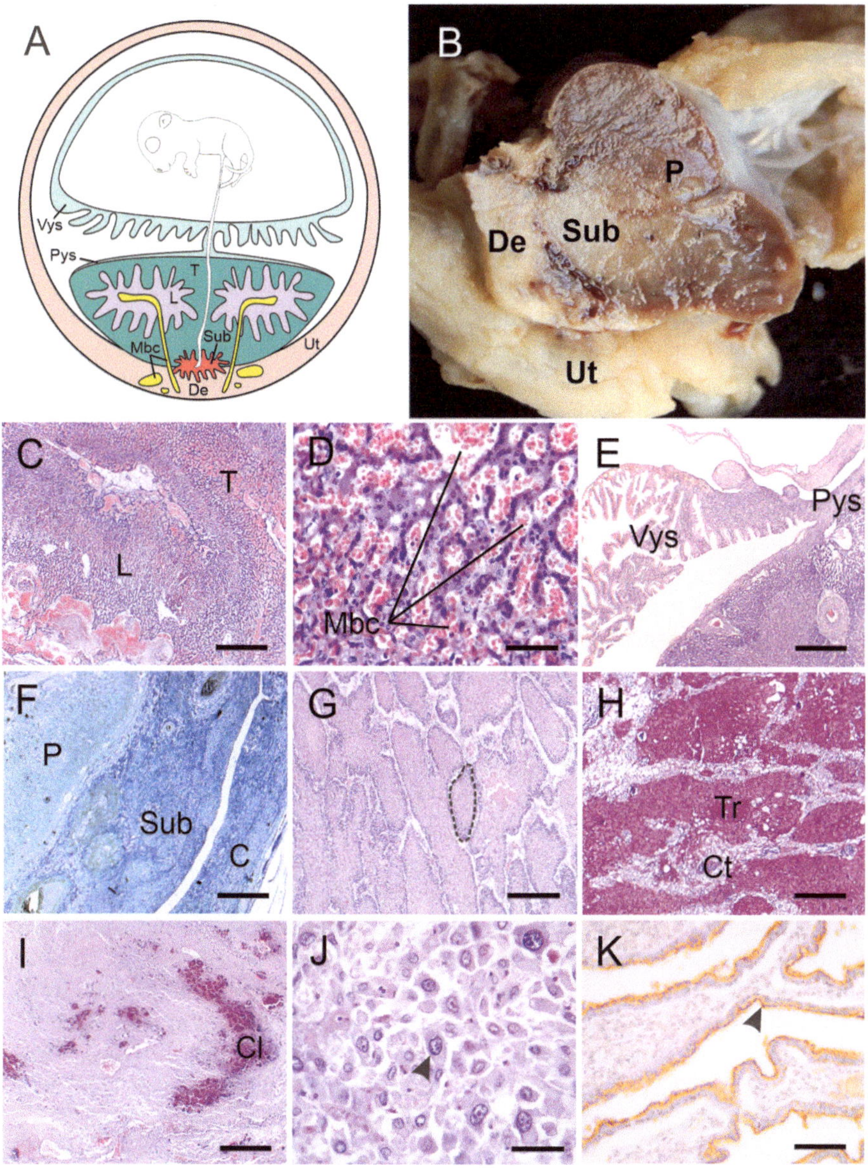

Fig. 13.1 (**a**) Schematic representation of the placental morphotype in caviomorph rodents. (**b**) Macroscopic view of the chorioallantoic placenta of *L. maximus*. (**c**) Main chorioallantoic placenta, with trophospongium and labyrinth. (**d**) Detail of the maternal blood chambers. (**e**) Yolk sac: visceral and parietal layers. (**f**) Connective tissue capsule, in relation to the main placenta and the subplacenta. Reticulin of Gomori (**g**) Subplacenta. The dotted line encloses a lobulation. (**h**) Subplacenta, trophoblast (magenta), and connective tissue. Periodic acid-Schiff reagent histochemical technique. (**i**) Dystrophic calcification in the decidua. (**j**) Polygonal and voluminous cells (arrowhead) of the decidua. (**k**) Labeling of the epithelium that covers the projections of the visceral yolk sac, lectin *Arachis hypogaea* agglutinin (PNA). Abbreviations: C, capsule; Cl, dystrophic calcification; Ct, connective tissue; De, decidua; L, labyrinth; Mbc, maternal blood chamber; P, placenta; Pys, parietal yolk sac; Sub, subplacenta; T, trophospongium; Tr, trophoblast; Ut, uterus; Vys, visceral yolk sac. (**c–e**) and (**i**, **j**) hematoxylin-eosin staining. Scale bar: 500 μm (**c**, **e**, **f**), 200 μm (**g–i**), 100 μm (**d**, **j**, **k**). (**a**, **b**, **f**, **j**): License Number 5625960935464; (**e**, **k**) License Number 5625961201745

The morphological characteristics of both have been extensively studied by different authors, mainly from South America (Bonatelli et al. 2005; Kanashiro et al. 2009; Mess 1999, 2001, 2003, 2007a, b, c, d, 2011; Miglino et al. 2002, 2004; Oliveira et al. 2008). Many of these features are similar across the clade, suggesting that the presence of the subplacenta is an ancestral feature in the Hystricognathi clade (Oliveira et al. 2006).

In hystricomorphs, the main placenta is lobed, (Mess 2007a) each lobe contains fetal blood vessels and maternal blood chambers, formed from blood vessels that lost their endothelium during trophoblast invasion. From the center to the periphery, sheets of syncytiotrophoblast (STB) delimit the maternal blood chambers, which are also in contact with fetal capillaries. These structures form the labyrinth, the site of the greatest exchange between the female and the embryo/fetus (Fig. 13.1c, d). In the placenta at term, there is only one layer of STB in direct contact with maternal blood, which makes this placenta hemomonochorial, like that of other hystricomorphs and that of humans. This type of placenta is considered highly invasive (da Anunciação et al. 2017).

The trophospongium occupies the interlobular and marginal area of the placenta (Fig. 13.1c). It is composed of maternal blood chambers surrounded by STB. In turn, the subplacenta presents both STB and cytotrophoblast (CTB) in the form of lamellae (Miglino et al. 2002, 2004, 2021; Bonatelli et al. 2005; Oliveira et al. 2006; Mess 2007a; Bezerra et al. 2018).

The yolk sac of hystricomorphs is divided into a parietal yolk sac that surrounds the chorioallantoic placenta and a visceral yolk sac that forms an inverted yolk placenta that, as in the other members of the studied group, is maintained until the end of gestation (Barbeito et al. 2021) (Fig.3.1e). A fibrovascular ring appears in the inverted yolk sac in all hystricomorph analysed (Mess 2005). Between the inverted yolk placenta and the chorioallantoic placenta, a connective tissue capsule of maternal origin is described, which is present until mid-gestation (Flamini et al. 2011) (Fig. 13.1f). This characteristic is observed in some hystricomorphs such as the Canadian porcupine *Erethizon dorsatum* (Perrotta 1959) but not in all members of the group. It appears mainly in larger body size representatives such as *Hydrochoerus hydrochaeris* and *Lagostomus maximus* (Kanashiro et al. 2009; Miglino et al. 2008; Barbeito et al. 2021).

13.1.2 **Lagostomus maximus *Placenta***

In the vizcacha, like other studied hystricomorphs, implantation occurs inside the endometrium, a process known as interstitial implantation. In each of the implantation sites (IS) (5–6/uterine horn) of the vizcacha, trophoblast ramifications directed towards the decidua at 26 days post-coitus (dpc) are observed (Acuña et al. 2021), although the first studies indicated that placentation began at 35 dpc (Weir 1971; Roberts and Weir 1973; Roberts and Perry 1974). At 46 dpc, the caudal IS, which are viable, show evident development of the placenta. Not so in those that are in resorption (Acuña et al. 2022a). During implantation, changes occur in the connective tissue of the endometrium, known as the decidual reaction (Roberts and Weir 1973). At 26 dpc, decidual cells are observed, which are positive to the PAS histochemical technique, as well as to vimentin antibodies and anti-progesterone and anti-prolactin receptors antibodies (Acuña et al. 2021). In caudal IS with 46 dpc, a differentiated decidua is observed in *basalis* and *parietalis* (Acuña et al. 2022a).

The general characteristics of the main placenta of the vizcacha are very similar to those of other hystricomorphs. The labyrinth forms lobes, and the trophospongium is distributed both interlobular and at the outer limit of the organ. From the center of each lobe, separate maternal blood chambers and fetal capillaries radiate, which are initially separated by CTB cells exhibiting mitotic activity, providing a source of new cell formation. As gestation progresses, the placenta is composed mainly of STB, while the CTB decreases (Flamini et al. 2011). The trophospongium is initially the most abundant component of the placenta, although it then decreases in relation to the labyrinthine lobes. Large clusters of CTB cells can be found at the edges of the main vizcacha placenta during early and mid-gestation. They are not found in all hystricomorphs (Barbeito et al. 2021), although it was found in *Octodon degus* (Bosco et al. 2007). These cells have proliferative capacity, although their function has not been fully elucidated. However, it has been observed that in other groups of rodents, they are related to invasiveness (Favaron and Carter 2023). The main placenta of the vizcacha is conical in shape and is attached to the decidua by a pedicel (Flamini et al. 2011; Acuña et al. 2023a). This structure, like the capsule, is not present in all hystricomorphs but has been described in some species such as *Myocastor coypus* (Hillemann et al. 1961), *Cuniculus paca* (Bonatelli et al. 2005) and the Canadian porcupine (Perrotta 1959).

During the early stages of gestation, the vizcacha subplacenta has a mesenchyme axis surrounded by CTB, forming about 5 layers. Initially, these cells are in contact with maternal vessels, while later, during advanced pregnancy, this structure is invaded by fetal capillaries that become the only blood vessels present in the subplacenta (Flamini et al. 2011; Acuña et al. 2023a) (Fig. 13.1g, h). The specific function of the subplacenta in the vizcacha has not yet been established, but it has been shown that in other hystricomorphs it has hormonal activity (Rodrigues et al. 2006).

In the subplacenta of vizcachas with advanced gestation, areas of necrosis with dystrophic calcification were observed, which is related to a remarkable reduction in size during the last third of gestation (Giacchino et al. 2020; Barbeito et al. 2021) (Fig. 13.1i). From the subplacenta, the trophoblast invades the uterine spiral arteries, replacing components of their wall (endovascular trophoblast). The invasiveness of this trophoblast is also evidenced by the presence of trophoblast cells in the lungs of early pregnant females (Acuña et al. 2022b; Carter et al. 2023), a phenomenon that was observed in other species with invasive placentas (La Perle et al. 2014) as the human being (Burton 2011) and *Chinchilla laniger* (Billington and Weir 1967).

In the vizcacha decidua (Fig. 13.1j), the presence of uterine natural killer cells (*uNK*), a cell type related to the invasion and remodeling of maternal blood vessels was observed. These cells have very different characteristics from circulating NK cells present in other locations, although they may have the same origin. *uNK* are recognized by their labeling with PAS and DBA lectin. The presence of these cells was also reported in other mammals such as humans, rats, and mice. For some authors, they are one of the features that characterize the Eutherian ancestral placenta (Fu et al. 2017; Vacca et al. 2019). Up to now, *uNK* cells have not been described in other hystricomorphs, except for an unconfirmed reference in guinea pigs (Debout et al. 1995). In the vizcacha, the presence of these cells was observed in the decidua of 26 dpc IS, close to the uterine spiral artery (Acuña et al. 2021). In a study with early, intermediate, and term placentas, the amount of *uNK* cells was observed to decrease towards the end of pregnancy (Flamini et al. 2011).

In the vizcacha placenta, variations in the glycosylation pattern of the different structures have been observed throughout gestation. The use of biotinylated lectins to label different components of the placenta, such as the subplacenta and the decidua, has shown differences in the glycosidic residues present between early, mid, and term stages of gestation (Fig. 13.1k). These variations could be related to hormonal changes during pregnancy (Acuña et al. 2023b). In this same study (Acuña et al. 2023b), the placental glycotype was defined, that is, the specific set of sugars present in the mature placenta. The glycotype has been associated with the function of sugars during trophoblast invasion and parturition (Jones et al. 1997). Part of the glycotype identified in the vizcacha is shared with other members of *Placentalia* (Acuña et al. 2023b).

In addition to the placenta and subplacenta, the vizcacha has an additional structure called the mesoplacenta. This structure originates from the decidua and is connected to the endometrium at the mesometrial end of the IS (Acuña et al. 2023a). This structure was first reported by Strahl (1905) from the observation of the placenta of the hystricomorph *Dasyprocta azarae*. Later, it was described in other members of the same taxa, such as *C. lanigera* (Tibbitts and Hillemann 1959), *Dasyprocta agouti* (agouti), and paca (Miglino et al. 2002). Uhlendorf and Kaufmann (1979) suggested that the function of the mesoplacenta is to restrict the damage inflicted on the uterus during parturition, allowing rapid regeneration of the organ and facilitating the possibility of a new pregnancy after postpartum mating. This function could be similar in the vizcacha, considering its type of placenta and the occurrence, in some females, of postpartum estrus (Acuña et al. 2023a). The visceral and parietal yolk sac are similar to others hystricomorphs including the existence of a fibrovascular ring (Barbeito et al. 2021). A connective capsule of matern origin is found until mid-gestation between inverted yolk sac and main placenta (Flamini et al. 2011).

13.2 Conclusions

The placenta of different species of hystricomporphs are similar. Some characters found in *L. maximus*, as the division in main placenta and subplacenta and the existence of a fibrovascular ring, were considerated as characteristics common of the hytricomporphs and are utilized to reinforce the hypothesis of the monophyletic origin of the group. The gestation of hystricomorphs is characterized by being longer as compared to other rodents. This has been related to the slow growth rate of embryos during early gestation, where the gestational process is primarily focused on placental establishment, rather than embryo development (Roberts and Perry 1974). In recent studies, carried out in 46–53 dpc vizcachas, it was observed that the rate of development of the placenta and subplacenta exceeds that of the embryo. In addition, it was observed that, during this period, the placenta exhibits a considerable exchange area, which could be related to the development and growth of precocial pups (Acuña et al. 2023a). There is a widely accepted hypothesis in the scientific community that associates less invasive placentations with the development of precocial offspring and more invasive ones with altricial offspring (Smith 2014). However, the characteristics of the hemomonochorial placenta and the precocial offspring observed in hystricomorphs contradict this hypothesis. Therefore, comparative morphological and molecular studies are required to test other hypotheses and consider additional variables such as gestation length and reproductive bioenergetics, among others.

Acknowledgments We would like to thank ECAS (Estación de Cría de Animales Silvestres) from the Ministry of Agroindustry of the Province of Buenos Aires (Argentina), for capturing the animals, and the Universidad Nacional de La Plata and CONICET for their support.

References

Acuña F, Barbeito CG, Portiansky EL, Miglino MA, Flamini MA (2021) Prenatal development in Lagostomus maximus (Rodentia, Chinchillidae): a unique case among eutherian mammals of physiological embryonic death. J Morphol 282(5):720–732

Acuña F, Barbeito CG, Portiansky EL, Ranea G, Miglino MA, Flamini MA (2022a) Spontaneous embryonic death in plains viscacha (*Lagostomus maximus* – Rodentia), a species with unique reproductive characteristics. Theriogenology 185:88–96

Acuña F, Castro GG, Gómez DG, Terraza A, Flamini MA, Barbeito C (2022b) A preliminary study of the possible trophoblast deportation to the lungs in the intermediate gestation of *Lagostomus maximus* (Rodentia) [Abstract]. Placenta 122:29

Acuña F, Portiansky EL, Miglino MA, Flamini MA, Barbeito CG (2023a) Embryonic-placental relationship in *Lagostomus maximus* as compared to other hystricognath rodents and eutherian mammals. Zoology 158:126082

Acuña F, Carril J, Portiansky EL, Flamini MA, Miglino MA, Barbeito CG (2023b) Placental glycotype of the caviomorph rodent *Lagostomus maximus* and its evolution within Eutheria. J Morphol 284(3):e21566

Barbeito CG, Acuña F, Miglino MA, Portiansky EL, Flamini MA (2021) Placentation and embryo death in the plains viscacha (*Lagostomus maximus*). Placenta 108:97–102

Bezerra FVF, Favaron PO, Mess AM, Araújo Júnior HN, Oliveira GB, Pereira AF, Miglino MA, Oliveira MF (2018) Subplacental development in *Galea spixii*. Pesqui Vet Bras 38(11):2175–2182

Billington WD, Weir BJ (1967) Deportation of trophoblast in the chinchilla. J Reprod Fertil 13(3):593–595

Blackburn DG (2015) Evolution of vertebrate viviparity and specializations for fetal nutrition: a quantitative and qualitative analysis: viviparity and fetal nutrition. J Morphol 276(8):961–990

Bonatelli M, Carter AM, Machado MRF, De Oliveira MF, de Lima MC, Miglino MA (2005) Placentation in the paca (*Agouti paca* L). Reprod Biol Endocrinol 3(1):9

Bosco C, Buffet C, Bello MA, Rodrigo R, Gutierrez M, García G (2007) Placentation in the degu (*Octodon degus*): analogies with extra subplacental trophoblast and human extravillous trophoblast. Comp Biochem Physiol A Mol Integr Physiol 146(4):475–485

Brawand D, Wahli W, Kaessmann H (2008) Loss of egg yolk genes in mammals and the origin of lactation and placentation. PLoS Biol 6(3):e63

Burton GJ (2011) Deportation of syncytial sprouts from the term human placenta. Placenta 32(1):96–98

Carter AM, Mess AM (2017) Placentation in the colugos *Cynocephalus volans* and *Galeopterus variegatus* (Dermoptera) and the transition from labyrinthine to villous placentation in primates. Placenta 55:47–53

Carter AM, Acuña F, Barbeito CG (2023) I W Rowlands, Barbara J Weir and the biology of the hystricomorph rodents. Reproduction (Cambridge, England), 166(1):R1–R13. https://doi.org/10.1530/REP-23-0028

Chavatte-Palmer P, Tarrade A (2016) Placentation in different mammalian species. Ann Endocrinol 77(2):67–74

da Anunciação A, Mess AM, Orechio D, Aguiar BA, Favaron PO, Miglino MA (2017) Extracellular matrix in epitheliochorial, endotheliochorial and haemochorial placentation and its potential application for regenerative medicine. Reprod Domest Anim 52(1):3–15

Debout C, Taouji S, Izard J (1995) Increase of a guinea pig natural killer cell (Kurloff cell) during leukemogenesis. Cancer Lett 97:117–122

Diessler ME, Hernández R, Gomez Castro G, Barbeito CG (2023). Decidual cells and decidualization in the carnivoran endotheliochorial placenta. Front Cell Dev Biol. 16;11:1134874. https://doi.org/10.3389/fcell.2023.1134874. PMID: 37009475; PMCID: PMC10060884.

Favaron PO, Carter AM (2023) The trophoblast giant cells of cricetid rodents. Front Cell Dev Biol 10:1097854

Flamini MA, Portiansky EL, Favaron PO, Martins DS, Ambrósio CE, Mess AM, Miglino MA, Barbeito CG (2011) Chorioallantoic and yolk sac placentation in the plains viscacha (*Lagostomus maximus*) – a caviomorph rodent with natural polyovulation. Placenta 32(12):963–968

Fu B, Zhou Y, Ni X, Tong X, Xu X, Dong Z, Sun R, Tian Z, Wei H (2017) Natural killer cells promote fetal development through the secretion of growth-promoting factors. Immunity 47(6):1100–1113.e6

Giacchino M, Claver JA, Inserra PI, Lange FD, Gariboldi MC, Ferraris SR, Vitullo DA (2020) Nutritional deficiency and placenta calcification underlie constitutive, selective embryo loss in pregnant South American plains viscacha, *Lagostomus maximus* (Rodentia, Caviomorpha). Theriogenology 155:77–87

Griffith OW (2021) Novel tissue interactions support the evolution of placentation. J Morphol 282(7):1047–1053

Hillemann HH, GAYNOR AI (1961) The definitive architecture of the placentae of nutria, Myocastor coypus (Molina). Am J Anat 109:299–317. https://doi.org/10.1002/aja.1001090306

Jones CJ, Dantzer V, Leiser R, Krebs C, Stoddart RW (1997) Localisation of glycans in the placenta: a comparative study of epitheliochorial, endotheliochorial, and haemomonochorial placentation. Microsc Res Tech 38:100–114

Jones CJ, Carter AM, Aplin JD, Enders AC (2007) Glycosylation at the fetomaternal interface in hemomonochorial placentae from five widely separated species of mammal: is there evidence for convergent evolution? Cells Tissues Organs 185(4):269–284

Kanashiro C, Santos TC, Miglino MA, Mess AM, Carter AM (2009) Growth and development of the placenta in the capybara (*Hydrochaeris hydrochaeris*). Reprod Biol Endocrinol 7:57

La Perle KM, Green MG, Niewiesk S (2014) Trophoblast deportation to the lungs of cotton rats (*Sigmodon hispidus*). Comp Med 64(6):448–455
Mess A (1999) Plazentation von Petromus typicus, ein Beitragzur Reproduktionsbiologie und evolutiven Differenzierung der Hystricognathi. Courier Senckenberg 215:153–157
Mess A (2001) Evolutionary differentiation of placental organisation in hystricognath rodents. In: Denys C, Granjon L, Poulet A (eds) African small mammals. IRD Éditions, Collection Colloques et Séminaires, Paris, pp 279–292
Mess A (2003) Evolutionary transformations of chorioallantoic placental characters in Rodentia with special reference to hystricognath species. J Exp Zool 299A:78–98
Mess A. (2005) The fibrovascualar ring: A synapomorphy of hystricognath Rodentia newly described in Petromus typicus and Octodon degus. Belg. J. Zool., 135 (supplement): 31–38
Mess A (2007a) The subplacenta in *Octodon degus* and *Petromus typicus*: two hystricognath rodents without significant placental lobulation. J Exp Zool 308B:172–188
Mess A (2007b) Development of the chorioallantoic placenta in *Octodon degus*: a model for growth processes in caviomorph rodents? J Exp Zool 308B:371–383
Mess A (2007c) The guinea pig placenta: model of placental growth dynamics. Placenta 28:812–815
Mess A (2007d) Chorioallantoic and yolk sac placentation in the dassie rat *Petromus typicus* and its bearing to the evolution of hystricognath Rodentia. Placenta 11:1229–1233
Mess AM (2011) Character transformations and their functional significance as a key to the evolution of hystricognath Rodentia. Pesq Vet Bras 31(12):1108
Mess AM, Favaron PO, Pfarrer C, Osmann C, Melo AP, Rodrigues RF, Ambrósio CE, Bevilacqua E, Miglino MA (2012) Placentation in the anteaters *Myrmecophaga tridactyla* and *Tamandua tetradactyla* (Eutheria, Xenarthra). Reprod Biol Endocrinol 10:102
Miglino MA, Carter AM, dos Santos Ferraz RH, Fernandes Machado MR (2002) Placentation in the capybara (*Hydrochoerus hydrochaeris*), Agouti (*Dasyprocta aguti*) and paca (*Agouti paca*). Placenta 23(5):416–428
Miglino MA, Carter AM, Ambrosio CE, Bonatelli M, De Oliveira MF, Dos Santos Ferraz RH, Rodrigues RF, Santos TC (2004) Vascular organization of the hystricomorph placenta: a comparative study in the agouti, capybara, guinea pig, paca and rock cavy. Placenta 25(5):438–448
Miglino MA, Franciolli AL, de Oliveira MF, Ambrósio CE, Bonatelli M, Machado MR, Mess A (2008) Development of the inverted visceral yolk sac in three species of caviids (Rodentia, Caviomorpha, Caviidae). Placenta 29(8):748–752
Miglino MA, de SáSchiavo MG, Rigoglio NN, Borghesi J, de Castro Sasahara TH, del Portal MJI, Granado GS, Caceres Ramos SC, de Oliveira MF, Conley AJ (2021) The comparative aspects of hystricomorph subplacenta: potential endocrine organ. BMC Zool 6(1):16
Mossman HW (1937) The comparative morphogenesis of the fetal membranes and accessory uterine structures. Contrib Embryol 26:129
Mossman HW (1987) Vertebrate fetal membranes. Comparative ontogeny and morphology, evolution, phylogenetic significance, basic function, research opportunities. Rutgers University Press, New Brunswick
Mossman HW (1991) Classics revisited: comparative morphogenesis of the fetal membranes and accessory uterine structures. Placenta 12(1):1–5
Oliveira MF, Carter AM, Bonatelli M, Ambrosio CE, Miglino MA (2006) Placentation in the rock cavy, *Kerodon rupestris* (Wied). Placenta 27(1):87–97
Oliveira MF, Mess A, Ambrósio CE, Dantas CA, Favaron PO, Miglino MA (2008) Chorioallantoic placentation in *Galea spixii* (Rodentia, Caviomorpha, Caviidae). Reprod Biol Endocrinol 6:39
Perrotta CA (1959) Fetal membranes of the Canadian porcupine, *Erethiz ondorsatum*. Am J Anat 104:35–59
Rezende LC, Barbeito CG, Favaron PO, Mess A, Miglino MA (2012) The fetomaternal interface in the placenta of three species of armadillos (Eutheria, Xenarthra, Dasypodidae). Reprod Biol Endocrinol 10:38
Roberts CM, Perry JS (1974) Hystricomorph embryology. In: Rowlands IW, Weir BJ (eds) The biology of hystricomorph rodents. Academic Press, New York, pp 333–360

Roberts CM, Weir BJ (1973) Implantation in the plains viscacha, *Lagostomus maximus*. J Reprod Fertil 33(2):299–307

Roberts RM, Green JA, Schulz LC (2016) The evolution of the placenta. J Reprod Fertil 152(5):R179–R189

Rodrigues RF, Carter AM, Ambrosio CE, dos Santos TC, Miglino MA (2006) The subplacenta of the red-rumped agouti (*Dasyprocta leporina* L). Reprod Biol Endocrinol 4:31

Smith KK (2014) Evolution of the placenta in therian mammals. In: Dial KP, Shubin N, Brainerd EL (eds) Great transformations in vertebrate evolution. University of Chicago Press, Chicago, pp 205–225

Strahl H (1905) Eine Placenta miteinem Mesoplacentarium. Anat Anz 26:524–528

Tibbitts FD, Hillemann HH (1959) The development and histology of the chinchilla placentae. J Morphol 105:317–365

Uhlendorf B, Kaufmann P (1979) Die Entwicklung des Plazentastielesbeim Meerschweinchen. Anat Histol Embryol 8(3):233–247

Vacca P, Chiossone L, Mingari MC, Moretta L (2019) Heterogeneity of NK cells and other innate lymphoid cells in human and murine decidua. Front Immunol 10:170

Van Dyke JU, Brandley MC, Thompson MB (2014) The evolution of viviparity: molecular and genomic data from squamate reptiles advance understanding of live birth in amniotes. J Reprod Fertil 147(1):R15–R26

Wagner GP (2018) Comparative placentation-mammals. In: Skinner MK (ed) Encyclopedia of reproduction. Academic Press, Elsevier, Oxford, Waltham, pp 455–461

Weir BJ (1971) The reproductive physiology of the plains viscacha, *Lagostomus maximus*. J Reprod Fertil 25(3):355–363

Wilsher S, Stansfield F, Allen WT (2019) Placentation in the blue wildebeest (*Connochaetes taurinus*). Placenta 82:46–56

Wooding P, Burton G (2008) Comparative placentation: structures, functions and evolution, 8th edn. Springer, Germany

Chapter 14
Neuroanatomy and Brain Histological Atlas of the South American Plains Vizcacha

Alejandro Raúl Schmidt, Alfredo Daniel Vitullo, and Verónica Berta Dorfman

Abstract Brains could be classified as lissencephalic or gyrencephalic according to the presence of sulci and gyri. In recent years, studies of the physiology and anatomy of wild species like the South American plains vizcacha, *Lagostomus maximus*, a hystricomorph rodent that inhabits Argentina, Bolivia, and Paraguay, have shown an increased interest. However, very few reports have investigated its neuroanatomy. This chapter describes the neuroanatomy and histology of the brain of the plains vizcacha and establishes the location of its main brain areas and nuclei. Adult vizcacha shows a 14 g weight, 3.8 cm length, and 3.2 cm width gyrencephalic brain, with a dorsoventrally flattened pyramidal shape. It shows a deep longitudinal fissure between both hemispheres and three pairs of sulci in the cortex (the rostral-lateral, the intraparietal, and the transversal sulci). For a histological inspection of the location of different areas and nuclei, the brain was divided into 12 coronal sections with a rostrocaudal orientation (according to Sigman's plates), and Nissl and Klüver Barrera, together with immunohistochemistry and immunofluorescence histological techniques, were used. Around the 4th coronal section, the rostral diencephalic septum plate has the preoptic area and regionalized striatum with the presence of an internal capsule, among other areas and nuclei. Then, in the 6th coronal section, the rostral diencephalon plate has the thalamic reticular and paraventricular nuclei and the hypothalamic median eminence among others. More caudally, in the 7th coronal section, the caudal diencephalon plate has the midbrain occupying the largest portion of this section, together with the mammillary structures and the mid-hippocampus formation with well-developed subiculum, horns of Ammon, and dentate gyrus. Finally, at the level of the 8th coronal section, the mesencephalon main dopaminergic structures substantia nigra and ventral tegmental

A. R. Schmidt · A. D. Vitullo · V. B. Dorfman (✉)
Centro de Estudios Biomédicos Básicos, Aplicados y Desarrollo (CEBBAD), Ciudad Autónoma de Buenos Aires, Argentina

Consejo Nacional de Investigaciones Científicas y Técnicas (CONICET), Ciudad Autónoma de Buenos Aires, Argentina
e-mail: dorfman.veronica@maimonides.edu

L. L. Rasia et al. (eds.), *Plains Vizcachas*, https://doi.org/10.1007/978-3-031-49487-1_14

area, connected by the interpeduncular nucleus, are found. The present chapter represents a significant advancement in our understanding of the neuroanatomy and histology of the South American plains vizcacha being a valuable tool for future neurological studies. In addition, the findings shed light on distinctive structural features of the brain, which are closely linked to its colonial and burrowing way of life.

Keywords Brain · Plains vizcacha · Neuroanatomy · Histology

14.1 Introduction

The anatomical macrostructure of the mammalian brain exhibits two distinct morphologies: lissencephaly, which is characterized by a smooth cortex, and gyrencephaly, featuring a folded cortex. In gyrencephalic brains, gyri and sulci display a high degree of conservation across species of the same order, with only slight variations being species-specific or individual in nature (Welker 1990; Pillay and Manger 2007). Conversely, lissencephalic brains are typically found in small mammals, particularly in small rodents such as mice and rats. The level of convolutedness in gyrencephalic brains is highly variable and enhances in large mammals with increasing brain mass (Pillay and Manger 2007; Kelava et al. 2013; Zilles et al. 2013; Triarhou 2017). Despite being confined by the skull, cortical folding has facilitated brain growth during evolution, leading to an increase in cortical surface area proportional to the degree of gyrification. Most of our comprehension of the morphological variability of the brain has been based on investigations conducted on a limited number of mammalian species, predominantly on the so-called laboratory models like rats and mice, and a handful of species of primates, ungulates, carnivores, and cetaceans (Triarhou 2017). Among rodents, except the brain of some murids like mice and rats, the brains of hystricomorphs and sciurus have not been extensively elucidated in academic literature. Elucidating the species-specific composition of the brain may contribute to expanding our understanding of brain anatomy, ultimately contributing to a better understanding of neural function.

The South American plains vizcacha, a hystricomorph rodent that inhabits Argentina, Bolivia, and Paraguay, has very few reports investigating its neuroanatomy, which have been exclusively produced by our research group over the course of the last dozen years (Dorfman et al. 2011; Inserra et al. 2017; Schmidt et al. 2019, 2020). We trust that the availability of the histological atlas of the brain of the plains vizcacha, which is described in this chapter, would contribute to exponentially increasing the interest in the neural knowledge of this species.

14.2 Neuroanatomy

The South American plains vizcacha exhibits a voluminous and pyramidal shaped brain with a dorsoventrally flattened structure. Regarding its morphometry, the species shows a length of 3.8 ± 0.4 cm and a width of 3.2 ± 0.5 cm (Fig. 14.1a), with a weight of 14.29 ± 0.98 g. At the rostral pole, it has a triangular shape with rounded vertices, finding two large olfactory bulbs that are connected to the brain via the olfactory peduncles, and a caudal pole laterally expanded and linked to the cerebellum (Fig. 14.1a). The cerebellum measures 1 cm in length and 1.8 cm in width (Fig. 14.1a). Regarding the dorsal surface of the brain, several fissures are evident. A sagittal longitudinal fissure divides the brain into two hemispheres. The depth of this longitudinal fissure reaches until it meets the corpus callosum. In addition, three lateral sulci with bilateral symmetry are observed including one pair of rostral lateral sulci, one pair of transversal sulci, and one pair of intraparietal sulci (Fig. 14.1a). These cortical folding characteristics allow to infer and place the vizcacha in the group of gyrencephalic species, according to the classification established by Beccari (1943) and Dozo (1997). Concerning the ventral surface of the brain, in a rostrocaudal direction, may be observed the projections of the olfactory tract in the anterior region followed by the optic nerves, the optic chiasm, the optic tracts, and the oculomotor nerves most caudally located (Fig. 14.1b). The optic chiasm and the optic tracts are situated over the hypothalamus, located in the central region of the ventral brain. Additionally, the infundibulum and the mammillary bodies can be externally identified with central and caudal location, respectively, in the ventral surface of the hypothalamus (Fig. 14.1b). Moreover, the sulcus rhinalis divides the piriform lobe from the temporoparietal lobe (Fig. 14.1b). In addition to the aforementioned encephalic structures, the vessels that supply the brain can be appreciated in this ventral surface (Fig. 14.1b).

It has been agreed that the extent of cortical convolutions depends not exclusively upon the brain mass or volume, but additionally upon the thickness of the cortex (Hofman 1985). In order to investigate the neuroanatomical structure, quantitative parameters such as the Encephalization Quotient (EQ) developed by Jerison (1973, 1977) and the gyrification index (GI) developed by Zilles et al. (2013) are commonly used. The calculation of the EQ considers values of brain mass, body mass, and the allometric exponent, while the GI is determined as the quotient between the complete pial contour (CPC) and the outer cortical contour (OCC) of coronal brain slices in the rostral to the caudal temporoparietal cortex. These parameters ultimately result in a value that characterizes the degree of cortical folding for the species and are useful for the classification of the brains in phyletic lineage studies. In accordance with this, we have recently ascertained the neuroanatomical values of EQ and GI for the plains vizcacha and have introduced a novel metric parameter denoted as Minimum Cortical Thickness (MCT). The value of the MCT is defined as the distance between the outer boundary of the cortex and the point of contact with the corpus callosum, measured at the rostral lateral sulcus of the temporoparietal cortex where the neopallium exhibits its least thickness (Schmidt et al.

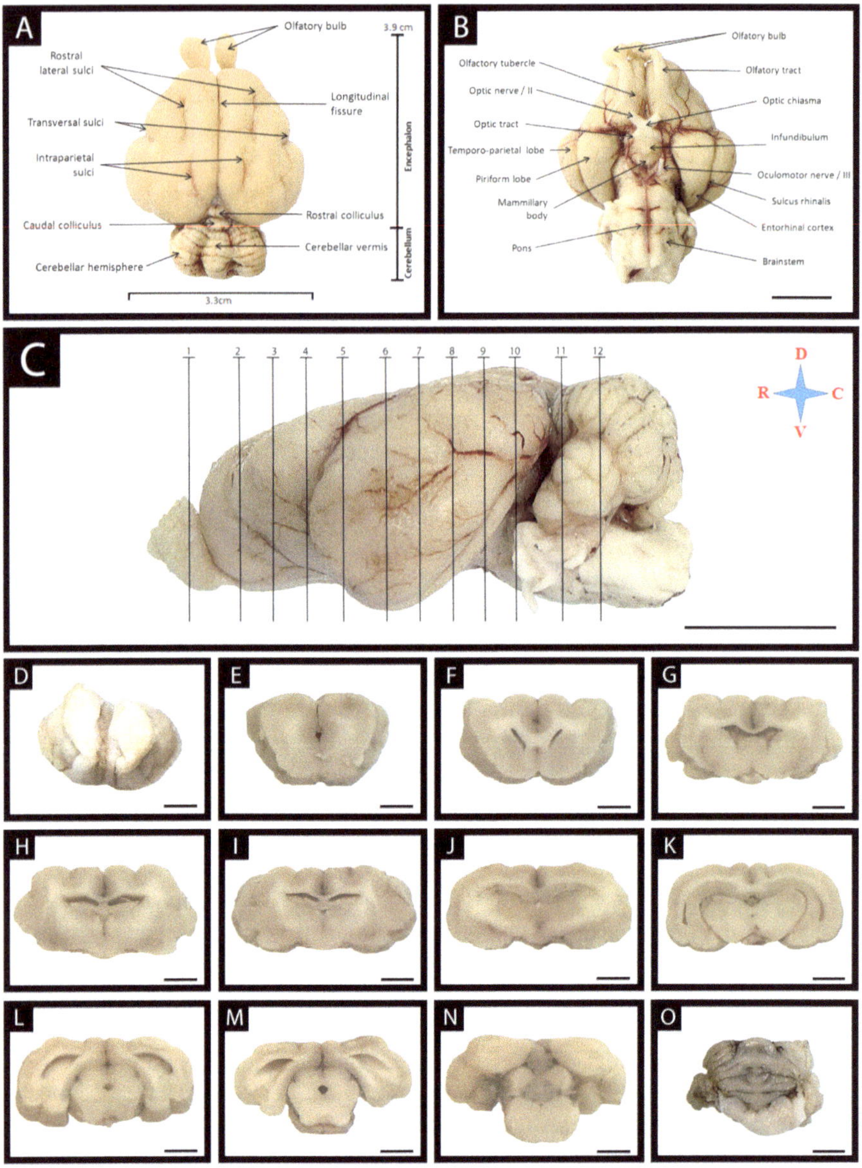

Fig. 14.1 Neuroanatomy of the plains vizcacha, *L. maximus*. (**a**) Dorsal view of the brain of the vizcacha, *L. maximus*, with the indication of its different structures; (**b**) ventral view of the brain of the plains vizcacha, *L. maximus*, with the indication of its different structures. Scale bar: 1 cm; (**c**) lateral view of the brain of the plains vizcacha, *L. maximus*. The 12 coronal sections of Sidman are outlined. Scale bar: 1 cm; (**d**) olfactory bulb (section 1); (**e**) pyriform cortex (section 2); (**f**) strial septum(section 3); (**g**) rostral diencephalic septum (section 4); (**h**) caudal diencephalic septum (section 5); (**i**) rostral diencephalon (section 6); (**j**) caudal diencephalon (section 7); (**k**) rostral midbrain (section 8); (**l**) midbrain (section 9); (**m**) caudal midbrain (section 10); (**n**) rostral cerebellum (section 11); (**o**) caudal cerebellum (section 12). Scale bar **d**–**o** = 5 mm. Images of each fresh section (**d**–**o**) were taken using a Nikon stereotaxic magnifying lens equipped with 0.5× and 1× magnification (Nikon, SMZ800, Nikon Corp)

2020). The calculation of the EQ for the plains vizcacha, adjusted for rodents by Herculano-Houzel (2007), resulted in a value of 1.26 ± 0.06, the GI 1.11 ± 0.02, and the MCT 0.63 mm (Schmidt et al. 2020). The EQ value for plains vizcacha falls within the expected range for extant caviomorphs (Bertrand and Silcox 2016). The GI determined for the plains vizcacha revealed an 11% increment in the cortical invaginations related to lissencephalic species, leading to the consequent increase in its cortical surface. This may indicate that cortical folds and fissures would facilitate the packing of the encephalon of the vizcacha inside the skull, as previously described for other species (Welker 1990; Rakic 1995; Albert and Huttner 2015). Moreover, the increment in the GI value of the vizcacha reflects a well-developed temporoparietal cortex where the brain structures associated with touch and hearing senses reside. A well development of both senses is crucial in recognizing the sounds of the individuals of the colony and the potential predators in animals of social and subterranean life such as the plains vizcacha (Contreras 1984). Finally, the relation between GI and MCT contributes to supporting the idea that the thinner the brain cortex the greater the folding of it, allowing a smaller and greater number of gyri.

14.3 Histological Characterization and Brain Atlas

Onwards in this chapter, it will be described for the very first time the complete histological atlas of the brain of the plains vizcacha. To ensure optimal organization and clarity, and to facilitate the analysis of the different areas and nuclei of the vizcacha brain atlas, its brain was divided into 12 sections with a rostrocaudal organization according to Sidman's plates (1971) (Fig. 14.1c): olfactory bulb (Fig. 14.1d), pyriform cortex (Fig. 14.1e), strial septum (Fig. 14.1f), rostral diencephalic septum (Fig. 14.1g), caudal diencephalic septum (Fig. 14.1h), rostral diencephalon (Fig. 14.1i), caudal diencephalon (Fig. 14.1j), rostral midbrain (Fig. 14.1k), midbrain (Fig. 14.1l), caudal midbrain (Fig. 14.1m), rostral cerebellum (Fig. 14.1n), and caudal cerebellum (Fig. 14.1o). Brain microsliced coronal sections were stained with Nissl, Klüver Barrera, or immunohistochemistry, and images were captured using a microscope or magnifier coupled to a video camera. Diagrams of coronal brain sections were generated elucidating by neuroanatomical comparison of the stained brain sections with images from brain atlases of evolutionarily related species (*Mus agouti, Cavia porcellus, Mesocricetus auratus, Hydrochoerus hydrochaeris, Chinchilla lanigera, Rattus norvegicus,* and *Mus musculus*). The location of regions, areas, and nuclei in the brain of the plains vizcacha were determined.

The following describes the different coronal plates of the 12 sections according to Sidman in a rostrocaudal organization.

In the most rostral region of the brain, the first section has the olfactory bulbs formed by concentrically arranged layers, organized into granular, internal plexiform, mitral, external plexiform, and glomerular layers (Figs. 14.1d and 14.2a, b). Caudally, in the piriform cortex section, the anterior olfactory nucleus stands out

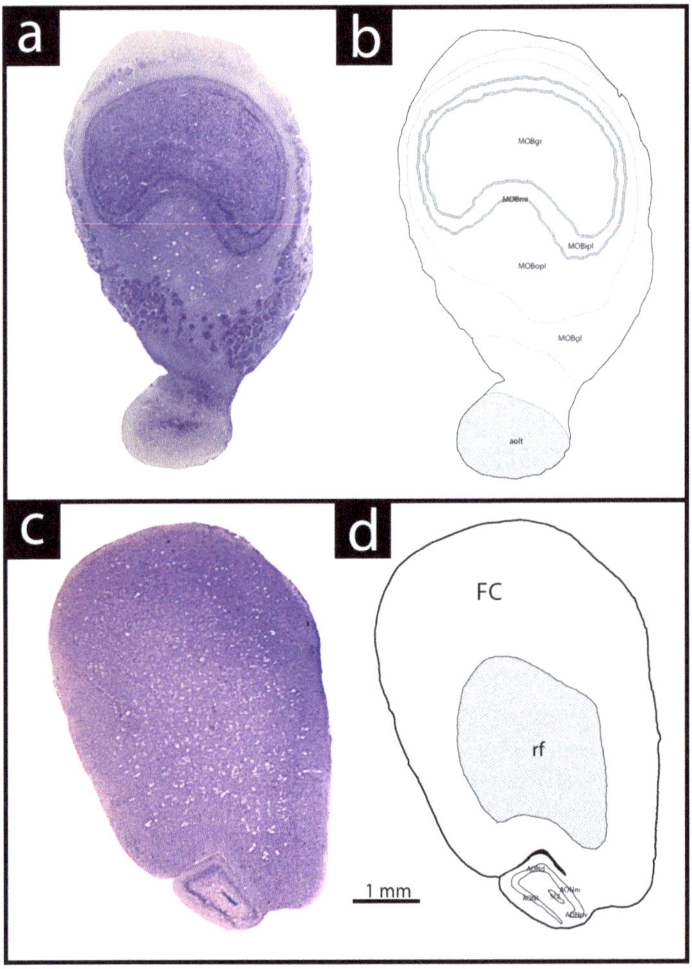

Fig. 14.2 (**a**, **b**) Coronal section of the olfactory bulb of the plains vizcacha, *L. maximus*. (**a**) Nissl stain in the olfactory bulb. (**b**) Diagram of the olfactory bulb and its structures. Nomenclature. aolt: accessory olfactory tract; MOBgl: main olfactory bulb, glomerular layer; MOBgr: main olfactory bulb, granule layer; MOBipl: main olfactory bulb, inner plexiform layer; MOBmil: main olfactory bulb, mitral layer; MOBopl: main olfactory bulb, outer plexiform layer. (**c**, **d**) Coronal plane of the piriform cortex section of the vizcacha, *L. maximus*. (**c**) Nissl staining in the piriform cortex section of vizcacha. (**d**) Diagram of the piriform cortex section and its structures. AON: Anterior olfactory nucleus; AONd: anterior olfactory nucleus, dorsal part; AONl: anterior olfactory nucleus, lateral part; AONm: anterior olfactory nucleus, medial part; AONpv: anterior olfactory nucleus, posteroventral part; FC: frontal cortex; rf: rostral forceps; SEZ: subependymal zone. Scale bar: 1 mm. Coronal sections of 5 μm thickness obtained with microtome (Leica brand, model RM 2125) and stained with Nissl were captured using a Nikon stereotaxic magnifying lens with 0.5× and 1× magnification (Nikon, SMZ800, Nikon Corp). The striped area refers to the white matter, the white area refers to the gray matter, and the dark area refers to the ventricles

ventrally with its four layers surrounding the subependymal zone (Figs. 14.1e and 14.2c, d). In addition, the frontal cortex with the rostral forceps in the central region, over the anterior olfactory nucleus, is observed (Fig. 14.2c, d). Then, caudally in the same section, the anterior commissure in the olfactory structures with a ventral location (Fig. 14.3a) and the rostral region of the caudo-putamen in the central portion of the brain section below the rostral forceps (Fig. 14.3b) are observed.

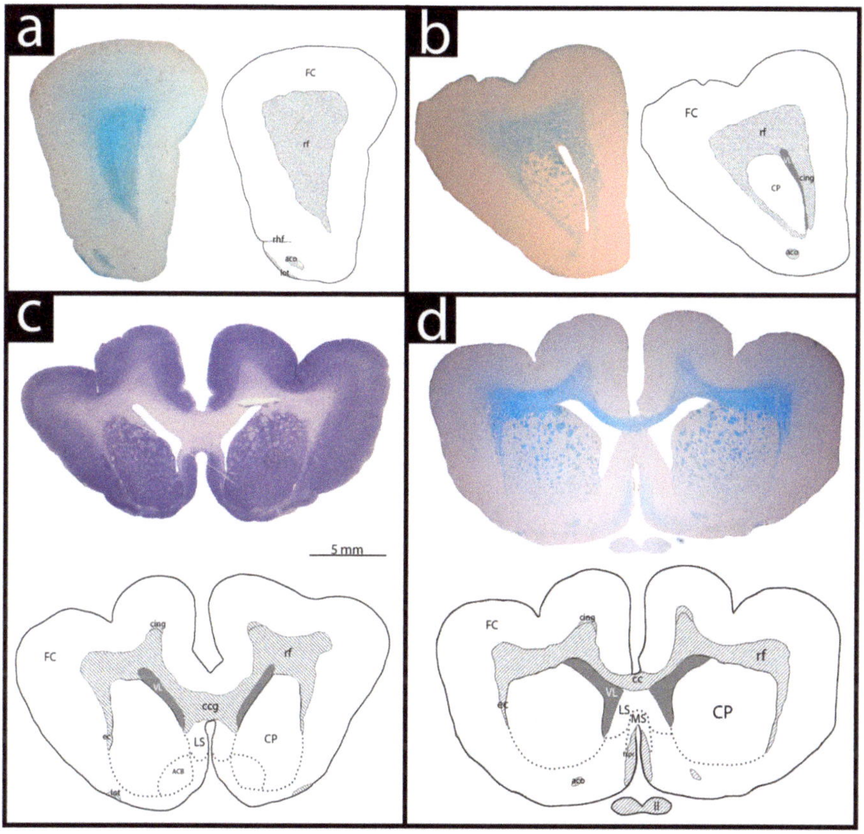

Fig. 14.3 (**a**, **b**) Coronal section of the piriform cortex of the vizcacha, *L. maximus*, with Klüwer-Barrera staining and its representative scheme. (**c**, **d**) Coronal section of the strial septum of the brain of the vizcacha, *L. maximus*. (**c**) With Nissl staining and its representative scheme. (**d**) With Klüwer-Barrera staining and its representative scheme. Nomenclature. ACB: Nucleus accumbens; aco: anterior commissure, olfactory limb; cc: corpus callosum; ccg: genu; cing: cingulum bundle; CP: caudo-putamen; rhf: rhinal fissure; ec: external capsule; FC: frontal cortex; lot: lateral olfactory tract; fxpr: precommissural fornix; LS: lateral septal nucleus; VL: lateral ventricle; MS: medial septal nucleus; rf: rostral forceps; II: optic nerve. Scale bar: 5 mm. Coronal sections of 5 μm thickness obtained with microtome (Leica brand, model RM 2125) and stained with Nissl (**c**) or Klüver Barrera (**a**, **b**, **d**) were captured using a Nikon stereotaxic magnifying lens with 0.5× and 1× magnification (Nikon, SMZ800, Nikon Corp). The striped area refers to the white matter, the white area refers to the gray matter, and the dark area refers to the ventricles

Nextly, in the strial septum section (Fig. 14.1f), the rostral-striatum with the caudo-putamen has a striosome-matrix compartmentalization (Fig. 14.3c, d). The rostral-striatum is paired located, laterally delimitated by the lateral ventricle and the external capsule and dorsally by the bundles of white matter arranged through the rostral forceps between the cingulate (Fig. 14.3c, d). In addition, the septal, medial, and lateral structures are located below the corpus callosum (Fig. 14.3c, d). Regarding the extra-encephalic structures, in the rostral part of the rostral diencephalic septal section, the optic nerves (pair II) are observed ventrally (Fig. 14.3d).

Caudally later, the rostral diencephalic septum section is characterized by the presence of the mid-striatum with a regionalization pattern where the three areas caudate, putamen, and globus pallium are differentiated (Figs. 14.1g and 14.4a, b).

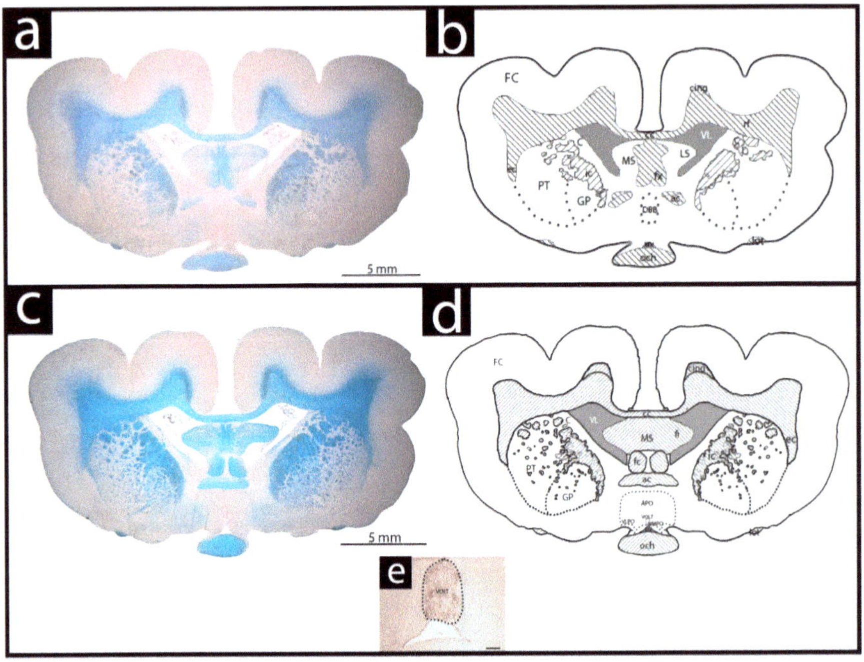

Fig. 14.4 (**a–d**) Coronal section of the rostral diencephalic septum of the vizcacha, *L. maximus*, with Klüwer-Barrera staining and its representative scheme. (**e**) Immunohistochemistry for gonadotropin-releasing hormone (GnRH) in the vascular organ of lamina terminalis (VOLT). Nomenclature. ac: Anterior commissure; C: caudate nucleus; cc: corpus callosum; cing: cingulum bundle; DBB: diagonal band of Broca; ec: external capsule; FC: frontal cortex; fc: fornix columns; fi: fimbria; fx: fornix; GP: globus pallidus; ic: internal capsule; IIIV: third ventricle; lot: lateral olfactory tract; LS: lateral septal nucleus; VL: Lateral ventricle; MS: medial septal nucleus; och: optic chiasm; POA: preoptic area; PT: putamen; rf: rostral forceps; VLPO: ventrolateral preoptic nucleus; VMPO: ventromedial preoptic nucleus; VOLT: vascular organ of the lamina terminalis. Scale bar: 5 mm. Coronal sections of 5 μm thickness obtained with microtome (Leica brand, model RM 2125) and stained with Klüver Barrera (**a**, **c**) were captured using a Nikon stereotaxic magnifying lens with 0.5× and 1× magnification (Nikon, SMZ800, Nikon Corp). The striped area refers to the white matter, the white area refers to the gray matter, and the dark area refers to the ventricles

The internal capsule appears separating the caudate from the putamen and the globus pallidus (Fig. 14.4a, b). The greater degree of regionalization provided by the presence of the internal capsule suggests an adaptive advantage in terms of motor and cognitive abilities allowing the division of striatal functions favoring reaction time in response to predators (Branch 1993; Jackson et al. 1996; Schmidt et al. 2019). In addition, this could be interpreted as an evolutionary retention that enables the lifestyle of the vizcacha, since it allows an increase in motor skills for the construction of their burrows and the manipulation of food (Nevo 1979; Contreras 1981; Bee de Speroni and Pellegrini de Gastaldo 1988; Schmidt et al. 2019). Below the corpus callosum appears the fornix column, and below the Diagonal Band of Broca (DBB) surrounded laterally by two anterior commissure structures (Fig. 14.4a, b). Some micrometers are caudally in the same rostral diencephalic septum section, the fornix columns are separated in two structures, and the anterior commissure is united in a single one (Fig. 14.4c, d). Under this structure, the anterior hypothalamus with the preoptic area (POA), ventrolateral and ventromedial nuclei, is established (Fig. 14.4c, d). In this rostral region of the hypothalamus, the vascular organ of the lamina terminalis (VOLT, belonging to the circumventricular organs) is basally located by the immunolocalization of the gonadotropin-releasing hormone (GnRH) (Fig. 14.4e). In the caudal side of the rostral diencephalic septum section, the neocortex becomes the temporoparietal cortex due to the composition of the granular heterotypic isocortex with a predominance of granule-type neurons in an extensive layer IV, associated with the sensory function (Fig. 14.5a) (Garey 2007). In addition, the fusion of the lateral ventricles forms the third ventricle (Fig. 14.5a) and separates the remaining upper septal structures (triangular nucleus of the septum, TRS) which express calretinin (Fig. 14.5b). Ventrally in this section, the preoptic area (POA) of the hypothalamus depicts the anteroventral periventricular nucleus (AVPV) with the expression of kisspeptin peptide hormone (Fig. 14.5c). In the following coronal section (caudal diencephalic septum) (Figs. 14.1h and 14.5d), the hypothalamic suprachiasmatic (SCN) and supraoptic (SON) nuclei are present over the optic chiasm (och) with calretinin and kisspeptin expression respectively (Fig. 14.5e, f). Additionally, the incipient structures of the hippocampal formation (HPF) are detected with dorsal location (Fig. 14.5d).

Further back in the subsequent plate, in the rostral diencephalon section (Figs. 14.1i and 14.6), the caudal-striatum appears laterally displaced and confined to a limited zone between the internal and external capsules (Fig. 14.6a–d). The large region in the center of the brain is occupied by the thalamus that is bounded dorsally by the third ventricle, ventrally by the imaginary line established between the internal capsule and the fornix columns, and laterally by the internal capsule (Fig. 14.6a–d). Adjacent to the internal capsule, the reticular nucleus of the thalamus (RT) is situated as a paired structure and has parvalbumin expression (Fig. 14.6b, d, e). In addition, in the rostral diencephalon section (Fig. 14.1i), in the midline of the brain, is situated the paired structure called paraventricular nuclei of the thalamus (PVT) which depicts tyrosine hydroxylase expression (Fig. 14.6d, f). Moreover, in this plane, the mediobasal hypothalamus formed by the arcuate nucleus (ARC) and the median eminence (ME) with its inner and outer regions surround the

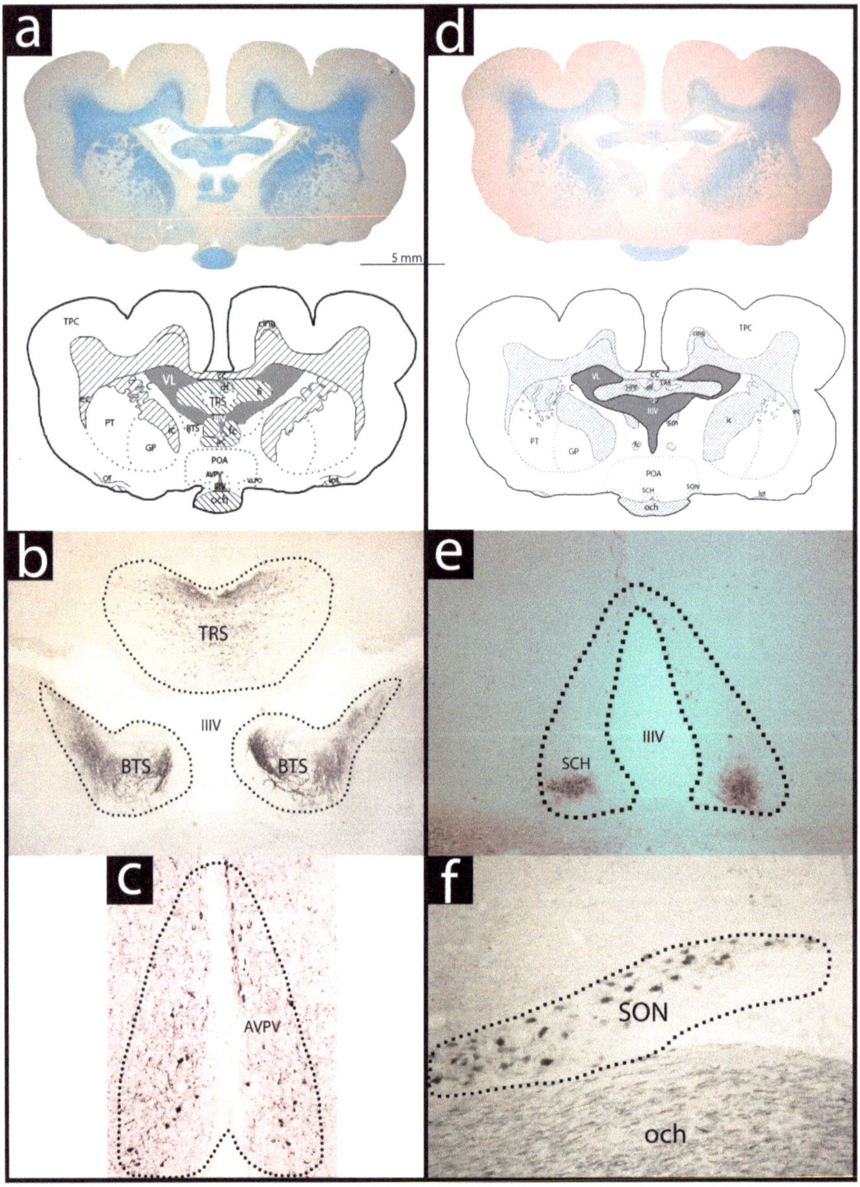

Fig. 14.5 (**a**) Coronal section of the rostral diencephalic septum (caudal part) of the vizcacha, *L. maximus*, with Klüwer-Barrera staining and its representative scheme. (**b**) Immunohistochemistry for calretinin in the triangular nucleus of the septum (TRS) and nucleus of the bed of the stria terminalis (BTS). (**c**) Immunohistochemistry for kisspeptin in the anteroventral periventricular nucleus (AVPV). (**d**) Coronal section of the caudal septo-diencephalon of the vizcacha, *L. maximus*, with Klüwer-Barrera stain and its representative scheme. (**e**) Immunohistochemistry for calretinin in the suprachiasmatic nucleus (SCH). (**f**) Immunohistochemistry for kisspeptin in the supraoptic nucleus (SON). Nomenclature. ac: Anterior commissure; AVPV: anteroventral periventricular nucleus; BTS: bed

(continued)

mammillary recessus of the third ventricle above and below, respectively (Fig. 14.6d, g). This marks the initial point of the infundibulum, which connects the hypothalamus to the pituitary gland.

Then, in the caudal diencephalon section (Figs. 14.1j and 14.7) and in the rostral midbrain section is found the hippocampal formation (HPF) inside the temporoparietal cortex (Figs. 14.1k and 14.8). In the caudal diencephalon section is possible to distinguish the dorsal and ventral hippocampal formations; the dorsal HPF with the Ammon's Horns CA1, CA2, and CA3; the dentate gyrus (DG) structures; and the ventral HPF with the presence of CA3 and DG, all of which express the neuron-specific protein NeuN (Fig. 14.7b, c). This section also highlights that the midbrain begins to appear in the center of the brain, displacing the dorsal thalamus to a lateral position next to the fimbria (Fig. 14.7b). Ventrally, the medial mammillary nuclei (MM) and their supramammillary nuclei (SUM, with tyrosine hydroxylase expression) are located (Fig. 14.7d). Then, in the rostral midbrain section (Fig. 14.1k), the fully developed hippocampal formation, with a complete dorsoventral extension, is observed (Fig. 14.8a, b). From the caudal side of the caudal diencephalic section up to the rostral midbrain section, the midbrain is found (Fig. 14.1j, k). Dorsally, in the cavity of the III ventricle, between the midbrain and the hippocampal formation, the pineal gland (an encephalic circumventricular organ) is placed (Figs. 14.1k and 14.8a–c). In the ventral region of the midbrain, the cerebral peduncles are found (Fig. 14.8b). In the midline, the interpeduncular nucleus (IPN) is located (Fig. 14.8d), laterally limited by the ventral tegmental area (VTA) (Fig. 14.8f) and externally bounded by the nuclei substantia nigra compact and reticular part (SNc and SNr, respectively) (Fig. 14.8e), all of them dopaminergic structures with tyrosine hydroxylase expression (Fig. 14.8d–f).

In the following mesencephalic plates (Fig. 14.1l, m and 14.9), that including midbrain and caudal midbrain sections, the red nuclei (RD), the superior colliculus motor related (SCm), the superior colliculus sensory related (SCs), and the periaqueductal gray (PAG) can be devised in the midbrain. In addition, the caudal thalamus and the caudal hippocampal formation that is surrounded by the external capsule and the alveus are observed in the midbrain section (Fig. 14.1l and 14.9a, b). Towards the caudal midbrain (Figs. 14.1m and 14.9c, d), the landscape of the

Fig. 14.5 (Continued)
nuclei of the stria terminalis; C: caudate nucleus; cc: corpus callosum; cing: cingulum bundle; ec: external capsule; df: dorsal fornix; fc: fornix columns; GP: globus pallidus; HPF: hippocampal formation; ic: internal capsule; IIIV: third ventricle; lot: lateral olfactory tract; VL: lateral ventricle; och: optic chiasm; OT: olfactory tubercle; POA: preoptic area; sm: stria medullaris; SON: supraoptic nucleus; SFO: subfornical organ; TPC: temporoparietal cortex; TRS: triangular nucleus of septum; VLPO: ventrolateral preoptic nucleus. Scale bar: 5 mm. Coronal sections of 5 μm thickness obtained with microtome (Leica brand, model RM 2125) and stained with Klüver Barrera (**a**, **d**) or immunohistochemistry were performed, and images were captured using an Olympus microscope (BX40, Olympus Optical Corporation) coupled to a video camera (390CU 3.2 Megapixel CCD Camera, Micrometrics) or a Nikon stereotaxic magnifying lens with 0.5× and 1× magnification (Nikon, SMZ800, Nikon Corp). The striped area refers to the white matter, the white area refers to the gray matter, and the dark area refers to the ventricles

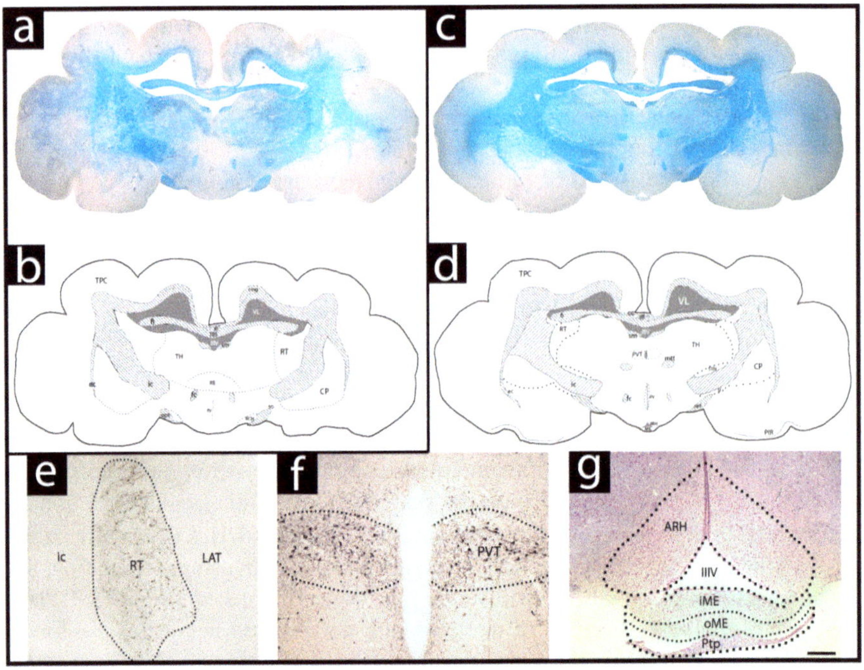

Fig. 14.6 Coronal section of the rostral diencephalic septum of the vizcacha, *L. maximus*. (**a**, **c**) Klüwer-Barrera stain. (**b**, **d**) Representative diagram of the section. (**e**) Immunohistochemistry for parvalbumin in the reticular nucleus of the thalamus (RT). (**f**) Immunohistochemistry for tyrosine hydroxylase in paraventricular thalamic nucleus (PVT). (**g**) Nissl stain of arcuate nucleus (ARH) and median eminence (ME). Nomenclature. ARH: Arcuate nucleus; cc: corpus callosum; cing: cingulum bundle; CP: caudo-putamen; ec: external capsule; df: dorsal fornix; fc: fornix columns; fi: fimbria; ic: internal capsule; IIIV: third ventricle; iME: inner median eminence; LAT: lateral group of thalamus; VL: lateral ventricle; ME: median eminence; mtt: mammillothalamic tract; oME: outer median eminence; opt: optic tract; PIR: piriform area; Ptp: pituitary tuberal part; PV: paraventricular nucleus; PVT: paraventricular thalamic nucleus; RCH: retrochiasmatic area; RE: nucleus of reuniens; RT: reticular nucleus of the thalamus; Si: substantia innominate; SON: supraopticnucleus; sm: stria medullaris; STN: subthalamic nucleus; TH: thalamus; TPC: temporoparietal cortex; TRS: triangular nucleus of septum. Scale bar: 5 mm. Coronal sections of 5 μm thickness obtained with microtome (Leica brand, model RM 2125) and stained with Nissl, Klüver Barrera, or immunohistochemistry were captured using an Olympus microscope (BX40, Olympus Optical Corporation) coupled to a video camera (390CU 3.2 Megapixel CCD Camera, Micrometrics) or a Nikon stereotaxic magnifying lens with 0.5× and 1× magnification (Nikon, SMZ800, Nikon Corp). The striped area refers to the white matter, the white area refers to the gray matter, and the dark area refers to the ventricles

mesencephalic areas is substantially modified. Two large paired nuclei, the reticular nucleus of the midbrain and the pontine reticular nucleus (PNR), as well as midline structures rostral linear nucleus raphe (CLI), the superior central nucleus raphe (CSr), and the pontine central gray (PG) are observed (Fig. 14.9d). Regarding the cerebellum, it is structured in the central vermis and two lateral cerebellar hemispheres. In the rostral cerebellum section, in addition to this structure, it is possible

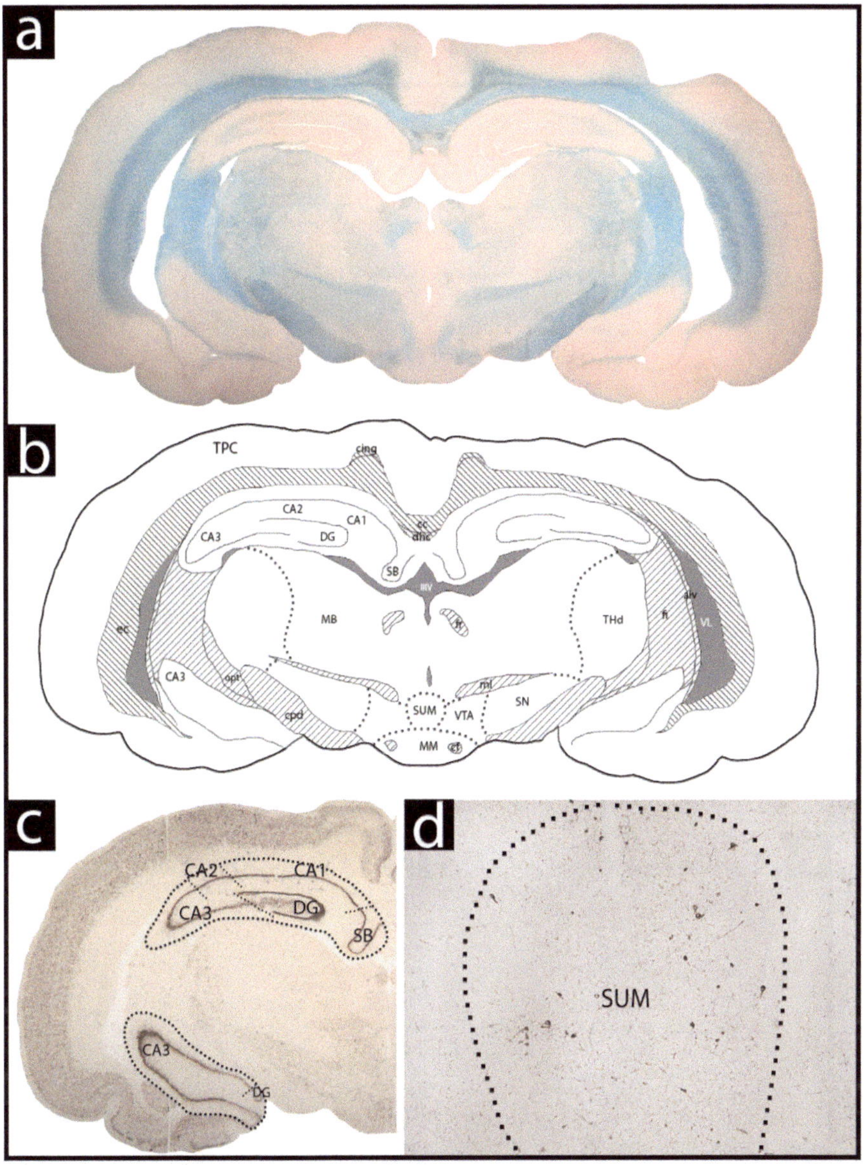

Fig. 14.7 Coronal section of the caudal diencephalon of the vizcacha, *L. maximus*. (**a**) Klüwer-Barrera stain. (**b**) Representative diagram of the section. (**c**) Immunohistochemistry for NeuN in the hippocampal formation (HPF). (**d**) Immunohistochemistry for tyrosine hydroxylase in supramammillary nucleus (SUM). Nomenclature. alv: Alveus; CA1: Ammon's horn 1; CA2: Ammon's horn 2; CA3: Ammon's horn 3; cing: cingulum bundle; fi: fimbria; cpd: cerebral peduncle; DG: dentate gyrus; dhc: dorsal hippocampal commissure; ec: external capsule; fc: fornix columns; fr: fasciculus retroflexus; IIIV: third ventricle; VL: lateral ventricle; MB: midbrain; ml: medial lemniscus; MM: medial mammillary nucleus; opt: optic tract; SB: subiculum; SN: substantia nigra; SUM: supramammillary nucleus; THd: dorsal thalamus; TPC: temporoparietal cortex; VTA: ventral tegmental area. Scale bar: 5 mm. Coronal sections of 5 μm thickness obtained with microtome (Leica brand, model RM 2125) and stained with Nissl, Klüver Barrera, or immunohistochemistry were captured using an Olympus microscope (BX40, Olympus Optical Corporation) coupled to a video camera (390CU 3.2 Megapixel CCD Camera, Micrometrics) or a Nikon stereotaxic magnifying lens with 0.5× and 1× magnification (Nikon, SMZ800, Nikon Corp). The striped area refers to the white matter, the white area refers to the gray matter, and the dark area refers to the ventricles

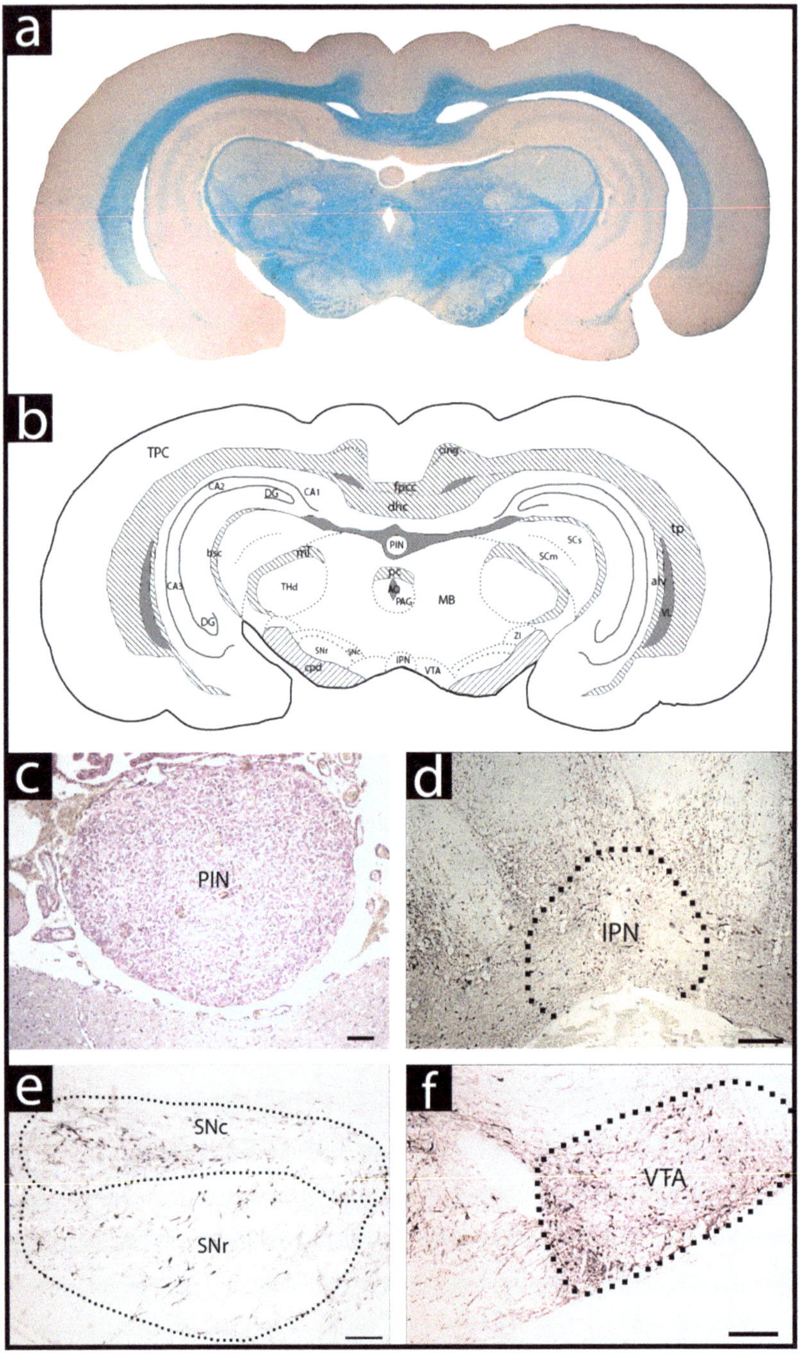

Fig. 14.8 Coronal section of the rostral midbrain of the vizcacha, *L. maximus*. (**a**) Klüwer-Barrera stain. (**b**) Representative diagram of the section. (**c**) Nissl stain of pineal gland (P). (**d–f**) (continued)

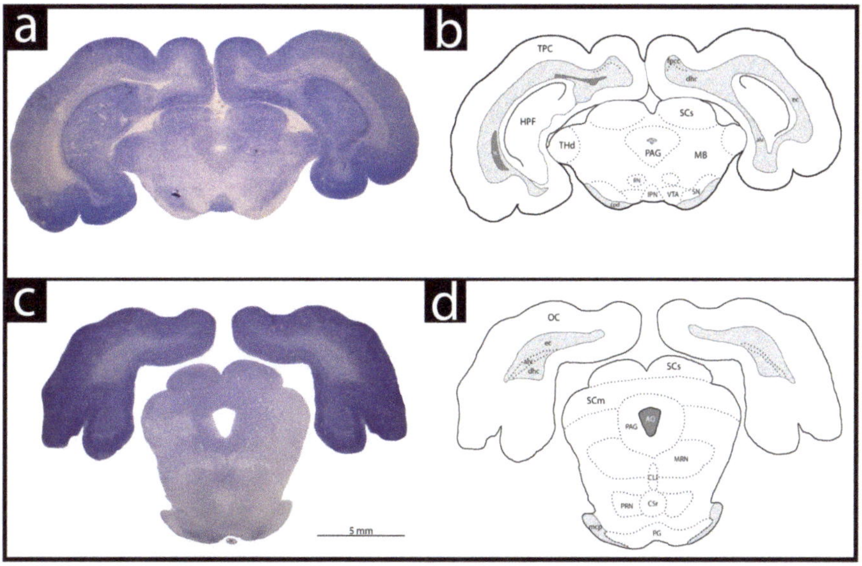

Fig. 14.9 (**a**, **b**) Coronal section of the midbrain of the vizcacha, *L. maximus*, with Nissl staining and its representative scheme. (**c**, **d**) Coronal section of the caudal midbrain of the vizcacha, *L. maximus*, with Nissl staining and its representative scheme. Nomenclature. alv: Alveus; AQ: cerebral aqueduct; arb: CLI: rostral linear nucleus raphe; cpd: cerebral peduncle; CSr: superior central nucleus raphe; dhc: dorsal hippocampal commissure; ec: external capsule; fpcc: posterior fórceps of corpus callosum; HPF: hippocampal formation; IPN: interpeduncular nucleus; VL: lateral ventricle; MB: midbrain; MRN: midbrain reticular nucleus; OC: occipital cortex; PAG: periaqueductal gray; PG: pontine central gray; mcp: middle cerebral peduncle; PNR: pontine reticular nucleus; RN: red nucleus; SCm: superior colliculus, motor related; SCs: superior colliculus, sensory related; SN: substantia nigra; THd: dorsal thalamus; TPC: temporoparietal cortex; VTA: ventral tegmental area. Scale bar: 5 mm. Coronal sections of 5 µm thickness obtained with microtome (Leica brand, model RM 2125) and stained with Nissl or Klüver Barrera were captured using a Nikon stereotaxic magnifying lens with 0.5× and 1× magnification (Nikon, SMZ800, Nikon Corp). The striped area refers to the white matter, the white area refers to the gray matter, and the dark area refers to the ventricles

Fig. 14.8 (Contiunued)

Immunohistochemistry for tyrosine hydroxylase in interpeduncular nucleus (IPN), substantia nigra (SN), and ventral tegmental area (VTA), respectively. Nomenclature. alv: Alveus; AQ: cerebral aqueduct; bsc: brachium of the superior colliculus; CA1: Ammon's horn 1; CA2: Ammon's horn 2; CA3: Ammon's horn 3; cing: cingulum bundle; fpcc: posterior fórceps of corpus callosum; cpd: cerebral peduncle; DG: dentate gyrus; dhc: dorsal hippocampal commissure; IPN: interpeduncular nucleus; VL: lateral ventricle; MB: midbrain; ml: medial lemniscus; PIN: pineal gland; PAG: periaqueductal gray; pc: posterior commissure; SCm: superior colliculus, motor related; SCs: superior colliculus, sensory related; SNc: substantia nigra, compact part; SNr: substantia nigra, reticular part; THd: dorsal thalamus; tp: tapetum; TPC: temporoparietal cortex; VTA: ventral tegmental area; ZI: zona incerta. Scale bar: 5 mm. Coronal sections of 5 µm thickness obtained with microtome (Leica brand, model RM 2125) and stained with Nissl were captured using a Nikon stereotaxic magnifying lens with 0.5× and 1× magnification (Nikon, SMZ800, Nikon Corp). The striped area refers to the white matter, the white area refers to the gray matter, and the dark area refers to the ventricles

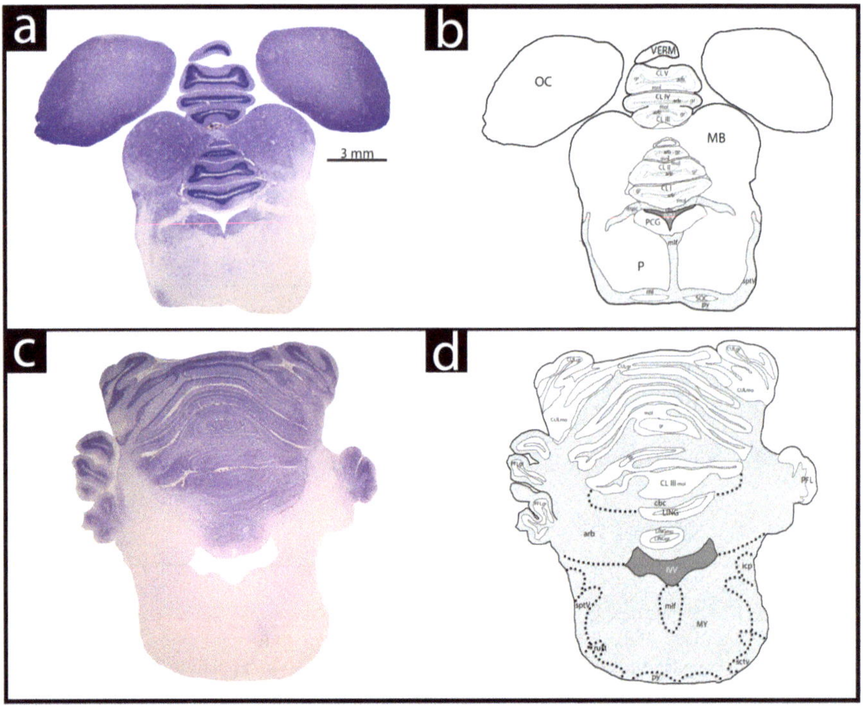

Fig. 14.10 (**a**, **b**) Coronal section of the rostral cerebellum of the vizcacha, *L. maximus*, with Nissl staining and its representative scheme. (**c**, **d**) Coronal section of the caudal cerebellum of the vizcacha, *L. maximus*, with Nissl staining and its representative scheme. Nomenclature. arb: Arbor vitae; cbc: cerebellar commissure; CL I: central lobule I; CL II: central lobule II; CL III: central lobule III; CL IV: central lobule IV; CL V: central lobule V; CULgr: culmen granular layer; CULmo: culmen molecular layer; gr: cerebellar granule layer; icp: inferior cerebellar peduncle; IVV: fourth ventricle; LING: lingula; LINGgr: lingula granule layer; LINGmo: lingula molecular layer; MB: midbrain; ml: medial lemniscus; mlf: medial longitudinal fascicle; mol: cerebellar molecular layer; MY: medulla; OC: occipital cortex; P: Pons; PCG: pontine central gray; PFL: paraflocculus; PFLgr: paraflocculus granular layer; py: pyramid; rust: rubrospinal tract; SOC: superior olivary complex; sctv: ventral spinocerebellar tract; sptV: spinal tract of the trigeminal nerve; VERM: vermis. Scale bar: 5 mm. Coronal sections of 5 μm thickness obtained with microtome (Leica brand, model RM 2125) and stained with Nissl were captured using a Nikon stereotaxic magnifying lens with 0.5× and 1× magnification (Nikon, SMZ800, Nikon Corp). The striped area refers to the white matter, the white area refers to the gray matter, and the dark area refers to the ventricles

to find the caudal region of the occipital cortex and the midbrain above the pons (Fig. 14.10a, b). In this section, the rostral portion of the vermis can be seen, in which the central lobes (CL) are located with their granular and molecular layers, and the arbor vitae (arb) (Fig. 14.10a, b). Finally, in the caudal cerebellum section, the central lobe appears as a continuous structure, and the culmen (CUL) structures are added in the dorsal portion and the lingula (LING) towards the base of the cerebellum. Furthermore, the paraflocculus structure (PFL) can be observed as part of the flocculonodular lobe, on the sides of the section (Fig. 14.10c, d).

14.4 Final Conclusions

The present chapter significantly advances our understanding of the neuroanatomy and histology of the South American plains vizcacha, *Lagostomus maximus*. The telencephalic morphology allowed the establishment of the species as gyrencephalic, evidencing an increase in the cortical surface due to the conjunction of the factors size, invagination, and cortical thickness, particularly benefiting the temporoparietal lobe. Related to the internal structures of the brain of this species, they exhibit distinctive features, highlighting the middle striatum's structural variation that showed regionalization due to the presence of an internal capsule that separates it into the caudate nucleus and the putamen, both with well-defined limits. Together, the enhanced development of the temporoparietal lobes and the regionalization of the striatal body provide the species with adaptive advantages that seem to be suited to its lifestyle. These adaptations allow greater development of the touch and hearing senses and facilitate the segregation of motor and cognitive abilities. Collectively, these neuroanatomical features may contribute to the species' improved recognition of sounds emitted by the members of the colony and to identify the presence or closeness of potential predators. Furthermore, they are useful for the construction of their burrow systems, called *vizcacheras,* and for the manipulation of food resources. Present findings shed light on the distinctive structural characteristics of the brain of the vizcacha, which are closely linked to its colonial and burrow dweller way of life.

Finally, from an evolutionary perspective, our research suggests that both the degree of encephalization and girencephalization, as well as the regionalization of the striatum, would have presented an early evolutionary/phylogenetic restriction process that would have favored the retention of both characteristics during the process of speciation of the plains vizcacha.

References

Albert M, Huttner WB (2015) Clever space saving—how the cerebral cortex folds. EMBO J 34(14):1845–1847

Beccari N (1943) Neurologia comparata, anatomo-funzionale dei vertebrati, compresol'uomo. Sansoni Edizioni Scientifiche, Firenze

Bee de Speroni N, Pellegrini de Gastaldo A (1988) Encefalización y composición cerebral en tres roedores sudamericanos (*Dolichotis patagonum, Lagostomus maximus y Calomys musculinus*). Physis 46(111):31–39

Bertrand OC, Silcox MT (2016) First virtual endocasts of a fossil rodent: *Ischyromys typus* (Ischyromyidae, Oligocene) and brain evolution in rodents. J Vertebr Paleontol 36:1–19

Branch LC (1993) Seasonal patterns of activity and body mass in plains vizcacha, Lagostomus maximus (family Chinchillidae). Can J Zool 71:1041–1045

Contreras JR (1981) El tunduque: Un modelo de ajuste adaptativo. Serie Científica:22–25

Contreras JR (1984) La vizcacha. Fauna Argentina 45:1–32

Dorfman VB, Fraunhoffer N, Inserra PIF, Loidl CF, Vitullo AD (2011) Histological characterization of gonadotropin-releasing hormone (GnRH) in the hypothalamus of the South American plains vizcacha (Lagostomus maximus). J Mol Histol 42:311–321

Dozo MT (1997) Paleoneurologia de Dolicaviaminúscula (Rodentia, Caviidae) y Paedotherium insigne (Notoungulata, hegetotheriidae) delpliocenio de Buenos Aires, Argentina. Ameghiniana Rev Paleontol Argent 34:427–435
Garey LJ (2007) Brodmann's localisation in the cerebral cortex: the principles of comparative localisation in the cerebral cortex based on cytoarchitectonics. Springer, Inglaterra, 298 pages
Herculano-Houzel S (2007) Encephalization, neuronal excess, and neuronal index in rodents. Anat Rec 290(10):1280–1287
Hofman MA (1985) Size and shape of the cerebral cortex in mammals. I. The cortical surface. Brain Behav Evol 27(1):28–40
Inserra PIF, Charif SE, Di Giorgio NP, Saucedo L, Schmidt AR, Fraunhoffer N, Halperin J, Gariboldi MC, Leopardo NP, Lux-Lantos V, Gonzalez CR, Vitullo AD, Dorfman VB (2017) ERα and GnRH co-localize in the hypothalamic neurons of the South American plains vizcacha, Lagostomus maximus (Rodentia, Caviomorpha). J Mol Histol 48(3):259–273
Jackson JE, Lyn C, Villarreal D (1996) Lagostomus maximus. Mamm Species 543:1–6
Jerison HJ (1973) Evolution of the brain and intelligence. Academic Press, New York
Jerison HJ (1977) The theory of encephalization. Ann N Y Acad Sci 299:146–160
Kelava I, Lewitus E, Huttner WB (2013) The secondary loss of gyrencephaly as an example of evolutionary phenotypical reversal. Front Neuroanat 7:16
Nevo E (1979) Adaptive convergence and divergence of subterranean mammals. Annu Rev Ecol Evol Syst 10:269–308
Pillay P, Manger PR (2007) Order-specific quantitative patterns of cortical gyrification. Eur J Neurosci 25:2705–2712
Rakic P (1995) A small step for the cell, a giant leap for mankind: a hypothesis of neocortical expansion during evolution. Trends Neurosci 18:383–388
Schmidt AR, Inserra PIF, Cortasa SA, Charif SE, Proietto S, Corso MC, Villarreal F, Halperin J, Loidl CF, Vitullo AD, Dorfman VB (2019) Structural organization, GABAergic and tyrosine hydroxylase expression in the striatum and globus pallidus of the South American plains vizcacha, Lagostomus maximus (Rodentia, Caviomorpha). J Mol Histol 50(6):515–531
Schmidt AR, Gariboldi MC, Cortasa SA, Proietto S, Corso MC, Inserra PIF, Jaime VS, Halperin J, Vitullo AD, Dorfman VB (2020) Neocortical anatomy in the South American Plains Vizcacha, Lagostomus maximus, reveals different strategies in encephalic development among hystricomorpha and myomorpha rodents. Brain Behav Evol 95:318–329
Sidman RL, Angevine JB, Taber Pierce E (1971) Atlas of the mouse brain and spinal cord. Harvard University Press, Cambridge
Triarhou LC (2017) The comparative neurology of neocortical gyration and the quest for functional specialization. Front Syst Neurosci 11:96
Welker W (1990) Why does cerebral cortex fissure and fold? A review of determinants of gyri and sulci, Chapter 10 (8B):3–136
Zilles K, Palomero-Gallagher N, Amunts K (2013) Development of cortical folding during evolution and ontogeny. Trends Neurosci 36:275–284

Chapter 15
Post-implantation Embryo Development in the South American Plains Vizcacha, *Lagostomus maximus* (Rodentia, Caviomorpha)

Noelia Paola Leopardo, Pablo Ignacio Felipe Inserra, and Alfredo Daniel Vitullo

Abstract The early embryo development after implantation varies among mammals. In muroid rodents, an egg cylinder is formed by folding during gastrulation; however, in most mammals, a flat disc develops after the embryo implants in the uterine wall. In contrast to what occurs in muroids, the embryo of the caviomorph *Lagostomus maximus* develops from a bilaminar disc by cavitation rather than folding. The flat disc morphology of the epiblast is shared with non-rodent mammals such as rabbits, pigs, cows, and humans and is considered the mammotypic embryonic morphology. Nevertheless, the flat embryo of *L. maximus* is a somewhat intermediate situation between the mammotypical bilaminar disc and the egg cylinder of muroids. In muroid rodents, the extraembryonic ectoderm is in close contact with the epiblast as a consequence of embryo folding during gastrulation. In contrast, the typical flat disc of mammals does not show extraembryonic ectoderm. *L. maximus* embryo develops extraembryonic ectoderm homologous to that found in muroids; however, it remains far apart from the epiblast by the visceral endoderm. Here we describe the embryonic and fetal development throughout gestation in *L. maximus* by discussing similarities and differences with other mammals from an evolutionary and phylogenetic perspective.

Keywords Bilaminar disk · Egg cylinder · *L. maximus* · Germ cells

---

N. P. Leopardo (✉) · P. I. F. Inserra · A. D. Vitullo
Centro de Estudios Biomédicos Básicos, Aplicados y Desarrollo (CEBBAD), Universidad Maimónides, Ciudad Autónoma de Buenos Aires, Argentina

Consejo Nacional de Investigaciones Científicas y Técnicas (CONICET), Ciudad Autónoma de Buenos Aires, Argentina
e-mail: leopardo.noelia@maimonides.edu

L. L. Rasia et al. (eds.), *Plains Vizcachas*, https://doi.org/10.1007/978-3-031-49487-1_15

Abbreviations

A	Anterior
Al	Allantois
Am	Amnion
AmC	Amniotic cavity
AnMe	Antimesometrial
Ar	Archenteron
D	Dorsal
De	Decidua
EcTr	Ectoplacental trophoblast
ED	Embryonic disc
EmEc	Embryonic ectoderm
EmEn	Embryonic endoderm
EmMe	Embryonic mesoderm
Ep	Epiblast
EpaC	Epamniotic cavity
EPC	Ectoplacental cone
EPGC	Ectoplacental giant cells
ExC	Exocoelomic cavity
ExE	Extraembryonic ectoderm
ExM	Extraembryonic mesoderm
HF	Head fold
Hp	Hindgut pocket
IC	Implantation chamber
ICM	Inner cell mass
IS	Implantation site
Me	Mesometrial
No	Node
P	Posterior
PEn	Primitive endoderm
PS	Primitive streak
SpMe	Splanchnic mesoderm
Tr	Trophectoderm
Trd	Trophoblast-derived cells
V	Ventral
VEn	Visceral endoderm

15.1 Introduction

Embryogenesis in mammals has been extensively studied through the use of laboratory animal models (Rossant and Tam 2002). The use of laboratory animals in the exploration of embryonic development has significant advantages, as they are generally small, cost-effective, and easy to manage. In addition, these animals have short gestation periods and large litters and have even undergone genetic manipulation to study gene expression without encountering major complications (Eakin and Behringer 2004). Nevertheless, constraining the comparison of embryonic development to a restricted group of species inevitably limits our understanding of the comprehensive nature of developmental processes in mammals.

The population of mammalian species on Earth is dominated by a minuscule percentage of mice, comprising less than 0.02% of the total population (Nowak 1999). Significant variability in reproductive strategies exists among species, as evidenced by disparities in implantation mechanisms, uterine and placental morphologies, gestation length, and offspring production numbers, as well as early and post-implantation embryo development (Tyndale-Biscoe and Renfree 1987). This variability among species implies that the mechanisms that regulate embryogenesis are susceptible to variation, even within the same mammalian order. Post-implantation development in rodents reveals a common phenomenon known as the inversion of germ layers (Copp and Cockroft 1990; Carlson 1999; Guzeloglu-Kayisli et al. 2007), which, nonetheless, manifests differently among various rodent species. While most rodent embryos, including rats, mice, and hamsters, develop into egg cylinders, other mammalian embryos, such as rabbits, guinea pigs, and even humans, evolve into bilaminar embryonic discs (O'Rahilly and Müller 2010; Bianchi et al. 1993; Eakin and Behringer 2004). The study of early embryo development provides valuable insights into a range of mechanisms, including germline specification. In mice, gene induction regulates the differentiation of the mammalian germline, with sequential gene expression relying on intricate cell-cell interactions among germ layers such as ectoderm, endoderm, and mesoderm as well as both embryonic and extraembryonic components. These interactions occur during the early post-implantation stage. However, due to the diverse nature of early embryonic development among different mammalian species, gene expression patterns also exhibit variations (Leopardo and Vitullo 2017). Therefore, a deeper understanding of embryonic development across diverse mammals can provide enhanced insights into the fundamental events occurring during this critical stage.

Lagostomus maximus, commonly known as the South American plains vizcacha, is a hystricognathe with unusual developmental and reproductive characteristics. The existing embryological bibliography on hystricomorphs is mainly confined to the domestic guinea pig (Sansom and Hill 1931) and chinchilla (Tibbitts and Hillemann 1959; Ilgren 1981). Reproduction in hystricomorph rodents is characterized by a long gestation, a long estrous cycle, a tendency to form accessory corpora lutea, and a vaginal closure membrane that is present until estrus (Weir and Rowlands 1973, 1974; Weir 1974). However, there are limited reports on post-implantation embryo development in these species (Leopardo and Vitullo 2017; Carter et al. 2023; Acuña et al. 2023). It is noteworthy that the plains vizcacha is considered the mammal with the highest ovulation rate known to date, releasing between 200 and 800 oocytes at each estrus cycle (two cycles per year) (Weir 1970, 1971a, b). Only a few 6–12 oocytes are successfully implanted after 18 days of pre-implantation

Table 15.1 *Lagostomus maximus* embryonic stages, according to capture time

Capture time	Embryonic stage
Late March to mid-April	Pre-somite
Early-April	1–7 somites
Mid-April	8–12 somites
Late-April	13–20 somites
Late April to early-May	21–29 somites
Early- May	30–34 somites
Early–mid-May	35–39 somites
Mid-May	>40 somites (7 weeks)
Mid-May to early-August	(>8 gestation weeks)

development from this massive ovulation (Weir 1971a, b). During the early post-implantation development, the embryos suffer developmental detention and subsequent reabsorption (Acuña et al. 2020). Therefore, only two embryos implanted nearest to the cervix (one per uterine horn) survive, avoiding the natural selective abortion mechanism (Roberts and Weir 1973). After 154 days of gestation, the species gives birth to one or two pups (Weir 1971a, b; Leopardo and Vitullo 2017). Although embryonic resorption is observed in other species such as chinchilla (Weir 1967) and coypu (Newson 1966), it is not as extensive as in the plains vizcacha (Carter et al. 2023).

In this chapter, we shall undertake a comprehensive review of the present state of understanding regarding the process of post-implantation development observed in the plains vizcacha. Additionally, we shall present a novel set of information related to the organization and development of various organs during embryonic stages.

15.2 Gestational Season of the Plains Vizcacha

The plains vizcacha is known to have a single breeding season annually, commencing in late March (Llanos and Crespo 1952; Leopardo and Vitullo 2017) or early April (Hudson 1872). Nevertheless, some authors have suggested that around 36% of females may exhibit a second pregnancy later in the year (Llanos and Crespo 1952), and it has been documented that vizcacha may breed in both autumn and spring (Gibson 1877). The possibility of a second pregnancy appears to be more prevalent in northern populations or in instances where the initial litter is lost (Weir 1971a, b). The gestation period can fluctuate between 145 and 166 days (Weir 1971a, b; Leopardo and Vitullo 2017). A detailed account of the embryonic stages of the plains vizcacha, including external morphology, somite number, and capture time in Buenos Aires, Argentina, is presented in Table 15.1 and Fig. 15.1.

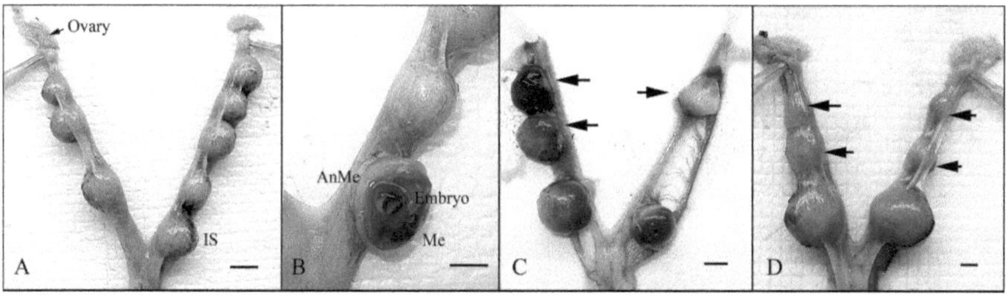

Fig. 15.1 Implantation site in the bicornuate uterus of *Lagostomus maximus*. (**a**) Uterine horns with eight implantation sites (IS) at an early stage of pregnancy (22–26 days gestation). (**b**) Cutaway view of the uterus showing the embryo implanted with inner cell mass towards the mesometrial region. (**c**) Uterine horns with necrotic embryos (arrows) at early-somites stage embryo (40–45 days gestation). (**d**) Uterine horns with necrotic embryos (arrows) at early-somites stage embryo (46–50 days gestation). Me: mesometrial, AnMe: antimesometrial. Scale bar: 2 mm

15.3 Morphological Stages of Post-implantation Embryonic Development

15.3.1 Formation of the Early Egg Cylinder: 22–26 Days of Gestation

At a gestational age of 22–26 days, the plains vizcacha exhibited the presence of multiple implantation sites (IS) in both uterine horns, varying from 6 to 12. These intrauterine structures were externally observable as spherical enlargements measuring approximately 4 cm in diameter (Table 15.2 and Fig. 15.1), at equivalent stages of development. The implantation pattern displayed an interstitial nature, with the embryo oriented in an anti-mesometrial direction (Fig. 15.1b). The embryo displayed a bilaminar structure, consisting of an inner cell mass (ICM) comprised of ectoderm cells, also known as the epiblast, and an outer layer of endoderm cells referred to as the visceral endoderm (VEn). The VEn closely interacted with the ICM, elongating and separating the embryo from the trophoblast. Additionally, the trophectoderm was in direct contact with the VEn (Fig. 15.2a).

15.3.2 The Pre-primitive-Streak Stage: 26–31 Days of Gestation

During the gestational period spanning from 26 to 31 days, observable transformations in the differentiation of the plains vizcacha embryo were observed. The IS, which measured about 7 mm in diameter (Table 15.2), exhibited a consistent

Table 15.2 Characterization and morphometrics of *Lagostomus maximus* embryonic development

Gestation (day)	Post implantation (week)	Number embryos/ fetus	Pairs of somites	IS (mm)	Cephalocaudal length (mm)	Weight (gr)
25	1	9	–	4 ± 0.02	–	–
		7	–	6978 ± 0.02	–	–
32	2	7	–	9,5 ± 0.72	–	–
		8	–	9,6 ± 0.71	–	–
39	3	6	–	11,17 ± 0.40	2.56 ± 0.29	–
		9	1–7	14,33 ± 0.84	3.02 ± 0.14	–
46	4	8	8–12	14,4 ± 0.40	3.41 ± 0.07	–
		9	13–20	16,2 ± 0.66	3.61 ± 0.26	–
53	5	7	21–29	16,5 ± 0.65	4.50 ± 0.39	–
		7	30–34	16,6 ± 0.49	5.80 ± 0.38	–
60	6	8	35–39	19,75 ± 0.47	8.47 ± 0.39	–
		8	40–44	20,33 ± 0.16	9.58 ± 0.45	–
67	7	8	>45	21,64 ± 0.43	10.33 ± 0.45	0.3 ± 0.1
			–	–	13.76 ± 0.53	0.7 ± 0.1
74	8	9	–	–	19.11 ± 1.63	0.9 ± 0.1
81	9	9	–	–	29.79 ± 1.46	2.8 ± 0.4
88	10	8	–	–	45.38 ± 1.59	7.7 ± 0.6
95	11	7	–	–	51.71 ± 1.84	10.9 ± 1.0
102	12	6	–	–	57.90 ± 2.04	14.9 ± 1.5
107	13	9	–	–	68.96 ± 3.24	24.2 ± 1.9
112	14	8	–	–	86.33 ± 3.09	49.0 ± 4.8
119	15	8	–	–	103.80 ± 2.14	81.9 ± 5.8
126	16	7	–	–	112.80 ± 2.22	95.8 ± 5.6
133	17	9	–	–	122.40 ± 2.34	114.3 ± 6.1
140	18	9	–	–	134.60 ± 2.23	165.8 ± 5.3
147	19	8	–	–	153.11 ± 0.74	216.3 ± 4.47

Gestation days were calculated based on an 18-day pre-implantation development according to Roberts and Weir (1973). Different letters in the same column indicate statistical differences ($p < 0.05$)

spacing of the embryo. No signs of embryonic regression were observed. The embryo underwent an enlargement and assumed a cone-like bulging shape into the segmentation cavity (Figs. 15.2b and 15.3b, c). At this pre-streak stage, both the embryonic axis and the differentiation between embryonic and extraembryonic tissues (ExE) were evident (Figs. 15.2b, c and 15.3b, c). The cone-like structures were primarily composed of unilaminar visceral endoderm (VEn) providing structural support to the inner cell mass (ICM). The VEn formed a continuous epithelial layer extending from beneath the ICM to the ectoplacental trophoblast (Figs. 15.2b, c and 15.3b, c). The ICM remained distinct from the ExE, resulting in a cup-shaped structure with the VEn acting as a barrier (Fig. 15.2b–e). The elongated VEn contributed to the formation of the exocoelomic cavity, while the ICM flattened (Fig. 15.3b, c).

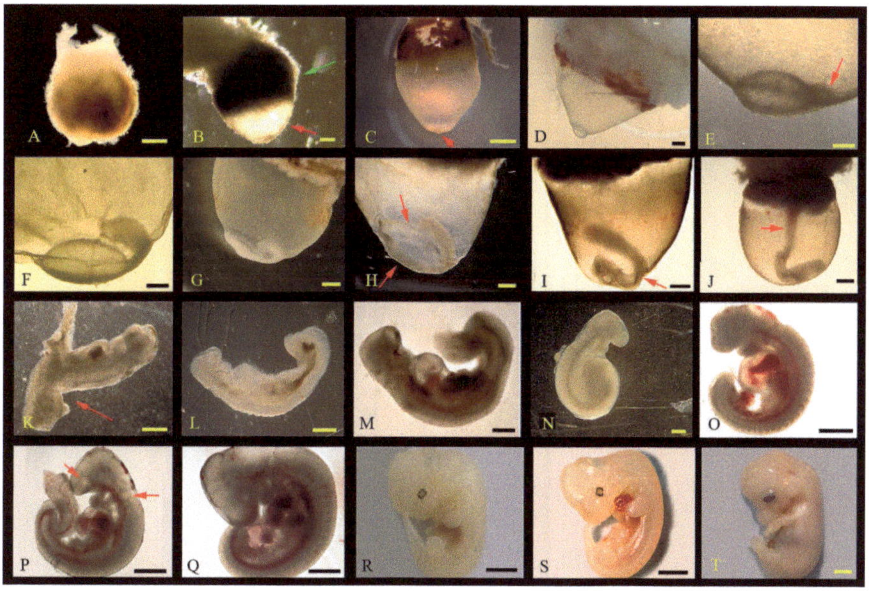

Fig. 15.2 General morphogenesis on fresh post-implantation conceptuses of *Lagostomus maximus*. (**a**) Newly implanted embryos (22–26 days) are round-shaped, (**b**) soon differentiating a pro-amniotic cavity (red arrow) and a trophoblast (green arrow). (**c**) At pre streak stage, the embryo acquires a cup-shaped aspect with visible inner cell mass (red arrow). (**d**) At the beginning of gastrulation, the first evidence of embryonic axis appears and the amniotic cavity develops, (**e**) soon followed by an allantoid bud appearance (red arrow) at the neural plate stage. (**f**) At the early head-fold pre-somite stage, the allantoids bud enlarges and (**g, h**) the cephalic region begins to turn. (**i, j**) With the appearance of the first somites, the allantoic extends through the exocoelomic cavity until it contacts the chorion; note the foregut pocket (red arrow in **j**). (**k**) Embryo turning initiates (red arrow) and (**l–o**) turned embryo progressively increases in somite number. (**p**) At 30–40 pairs of somites, optic and otic vesicles are evident (red arrows). (**q**) Umbilical hernia is visible. (**r–t**) From 45–60 pairs of somites lens vesicles close, the eye is defined, the tail elongates and individual fingers are visible. Scale bars: 100 μm (**a, b**); 250 μm (**c, d, f**); 500 μm (**e, g, h, k–m**), 1.000 μm (**i, j, n–q**); 2.000 μm (**r–t**)

The embryo sac consisted of the ectoplacental trophoblast enveloping the epamniotic cavity (Fig. 15.3e). The outer lamina of this cavity exhibited a simple epithelium, while the inner lamina appeared to vary in thickness, ranging from two to three cells (Fig. 15.3b, e). No ectoplacental cone (EPC) was identified, and the embryo displayed a bilaminar disc morphology (Fig. 15.3d), accompanied by a rim of mesometrially located ectoplacental giant cells (EPGC) (Figs. 15.2d, e and 15.3c, d). There was no histological evidence of a parietal yolk sac layer (Reichert's membrane). The egg cylinder was attached mesometrially within the mesometrial portion of the implantation chamber. The anti-mesometrial chamber contained maternal blood freely (Fig. 15.3c). This material has been previously published (Leopardo and Vitullo 2017).

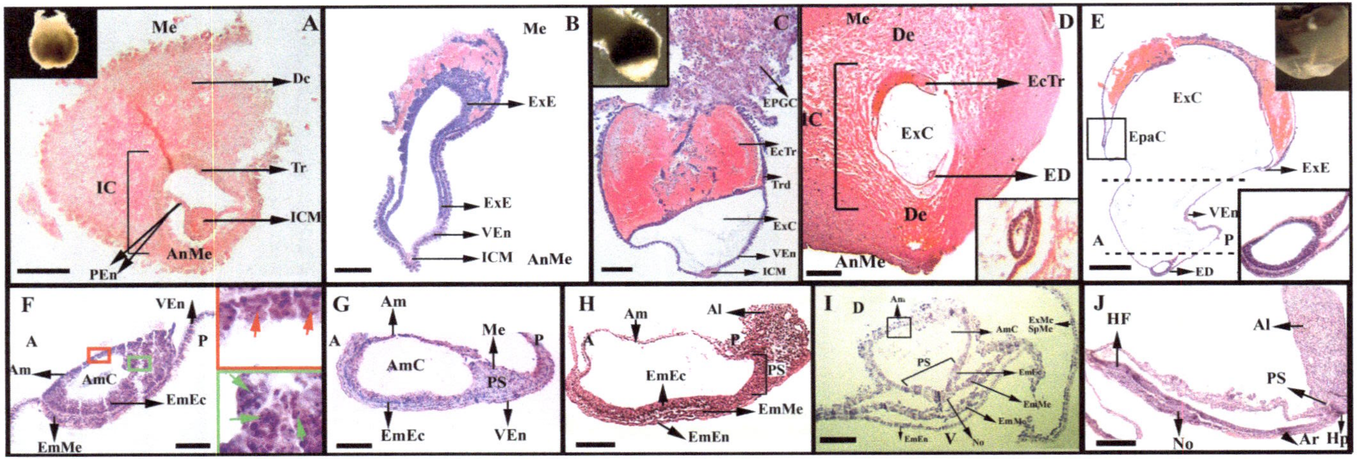

Fig. 15.3 Sagittal sections of the early post-implantation embryo of *Lagostomus maximus*. (**a**) Newly implanted embryo (22–26 days gestation). (**b**) Pre-streak stage, elongated embryo; note that the embryonic and extra-embryonic tissues are far apart. (**c**) Differentiation of the embryo (26–32 days gestation) with the ectoplacental trophoblast (EcTr) invaded by maternal blood; exocoelomic cavity (ExC) formed. Note the mesometrially situated, ectoplacental giant cells (EPGC). (**d**) General view of the implantation chamber, with the embryo disc at the anti-mesometrial pole (26–32 days gestation). Inset: detail of the embryonic disc (ED). (**e**) ED with allantoic bud in the neural plate stage. Note the extraembryonic ectoderm (ExMe) far apart from the cavitated epiblast, separated by the visceral endoderm (VEn) (dotted lines). Inset: detail of the ED. (**f**) Embryonic cavitation; upper inset: cells undergoing apoptosis in the amniotic cavity (AmC) (red arrows); lower inset: cells in apoptosis in the epiblast (green arrows). (**g**) Primitive streak (PS) stage: the appearance of some mesoderm cells in the posterior region of the epiblast that lines the VEn. (**h**, **i**) Neural plate stage (32–39 days gestation): the appearance of the node (No). (**j**) Late head-fold stage: foregut invagination, the extension of the allantois (Al). A: anterior; Am: amnion; AnMe: antimesometrial; Ar: archenteron; D: dorsal; De: decidua; EmEc: embryonic ectoderm; EmEn: embryonic endoderm; EmMe: embryonic mesoderm; EpaC: epamniotic cavity; ExE: extraembryonic ectoderm; HF: head fold, Hp: hindgut pocket; IC: implantation chamber; ICM: inner cell mass; Me: mesometrial; P: posterior; PEn: primitive endoderm; SpMe: splanchnic mesoderm; Tr: trophectoderm; Trd: trophoblast-derived cells; V: ventral. Scale bars: 100 µm. (Adapted from Leopardo and Vitullo (2017))

15.3.3 *Cavitation, Gastrulation, and Formation of the Amniotic Cavity: 32–39 Days of Gestation*

At an advanced stage, approximately 39 days post-implantation, the implantation sites exhibited a diameter of approximately 11 mm (Table 15.2), with embryos evenly spaced. The embryo itself presented as a tubular egg cylinder, maintaining a similar orientation to the preceding stage. The amino-embryonic mass appeared as a flattened cluster of undifferentiated cells, featuring a central cleft that initiated the formation of an incipient amniotic cavity (Figs. 15.2d and 15.3d–g). Cavitation was responsible for the development of the amniotic cavity, accompanied by apoptosis of the epiblast cells (Fig. 15.3f). In plains vizcacha, the amino-embryonic mass underwent cavitation and remained separate from the extraembryonic ectoderm. A central lumen was observed, and the proamniotic cavity emerged before gastrulation, shaping the embryo into a characteristic bilaminar disc (Figs. 15.2 and 15.3d, e). Prior to the emergence of the allantoic bud, the embryo established ectoplacental, amniotic, and exocoelomic cavities (Figs. 15.2d and 15.3e). All embryonic cavities and membranes in plains vizcacha originated from the cavitated embryo, leading to a well-structured development progression.

At the primitive streak stage, following cavitation, there was limited evidence of differentiation among the cells within the embryonic disc. Nevertheless, mesoderm cells displayed signs of differentiation in the posterior region of the embryo, signifying the commencement of gastrulation (Figs. 15.2e and 15.3g). At the neural plate stage, significant morphological changes included a prominent swelling of the allantois at one edge of the disc and the folding of the head at the opposite edge (Fig. 15.2f). This stage marked the establishment of a trilaminar embryonic disc encompassing the three primary germ layers: ectoderm, mesoderm, and endoderm (Fig. 15.3h). Additionally, the primitive streak, the notochord, and the neural tube were formed (Fig. 15.2g). Mesoderm emerged, spreading between the ectoderm and the endoderm of the embryonic disc. Somatic mesoderm covered the amniotic ectoderm, while splanchnic mesoderm (ExM) extended over the visceral endoderm (yolk sac). Simultaneously, the ectodermal tissue thickened and flattened into the neural plate (Fig. 15.3i). Finally, at the early head fold pre-somite stage, an invagination of the hindgut was observed, accompanied by the extension of the allantois and head folding (Figs. 15.2h and 15.3j). This material has been previously published (Leopardo and Vitullo 2017).

15.3.4 *Early Somites Stage Embryo: 40–45 Days of Gestation*

At the early somites stage, the implantation site dimensions were around 14 cm in diameter, as indicated in Table 15.2. Furthermore, signs of embryonic regression were evident in the uterine horns. Notably, the implantation sites closest to the ovary were necrotic (Fig. 15.1c, d). At this stage, the allantoic primordium underwent further development, forming an elongated mass (Figs. 15.2i and 15.4). Additionally, the differentiation of the embryonic disc led to an increase in the cephalic neural folds size, representing the most significant advances in the embryo's development. Neurulation is initiated during this stage, with four or five pairs of somites observed (Fig. 15.4).

The demarcation of the embryo's anterior portion was defined by the undercutting of the amnion, where the ectodermal and mesodermal layers of the amnion were loosely attached. Posteriorly, the mesoderm of the amnion was continuous with that of the allantois primordium, extending into the exocoelomic space as a loosely aggregated mass. This mass was constituted by a series of irregular blind pouches formed through multiple invaginations of the allantoic mesothelium (Fig. 15.4). In sagittal section, the embryo displayed a U-shaped configuration, with the surface and neural ectoderm located within the concavity of the U-shaped structure. The endoderm, corresponding to the prospective midgut region or the exposed part of the gut, was situated on the outer aspect of the convex portion of the U (Fig. 15.4). The embryo's trunk region underwent a reversal of curvature through a process of turning or axial rotation, which commenced when the embryo had 8–10 pairs of somites (Fig. 15.2j, k), and was completed when it had 14–16 pairs of somites (Fig. 15.2l). Throughout this turning process, the embryo became invaginated and enveloped by the amnion and yolk sac, leading to the inversion of its germ layers. This material has been previously published (Leopardo and Vitullo 2017).

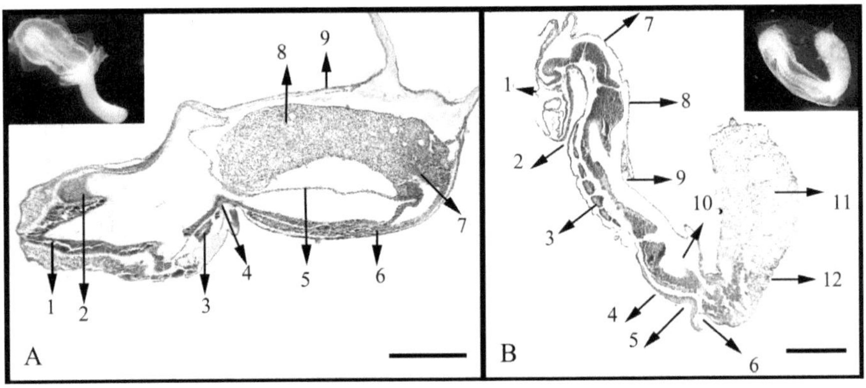

Fig. 15.4 Sagittal histological sections of *Lagostomus maximus* early embryonic development (40–45 days gestation). (**a**) Unrotated embryo with 5–6 pairs of somites in the neural plate stage. Allantois extends, first branchial arch, heart starts to form, foregut pocket visible, preotic sulcus. (1): neural plate, (2): cephalic neural fold, (3): somites, (4): neurulation initiation site, (5): amnios, (6): embryonic endoderm, (7): allantois base, (8): allantois, (9): visceral endoderm. (**b**) Turning off the embryo with 8–10 pairs of somites. First branchial arch has maxillary and mandibular components; the second arch is present. (1): oral plate, (2): anterior intestinal pocket, (3): somites, (4): embryonic endoderm, (5): archenteron, (6): posterior intestinal pocket, (7): mesencephalon, (8): rhombencephalon, (9): neural contour, (10): primitive line, (11): allantois, (12): anterior intestinal pocket. Scale bars: 250 μm (**a**) and 500 μm (**b**)

15.4 Advanced Post-implantation Development

There are notable interspecific disparities in the processes of implantation and early implantation stages in mammals. Nonetheless, similarities can be observed regarding overall changes that occur during advanced and late implantation phases across all mammalian species. Figures 15.5, 15.6, and 15.7 depict the most prominent developmental features and organ system development during advanced implantation in the vizcacha, which occurs from the 29 somites stage until approximately mid-gestation (74 gestation).

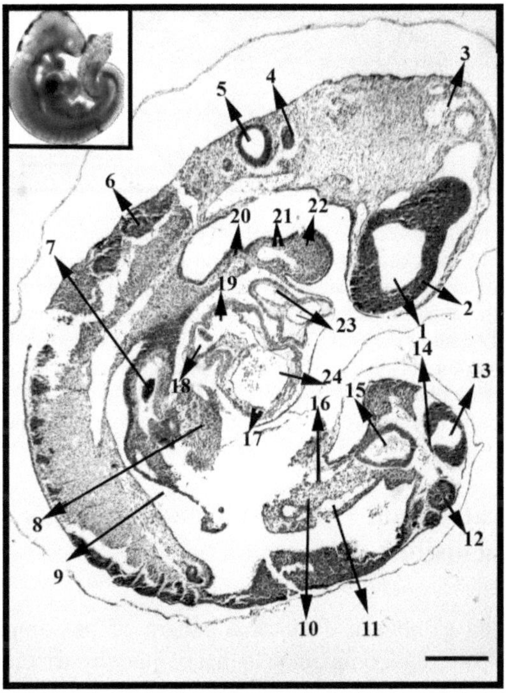

Fig. 15.5 Sagittal histological section through a *Lagostomus maximus* embryo (53–59 days gestation). Rotated embryo with 29 pairs of somites in the neuropore formation stage. (1): forebrain, (2): telencephalon, (3): mesencephalon, (4): acoustic fascia, (5): otic vesicle, (6): somite, (7): dorsal aorta, (8): junctions of the anterior and middle gut, (9): coelomic cavity, (10): dorsal mesentery of the gut, (11): right umbilical artery, (12): somite, (13): neural tube, (14): notochord, (15): dorsal aorta, (16): left umbilical artery, (17): ventricular wall, (18): aortic sac, (19): cardiac bulbs, (20): third aortic arch, (21): pharyngeal region, (22): first-gill arch, (23): arteriosus truncus, (24): developing heart, (25): amnion. Scale bar: 500 μm

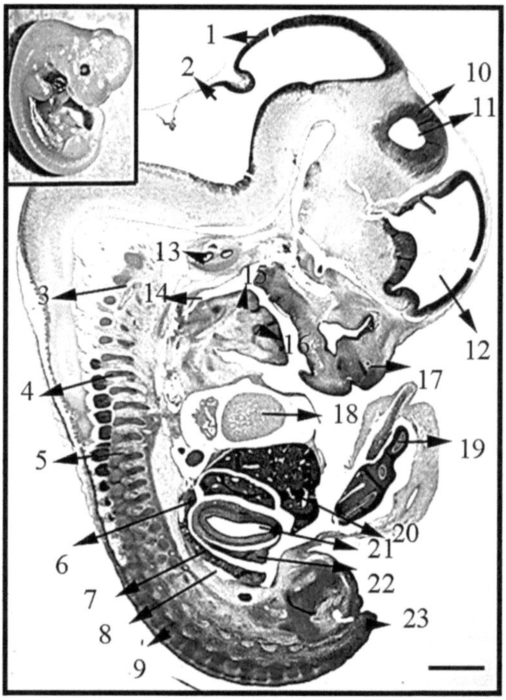

Fig. 15.6 Sagittal histological section of a female *Lagostomus maximus* embryo (67 days gestation). Female embryo of approximately 55 somites and 12 mm in length. Earliest sign of fingers, (splayed-out), posterior footplate apparent, retina pigmentation apparent, tongue well-defined, brain vesicles clear. Absent hair follicles, fingers separate distally. (1): mesencephalon, (2): fourth ventricle, (3): atlas, (4): ganglia cervical, (5): somite, (6): kidney, (7): genital ridge, (8): aorta, (9): somite, (10): primary cortex, (11): left cerebral hemisphere, (12): ventriculus lateralis, (13): ductus cochlearis, (14): future esophagus, (15): tongue, (16): Meckel's cartilage, (17): nasal cavity, (18): heart, (19): umbilical vein, (20): liver, (21): stomach, (22): pancreas, (23): lower end of spinal cord. Scale bar: 2000 μm

15.5 Comparative and Evolutive Considerations and Conclusions

The plains vizcacha exhibits a distinctive pattern of post-implantation embryo development. This pattern is comparable to that of other hystricomorph rodents, like guinea pig and chinchilla. The gestation period for plains vizcachas spans approximately 147 ± 5 days, which is notably longer than that of other rodents, including caviomorphs. For instance, guinea pig has a gestation period of 68 days, while chinchilla has a gestation period of 109–111 days. The sole species of caviomorph that appears to exhibit a more extended gestation period than that of vizcacha is the pacarana, *Dinomys branickii*, which can attain a duration of 222–283 days (Grzimek 1975).

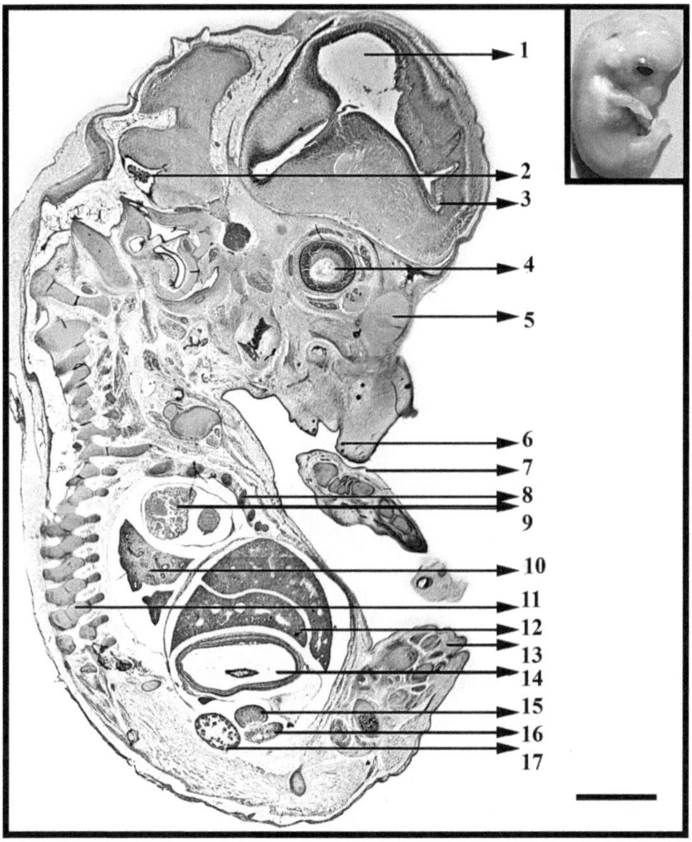

Fig. 15.7 Sagittal histological section of a female *Lagostomus maximus* embryo (more than 74 days gestation). Female embryo of 19 mm in length. The most prominent feature that defines this period is the separation of toes and fingers. Follicles on the nose showed an advanced degree of organogenesis. The eyelids are open. The 60 somites cannot be detected by external inspection. (1): ventricle, (2): fourth ventricle (3): encephalon, (4): eye, (5): olfactory lobe, (6): hair follicles, (7): upper limb, (8): sternum, (9): heart, (10): lung, (11): somite, (12): liver, (13): leg, (14): gut, (15): ovary, (16): rete ovarii, (17): kidney, (18): future brain. Scale bar: 2500 μm

When we compare the developmental stages of plains vizcacha with those of guinea pigs and chinchillas, it becomes apparent that the chinchilla's developmental trajectory exhibits an acceleration of approximately 10–12 days relative to plains vizcacha. In the initial stages of implantation, the pre-gastrulating embryo manifests as an undifferentiated ICM ball situated at the distal part of the conceptus. There is a significant advancement of extra-embryonic tissues during this stage. During analysis of the egg cylinder stage, it was observed that the plains vizcacha manifested the formation of an ectoplacental cavity that was replete with extravasated maternal blood. The embryonic tissue, on the other hand, persisted as an undifferentiated and compact inner cell mass. It is possible that, during the preimplantation period of 18 days, while the blastocyst was in the uterine lumen, it developed a cone-shaped structure on the antithetical side of the inner cell mass, potentially indispensable for penetrating the luminal epithelium. Nevertheless, the specific occurrences during the pre-implantation development days remain incompletely comprehended.

It is worth noting that the embryonic development of the plains vizcacha is similar to that of other mammals, such as rabbits, pigs, and even humans. Specifically, the epiblast and extraembryonic tissues develop from a bilaminar disc or cavitated embryo. However, this differs from other species within the same order, such as mice or rats, where the extraembryonic tissue develops via folding. Unlike some other hystricomorph rodents, the plains vizcacha does not undergo the development of an ectoplacental cone. Instead, it demonstrates an egg cylinder with a rim of mesometrially situated ectoplacental giant cells. Interestingly, the plains vizcacha exhibits a deviant pattern in contrast to most rodents studied, despite displaying an egg cylindrical embryo shape. Unlike its counterparts, the plains vizcacha's embryonic disc is the source of the epiblast-hypoblast and does not elongate, as is commonly observed. This distinctive feature implies that the early embryonic development of the plains vizcacha is unique and represents a combination of two evolutionarily types of embryonic development (Downs et al. 2009).

The egg cylinder morphology of the mouse has long been considered the archetypal early post-implantation rodent embryo morphology (Kaufman 1992; Nowak 1999; Rossant and Tam 2002). However, it is important to note that this generalization does not apply to all rodents, as both the plains vizcacha and the guinea pig exhibit deviations from this pattern. It is plausible to hypothesize that the developmental traits observed in plains vizcachas can be attributed to their evolutionary history (Leopardo et al. 2018). Notably, there exist significant differences between Muridae and Hystricognathi, including the gestation period, litter size, and pre- and post-implantation embryo development. Murid (such as rats and mice) and cricetid rodents (e.g., hamsters) have adapted for species survival by maintaining short gestation periods and producing a high number of poorly developed offspring, a so-called *r*-reproductive strategy (Copp and Cockroft 1990; Nagy et al. 2003). Conversely, caviomorphs, such as chinchillas and plains vizcacha, have retained a *k*-reproductive strategy typical from most mammalian orders, with long gestation periods, giving birth to one or two well-developed pups and displaying a long pre-implantation phase with a bilaminar embryo disc in the egg cylinder, which exemplifies the retention of ancestral traits in its developmental process (Tyndale-Biscoe and Renfree 1987).

Hystricognathi, along with Sciuromorpha, represents the first branches in the rodent phylogeny subsequent to the separation of Lagomorpha and Rodentia. Conversely, the most recently evolved rodents are those belonging to the Myomorpha, with the Muroidea clade, which is related to the mouse. The basal plains vizcacha's planar epiblast morphology lends support to the notion that the egg cylinder, in which the ExE comes into proximity with the epiblast, is an innovation that was introduced by more recently evolved rodents, such as murids, as opposed to being a characteristic of the entire order. The evolutionary divergence between the clades Hystricomorpha and Sciuromorpha, which represents a profound dichotomy within Rodentia, holds the potential to shed light on the unique characteristics that are observed in plains vizcacha. Recent studies have recognized this dichotomy as one of the earliest divisions within the rodent evolutionary tree (Montgelard et al. 2008).

Finally, it is noteworthy to mention that the origin of primordial germ cells, which will ultimately generate oocytes and spermatozoa, in the plains vizcacha is facilitated by a gene pathway that is partially shared with that of the laboratory mouse. The origin and specification of germ cells in mice have been extensively established and are considered the model of germline development. However, in mammals other than mice, such as rabbits, pigs, or even humans, many disparities have been identified in gene expression during primordial germ cell specification, as compared to the mouse model. Furthermore, in humans, where the origin of primordial germ cells remains elusive, the data accumulated thus far indicate that the germline gene pathway of the plains vizcacha is more akin to that of humans than that of the laboratory mouse (Leopardo and Vitullo 2017).

Acknowledgments We especially thank the personnel of ECAS for their fieldwork. We would like to thank Ms. María Grisel Clausi-Schettini for her technical assistance in embryo sectioning. This study was supported by the intramural grant program from Universidad Maimónides, Buenos Aires, Argentina.

References

Acuña F, Barbeito CG, Portiansky EL, Ranea G, Nishida F, Miglino MA, Flamini MA (2020) Early and natural embryonic death in *Lagostomus maximus*: association with the uterine glands, vasculature, and musculature. J Morphol 281:710–724. https://doi.org/10.1002/jmor.21127

Acuña F, Portiansky EL, Miglino MA, Flamini MA, Barbeito CG (2023) Embryonic-placental relationship in *Lagostomus maximus* as compared to other hystricognath rodents and eutherian mammals. Zoology 158:126082. https://doi.org/10.1016/j.zool.2023.126082. ISSN 0944-2006

Bianchi DW, Wilkins-Haug LE, Enders AC, Hay ED (1993) Origin of extraembryonic mesoderm in experimental animals: relevance to chorionic mosaicism in humans. Am J Med Genet 46:542–550. https://doi.org/10.1002/ajmg.1320460517

Carlson BM (1999) Human embryology and developmental biology, 2nd edn. Mosby, St. Louis. ISBN: 0815114583

Carter AM, Acuña F, Barbeito CG (2023) IW Rowlands, Barbara J Weir and the biology of the hystricomorph rodents. Reproduction 166(1):R1–R13. https://doi.org/10.1530/REP-23-0028

Copp AJ, Cockroft DL (1990) Postimplantation mammalian embryos: a practical approach. IRL Press, Oxford. ISBN: 0199630895

Downs KM, Inman KE, Jin DX, Enders AC (2009) The allantoic core domain: new insights into the development of the murine allantois and its relation to the primitive streak. Dev Dyn 238:532–553. https://doi.org/10.1002/dvdy.21862

Eakin GS, Behringer RR (2004) Diversity of germ layer and axis formation among mammals. Semin Cell Dev Biol 15(5):619–629. https://doi.org/10.1016/j.semcdb.2004.04.008

Gibson E (1877) On the viscacha (*Lagostomus trichodactylus*) a South American rodent. Proc Nat Hist Soc Glasg 3:136

Grzimek B (1975) Grzimek's animal life encyclopedia. Van Nostrand Reinhold Co., London

Guzeloglu-Kayisli O, Basar M, Arici A (2007) Basic aspects of implantation. Reprod Biomed Online 15(6):728–739. https://doi.org/10.1016/s1472-6483(10)60541-x

Hudson WH (1872) On the habits of the vizcacha (*Lagostomus trichodactylus*). Proc Zool Soc London 1872:822–833

Ilgren EB (1981) Placental polyploidization: a comparative study in the mouse and the guinea pig. Placenta 2(4):333–342. https://doi.org/10.1016/s0143-4004(81)80029-x

Kaufman MH (1992) The atlas of mouse development. Academic Press, London. ISBN: 0-12-402035-6

Leopardo NP, Vitullo AD (2017) Early embryonic development and spatiotemporal localization of mammalian primordial germ cell-associated proteins in the basal rodent *Lagostomus maximus*. Sci Rep 7(1):594. https://doi.org/10.1038/s41598-017-00723-6

Leopardo NP, Inserra PIF, Vitullo AD (2018) Challenging the paradigms on the origin, specification and development of the female germ line in placental mammals. In: Ahmed RG (ed) Germ cells. InTech Open, Egypt. https://doi.org/10.5772/intechopen.71559

Llanos AC, Crespo JA (1952) Ecología de la vizcacha (*Lagostomus maximus maximus* Blainv.) en el nordeste de la Provincia de Entre Ríos. Rev de Investig Agríc 6:289–378

Montgelard C, Forty E, Arnal V, Matthee CA (2008) Suprafamilial relationships among Rodentia and the phylogenetic effect of removing fast-evolving nucleotides in mitochondrial, exon and intron fragments. BMC Evol Biol 8:321. https://doi.org/10.1186/1471-2148-8-321

Nagy A, Gertsenstein M, Vintersten K, Behringer R (2003) Manipulating the mouse embryo: a laboratory manual, 3rd edn. Cold Spring Harbor Laboratory Press, New York

Newson RM (1966) Reproduction in the feral coypu (*Myocastor coypus*). Symp Zool Soc Lond 15:323–334

Nowak RM (1999) Walker's mammals of the world, vol 2, 6th edn. Johns Hopkins University Press, London. ISBN 13: 9780801857898

O'Rahilly R, Müller F (2010) Developmental stages in human embryos: revised and new measurements. Cells Tissues Organs 192(2):73–84. https://doi.org/10.1159/000289817

Roberts CM, Weir BJ (1973) Implantation in the plains viscacha, *Lagostomus maximus*. J Reprod Fertil 33:299–307. https://doi.org/10.1530/jrf.0.0330299

Rossant J, Tam P (2002) Mouse development: patterning, morphogenesis, and organogenesis. Academic Press, San Diego. https://doi.org/10.1016/B978-0-12-597951-1.X5000-8

Sansom GS, Hill JP (1931) Observations of the structure and mode of implantation of the blastocyst of Cavia. Trans Soc Zool 21:295–354. https://doi.org/10.1111/j.1096-3642.1931.tb00451.x

Tibbitts FD, Hillemann HH (1959) The development and histology of the chinchilla placentae. J Morphol 105:317–365. https://doi.org/10.1002/jmor.1051050205

Tyndale-Biscoe CH, Renfree MB (1987) Reproductive physiology of marsupials. Cambridge University Press, Cambridge. https://doi.org/10.1017/CBO9780511623493

Weir BJ (1967) The care and management of laboratory hystricomorph rodents. Lab Anim 1:95–104. https://doi.org/10.1258/002367767781035521

Weir BJ (1970) The management and breeding of some more hystricomorph rodents. Lab Anim 4:83–97. https://doi.org/10.1258/002367770781036562

Weir BJ (1971a) The reproductive organs of the female plains viscacha, *Lagostomus maximus*. J Reprod Fertil 25:365–373. https://doi.org/10.1530/jrf.0.0250365

Weir BJ (1971b) The reproductive physiology of the plains viscacha, *Lagostomus maximus*. J Reprod Fertil 25:235–373. https://doi.org/10.1530/jrf.0.0250355

Weir BJ (1974) Reproductive characteristics of hystricomorph rodents. In: Rowlands IW, Weir BJ (eds) The biology of hystricomorph rodents. Academic Press, London

Weir BJ, Rowlands IW (1973) Reproductive strategies of mammals. Annu Rev Ecol Syst 4:139–163. https://doi.org/10.1146/annurev.es.04.110173.001035

Weir BJ, Rowlands IW (1974) Functional anatomy of the hystricomorph ovary. In: Rowlands IW, Weir BJ (eds) The biology of hystricomorph rodents. Academic Press, London

Chapter 16
Reproductive Neuroendocrinology of the Female South American Plains Vizcacha, *Lagostomus maximus*

Verónica Berta Dorfman, Pablo Ignacio Felipe Inserra, Alfredo Daniel Vitullo, and Julia Halperin

Abstract Female plains vizcachas have several peculiar reproductive characteristics as a true reactivation of the hypothalamic–pituitary–ovarian axis during pregnancy among others. This axis-reactivation is enabled by precise regulation of the hypothalamic gonadotropin-releasing hormone (GnRH) neurons allowing the pulsatile delivery of GnRH at mid-gestation. Such an event is followed by sequential secretion of estradiol (E2), follicle-stimulating hormone (FSH), and luteinizing hormone (LH), ultimately leading to a reboot of luteal steroidogenesis and a consequent boost of progesterone (P4) that enables the successful delivery of (at most) two well-developed newborns. The regulation of the hypothalamic GnRH neurons results from the combination of different endocrine pathways. GnRH neurons located in the preoptic area express progesterone and estrogens receptors, and aromatase, the estradiol-synthesizing enzyme. This local expression points to a mechanism of direct hormonal regulation of GnRH neurons that would synergize with the classic feedback pathways driven from the ovaries. Furthermore, the conspicuous organization of the kisspeptin system and the neuromodulators prolactin and dopamine observed in this area strongly suggests their involvement in the modulation of GnRH neurons. All these hormones, receptors, enzymes, and neurotransmitters coordinately regulate the functioning of this distinctive mechanism of which many details remain to be elucidated. In addition, photoperiod information transduced through melatonin hormone would coordinate plains vizcacha gestation with the most favorable moment for offspring surviving. To our knowledge, plains vizcacha is the unique rodent species that has a true reactivation of the reproductive axis during pregnancy. This trait is likely related to a reproductive strategy of plains vizcacha favoring offspring success.

V. B. Dorfman (✉) · P. I. F. Inserra · A. D. Vitullo · J. Halperin
Centro de Estudios Biomédicos Básicos, Aplicados y Desarrollo (CEBBAD), Universidad Maimónides, Ciudad Autónoma de Buenos Aires, Argentina

Consejo Nacional de Investigaciones Científicas y Técnicas (CONICET), Buenos Aires, Argentina
e-mail: dorfman.veronica@maimonides.edu

L. L. Rasia et al. (eds.), *Plains Vizcachas*, https://doi.org/10.1007/978-3-031-49487-1_16

Keywords GnRH · Plains vizcacha · Neuroendocrinology · Reproduction

Abbreviations

17β-HSD	17β-hydroxysteroid dehydrogenase
3β-HSD	3β-hydroxysteroid dehydrogenase
A4	Androstenedione
AANAT	Arylalkylamine N-acetyltransferase
ARC	Arquate nucleus
AVPV	Anteroventral periventricular
CYP19A1	Aromatase
D2R	Dopamine type 2 receptors
DA	Dopamine
DBB	Diagonal band of Brocca
dmARC	Dorsomedial arcuate nucleus
Dyn	Dynorphin A
E1	Estrone
E2	Estradiol
EP	Early pregnant
ERE	Estrogen response elements
ERα	Estrogen receptor alpha
ERβ	Estrogen receptor beta
FSH	Follicle-stimulating hormone
GAP	GnRH associated protein
GnRH	Gonadotropin-releasing hormone
HPO	Hypothalamic–pituitary–ovarian
KISS	Kisspeptin
KISSR	Kisspeptin receptor
KNDy	Kisspeptin, neurokinin B, and dynorphin neurons
LH	Luteinizing hormone
ME	Median eminence
MEL	Melatonin
mGluR1a	Metabotropic glutamate receptor 1a
MP	Midpregnant
MT1	Melatonin receptor 1
MT2	Melatonin receptor 2
neuroP	Neuroprogesterone
NK3R	Neurokinin 3 receptor
NKB	Neurokinin B
P4	Progesterone
PeVN	Periventricular nucleus
PIF	Prolactin inhibitory factor
POA	Preoptic area

PR	Progesterone receptor
PRL	Prolactin
PRLR	Prolactin receptor
RM	Recessus mammillaris
SCN	Suprachiasmatic nucleus
S-NAT	Serotonin N-acetyltransferase
SON	Supraoptic nucleus
StAR	Steroidogenic acute regulatory protein
TH	Tyrosine hydroxylase
TIDA	Tuberoinfundibular dopaminergic
TP	Late-pregnant females
TSPO	Translocator protein
TT	Testosterone
VMN	Ventromedial nucleus

16.1 Introduction

The South American plains vizcacha, *Lagostomus maximus*, is a hystricomorph caviomorph fossorial rodent native from Argentina, Bolivia, and Paraguay (Jackson et al. 1996). The plains vizcacha belongs to the family Chinchillidae which also comprises two other living species, the chinchilla (*Chinchilla lanigera*) and the mountain viscacha (*Lagidium viscacia*). Chinchillidae is one of the more recently evolved families from Hystricomorpha, a clade that radiated early in rodent phylogeny. Plains vizcacha, chinchilla, and mountain viscacha have gyrencephalic neuroanatomy with a longitudinal fissure, two transverse sulci, and two lateral sulci (Schmidt et al. 2020).

Female plains vizcachas have nocturnal foraging outings and a biannual seasonal reproduction. Its estrous cycle lasts 40–45 days releasing 200–800 oocytes, the highest polyovulation rate recorded so far for a mammal (Weir 1971a, b). However, just five or six embryos will get implanted in each uterine horn, and only two of them (the most caudally located) will be born after 155 days of gestation (Acuña et al. 2022; Barbeito et al. 2021; Weir 1971a, b). The main reproductive season occurs from March to April and births from the end of August to the beginning of September (Branch et al. 1993; Spotorno and Patton 2015; Weir 1970). In females that have lost their progeny, a second breeding season takes place from October to February (Llanos and Crespo 1952).

This chapter discusses the state of the knowledge of the reproductive neuroendocrinology of the female plains vizcacha that has some peculiar features compared with other mammalian species. The specific modulation of the gonadotropin-releasing hormone (GnRH) expression and delivery by hormones, neuropeptides, and neurotransmitters will be examined.

16.2 Endocrinology of the Plains Vizcacha

Female plains vizcachas have several peculiar reproductive characteristics. One of them is a true reactivation of the hypothalamic–pituitary–ovarian (HPO) axis during pregnancy. From an endocrine point of view, two well-defined phases separated around mid-gestation could be recognized during the gestation of the plains vizcacha. The endocrinology and neuroendocrinology during the first phase resemble that of most mammals with an early increase in serum progesterone (P4) until around 70-day of pregnancy (~50% of gestation progression) and minimum levels of GnRH expression and delivery (Dorfman et al. 2013; Fraunhoffer et al. 2017) (Fig. 16.1). During this period, despite a successful embryo implantation, their

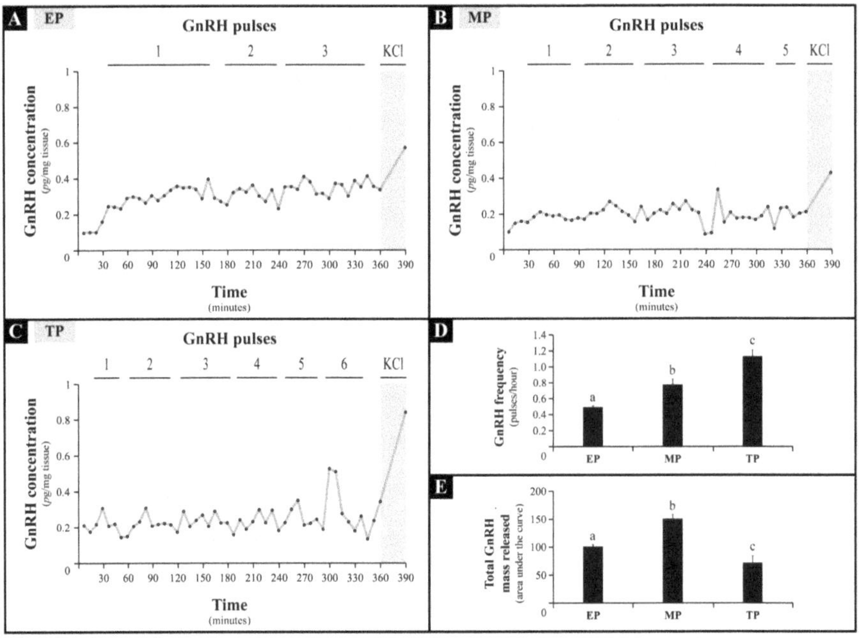

Fig. 16.1 Delivery of gonadotropin-releasing hormone (GnRH) during gestation in the female plains vizcacha, *Lagostomus maximus*. The delivery of GnRH was determined using ex vivo explants of hypothalami dissected from pregnant vizcachas. Hypothalami were separately incubated for 6 hours with Krebs-Ringer buffer as described in Charif et al. (2016) and Inserra et al. (2020). (**a–c**) Representative images of the pulsatile pattern of GnRH in early-pregnant females (EP) (**a**), midpregnant females (MP) (**b**), and late-pregnant females (TP) (**c**). The numbers over each graphic indicate the pulses of GnRH delivery. (**d**) Significant variations in the GnRH pulsatile frequency throughout the pregnancy are measured as the number of pulses per hour. (**e**) Significant increment of the total mass of GnRH released at MP related to EP and TP. (**d, e**) Results are expressed as the mean ± standard deviation per group ($n = 4$–5 per group). One-way ANOVA followed by the Tukey test was employed to determine significant differences (software Prism 4.0, GraphPad Software Inc.). Different letters over the bars indicate significant differences among groups with $p < 0.05$

growth stops around 50-embryonic day, and a sequential and progressive reabsorption is triggered. Around 70-embryonic day, only the 2 embryos from each uterine horn located closest to the cervix are still viable. From this day up to around 90-embryonic day, serum P4 gradually decreases because of the decay in the steroidogenic activity of the primary corpora lutea reaching its minimum level (Cortasa et al. 2022; Fraunhoffer et al. 2017). However, despite the decrease of P4, the two surviving embryos are still too immature to be born as would occur in most mammals, and a hormonal boost becomes essential to continue gestation. At this moment, a particular reproductive strategy takes place in this species, and the second phase of gestation begins. The minimum level of P4 that has been reached at this point seems to enable the reactivation of the hypothalamic activity around 100 days of pregnancy recovering the expression levels and pulsatile delivery of GnRH (Fig. 16.1). This milestone triggers pituitary activity (Dorfman et al. 2013; Inserra et al. 2017; Proietto et al. 2019), and the systemic increment of the pituitary follicle-stimulating hormone (FSH) secretion increases ovary estradiol (E2) secretion and stimulates the surge and pulsatility of the pituitary luteinizing hormone (LH) (Dorfman et al. 2013; Fraunhoffer et al. 2017; Proietto et al. 2019). As result, the reactivation of the steroidogenic activity of the primary corpora lutea and the development of a new set of smaller corpora lutea with oocyte retention called secondary corpora lutea or accessory corpora lutea was registered (Cortasa et al. 2022; Dorfman et al. 2013; Fraunhoffer et al. 2017). This process was called "pseudo-ovulation" (Dorfman et al. 2016; Halperin et al. 2013) since although the formation of active corpora lutea occurs, follicle rupture with the release of the oocyte does not. During this process of HPO reactivation, the boost of P4 necessary for fully developed fetuses to be born occurs and is sufficient to enable the two surviving fetuses to reach successful to term favoring offspring success (Cortasa et al. 2022; Dorfman et al. 2016). Some of these peculiar reproductive characteristics of female plains vizcachas as the accessory corpora lutea formation during pregnancy have been described in other hystricomorphs as the Zambian mole-rat (*Fukomys amatus*) and the brush-tailed porcupines (*Atherurus africanus*) (Jori et al. 2002; Willingstorfer et al. 1998). However, to our knowledge, the reactivation of the reproductive axis during pregnancy has not been described to date in any other rodent.

The amazing recovery of the HPO axis described so far is achieved, thanks to the precise, fine, and rapid reactivation of hypothalamic activity with a particular modulation that points to the triggering of GnRH secretion.

16.3 GnRH: The Key Hormone for Reproduction

GnRH is the decapeptide involved in the modulation of fertility in mammals. GnRH is synthesized as a prepro-gonadotropin, including a signal peptide and an enzymatic cleavage site that separates GnRH from another sequence termed GnRH-associated protein (GAP), originally proposed as prolactin inhibitory factor (PIF) (Adelman et al. 1986). Most vertebrates possess at least two, usually three, variants of GnRH that differ in their amino acid sequence, localization, and embryonic origin (Cheng and Leung 2005). According to its amino acid sequence composition, function, localization, and embryonic origin, 30 GnRH peptides have been identified in nervous tissues from protochordates to vertebrates, with highly NH2– and COOH– terminal conserved sequences (Lethimonier et al. 2004; Millar 2005; Roch et al. 2014). The first identified form of GnRH, named mammalian GnRH (mGnRH or GnRH1), was initially isolated from porcine and ovine brains (Burgus et al. 1972; Matsuo et al. 1971) and then was also detected in chicken (King and Millar 1982). In addition, a different variant of GnRH1, originally termed gpGnRH, was first described in guinea pig (*Cavia porcellus*) (Jimenez-Liñan et al. 1997) and also reported in capybara (*Hydrochoerus hydrochaeris*) (Montaner et al. 2002). A second variant namely cGnRH or GnRH2 was later identified in chicken too (Miyamoto et al. 1984). Different biological functions for GnRH1 and GnRH2 were described (Cheng and Leung 2005). Another form of GnRH (sGnRH or GnRH3) firstly isolated from salmon is present in all teleosts (Sherwood et al. 1986; Somoza et al. 2002; Zohar et al. 2010). Comparing the obtained sequence of the GnRH prepro-gonadotropin mRNA in the plains vizcacha with that among species, we deduced the three classical components of the GnRH prepro-gonadotropin: the signal peptide, the GnRH decapeptide, and GAP (Charif et al. 2016). In addition, its composition has four exons as for other mammals (Tsai and Zhang 2008). It is likely that the first exon would contain the 5′UTR, the second exon would encode for GnRH and the 5′end of GAP, the third exon would encode for GAP, and the fourth exon would encode for the resting sequence of GAP followed by the 3′UTR (Charif et al. 2016). The high conservation of GnRH among species confirms once again the relevance of this hormone, among other factors to guarantee reproduction.

In the mammalian postnatal brain, GnRH is synthesized by a discrete specialized group of neurons scattered throughout the preoptic area of the hypothalamus in the basal forebrain, the ventromedial nucleus, and the arcuate nucleus (Silverman et al. 1987; Urbanski et al. 1991). Most hypothalamic GnRH-secreting neurons project towards the median eminence releasing GnRH into the hypothalamic–pituitary portal circulation up to the anterior pituitary gland where it binds to specific receptors and stimulates the synthesis and release of the gonadotropins LH and FSH (Belchetz et al. 1978; Krey and Silverman 1978; Silverman and Witkin 1994; Silverman et al. 1987). We have described that the distribution of GnRH in the hypothalamus of the plains vizcacha is mostly comparable to other mammalian species like mice, rats, rabbits, guinea-pigs, and long-tailed chinchillas among others (Dorfman et al. 2011; Inserra et al. 2017). GnRH immunoreactive neurons are scattered around the third ventricle and below the recessus mammillaris (RM) of the third ventricle, with localization in the diagonal band of Brocca (DBB), preoptic area (POA), the ventromedial nucleus (VMN), the periventricular nucleus (PeVN), and the arquate nucleus (ARC) (Dorfman et al. 2011; Inserra et al. 2017). In the POA, two subgroups of GnRH neurons were detected according to their labeling intensity (high or low) of GnRH by immunohistochemistry (Charif et al. 2017) as previously described in rats (Wray et al. 1989). GnRH dendrites and axons with a beaded pattern of GnRH immunoreactive varicosities are distributed all throughout the rostral region of the hypothalamus and reach the external borders of the median eminence (ME) where they acquire a radially orientated pattern with respect to the RM. GnRH immunoreactive neurosecretory terminals surrounding the primary plexus of the hypothalamic–pituitary portal vessels were described (Dorfman et al. 2011; Inserra et al. 2017). Furthermore, the ventral preoptic area of the hypothalamus of the plains vizcacha showed GnRH immunoreactive neurons localized in the supraoptic nucleus (SON) and in the suprachiasmatic nucleus (SCN) (Dorfman et al. 2013; Inserra et al. 2017). In addition, we have described GnRH immunolocalization at cellular and ultracellular levels by electron microscopy (Dorfman et al. 2011). GnRH distribution was found in dendrites restricted to varicosities containing GnRH immunoreactive core and non-core vesicles and in somas localized in cytoplasmic vesicles associated with the Golgi apparatus (Dorfman et al. 2011) as is also described in other species (Naik 1975; Silverman et al. 1990).

The normal reproductive development and fertility depends in part on the frequency of GnRH pulses with variations across the reproductive cycle that differentially stimulates the release of FSH or LH. While a low frequency of GnRH pulse stimulates FSH release, a high frequency of GnRH pulse stimulates LH surge (Burger et al. 2008; Ciccone et al. 2010; Wildt et al. 1981). Studying cultured hypothalamic explants, we described GnRH secretion of the plains vizcacha. Surprisingly, the pulsatile pattern of GnRH showed 5 pulses of GnRH during the 6 hours of the experiment at both follicular and luteal phases (Charif et al. 2016, 2022); however, a double amount of GnRH mass was secreted during the follicular phase compared with the luteal phase (Charif et al. 2022). Why the frequency of GnRH delivery did not show differences between astral phases could be the result of the isolation of the hypothalamus from the whole animal with explants blind to the ovarian negative feedback; however, differences in the total mass secretion may represent the differences of hypothalamic stimulation over the pituitary with differences on gonadotropins secretion.

In mammalian females, GnRH is delivered in a pulsatile manner throughout reproductive life except during pregnancy due to negative feedback exerted by progesterone inhibiting GnRH release (Belchetz et al. 1978). In contrast, the plains vizcacha has variations of GnRH expression (Charif et al. 2017; Dorfman et al. 2013; Inserra et al. 2017) and delivery (Fig. 16.1) even during pregnancy. This particular modulation of the hypothalamic GnRH neurons in the plains vizcacha would result from the coordinated regulation of different endocrine pathways that involves hormones, neuropeptides, and neurotransmitters with a distinctive mechanism that enables the fine and rapid delivery of GnRH during pregnancy.

16.3.1 GnRH Regulation by Estrogens and Progesterone

The biosynthesis of estrogens results from the aromatization of the androgens testosterone (TT) and androstenedione (A4) by aromatase (CYP19A1), a monooxygenase enzyme that belongs to the cytochrome P450 superfamily, located in the endoplasmic reticulum and mitochondria. So, TT and A4 are converted into E2 and estrone (E1), respectively by aromatase. In addition, E1 and A4 are converted into E2 and TT through the enzyme 17β-hydroxysteroid dehydrogenase (17β-HSD), respectively (Conley and Hinshelwood 2001; Luu-The et al. 1995). E2 is mainly produced by the ovaries and it crosses the hematoencephalic barrier; however, brain neuroendocrine areas like the hypothalamus in mammals, fishes, and birds have also shown aromatase expression and activity (Naftolin et al. 1972; Roselli et al. 1985, 2009; Selmanoff et al. 1977). These findings indicate the importance of the local synthesis of E2, called neuroestradiol, on reproduction.

In this context, the plains vizcacha is not an exception. We have described aromatase expression in neurons distributed in the POA of the hypothalamus of pregnant and nonpregnant females with significant variations in its expression during pregnancy (Charif et al. 2017). Besides, the concordant pattern of aromatase and GnRH expression, and their colocalization in neurons of the POA throughout gestation, suggests the association of neuroestradiol with GnRH. Strikingly, we noted that their colocalization was exclusively detected in the low-stained GnRH neurons which may be related to differences in the modulation of pulse or surge of GnRH by E2 action (Charif et al. 2017). On the other hand, the action of neuroestradiol-modifying GnRH release has been previously shown using microdialysis method in monkeys (Kenealy et al. 2013). In plains vizcacha, studying GnRH secretion in hypothalamic explants culture, we determined changes in GnRH pulsatile parameters induced by aromatase inhibition. This local expression points to a mechanism of direct hormonal regulation of GnRH neurons by neuroestradiol that would synergize with the classic feedback pathways driven from the ovaries (Charif et al. 2017).

Estrogens have a bimodal effect on the hypothalamus, with either an inhibitory or stimulatory action on GnRH delivery. A rapid E2 increase at the end of the follicular phase triggers a stimulatory effect on GnRH surge resulting in the secretion of FSH and LH from the pituitary gland, which is essential for follicular maturation and ovulation, respectively (Abraham et al. 1972; Simerly 2002). In contrast, during the luteal phase, the gentle rise of E2 produces negative feedback over GnRH secretion (March et al. 1979). Two nuclear estrogen receptor isoforms, alpha (ERα) and beta (ERβ), have been described to bind E2 with similar affinity (Kuiper et al. 1996). Both receptors bind to estrogen response elements (ERE) localized in the GnRH gene promoter activating GnRH synthesis (Radovick et al. 1991). On the other hand, the promoter of the progesterone receptor (PR) gene also shows the ERE sequence where ER-E2 binds and activates GnRH synthesis (Quadros and Wagner 2008).

In the plains vizcacha, we have determined significant variations in the hypothalamic ERα expression throughout pregnancy at both transcription and protein translation levels, and this correlates with serum E2 levels (Inserra et al. 2017). However, hypothalamic ERβ mRNA or protein expression changes were not observed (Inserra et al. 2017; Inserra 2016). Moreover, the increment in hypothalamic expression of ERα was correlated with increases in hypothalamic GnRH and serum LH at mid-pregnancy supporting the involvement of ERα in GnRH expression (Inserra et al. 2017). These observations were later confirmed by studying GnRH and LH secretion in ex vivo assays using explants of hypothalamic and pituitary glands respectively incubated with agonists and antagonists of ERα and ERβ (Inserra et al. 2020; Proietto et al. 2019).

On the other hand, we described the distribution of ERα in hypothalamic neurons and pituitary gonadotrophs of the plains vizcacha (Inserra et al. 2017; Proietto et al. 2019). We determined the immunolocalization of ERα in neurons distributed throughout the POA (including DBB, PeVN, mPOA, SCN, and SON), VMN, ARC, and ME with both nuclear and cytoplasmic localization, with a predominant nuclear signal. This distribution was comparable with that described in other mammalian species like mouse, rat, guinea pig, sheep, mink, and macaque (Gonzales et al. 2008; Gundlah et al. 2000; Scott et al. 2000a; Shughrue et al. 1997; Warembourg and Leroy 2004; Warembourg et al. 1998).

GnRH hypothalamic neurons of most mammalian species like mink, sheep, ewe, guinea pig, mouse, and rat do not express ERα (Herbison et al. 1993; Warembourg et al. 1998; Watson Jr et al. 1992; Wintermantel et al. 2006). This determines an indirect model of regulation by E2 where this steroid regulates GnRH neurons through afferent neurotransmitter and neuromodulatory systems (Wintermantel et al. 2006). However, the controversy was established when ERα expression was evidenced in 17% of GnRH-expressing neurons of the POA of rats (Butler et al. 1999). The localization of ERα and GnRH in the SON of the plains vizcacha was also confirmed at mRNA level by laser microdissection of SON and later ERα and GnRH mRNA amplification (Inserra et al. 2017). In this context, the co-localization of ERα and GnRH in the hypothalamus of plains vizcacha suggests a direct modulation of ERα on GnRH expression in this species (Fig. 16.2). This may represent a peculiar strategy developed by the plains vizcacha that favors ovulation during gestation. The "direct model" of GnRH regulation, opposed to the "indirect model" established for most mammals, may support the classical genomic mechanism for ERα action as a transcription factor binding to the ERE sequence localized in the GnRH promoter to stimulate GnRH synthesis (Radovick et al. 1991).

Regarding P4, this hormone modulates the activity of the HPO axis during pregnancy in most mammals by inhibiting the secretion of GnRH and the expression of pituitary gonadotropins. This would block follicular recruitment and ovulation and maintain embryo's viability (Skinner et al. 1998). P4 increases during early-gestation and remains elevated until the end of gestation when its decrease allows the delivery (Challis et al. 1971; Leavitt and Blaha 1970; McCormack and Greenwald 1974). As described above, the plains vizcacha shows two waves of serum P4 increments during gestation, enabling the embryos to complete their development successfully until birth. This behavior of P4 contrasts with that described in other caviomorphs such as *Cavia porcellus* and *Cavia aperea*, which show a single peak of P4 at mid-gestation (Heap et al. 1981; Tam 1974). However, these species do not have HPO axis reactivation during pregnancy.

It has been recently described that P4 is also synthesized de novo by hypothalamic astrocytes (called neuroprogesterone, neuroP) (Sinchak et al. 2020). The production of neuroP is induced by the increment of E2 during the proestrus stage. Then, the binding of E2 to the ERα subtype that is associated with the metabotropic

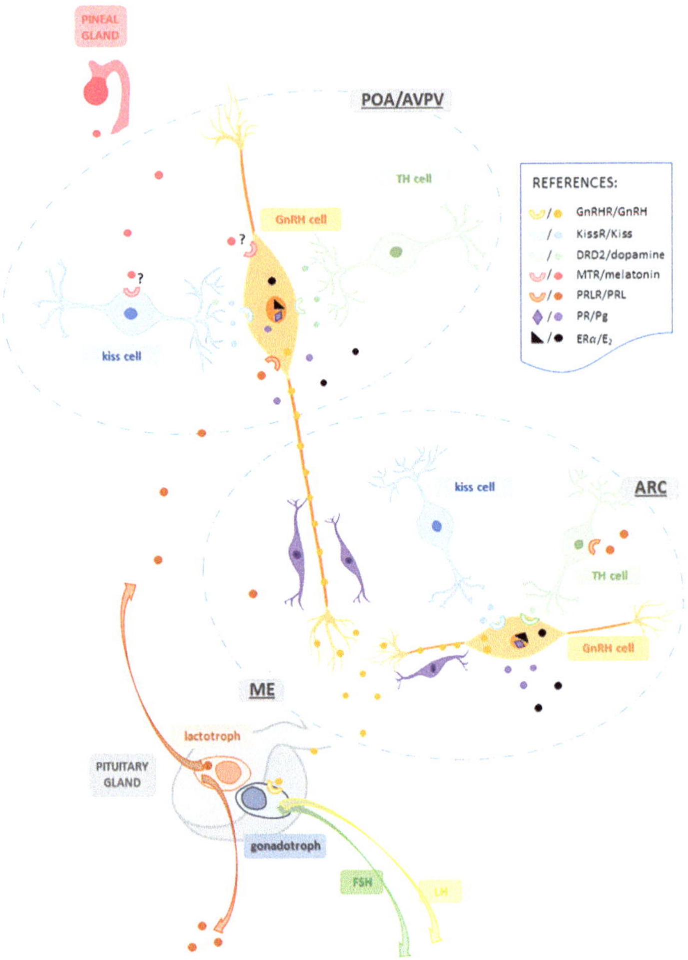

Fig. 16.2 Neuromodulation of gonadotropin-releasing hormone (GnRH) synthesis and delivery in the female plains vizcacha, *Lagostomus maximus*. GnRH expression and delivery are regulated by several neuromodulators like kisspeptin (KISS) and dopamine (DA) and hormones like prolactin (PRL), estradiol (E2), progesterone (P4), and melatonin (MEL) among others. In the preoptic area/periventricular anteroventral area (POA/AVPV) of the hypothalamus, GnRH cells coexist with Kiss and tyrosine hydroxylase (TH) cells that produce KISS and DA, respectively. Those cells send their products to GnRH cells modulating GnRH expression. At the same time, Kiss cells express estrogen receptor alpha (ERα) and progesterone receptors (PR) that enable both hormones to modulate GnRH through KISS. On the other hand, TH cells express PRLR. So, as a result of PRL increments, the increase of DA secretion inhibits GnRH expression. In addition, GnRH neurons of female plains vizcachas also express ERα, PR, and PRLR. This enables the hormonal direct modulation of GnRH expression. At the median eminence (ME), the site for GnRH secretion, KISS neuroterminals are in the close boundary with GnRH neuroterminals, which are also surrounded by PR expressing cells, especially at midpregnancy. In addition, MEL from the pineal gland also modulates GnRH secretion

glutamate receptor 1a (mGluR1a) in the membrane of astrocytes triggers a signaling cascade that induces the transport of cholesterol into the mitochondria (Mittelman-Smith et al. 2017). In the mitochondria, cholesterol is converted to pregnenolone by the CYP11A1 enzyme and then transformed into neuroP by the 3β-hydroxysteroid dehydrogenase (3β-HSD) enzyme (Mittelman-Smith et al. 2017; Sinchak et al. 2020). Thus, both the increase in E2 and the preovulatory signaling of neuroP are key precursors for the synthesis and release of GnRH and the LH surge (Mittelman-Smith et al. 2017). This background strongly indicates the possible intervention of neuroP in the hypothalamic and pituitary activity in the pregnant plains vizcacha.

Before ovulation, the binding of P4 to its specific receptors localized in GnRH neurons stimulates GnRH expression acting directly as a transcription factor or acting indirectly by specific interactions with co-regulator proteins (McKenna et al. 1999; Micevych and Sinchak 2008). After ovulation, the progressive increase of P4 secretion from corpora lutea produces negative feedback over GnRH secretion (Goodman and Karsch 1980; Skinner et al. 1998). Strikingly, although the significantly increased levels of serum P4 that were determined in midpregnant plains vizcachas may decrease GnRH expression in nonpregnant vizcachas, those increased levels of serum P4 are not able to show inhibitory effects on GnRH expression during mid-gestation (Dorfman et al. 2013). However, an imbalance in the FSH/LH ratio during midpregnancy could be responsible for the formation of accessory corpora lutea which, according to their size, could originate from secondary follicles more than from antral follicles as expected for primary corpora lutea (Cortasa et al. 2023). Curiously, this physiological change in the female vizcacha may represents a pathological alteration in other species like humans where changes in the FSH/LH ratio were determinded as the origin of cysts in the Polycystic Ovary Syndrome (McCartney et al. 2022).

We have described significantly increased levels of hypothalamic PR in midpregnant ovulating plains vizcachas concordant with high levels of serum P4 (Dorfman et al. 2013). In addition, the distribution of PR in the hypothalamus of the plains vizcacha is consistent with that previously described in other mammals such as a rat, mouse, guinea pig, primate, mink, cat, and ewe (Bayliss et al. 1991; Hagihara et al. 1992; Leranth et al. 1992; Moffatt et al. 1998; Scott et al. 2000b; Warembourg et al. 1989, 1998). PR distribution was found in neurons of the POA (DBB, PeVN, POA, SON, and SCN), VMN, ARC, and ME hypothalamic areas (Dorfman et al. 2013). The immunolocalization of PR in neurons of SON agrees with a previous report in the monkey brain (Bethea et al. 1992).

Genomic effects of P4 are mediated through its binding to nuclear PRs which act as ligand-dependent transcription factors after undergoing phosphorylation. This is assumed to be the primary pathway for P4 action in the brain. However, non-genomic actions of PR have been recently described (Mani and Blaustein 2012) and should be considered, specially taking in account the PR immunoreactivity in plains vizcacha at both nuclear and cytoplasmic compartments (Dorfman et al. 2013).

Finally, we determined the co-existence of PR and GnRH in neurons of POA and SON of pregnant and nonpregnant female plains vizcachas by confocal microscopy (Dorfman et al. 2013). In addition, regarding ME, PR cells were localized along

GnRH axons of midpregnant vizcachas (Dorfman et al. 2013) (Fig. 16.2). This observation suggested the key function of PR on GnRH delivery at this stage of the pregnancy. Notably, the hypothalamus of the guinea pig, which belongs to the same suborder of plains vizcacha and is a close evolutionary relative with a true luteal phase (Weir 1970), shows a small subpopulation of GnRH neurons containing PR (King et al. 1995). Later, these presumptions were supported by studying GnRH secretion in ex vivo assays using explants of hypothalami incubated with agonists and antagonists of PR where direct effects of P4 were observed on GnRH delivery (Inserra et al. 2020).

In summary, hypothalamic GnRH neurons in plains vizcacha express steroid hormone receptors as PR and ER that are not classically co-localized in most mammals. It is tempting to speculate that these peculiarities might enable a fine and rapid recovery of the activity of GnRH neurons during pregnancy.

16.3.2 GnRH Regulation by Kisspeptin

Kisspeptin (KISS) is a key hormone in the regulation of the HPO axis (Lehman et al. 2010). KISS1 gene codes for a 145-amino acid precursor peptide that is processed into a family of peptides known as kisspeptins (Kauffman 2010; Li et al. 2009). KISS has been predominantly localized in the hypothalamus at the anteroventral periventricular (AVPV) nucleus and neighboring rostral PeVN and in the ARC of several mammals such as rats, mice, hamsters, sheep, non-human primates, and humans (Clarkson et al. 2009; d'Anglemont de Tassigny et al. 2010; Gottsch et al. 2004; Kauffman et al. 2007; Shibata et al. 2007; Smith 2009). In the ARC, KISS neurons co-express neurokinin B (NKB) and dynorphin A (Dyn); therefore, these cells are named KNDy (Burke et al. 2006; Goodman et al. 2007; Navarro et al. 2009). KISS cells express both NK3R, the NKB receptor, and KOR, the Dyn-specific receptor. In addition, both peptides NKB and Dyn have opposite effects over KISS; whereas NKB induces KISS secretion, Dyn inhibits it (d'Anglemont de Tassigny and Colledge 2010; Helena et al. 2015; Krajewski et al. 2010). Functionally, KISS modulates the synthesis and release of GnRH (Gottsch et al. 2004; Irwig et al. 2004; Lehman et al. 2010; Navarro et al. 2005, 2009). KISS and KNDy neurons participate in the generation of GnRH/LH surge (Kauffman et al. 2007; Uenoyama et al. 2011; Yeo 2013). Understanding the role of KISS is critical to defining the underlying neuroendocrine mechanisms that regulate the process of reproduction in mammals. During the estrous cycle, the bimodal action of E2 over GnRH would be mainly mediated by KISS. KISS neurons show the presence of ERα and RP enabling the modulation of E2 and P4 over KISS expression (Smith and Clarke 2007). So, the bimodal action of E2 over GnRH, and the inhibiting action of P4 peaks, would be mediated by KISS action. In addition, different anatomical localization has been described for E2 modulation. Specifically, the E2-positive feedback on GnRH is observed in the AVPV, and the E2-negative feedback on GnRH is observed in the ARC, and both of them are mediated by KISS. In addition, the interconnected network of KNDy cells seems to be synchronized by NKB. In detail, NKB is involved in the positive feedback of E2 over KISS expression, whereas Dyn mediates the negative feedback of P4 (Foradori et al. 2002; Goodman et al. 2004; Goubillon et al. 2000; Lehman et al. 2010; Navarro et al. 2009).

Recently, we have characterized the expression of KISS, NKB, and Dyn in neurons of the hypothalamus of female plains vizcacha. Co-expression of KISS, NKB, and Dyn has been determined in the ARC supporting the existenceof KNDy cells in this species (Schmidt et al. 2022) (Fig. 16.2). This finding reinforces the highly conservation of this neurons throughout mammalian evolution. On the other hand, three different subpopulations of neurons were detected in the AVPV of plains vizcacha: one of them exhibiting NKB and KISS immunoreactivity, another one with NKB and Dyn immunoreactivity, and a third one expressing only NKB (Schmidt et al. 2022).

Considering that the AVPV region modulates the surge of LH, the absence of KNDy cells in this area suggests that the co-expression of KISS, NKB, and Dyn in the same cell is not required for the induction of LH surge. In addition, the presence of NK3R in KNDy cells of the ARC, as well as in the three subpopulations of NKB cells of the AVPV, points to the self-regulation of NKB. This suggests a possible cross-regulation among the different subpopulations of NKB neurons.

The distribution of NKB cells in the forebrain has been determined in humans (Hrabovszky et al. 2010). However, regarding Dyn cells distribution, the finding of Dyn immunoreactive cells in the AVPV of the plains vizcacha suggests a striking difference with mice since recent work has described that KISS cells of the AVPV are modulated by Dyn of ARC origin in that rodent (Helena et al. 2015). Previous reports described that both subpopulations of KISS neurons, those residing in the AVPV/PeVN and the KNDy neurons in the ARC, project their fibers towards GnRH neurons and terminals at both POA and ARC/ME and modulate the reproductive axis in rats, mice, and humans (Hrabovszky et al. 2010; Uenoyama et al. 2011; Yeo 2013). Similarly, we have detected KISS immunoreactive fibers making close appositions with GnRH cells (forming axo-somatic, axo dendritic, and axo-axonal contacts) (Schmidt et al. 2022) (Fig. 16.2). Moreover, both GnRH subpopulations of neurons, those at the POA and at the ARC, express GPR-54 confirming the intimate relation between KISS and GnRH in the plains vizcacha (Schmidt et al. 2022).

16.3.3 GnRH Regulation by Prolactin and Dopamine

Prolactin (PRL) is a peptide hormone mainly synthesized by the lactotroph cells of the anterior pituitary gland (McNeilly 1980, 2001). PRL is involved in the control of mammary gland development and milk synthesis (Bole-Feysot et al. 1998; Riddle et al. 1933; Tyson et al. 1972); however, a broad range of functions in the brain have been described (Bridges et al. 1985; Cohen-Becker et al. 1986; Freemark et al. 2001; Grattan 2015; Kokay et al. 2006; Lopez-Vicchi et al. 2020; Lucas et al. 1998; Shingo et al. 2003; Torner and Neumann 2002).

The tuberoinfundibular dopaminergic (TIDA) neurons, located in the dorsomedial arcuate nucleus (dmARC), are stimulated by increments of PRL acting through PRL receptors (PRLR) (Grattan and Kokay 2008; Kokay and Grattan 2005). In response to PRL, TIDA neurons secrete dopamine into the hypothalamic–pituitary portal vasculature. Then, in the pituitary gland, dopamine inhibits PRL secretion by acting through dopamine type 2 receptors (D2R) (Björklund et al. 1973; Freeman et al. 2000; Goudreau et al. 1995; MacLeod et al. 1970). In this way, PRL establishes a short-loop feedback by modulating TIDA neurons, resulting in the inhibition of its own secretion (Schuff et al. 2002). However, during lactation (the physiological state of hyperprolactinemia), PRLR favors enkephalin production instead of dopamine, so PRL secretion is not inhibited (Arbogast and Voogt 1991; Grattan and Kokay 2008; Yip et al. 2019).

On the other hand, PRLR expression has been also described in a small subpopulation of GnRH neurons of the POA (Grattan et al. 2007; Kokay et al. 2011). This suggests the direct regulation of PRL on GnRH expression. Moreover, GnRH neurons also receive projections from afferent regulatory neurons that express PRLR (Brown et al. 2014; Grattan et al. 2007). The expression of PRLR in most KISS neurons (Kokay et al. 2011; Li et al. 2011) suggests PRL involvement in GnRH regulation mediated by KISS. On the other hand, a subpopulation of GnRH neurons of the POA also expresses dopamine receptors D1R and D2R (Chen et al. 1989; Grattan et al. 2007; Horvath et al. 1993; Kuljis and Advis 1989). Tyrosine hydroxylase (TH), the rate-limiting enzyme in the synthesis of dopamine, is directly stimulated by PRL (Simerly et al. 1997). PRL modulates TH expression in the hypothalamus of mice and rats (Arbogast and Voogt 1991; Ma et al. 2005). Altogether, these data suggest that GnRH cells may be indirectly modulated by PRL acting through the dopamine pathway. This would constitute the so-called indirect PRL-TH/dopamine regulatory pathway on GnRH (Liu and Herbison 2013). However, this indirect regulatory pathway has been less characterized (Koike et al. 1991).

A significant increase of PRL has been detected in the pituitary glands of plains vizcachas from midpregnancy onward, with its maximum at term pregnancy (Corso et al. 2020; Proietto et al. 2018). Strikingly, the levels of hypothalamic transcription of PRLR did not follow a concurrent variation during pregnancy nor with PRL concentrations. However, the number of PRLR+ TH+ colocalizing cells was correlated with the increment of PRL from midpregnancy onward with a maximum increase of TIDA neurons at term-pregnancy. This may suggest that the increments of PRL at term-pregnancy may induce the expression of PRLR at TH cells. Nevertheless, the increment of TH was probably not sufficient to suppress the expression of PRL, which remained increased until the end of the pregnancy and after parturition (Proietto et al. 2018). This may represent the desensitization of the short loop feedback of PRL as has also been described during parturition and lactation in rats and other mammals (Arbogast and Voogt 1991; Grattan and Kokay 2008).

Taking in advance the extended gestation of plains vizcacha, this species enables to infer the possible effects of PRL at each specific stage of the gestation. While during midpregnancy, the increments of pituitary PRL may induce downregulation of PRLR expression in TH neurons of the POA, at term-pregnancy, the long-term increments of PRL may produce an increase in PRLR expression in those cells. Previous reports in mice suggested an indirect regulatory PRL-TH/dopamine pathway on GnRH expression, with negative effects of dopamine on GnRH (Liu and Herbison 2013; Ma et al. 2005). In this context, this indirect regulatory pathway may occur in the plains vizcacha. The decrease in PRLR expression in TH cells at midpregnancy would favor a decrease in dopamine synthesis. Considering the presence of TH fibers in close apposition to GnRH somas, this decrease of dopamine would allow the observed increase in GnRH expression and secretion previously described in this stage of the pregnancy (Charif et al. 2017; Dorfman et al. 2013; Inserra et al. 2017) (Fig. 16.2). Interestingly, the localization of D2R in GnRH neurons supports the action of dopamine on GnRH neurons. Moreover, the presence of TH immunoreactive fibers surrounding GnRH cells at the three gestational stages, together with the unvarying expression of D2R at GnRH cells during pregnancy, may pinpoint dopamine as a basal modulator of GnRH (Cortasa et al. 2023) (Fig. 16.2). Altogether, these results further highlight the relevance of an indirect regulatory PRL-TH/dopamine-GnRH pathway during gestation.

We determined the expression of PRLR in a substantial proportion of GnRH neurons of the rostral POA of early pregnant plains vizcachas (Cortasa et al. 2023). This has been rarely reported in other species. In mice, the presence of PRLR in a small subpopulation of GnRH neurons has been described (Grattan et al. 2007; Kokay et al. 2011). This may indicate a direct regulatory pathway during gestation by which PRL may prevent the expression of GnRH in early pregnancy. In this context, the sustained increase of pituitary PRL from midpregnancy onward probably induces a decrease of PRLR expression in GnRH neurons that would finally enable the increase of GnRH (Cortasa et al. 2023). These observations highlight the plains vizcacha as a possible model to let us learn about the direct PRL regulatory pathway on GnRH (Fig. 16.2). Furthermore, we developed a model of pharmacological induced hyperprolactinemia in female plains vizcachas (Corso et al. 2020; Cortasa et al. 2023). The significant increment of PRLR and GnRH colocalization in the POA of hyperprolactinemic plains vizcachas, together with the decrease of their GnRH and basal LH secretion (Cortasa et al. 2023), indicated a strong inhibitory effect of PRL on GnRH through a direct regulatory pathway as previously described in the hypothalamus of other species.

16.3.4 GnRH Regulation by Melatonin

The relative duration of the periods of dark and light is called photoperiod. This serves as seasonal indicator for resource availability. Based on the specific photoperiod length where each species reside, they are classified as short-day or long-day breeders and show different reproductive axis responses. In short-day breeder species like the sheep, the winter-like photoperiod triggers LH release (Bittman et al. 1985). In contrast, in long-day breeder species like the Syrian hamster, the winter-like photoperiod inhibits there productive cycle (Pratt and Goldman 1986; Seegal and Goldman 1975).

Environment light, captured by the retina, modulates the synthesis and release of melatonin from the pineal gland (Berson et al. 2002; Pandi-Perumal et al. 2008). Melatonin transduces the information of daylight duration into an endocrine signal allowing the translation of photoperiodical information into physiological and behavioral responses, being able to mediate seasonal reproduction events (Aleandri et al. 1996; Barret and Bolborea 2013; Nakane and Yoshimura 2014). The amino acid l-tryptophan is the initial substrate for the synthesis of melatonin (Simmoneaux and Ribelayga 2003). The enzyme arylalkylamine N-acetyltransferase (AANAT), also known as serotonin N-acetyltransferase (S-NAT), is considered the limiting step for the synthesis of melatonin. AANAT is directly modulated by the circadian rhythm (Borjigin et al. 1995; Klein and Weller 1972). Variations in AANAT protein expression linked to photoperiod have been reported in primates and ungulates with increases during the night (Ackermann et al. 2006; Ganguly et al. 2002; Stehle et al. 2001). On the other hand, the SCN is considered the residing place of the endogenous circadian pacemaker. Acting mainly through its two specific membrane G-coupled protein receptors, MT1 and MT2, melatonin synchronizes the rhythm of the SCN with the environment photoperiod information (Liu et al. 1997; Simmoneaux and Ribelayga 2003).

Melatonin is able to modulate the HPO axis, and the expression of MTs in hypothalamus and pituitary gland can mediate this event. Perhaps one of the most relevant reports is the direct modulation of GnRH expression and secretion by melatonin in the GT1–7 hypothalamic neuronal cell line that expresses MTs (Gillespie et al. 2003; Roy and Belsham 2002). In addition, in vivo and ex vivo evidence about melatonin downregulation of hypothalamic GnRH has been shown in rats, hamsters, and ewe (Buchanan and Yellon 1991; Rasmussen 1993; Viguie et al. 1995). In addition, melatonin is able to modulate puberty onset in mice and ewe (Carcangiu et al. 2012; Yang et al. 2021). These findings suggest that the combined actions of melatonin and GnRH may be relevant in the regulation of the timing and length of the reproductive cycle in seasonal breeders.

Histological changes have been observed in the pineal gland of plains vizcachas during pregnancy (Gil et al. 2005) and other species like mole, sow, and rat (Pevet 1980; Satodate et al. 1980; Wyrzykowski et al. 1987). In addition, it has been shown that melatonin is able to stimulate the pituitary gland cell proliferation in plains vizcachas altering its morphology in a seasonal fashion and according to the estrous cycle phase (Filippa and Mohamed 2010; Romera et al. 2010; Rosales et al. 2016). Furthermore, the maximum and minimum activity of the pineal gland during the winter and the summer, respectively, has been reported (Busolini et al. 2017; Cernuda-Cernuda et al. 2003; Pelzer and Guzman 1989).

We and other researchers have shown morphological variations in testis and spermatogenesis of plains vizcachas linked to changes in their gonadal activity according to the season confirming male plains vizcacha as a seasonal breeder (Chaves et al. 2012; Cruceño et al. 2012; González et al. 2018). On the other hand, we have recently reported the first pieces of evidence showing seasonal breeding in female plains vizcacha too (Charif et al. 2022). We determined progressively increased activity and expression of the pineal gland AANAT throughout the gestation of plains vizcacha, concordant with the pattern of serum melatonin levels, and together with significant variations in hypothalamic and pituitary gland MT1 and MT2 expression (Charif et al. 2022). These observations, together with our previously reported variations of GnRH and LH expression and delivery throughout gestation, may indicate a central leadership of melatonin in the reproduction of this species (Fig. 16.2).

Homogeneous distribution of AANAT in pinealocytes was described for plains vizcacha (Cernuda-Cernuda et al. 2003; Charif et al. 2022) and for other rodents like rat, hamster, and mouse (revised in (Hamada et al. 1992; Vivien-Roels et al. 1981)). In addition, both MT1 and MT2 expression was detected in cells of POA, PeVN, and SCN of female plains vizcacha (Charif et al. 2022) in agreement with the previously reported localization in other species such as the mole, rat, and Siberian hamster (Masson-Pevet et al. 1994; Pevet 2003). The progressively increased activity of AANAT during pregnancy suggests the involvement of melatonin during gestation. These variations of AANAT expression agree with the observations of Gil et al. (2005); however, they found increments in the content of melatonin in the pineal gland of pregnant plains vizcachas but not alterations throughout gestation. However, in their study, Gil et al. (2005) considered pregnancy as a homogeneous group possibly masking differences during pregnancy. Notably, variation in melatonin levels during pregnancy were reported in humans with the maximum level at gestational week 32 (Kivela 1991; McCarthy et al. 2019; Nakamura et al. 2001).

By employing ex vivo explants of hypothalamic and pituitary glands, we showed that melatonin is able to regulate GnRH and LH delivery, respectively, suggesting its key role on the reproductive axis activity of the plains vizcacha (Charif et al. 2022). In other rodents like the rat, the activity of the pineal gland has been reported to induce changes in the synthesis and delivery of GnRH and gonadal hormones during the last phase of gestation (Nir et al. 1979), while in the mouse, the exogenous administration of melatonin to pregnant females increases the number of corpora lutea, serum P4 and E2 levels, and the expression of MTs in the uterus (Guan et al. 2017; Zhang et al. 2017). Thus, the increase of melatonin in the plains vizcacha during pregnancy may be involved with the increment of the E2 and P4 levels required to guarantee the success of the gestation. Pro-gonadotropic effect of melatonin observed in plains vizcacha has been previously reported during anestrous in sheep with an increase of LH secretion during mimicked luteal phase (Bittman et al. 1985; Ciechanowska et al. 2017). However, the opposite has been observed in rats, where melatonin reduced LH and E2 levels during ovulation (Chuffa et al. 2011). All these reports may suggest a species-specific action of melatonin.

In conclusion, the changes in the seasonal photoperiod would directly impact in the plains vizcacha's melatonin levels through gestation, reflecting the progressive shortening of the days when gestation is near to term. This might stimulate the hypothalamic–pituitary activity at the most suitable season for breeding. This highlights the importance of its synchronicity and its relationship with the central melatoninergic components, focusing on the impact on reproductive axis activity and on the maintenance of the extensive gestational period of the plains vizcacha.

16.4 Final Conclusions

The decrease of P4 levels at the midpregnancy of the plains vizcacha elicits the reactivation of its HPO axis with the recovery of primary corpora lutea steroidogenic activity and a pseudo-ovulation event that recruits a new set of secondary corpora lutea during gestation. This would enable the continuity and success of gestation in this species which would never be possible without the precise disinhibition of the hypothalamic activity. According to our observations, this could be achieved by the concerted combination of direct and indirect action of hormones and neuropeptides on GnRH neurons which are definitively responsible for the trigger of the HPO axis. GnRH neurons of the POA of female plains vizcacha seem to have a greater ability than that of other mammals to express hormone receptors like REα, RP, and RPRL. Those receptors may confer GnRH neurons the capacity to directly sense the hormonal environment and, together with their indirect action and the concomitant effect of KISS, enable the fine and precise regulation of GnRH synthesis and delivery. Moreover, photoperiod, with melatonin as its chemical transducer, seems to coordinate the most adequate moment of the year for the reproduction of this species.

References

Abraham GE, Odell WD, Swerdloff RS, Hopper K (1972) Simultaneous radioimmunoassay of plasma FSH, LH, progesterone, 17-hydroxyprogesterone and estradiol-17b during the menstrual cycle. J Clin Endocrinol Metab 34:312–318. https://doi.org/10.1210/jcem-34-2-312

Ackermann K, Bux R, Rub U, Korf HW, Kauert G, Stehle JH (2006) Characterization of human melatonin synthesis using autoptic pineal tissue. Endocrinology 147:3235–3242. https://doi.org/10.1210/en.2006-0043

Acuña F, Barbeito CG, Portiansky EL, Ranea G, Miglino MA, Flamini MA (2022) Spontaneous embryonic death in plains viscacha (*Lagostomus maximus*, Rodentia), a species with unique reproductive characteristics. Theriogenology 185:88–96. https://doi.org/10.1016/j.theriogenology.2022.03.024

Adelman J, Mason AJ, Hayflick JS, Seeburg PH (1986) Isolation of the gene and hypothalamic cDNA for the common precursor of gonadotropin-releasing hormone and prolactin release-inhibiting factor in human and rat. Proc Natl Acad Sci USA 83:179–183. https://doi.org/10.1073/pnas.83.1.179

Aleandri V, Spina V, Moirini A (1996) The pineal gland and reproduction. Hum Reprod Update 2(3):225–235. https://doi.org/10.1093/humupd/2.3.225

Arbogast LA, Voogt JL (1991) Hyperprolactinemia increases and hypoprolactinemia decreases tyrosine hydroxylase messenger ribonucleic acid levels in the arcuate nuclei, but not the substantia nigra or zona incerta. Endocrinology 128:997–1005. https://doi.org/10.1210/endo-128-2-997

Barbeito CG, Acuña F, Miglino MA, Portiansky EL, Flamini MA (2021) Placentation and embryo death in the plains viscacha (*Lagostomus maximus*). Placenta 108:97–102. https://doi.org/10.1016/j.placenta.2021.03.006

Barret P, Bolborea M (2013) Molecular pathways involved in seasonal body weight and reproductive responses governed by melatonin. J Pineal Res 52(4):376–388. https://doi.org/10.1111/j.1600-079X.2011.00963.x

Bayliss DA, Seroogy KB, Millhorn DE (1991) Distribution and regulation by estrogen of progesterone receptor in the hypothalamus of the cat. Endocrinology 128:2610–2617. https://doi.org/10.1210/endo-128-5-2610

Belchetz PE, Plant TM, Nakai Y, Keogh EJ, Knobil E (1978) Hypophysial responses to continuous and intermittent delivery of hypothalamic gonadotropin-releasing hormone. Science 202:631–633. https://doi.org/10.1126/science.100883

Berson DM, Dunn FA, Takao M (2002) Phototransduction by retinal ganglion cells that set the circadian clock. Science 295:1070–1073. https://doi.org/10.1126/science.1067262

Bethea CL, Fahrenbach WH, Sprangers SA, Freesh F (1992) Immunocytochemical localization of progestin receptors in monkey hypothalamus: effect of estrogen and progestin. Endocrinology 130:895–905. https://doi.org/10.1210/endo.130.2.1733733

Bittman EL, Kaynard AH, Olster DH, Robinson JE, Yellon SM, Karsch FJ (1985) Pineal melatonin mediates photoperiodic control of pulsatile luteinizing hormone secretion in the ewe. Neuroendocrinology 40(5):409–418. https://doi.org/10.1159/000124106

Björklund A, Moore RY, Nobin A, Stenevi U (1973) The organization of tubero-hypophyseal and reticulo-infundibular catecholamine neuron systems in the rat brain. Brain Res 51:171–191. https://doi.org/10.1016/0006-8993(73)90371-5

Bole-Feysot C, Goffin V, Edery M, Binart N, Kelly PA (1998) Prolactin (PRL) and its receptor: actions, signal transduction pathways and phenotypes observed in PRL receptor knockout mice. Endocr Rev 19(3):225–268. https://doi.org/10.1210/edrv.19.3.0334

Borjigin J, Wang MM, Snyder SH (1995) Diurnal variation in mRNA encoding serotonin N-acetyltransferase in the pineal gland. Nature 378:783–785. https://doi.org/10.1038/378783a0

Branch LC, Villarreal D, Fowler GS (1993) Recruitment, dispersal, and group fusion in a declining population of the plains viscacha (*Lagostomus maximus*; Mammalia, Chinchillidae). J Mammal 74:9–20. https://doi.org/10.2307/1381901

Bridges RS, DiBiase R, Loundes DD, Doherty PC (1985) Prolactin stimulation of maternal behavior in female rats. Science 227:782–784. https://doi.org/10.1126/science.3969568

Brown RSE, Herbison AE, Grattan DR (2014) Prolactin regulation of kisspeptin neurones in the mouse brain and its role in the lactation induced suppression of kisspeptin expression. J Neuroendocrinol 26(12):898–908. https://doi.org/10.1111/jne.12223

Buchanan KL, Yellon SM (1991) Delayed puberty in the male Djungarian hamster: effect of short photoperiod or melatonin treatment on the GnRH neuronal system. Neuroendocrinology 54(2):96–102. https://doi.org/10.1159/000125857

Burger LL, Haisenleder DJ, Aylor KW, Marshall JC (2008) Regulation of intracellular signaling cascades by GnRH pulse frequency in the rat pituitary: roles for CaMK II, ERK, and JNK activation. Biol Reprod 79:947–953. https://doi.org/10.1095/biolreprod.108.070987

Burgus R, Butcher M, Amoss M, Ling N, Monahan M, Rivier J, Fellows R, Blackwell R, Vale W, Guillemin R (1972) Primary structure of the ovine hypothalamic luteinizing hormone-releasing factor (LRF). Proc Natl Acad Sci USA 69:278–282. https://doi.org/10.1073/pnas.69.1.278

Burke MC, Letts PA, Krajewski SJ, Rance NE (2006) Coexpression of dynorphin and neurokinin B immunoreactivity in the rat hypothalamus: morphologic evidence of interrelated function within the arcuate nucleus. J Comp Neurol 498(5):712–726. https://doi.org/10.1002/cne.21086

Busolini FI, Rosales GJ, Filippa VP, Mohamed FH (2017) A seasonal and age-related study of interstitial cells in the pineal gland of male vizcacha (*Lagostomus maximus maximus*). Anat Rec 300(10):1847–1857. https://doi.org/10.1002/ar.23621

Butler JA, Sjoberg M, Coen C (1999) Evidence for oestrogen receptor alpha-immunoreactivity in gonadotropin-releasing hormone expressing neurones. J Neuroendocrinol 11:331–335. https://doi.org/10.1046/j.1365-2826.1999.00347.x

Carcangiu V, Mura MC, Bini PP, Vacca GM, Daga C, Luridiana S (2012) Can advance of first lambing induced by melatonin implants influence the next lambing time in Sarda breed sheep? J Anim Sci 92(1):67–71. https://doi.org/10.4141/cjas2010-011

Cernuda-Cernuda R, Piezzi RS, Dominguez S, Alvarez-Uria M (2003) Cell populations in the pineal gland of the viscacha (Lagostomus maximus) seasonal variations. Histol Histopathol 18:827–836. https://doi.org/10.1152/10.14670/HH-18.827

Challis JRG, Heap RB, Illingworth DV (1971) Concentration of oestrogen and progesterone in the plasma of non-pregnant, pregnant and lactating guinea pigs. J Endocrinol 51:333–345. https://doi.org/10.1677/joe.0.0510333

Charif SE, Inserra PIF, Di Giorgio NP, Schmidt AR, Lux-Lantos V, Vitullo AD, Dorfman VB (2016) Sequence analysis, tissue distribution and molecular physiology of the GnRH preprogonadotrophin in the South American plains vizcacha (*Lagostomus maximus*). Gen Comp Endocrinol 232:174–184. https://doi.org/10.1016/j.ygcen.2015.12.012

Charif SE, Inserra PIF, Schmidt AR, Di Giorgio NP, Cortasa SA, Gonzalez CR, Lux-Lantos V, Halperin J, Vitullo AD, Dorfman VB (2017) Local production of neurostradiol affects gonadotropin-releasing hormone (GnRH) secretion at mid-gestation in *Lagostomus maximus* (Rodentia, Caviomorpha). Physiol Rep 5(16):e13439. https://doi.org/10.14814/phy2.13439

Charif SE, Inserra PIF, Schmidt AR, Cortasa SA, Proietto S, Corso MC, Halperin J, Di Giorgio NP, Lux-Lantos V, Vitullo AD, Dorfman VB (2022) Melatonin is involved in the modulation of the hypothalamic and pituitary activity in the South American plains vizcacha, *Lagostomus maximus*. J Comp Physiol B 192(1):141–159. https://doi.org/10.1007/s00360-021-01405-6

Chaves EM, Aguilera-Merlo C, Cruceño A, Fogal T, Piezzi R, Scardapane L, Dominguez S (2012) Seasonal morphological variations and age-related changes of the seminal vesicle of viscacha (*Lagostomus maximus maximus*): an ultrastructural and immunohistochemical study. Anat Rec 295(5):886–895. https://doi.org/10.1002/ar.22434

Chen WP, Witkin JW, Silverman AJ (1989) Gonadotropin releasing hormone (GnRH) neurons are directly innervated by catecholamine terminals. Synapse 3:288–290. https://doi.org/10.1002/syn.890030314

Cheng CK, Leung PCK (2005) Molecular biology of gonadotropin-releasing hormone (GnRH)-I, GnRH-II, and their receptors in humans. Endocr Rev 26:283–306. https://doi.org/10.1210/er.2003-0039

Chuffa LGA, Seiva FR, Favaro WJ, Teixeira GR, Amorim JP, Mendes LO, Fioruci BA, Pinheiro PFF, Fernandes AAH, Franc JAA, Delella FK, Martinez M, Martinez FE (2011) Melatonin reduces LH, 17 beta-estradiol and induces differential regulation of sex steroid receptors in reproductive tissues during rat ovulation. Reprod Biol Endocrinol 9(1):108. https://doi.org/10.1186/1477-7827-9-108

Ciccone NA, Xu S, Lacza CT, Carroll RS, Kaiser UB (2010) Frequency-dependent regulation of follicle-stimulating hormone beta by pulsatile gonadotropin-releasing hormone is mediated by functional antagonism of bZIP transcription factors. Mol Cell Biol 30:1028–1040. https://doi.org/10.1128/MCB.00848-09

Ciechanowska MO, Łapot M, Mateusiak K, Paruszewska E, Malewski T, Przekop F (2017) Biosynthesis of gonadotropin-releasing hormone (GnRH) and GnRH receptor (GnRHR) in hypothalamic–pituitary unit of anoestrous and cyclic ewes. Can J Physiol Pharmacol 95(2):178–184. https://doi.org/10.1139/cjpp-2016-0137

Clarkson J, d'Anglemont de Tassigny X, Colledge WH, Caraty A, Herbison AE (2009) Distribution of kisspeptin neurons in the adult female mouse brain. J Neuroendocrinol 21:673–682. https://doi.org/10.1111/j.1365-2826.2009.01892.x

Cohen-Becker IR, Selmanoff M, Wise PM (1986) Hyperprolactinemia alters the frequency and amplitude of pulsatile luteinizing hormone secretion in the ovariectomized rat. Neuroendocrinology 42:328–333. https://doi.org/10.1159/000124459

Conley A, Hinshelwood M (2001) Mammalian aromatases. Reproduction 121:685–695. https://doi.org/10.1530/rep.0.1210685

Corso MC, Cortasa SA, Schmidt AR, Proietto S, Inserra PIF, Fernandez MO, Di Giorgio N, Lux-Lantos V, Vitullo AD, Dorfman VB, Halperin J (2020) Mammary gland-specific regulation of GnRH and GnRH-receptor gene expression is likely part of a local autoregulatory system in female vizcachas (Rodentia: Chinchillidae). Gen Comp Endocrinol 296:113518. https://doi.org/10.1016/j.ygcen.2020.113518

Cortasa SA, Inserra PIF, Proietto S, Corso MC, Schmidt AR, Vitullo AD, Dorfman VB, Halperin J (2022) Achieving full-term pregnancy in the vizcacha relies on a reboot of luteal steroidogenesis in midgestation (*Lagostomus maximus*, Rodentia). PLoS One 17(7):e0271067. https://doi.org/10.1371/journal.pone.0271067

Cortasa SA, Schmidt AR, Proietto S, Corso MC, Inserra PIF, Di Giorgio NP, Lux-Lantos V, Vitullo AD, Halperin J, Dorfman VB (2023) Hypothalamic GnRH expression and pulsatility depends on a balance of prolactin receptors in the plains vizcacha, *Lagostomus maximus*. J Comp Neurol 531(7):720–742. https://doi.org/10.1002/cne.25457

Cruceño AM, de Rosas JC, Fóscolo M, Chaves EM, Scardapane L, Dominguez S, Aguilera-Merlo C (2012) Effects of photoperiod on epididymal and sperm morphology in a wild rodent, the viscacha (*Lagostomus maximus maximus*). ISRN Anat 2013:128921. https://doi.org/10.5402/2013/128921

d'Anglemont de Tassigny X, Colledge WH (2010) The role of kisspeptin signaling in reproduction. Physiology (Bethesda) 25(4):207–217. https://doi.org/10.1152/physiol.00009.2010

d'Anglemont de Tassigny X, Ackroyd KJ, Chatzidaki EE, Colledge WH (2010) Kisspeptin signaling is required for peripheral but not central stimulation of gonadotropin-releasing hormone neurons by NMDA. J Neurosci 30(25):8581–8590. https://doi.org/10.1523/JNEUROSCI.5486-09.2010

Dorfman VB, Fraunhoffer N, Inserra PIF, Loidl CF, Vitullo AD (2011) Histological characterization of gonadotropin-releasing hormone (GnRH) in the hypothalamus of the SouthAmerican plains vizcacha (*Lagostomus maximus*). J Mol Histol 42:311–321. https://doi.org/10.1007/s10735-011-9335-5

Dorfman VB, Saucedo L, Di Giorgio NP, Inserra PIF, Fraunhoffer N, Leopardo NP, Halperin J, Lux-Lantos V, Vitullo AD (2013) Variation in progesterone receptors and GnRH expression in the hypothalamus of the pregnant South American plains vizcacha, *Lagostomus maximus* (Mammalia, Rodentia). Biol Reprod 89(5):115–125. https://doi.org/10.1095/biolreprod.113.107995

Dorfman VB, Inserra PIF, Leopardo NP, Halperin J, Vitullo AD (2016) The South American plains vizcacha, *Lagostomus maximus*, as a valuable animal model for reproductive studies. JSM Anat Physiol 1(1):1004

Filippa VP, Mohamed FH (2010) Morphological and morphometric changes of pituitary lactotrophs of viscacha (*Lagostomus maximus maximus*) in relation to reproductive cycle, age and sex. Anat Rec 293(1):150–161. https://doi.org/10.1002/ar.21013

Foradori CD, Coolen LM, Fitzgerald ME, Skinner DC, Goodman RL, Lehman MN (2002) Colocalization of progesterone receptors in parvicellular dynorphin neurons of the ovine preoptic area and hypothalamus. Endocrinology 143(11):4366–4374. https://doi.org/10.1210/en.2002-220586

Fraunhoffer NA, Jensen F, Leopardo N, Inserra PIF, Abuelafia AM, Espinosa MB, Charif SE, Dorfman VB, Vitullo AD (2017) Hormonal behavior correlates with follicular recruitment at mid-gestation in the south American plains vizcacha, *Lagostomus maximus* (Rodentia, Caviomorpha). Gen Comp Endocrinol 250:162–174. https://doi.org/10.1016/j.ygcen.2017.06.010

Freeman ME, Kanyicska B, Lerant A, Nagy G (2000) Prolactin: structure, function, and regulation of secretion. Physiol Rev 80:1523–1631. https://doi.org/10.1152/physrev.2000.80.4.1523

Freemark M, Fleenor D, Driscoll P, Binart N, Kelly P (2001) Body weight and fat deposition in prolactin receptor-deficient mice. Endocrinology 142:532–537. https://doi.org/10.1210/endo.142.2.7979

Ganguly S, Coon SL, Klein DC (2002) Control of melatonin synthesis in the mammalian pineal gland: the critical role of serotonin acetylation. Cell Tissue Res 309:127–137. https://doi.org/10.1007/s00441-002-0579-y

Gil E, Calderon C, Pelzer L, Dominguez S, Fogal T, Scardapane L, Piezzi RS (2005) Morphological and biochemical study of the pineal gland of pregnant and non-pregnant female vizcachas (*Lagostomus maximus maximus*). Neuro Endocrinol Lett 26(5):575–580

Gillespie JM, Chan BP, Roy D, Cai F, Belsham DD (2003) Expression of circadian rhythm genes in gonadotropin-releasing hormone secreting GT1-7 neurons. Endocrinology 144(12):5285–5292. https://doi.org/10.1210/en.2003-0802

Gonzales KL, Tetel MJ, Wagner CK (2008) Estrogen receptor (ER) β modulates ERα response to estrogen in the developing rat ventromedial nucleus of the hypothalamus. Endocrinology 149(9):4615–4621. https://doi.org/10.1210/en.2008-0511

Goodman RL, Karsch FJ (1980) Pulsatile secretion of luteinizing hormone: differential suppression by ovarian steroids. Endocrinology 107:1286–1290. https://doi.org/10.1210/endo-107-5-1286

Goodman RL, Coolen LM, Anderson GM, Hardy SL, Valent M, Connors JM, Fitzgerald ME, Lehman MN (2004) Evidence that dynorphin plays a major role in mediating progesterone negative feedback on gonadotropin-releasing hormone neurons in sheep. Endocrinology 145(6):2959–2967. https://doi.org/10.1210/en.2003-1305

Goodman RL, Lehman MN, Smith JT, Coolen LM, de Oliveira CV, Jafarzadehshirazi MR, Pereira A, Iqbal J, Caraty A, Ciofi P, Clarke IJ (2007) Kisspeptin neurons in the arcuate nucleus of the ewe express both dynorphin A and neurokinin B. Endocrinology 148:5752–5760. https://doi.org/10.1210/en.2007-0961

Gottsch ML, Cunningham MJ, Smith JT, Popa SM, Acohido BV, Crowley WF, Seminara S, Clifton DK, Steiner RA (2004) A role for kisspeptins in the regulation of gonadotropin secretion in the mouse. Endocrinology 145:4073–4077. https://doi.org/10.1210/en.2004-0431

Goubillon ML, Forsdike RA, Robinson JE, Ciofi P, Caraty A, Herbison AE (2000) Identification of neurokinin B-expressing neurons as an highly estrogen-receptive, sexually dimorphic cell group in the ovine arcuate nucleus. Endocrinology 141:4218–4225. https://doi.org/10.1210/endo.141.11.7743

Goudreau JL, Falls WM, Lookingland KJ, Moore KE (1995) Periventricular-hypophysial dopaminergic neurons innervate the intermediate but not the neural lobe of the rat pituitary gland. Neuroendocrinology 62:147–154. https://doi.org/10.1159/000126999

Grattan DR (2015) 60 years of neuroendocrinology: the hypothalamo prolactin axis. J Endocrinol 226(2):T101–T122. https://doi.org/10.1530/JOE-15-0213

Grattan DR, Kokay IC (2008) Prolactin: a pleiotropic neuroendocrine hormone. J Neuroendocrinol 20:752–763. https://doi.org/10.1111/j.1365-2826.2008.01736.x

Grattan DR, Jasoni CL, Liu X, Anderson GM, Herbison AE (2007) Prolactin regulation of gonadotropin-releasing hormone neurons to suppress luteinizing hormone secretion in mice. Endocrinology 148(9):4344–4351. https://doi.org/10.1210/en.2007-0403

Guan S, Zie L, Ma T, Lv D, Jing W, Tian X, Song Y, Liu Z, Xiao X, Liu G (2017) Effects of melatonin on early pregnancy in mouse: involving the regulation of StAR, Cyp11a1 and Ihh expression. Int J Mol Sci 18:1637–1650. https://doi.org/10.3390/ijms18081637

Gundlah C, Kohama SG, Mirkes SJ, Garyfallou VT, Urbanski HF, Bethea CL (2000) Distribution of estrogen receptor beta (ERβ) mRNA in hypothalamus, midbrain and temporal lobe of spayed macaque: continued expression with hormone replacement. Mol Brain Res 76:191–204. https://doi.org/10.1016/s0006-8993(99)02475-0

Hagihara K, Hirata S, Osada T, Hirai M, Kato J (1992) Distribution of cells containing progesterone receptor mRNA in the female rat di- and telencephalon: an *in situ* hybridization study. Mol Brain Res 14:239–249. https://doi.org/10.1016/0169-328x(92)90179-f

Halperin J, Dorfman VB, Fraunhoffer N, Vitullo AD (2013) Estradiol, progesterone and prolactin modulate mammary glandmorphogenesis in adult female plains vizcacha (*Lagostomus maximus*). J Mol Histol 44(3):299–310. https://doi.org/10.1007/s10735-012-9477-0

Hamada T, Shimizu T, Ichikawa M, Mori Y (1992) Immunohistochemical study on gonadotropin-releasing hormone neurons in the Shiba goat brain. J Reprod Dev 38:133–142. https://doi.org/10.1262/jrd.38.133

Heap RB, Ackland N, Weir BJ (1981) Progesterone-binding proteins in plasma of guinea-pigs and other hystricomorph rodents. J Reprod Fertil 63:477–489. https://doi.org/10.1530/jrf.0.0630477

Helena CV, Toporikova N, Kalil B, Stathopoulos AM, Pogrebna VV, Carolino RO, Anselmo-Franci JA, Bertram R (2015) KNDy neurons modulate the magnitude of the steroid-induced luteinizing hormone surges in ovariectomized rats. Endocrinology 156(11):4200–4213. https://doi.org/10.1210/en.2015-1070

Herbison AE, Robinson JE, Skinner DC (1993) Distribution of estrogen receptor-immunoreactive cells in the preoptic area of the ewe: co-localization with glutamic acid decarboxylase but not luteinizing hormone-releasing hormone. Neuroendocrinology 57(4):751–759. https://doi.org/10.1159/000126433

Horvath TL, Naftolin F, Leranth C (1993) Luteinizing hormone releasing hormone and gamma-aminobutyric acid neurons in the medial preoptic area are synaptic targets of dopamine axons originating in anterior periventricular areas. J Neuroendocrinol 5:71–79. https://doi.org/10.1111/j.1365-2826.1993.tb00365.x

Hrabovszky E, Ciofi P, Vida B, Horvath MC, Keller E, Caraty A, Bloom SR, Ghatei MA, Dhillo WS, Liposits Z, Kallo I (2010) The kisspeptin system of the human hypothalamus: sexual dimorphism and relationship with gonadotropin-releasing hormone and neurokinin B neurons. Eur J Neurosci 31:1984–1998. https://doi.org/10.1111/j.1460-9568.2010.07239.x

Inserra PIF (2016) Circuitos neuromoduladores del eje hipotalámico-hipofisario-gonadal en la vizcacha de las llanuras, *Lagostomus maximus*: un mamífero con oogénesis continua y ovulación durante la gestación. Doctoral thesis. Facultad de Medicina, Universidad de Buenos Aires

Inserra PIF, Charif SE, Di Giorgio NP, Saucedo L, Schmidt AR, Fraunhoffer N, Halperin J, Gariboldi MC, Leopardo NP, Lux-Lantos V, Gonzalez CR, Vitullo AD, Dorfman VB (2017) ERα and GnRH co-localize in the hypothalamic neurons of the South American plains vizcacha, *Lagostomus maximus* (Rodentia, Caviomorpha). J Mol Histol 48(3):259–273. https://doi.org/10.1007/s10735-017-9715-6

Inserra PIF, Charif SE, Fidel V, Giacchino M, Schmidt AR, Villarreal FM, Proietto S, Cortasa SA, Corso MC, Gariboldi MC, Leopardo NP, Fraunhoffer NA, Di Giorgio NP, Lux-Lantos V, Halperin J, Vitullo AD, Dorfman VB (2020) The key action of estradiol and progesterone enables GnRH delivery during gestation in the South American plains vizcacha, *Lagostomus maximus*. J Steroid Biochem Mol Biol 200:105627. https://doi.org/10.1016/j.jsbmb.2020.105627

Irwig MS, Fraley GS, Smith JT, Acohido BV, Popa SM, Cunningham MJ, Gottsch ML, Clifton DK, Steiner RA (2004) Kisspeptin activation of gonadotropin releasing hormone neurons and regulation of KiSS-1 mRNA in the male rat. Neuroendocrinology 80:264–272. https://doi.org/10.1159/000083140

Jackson JE, Branch LC, Villarreal D (1996) *Lagostomus maximus*. Mamm Species 543:1–6. https://doi.org/10.2307/3504168

Jimenez-Liñan M, Rubin BS, King JC (1997) Examination of guinea pig luteinizing hormone-releasing hormone gene reveals a unique decapeptide and existence of two transcripts in the brain. Endocrinology 138:4123–4130. https://doi.org/10.1210/endo.138.10.5454

Jori F, López-Béjar M, Mayor P, López C (2002) Functional anatomy of the ovaries of wild brush-tailed porcupines (*Atherurus africanus*, Gray 1842) from Gabon. J Zool 256(1):35–43. https://doi.org/10.1017/S0952836902000055

Kauffman AS (2010) Coming of age in the kisspeptin era: sex differences, development, and puberty. Mol Cell Endocrinol 324(1–2):51–63. https://doi.org/10.1016/j.mce.2010.01.017

Kauffman AS, Gottsch ML, Roa J, Byquist AC, Crown A, Clifton DK, Hoffman GE, Steiner RA, Tena-Sempere M (2007) Sexual differentiation of Kiss1 gene expression in the brain of the rat. Endocrinology 148:1774–1783. https://doi.org/10.1210/en.2006-1540

Kenealy BP, Kapoor A, Guerriero KA, Keen KL, Garcia JP, Kurian JP, Ziegler TE, Terasawa E (2013) Neuroestradiol in the hypothalamus contributes to the regulation of gonadotropin releasing hormone release. J Neurosci 33:19051–19059. https://doi.org/10.1523/JNEUROSCI.3878-13.2013

King JA, Millar RP (1982) Structure of chicken hypothalamic luteinizing hormone-releasing hormone I. Structural determination on partially purified material. J Biol Chem 257:10722–10728

King JC, Tai DW, Hanna IK, Pfeiffer A, Haas P, Ronsheim PM, Mitchell SC, Turcotte JC, Blaustein JD (1995) A subgroup of LHRH neurons in guinea pigs with progestin receptors is centrally positioned within the total population of LHRH neurons. Neuroendocrinology 61:265–275. https://doi.org/10.1159/000126848

Kivela A (1991) Serum melatonin during human pregnancy. Acta Endocrinol (Norway) 124:233–237. https://doi.org/10.1530/acta.0.1240233

Klein DC, Weller JL (1972) Rapid light-induced decrease in pineal serotonin N-acetyltransferase activity. Science 177(4048):532–533. https://doi.org/10.1126/science.177.4048.532

Koike K, Miyake A, Aono T, Sakumoto T, Ohmichi M, Yamaguchi M, Tanizawa O (1991) Effect of prolactin on the secretion of hypothalamic GnRH and pituitary gonadotropins. Horm Res 35(1):5–12. https://doi.org/10.1159/000181921

Kokay IC, Grattan DR (2005) Expression ofmRNAfor prolactin receptor (long form) in dopamine and pro-opiomelanocortin neurones in the arcuate nucleus of non-pregnant and lactating rats. J Neuroendocrinol 17:827–835. https://doi.org/10.1111/j.1365-2826.2005.01374.x

Kokay IC, Bull PM, Davis RL, Ludwig M, Grattan DR (2006) Expression of the long form of the prolactin receptor in magnocellular oxytocin neurons is associated with specific prolactin regulation of oxytocin neurons. Am J Physiol Regul Integr Comp Physiol 290(5):R1216–R1225. https://doi.org/10.1152/ajpregu.00730.2005

Kokay IC, Petersen SL, Grattan DR (2011) Identification of prolactin-sensitive GABA and kisspeptin neurons in regions of the rat hypothalamus involved in the control of fertility. Endocrinology 152(2):526–535. https://doi.org/10.1210/en.2010-0668

Krajewski SJ, Burke MC, Anderson MJ, McMullen NT, Rance NE (2010) Forebrain projections of arcuate neurokinin B neurons demonstrated by anterograde tract-tracing and monosodium glutamate lesions in the rat. Neuroscience 166(2):680–697. https://doi.org/10.1016/j.neuroscience.2009.12.053

Krey LC, Silverman AJ (1978) The luteinizing hormone-releasing hormone (LHRH) neuronal networks of the guinea pig brain. II. The regulation of gonadotropin secretion and the origin of terminals in the median eminence. Brain Res 157:247–255. https://doi.org/10.1016/0006-8993(78)90027-6

Kuiper GG, Enmark E, Pelto-Huikko M, Nilsson S, Gustafsson JA (1996) Cloning of a novel receptor expressed in rat prostate and ovary. Proc Natl Acad Sci USA 93:5925–5930. https://doi.org/10.1073/pnas.93.12.5925

Kuljis RO, Advis JP (1989) Immunocytochemical and physiological evidence of a synapse between dopamine- and luteinizing hormone releasing hormone-containing neurons in the ewe median eminence. Endocrinology 124:1579–1581. https://doi.org/10.1210/endo-124-3-1579

Leavitt WW, Blaha GC (1970) Circulating progesterone levels in the golden hamster during the estrus cycle, pregnancy, and lactation. Biol Reprod 3:353–361. https://doi.org/10.1093/biolreprod/3.3.353

Lehman MN, Coolen LM, Goodman RL (2010) Minireview: kisspeptin/neurokinin B/dynorphin (KNDy) cells of the arucate nucleus: a central node in the control of GnRH secretion. Endocrinology 151:3479–3489. https://doi.org/10.1210/en.2010-0022

Leranth C, MacLusky NJ, Brown TJ, Chen EC, Redmond DE, Naftolin F (1992) Transmitter content and afferent connections of estrogen-sensitive progestin receptor-containing neurons in the primate hypothalamus. Neuroendocrinology 55:667–682. https://doi.org/10.1159/000126187

Lethimonier C, Madigou T, Muñoz-Cueto JA, Lareyre JJ, Kah O (2004) Evolutionarily aspects of GnRHs, GnRH neuronal systems and GnRH receptors in teleost fish. Gen Comp Endocrinol 135:1–16. https://doi.org/10.1016/j.ygcen.2003.10.007

Li D, Yu W, Liu M (2009) Regulation of KiSS1 gene expression. Peptides 30(1):130–138. https://doi.org/10.1016/j.peptides.2008.09.025

Li Q, Rao A, Pereira A, Clarke IJ, Smith JT (2011) Kisspeptin cells in the ovine arcuate nucleus express prolactin receptor but not melatonin receptor. J Neuroendocrinol 23(10):871–882. https://doi.org/10.1111/j.1365-2826.2011.02195.x

Liu X, Herbison AE (2013) Dopamine regulation of gonadotropin releasing hormone neuron excitability in male and female mice. Endocrinology 154(1):340–350. https://doi.org/10.1210/en.2012-1602

Liu C, Weaver DR, Jin X, Shearman LP, Pieschl RL, Gribkoff VK, Reppert SM (1997) Molecular dissection of two distinct actions of melatonin on the suprachiasmatic circadian clock. Neuron 19(1):91–102. https://doi.org/10.1016/S0896-6273(00)80350-5

Llanos AC, Crespo JA (1952) Ecología de la vizcacha (*Lagostomus maximus maximus* Blainv.) en el nordeste de la Provincia de Entre Ríos. Rev de Investig Agríc 6:289–378

Lopez-Vicchi F, Ladyman SR, Ornstein AM, Gustafson P, Knowles P, Luque GM, Grattan DR, Becu-Villalobos D (2020) Chronic high prolactin levels impact on gene expression at discrete hypothalamic nuclei involved in food intake. FASEB J 34(3):3902–3914. https://doi.org/10.1096/fj.201902357R

Lucas BK, Ormandy CJ, Binart N, Bridges RS, Kelly PA (1998) Null mutation of the prolactin receptor gene produces a defect inmaternal behavior. Endocrinology 139:4102–4107. https://doi.org/10.1210/endo.139.10.6243

Luu-The V, Zhang Y, Poirier D, Labrie F (1995) Characteristics of human types 1, 2 and 3 17beta-hydroxysteroid dehydrogenase activities: oxidation/reduction and inhibition. J Steroid Biochem Mol Biol 55:581–587. https://doi.org/10.1016/0960-0760(95)00209-x

Ma FY, Grattan DR, Goffin V, Bunn SJ (2005) Prolactin-regulated tyrosine hydroxylase activity and messenger ribonucleic acid expression in mediobasal hypothalamic cultures: the differential role of specific protein kinases. Endocrinology 146:93–102. https://doi.org/10.1210/en.2004-0800

MacLeod RM, Fontham EH, Lehmeyer JE (1970) Prolactin and growth hormone production as influenced by catecholamines and agents that affect brain catecholamines. Neuroendocrinology 6:283–294. https://doi.org/10.1159/000121933

Mani SK, Blaustein JD (2012) Neural progestin receptors and female sexual behavior. Neuroendocrinology 96:152–161. https://doi.org/10.1159/000338668

March CM, Goebelsmann U, Nakamura RM, Mishell DR Jr (1979) Roles of estradiol and progesterone in eliciting the mid cycle luteinizing hormone and follicle stimulating hormone surges. J Clin Endocrinol Metab 49:507–513. https://doi.org/10.1210/jcem-49-4-507

Masson-Pevet M, George D, Kalsbeek A, Saboureau M, Lakhdar-Ghazal N, Pevet P (1994) An attempt to correlate brain areas containing melatonin-binding sites with rhythmic functions: a study in five hibernator species. Cell Tissue Res 278:97–106. https://doi.org/10.1007/BF00305781

Matsuo H, Baba Y, Nair RM, Arimura A, Schally AV (1971) Structure of the porcine LH- and FSH-releasing hormone I. The proposed amino acid sequence. Biochem Biophys Res Commun 43:1334–1339. https://doi.org/10.1016/s0006-291x(71)80019-0

McCarthy R, Jungheim ES, Fay JC, Bates K, Herzog ED, England SK (2019) Riding the rhythm of melatonin through pregnancy to deliver on time. Front Endocrinol 10:616. https://doi.org/10.3389/fendo.2019.00616

McCartney CR, Campbell RE, Marshall JC, Moenter SM (2022) The role of gonadotropin-releasing hormone neurons in polycystic ovary syndrome. J Neuroendocrinol 34(5):e13093. https://doi.org/10.1111/jne.13093

McCormack JT, Greenwald GS (1974) Progesterone and oestradiol-17b concentration in the peripheral plasma during pregnancy in the mouse. J Endocrinol 62:101–107. https://doi.org/10.1677/joe.0.0620101
McKenna NJ, Lanz RB, O'Malley BW (1999) Nuclear receptor co-regulators: cellular and molecular biology. Endocr Rev 20:321–344. https://doi.org/10.1210/edrv.20.3.0366
McNeilly AS (1980) Prolactin and the control of gonadotrophin secretion in the female. J Reprod Fertil 58:537–549. https://doi.org/10.1530/jrf.0.0580537
McNeilly AS (2001) Lactational control of reproduction. J Reprod Fertil Dev 13:583–590. https://doi.org/10.1071/rd01056
Micevych P, Sinchak K (2008) Synthesis and function of hypothalamic neuroprogesterone in reproduction. Endocrinology 149:2739–2742. https://doi.org/10.1210/en.2008-0011
Millar RP (2005) GnRHs and GnRH receptors. Anim Reprod Sci 88:5–28. https://doi.org/10.1016/j.anireprosci.2005.05.032
Mittelman-Smith MA, Rudolph LM, Mohr MA, Micevych PE (2017) Rodent models of non-classical progesterone action regulating ovulation. Front Endocrinol 8:165. https://doi.org/10.3389/fendo.2017.00165
Miyamoto K, Hasegawa Y, Nomura M, Igarashi M, Kangawa K, Matsuo H (1984) Identification of the second gonadotropin-releasing hormone in chicken hypothalamus: evidence that gonadotropin secretion is probably controlled by two distinct gonadotropin-releasing hormones in avian species. Proc Natl Acad Sci USA 81:3874–3878. https://doi.org/10.1073/pnas.81.12.3874
Moffatt CA, Rissman E, Shupnik M, Blaustein JD (1998) Induction of progestin receptors by estradiol in the forebrain of estrogen receptor-a gene disrupted mice. J Neurosci 18:9556–9563. https://doi.org/10.1523/JNEUROSCI.18-22-09556.1998
Montaner AD, Mongiat L, Lux-Lantos VA, Warby C, Chewpoy B, Bianchi MS, Libertun C, Rivier JE, Sherwood NM, Somoza GM (2002) Guinea pig gonadotropin-releasing hormone: expression pattern, characterization and biological activity in rodents. Neuroendocrinology 75:326–338. https://doi.org/10.1159/000057342
Naftolin F, Ryan KJ, Petro Z (1972) Aromatization of androstenedione by the anterior hypothalamus of adult male and female rats. Endocrinology 90:295–298. https://doi.org/10.1210/endo-90-1-295
Naik DV (1975) Immuno-electron microscopic localization of luteinizing hormone-releasing hormone in the arcuate nuclei and median eminence of the rat. Cell Tissue Res 157:437–455. https://doi.org/10.1007/BF00222598
Nakamura Y, Tamura H, Kashida S, Takayama H, Yagamata Y, Karube A, Sugino N, Kato H (2001) Changes of serum melatonin level and its relationship to feto-placental unit during pregnancy. J Pineal Res 30:29–33. https://doi.org/10.1034/j.1600-079X.2001.300104.x
Nakane Y, Yoshimura T (2014) Universality and diversity in the signal transduction pathway that regulates seasonal reproduction in vertebrates. Front Neurosci 8:115. https://doi.org/10.3389/fnins.2014.00115
Navarro VM, Castellano JM, Fernandez-Fernandez R, Tovar S, Roa J, Mayen A, Barreiro ML, Casanueva FF, Aguilar E, Dieguez C, Pinilla L, Mena-Sempere M (2005) Effects of KiSS-1 peptide, the natural ligand of GPR54, on follicle-stimulating hormone secretion in the rat. Endocrinology 146(4):1689–1697. https://doi.org/10.1210/en.2004-1353
Navarro VM, Gottsch ML, Chavkin C, Okamura H, Clifton DK, Steiner RA (2009) Regulation of GnRH secretion by KISS1/dynorphin/neurokinin B neurons in the arcuate nucleus of the mouse. J Neurosci 29:11859–11866. https://doi.org/10.1523/JNEUROSCI.1569-09.2009
Nir I, Hirschmann N, Goldhaber G, Shani J (1979) Pinealectomy induced changes in blood and pituitary luteinizing and prolactin levels during the last phase of pregnancy in rats. Neuroendocrinology 28:44–51. https://doi.org/10.1159/000122843
Pandi-Perumal SR, Trakht I, Srinivasan V, Spence DW, Maestroni GJM, Zisapel N, Cardinali DP (2008) Physiological effects of melatonin: role of melatonin receptors and signal transduction pathways. Prog Neurobiol 85(3):335–353. https://doi.org/10.1016/jpneurobio.2008.04.001

Pelzer LE, Guzman JA (1989) Pineal serotonin N-acetyltransferase and hydroxyndole-O-methyltransferase activities in vizcacha (*Lagostomus maximus maximus*) maintained under various photoperiods. Acta Physiol Pharmacol Latinoam 39(3):299–305

Pevet P (1980) The pineal gland of the mole (*Talpa europaea* L.). IV. Fine structure of fetal pinealocytes. Cell Tissue Res 206:417–430. https://doi.org/10.1007/BF00237971

Pevet P (2003) Melatonin in animal models. Dialogues Clin Neurosci 5(4):343–352. https://doi.org/10.31887/DCNS.2003.5.4/ppevet

Pratt BL, Goldman BD (1986) Activity rhythms and photoperiodism of Syrian hamsters in a simulated burrow system. Physiol Behav 36(1):83–89. https://doi.org/10.1016/0031-9384(86)90078-8

Proietto S, Cortasa SA, Corso MC, Inserra PIF, Charif SE, Schmidt AR, Di Giorgio NP, Lux-Lantos V, Vitullo AD, Dorfman VB, Halperin J (2018) Prolactin is a strong candidate for the regulation of luteal steroidogenesis in vizcachas (*Lagostomus maximus*). Int J Endocrinol 2018:1910672. https://doi.org/10.1155/2018/1910672

Proietto S, Yankelevich L, Villarreal FM, Inserra PIF, Charif SE, Schmidt AR, Cortasa SA, Corso MC, Di Giorgio NP, Lux-Lantos V, Vitullo AD, Halperin J, Dorfman VB (2019) Pituitary estrogen receptor alpha is involved in luteinizing hormone pulsatility at midgestation in the South American plains vizcacha, *Lagostomus maximus* (Rodentia, Caviomorpha). Gen Comp Endocrinol 273:40–51. https://doi.org/10.1016/j.ygcen.2018.04.001

Quadros PS, Wagner CK (2008) Regulation of progesterone receptor expression by estradiol is dependent on age, sex and region in the rat brain. Endocrinology 149(6):3054–3061, R1216–R1225. https://doi.org/10.1210/en.2007-1133

Radovick S, Ticknor CM, Nakayama Y, Notides AC, Rahman A, Weintraub BD, Cutler GB Jr, Wondisford FE (1991) Evidence for direct estrogen regulation of the human gonadotropin-releasing hormone gene. J Clin Investig 88:1649–1655. https://doi.org/10.1172/JCI115479

Rasmussen DD (1993) Diurnal modulation of rat hypothalamic gonadotropin-releasing hormone release by melatonin *in vitro*. J Endocrinol Investig 16:1–7. https://doi.org/10.1007/BF03345819

Riddle O, Bates RW, Dykshorn SW (1933) The preparation, identification and assay of prolactin: a hormone of the anterior pituitary. Am J Physiol 105:191–216. https://doi.org/10.1152/ajplegacy.1933.105.1.191

Roch GJ, Busby ER, Sherwood NM (2014) GnRH receptors and peptides: skating backward. Gen Comp Endocrinol 209:118–134. https://doi.org/10.1016/j.ygcen.2014.07.025

Romera EP, Mohamed FH, Fogal T, Dominguez S, Piezzi R, Scardapane L (2010) Effect of the photoperiod and administration of melatonin on the pars tuberalis of viscacha (*Lagostomus maximus maximus*): an ultrastructural study. Anat Rec 293(5):871–878. https://doi.org/10.1002/ar.21083

Rosales GJ, Busolini FI, Mohamed FH, Filippa VP (2016) Effects of melatonin and gonadal androgens on cell proliferation in the pituitary of viscachas (*Lagostomus maximus maximus*). Cell Prolif 49(5):644–653. https://doi.org/10.1111/cpr.12280

Roselli CE, Horton LE, Resko JA (1985) Distribution and regulation of aromatase activity in the rat hypothalamus and limbic system. Endocrinology 117:2471–2477. https://doi.org/10.1210/endo-117-6-2471

Roselli CE, Liu M, Hurn PD (2009) Brain aromatization: classic roles and new perspectives. Semin Reprod Med 27:207–217. https://doi.org/10.1055/s-0029-1216274

Roy D, Belsham DD (2002) Melatonin receptor activation regulates GnRH gene expression and secretion in GT1-7 GnRH neurons. J Biol Chem 277(1):251–258. https://doi.org/10.1074/jbc.M108890200

Satodate R, Sasou S, Ota M, Takashio M, Yoshida Y (1980) An electron microscopical morphometrical study on pinealocytes of pregnant rats. Endokrinologie 75:285–291

Schmidt AR, Gariboldi MC, Cortasa SA, Proietto S, Clara MC, Inserra PIF, Jaime V, Halperin J, Vitullo AD, Dorfman VB (2020) Neocortical anatomy in the South American plains vizcacha, *Lagostomus maximus*, reveals different strategies in encephalic development among

Hystricomorpha and Myomorpha rodents. Brain Behav Evol 95(6):318–329. https://doi.org/10.1159/000515638

Schmidt AR, Inserra PIF, Cortasa SA, Proietto S, Fidel V, Halperin J, Vitullo AD, Dorfman VB (2022) Distribution of kisspeptin system and its relation with gonadotropin-releasing hormone in the hypothalamus of the South American plains vizcacha, *Lagostomus maximus*. Gen Comp Endocrinol 317:113974. https://doi.org/10.1016/j.ygcen.2021.113974

Schuff KG, Hentges ST, Kelly MA, Binart N, Kelly PA, Iuvone PM, Asa SL, Low MJ (2002) Lack of prolactin receptor signaling in mice results in lactotroph proliferation and prolactinomas by dopamine dependent and -independent mechanisms. J Clin Investig 110:973–981. https://doi.org/10.1172/JCI15912

Scott CJ, Tilbrook AJ, Simmons DM, Rawson JA, Chu S, Fuller PJ, Ing NH, Clarke IJ (2000a) The distribution of cells containing estrogen receptor-alpha (ERα) and ERβ messenger ribonucleic acid in the preoptic area and hypothalamus of the sheep: comparison of males and females. Endocrinology 141:2951–2962. https://doi.org/10.1210/endo.141.8.7622

Scott CJ, Pereira AM, Rawson JA, Simmons DM, Rossmanith WG, Ing NH, Clarke IJ (2000b) The distribution of progesterone receptor (PR) immunoreactivity (-ir) and mRNA in the preoptic area and hypothalamus of the ewe: up-regulation of PR and mRNA in the mediobasal hypothalamus by estrogen. J Neuroendocrinol 12:565–575. https://doi.org/10.1046/j.1365-2826.2000.00490.x

Seegal RF, Goldman BD (1975) Effects of photoperiod on cyclicity and serum gonadotropins in the Syrian hamster. Biol Reprod 12(2):223–231. https://doi.org/10.1095/biolreprod12.2.223

Selmanoff MK, Brodkin LD, Weiner RI, Siiteri PK (1977) Aromatization and 5a-reduction of androgens in discrete hypothalamic and limbic regions of the male and female rat. Endocrinology 101:841–848. https://doi.org/10.1210/endo-101-3-841

Sherwood NM, Sower SA, Marshak DR, Fraser BA, Brownstein MJ (1986) Primary structure of gonadotropin-releasing hormone from lamprey brain. J Biol Chem 261:4812–4819

Shibata M, Friedman RL, Ramaswamy S, Plant TM (2007) Evidence that down regulation of hypothalamic KISS-1 expression is involved in the negative feedback action of testosterone to regulate luteinising hormone secretion in the adult male rhesus monkey (*Macaca mulatta*). J Neuroendocrinol 19(6):432–438. https://doi.org/10.1111/j.1365-2826.2007.01549.x

Shingo T, Gregg C, Enwere E, Fujikawa H, Hassam R, Geary C, Cross JC, Weiss S (2003) Pregnancy-stimulated neurogenesis in the adult female forebrain mediated by prolactin. Science 299:117–120. https://doi.org/10.1126/science.1076647

Shughrue PJ, Lane MV, Merchenthaler I (1997) Comparative distribution of estrogen receptor-α and -β mRNA in the rat central nervous system. J Comp Neurol 388:507–525. https://doi.org/10.1002/(sici)1096-9861(19971201)388:4<507::aid-cne1>3.0.co;2-6

Silverman AJ, Witkin JW (1994) Biosynthesis of gonadotropin-releasing hormone during the rat estrous cycle: a cellular analysis. Neuroendocrinology 59:545–551. https://doi.org/10.1159/000126704

Silverman AJ, Jhamandas J, Renaud LP (1987) Localization of luteinizing hormone releasing hormone (LHRH) neurons that project to the median eminence. J Neurosci 7(8):2312–2319

Silverman AJ, Witkin JW, Millar RP (1990) Light and electron microscopic immunocytochemical analysis of antibodies directed against GnRH and its precursor in hypothalamic neurons. J Histochem Cytochem 38:803–813. https://doi.org/10.1177/38.6.2186087

Simerly RB (2002) Wired for reproduction: organization and development of sexually dimorphic circuits in the mammalian forebrain. Annu Rev Neurosci 25:507–536. https://doi.org/10.1146/annurev.neuro.25.112701.142745

Simerly RB, Zee MC, Pendleton JW, Lubahn DB, Korach KS (1997) Estrogen receptor-dependent sexual differentiation of dopaminergic neurons in the preoptic region of the mouse. Proc Natl Acad Sci USA 94:14077–14082. https://doi.org/10.1073/pnas.94.25.14077

Simmoneaux V, Ribelayga C (2003) Generation of the melatonin endocrine message in mammals: a review of the complex regulation of melatonin synthesis by norepinephrine, peptides, and other pineal transmitters. Pharmacol Rev 55(2):325–395. https://doi.org/10.1124/pr.55.2.2

Sinchak K, Mohr MA, Micevych PE (2020) Hypothalamic astrocyte development and physiology for neuroprogesterone induction of the luteinizing hormone surge. Front Endocrinol 11:420. https://doi.org/10.3389/fendo.2020.00420

Skinner DC, Evans NP, Delaleu B, Goodman RL, Bouchard P, Caraty A (1998) The negative feedback actions of progesterone on gonadotropin-releasing hormone secretion are transduced by the classical progesterone receptor. Proc Natl Acad Sci USA 95:10978–10983. https://doi.org/10.1073/pnas.95.18.10978

Smith JT (2009) Sex steroid control of hypothalamic KISS1 expression in sheep and rodents: comparative aspects. Peptides 30(1):94–102. https://doi.org/10.1016/j.peptides.2008.08.013

Smith JT, Clarke IJ (2007) Kisspeptin expression in the brain: catalyst for the initiation of puberty. Rev Endocr Metab Disord 8:1–9. https://doi.org/10.1007/s11154-007-9026-4

Somoza GM, Miranda LA, Strobl-Mazzulla P, Guilgur LG (2002) Gonadotropin releasing hormone (GnRH): from fish to mammalian brains. Cell Mol Neurobiol 22:589–609. https://doi.org/10.1023/a:1021888420271

Spotorno AE, Patton JL (2015) Superfamily Chinchilloidea Bennett 1833. In: Patton JL, Pardiñas UFJ, Elia GD (eds) Mammals of South America, Rodents, vol 2. University of Chicago Press, USA

Stehle JH, Von Gall C, Schomerus C, Korf HW (2001) Of rodents and ungulates and melatonin: creating an uniform code for darkness by different signaling mechanisms. J Biol Rhythm 16(4):312–325. https://doi.org/10.1177/074873001129002033

Tam WH (1974) The synthesis of progesterone in some hystricomorph rodents. Symp Zool Soc Lond 34:363–384

Torner L, Neumann ID (2002) The brain prolactin system: involvement in stress response adaptations in lactation. Stress 5:249–257. https://doi.org/10.1080/1025389021000048638

Tsai PS, Zhang L (2008) The emergence and loss of gonadotropin-releasing hormone in protostomes: orthology, phylogeny, structure, and function. Biol Reprod 79:798–805. https://doi.org/10.1095/biolreprod.108.070185

Tyson JE, Friesen HG, Anderson MS (1972) Human lactational and ovarian response to endogenous prolactin release. Science 177:897–900. https://doi.org/10.1126/science.177.4052.897

Uenoyama Y, Inoue N, Pheng V, Homma T, Takase K, Yamada S, Ajiki K, Ichikawa M, Okamura H, Maeda KI, Tsukamura H (2011) Ultrastructural evidence of kisspeptin-gonadotrophin-releasing hormone (GnRH) interaction in the median eminence of female rats: implication of axo-axonal regulation of GnRH release. J Neuroendocrinol 23(10):863–870. https://doi.org/10.1111/j.1365-2826.2011.02199.x

Urbanski HF, Doan A, Pierce M (1991) Immunocytochemical investigation of luteinizing hormone-releasing hormone neurons in Syrian hamsters maintained under long or short days. Biol Reprod 44:687–692. https://doi.org/10.1095/biolreprod44.4.687

Viguie C, Caraty A, Locatelli A, Caraty A (1995) Regulation of luteinizing hormone-releasing hormone (LHRH) secretion by melatonin in the ewe. I. Simultaneous delayed increase in LHRH and luteinizing hormone pulsatile secretion. Biol Reprod 52(5):1114–1120. https://doi.org/10.1095/biolreprod52.5.1114

Vivien-Roels DB, Pevet P, Dubois MP, Arendt J, Brown GM (1981) Immunohistochemical evidence for the presence of melatonin in the pineal gland, the retina and the Harderian gland. Cell Tissue Res 217:105–115. https://doi.org/10.1007/bf00233830

Warembourg M, Leroy D (2004) Comparative distribution of estrogen receptor α and β immunoreactivities in the forebrain and the midbrain of the female guinea pig. Brain Res 1002(1–2):55–66. https://doi.org/10.1016/j.brainres.2003.12.014

Warembourg M, Jolivet A, Milgrom E (1989) Immunohistochemical evidence of the presence of estrogen and progesterone receptors in the same neurons of the guinea pig hypothalamus and preoptic area. Brain Res 480:1–15. https://doi.org/10.1016/0006-8993(89)91561-8

Warembourg M, Leroy D, Peytevin J, Martinet L (1998) Estrogen receptor and progesterone receptor immunoreactive cells are not co-localized with gonadotropin-releasing hormone

in the brain of the female mink (*Mustela vison*). Cell Tissue Res 291(1):33–41. https://doi.org/10.1007/s004410050977
Watson RE Jr, Langub MC Jr, Landis JW (1992) Further evidence that most luteinizing hormone-releasing hormone neurons are not directly estrogen-responsive: simultaneous localization of luteinizing hormone-releasing hormone and estrogen receptor immunoreactivity in the guinea-pig brain. J Neuroendocrinol 4(3):311–317. https://doi.org/10.1111/j.1365-2826.1992.tb00173.x
Weir BJ (1970) The management and breeding of some more hystricomorph rodents. Lab Anim 4:83–97. https://doi.org/10.1258/002367770781036562
Weir BJ (1971a) The reproductive organs of the female plains viscacha, *Lagostomus maximus*. J Reprod Fertil 25:365–373. https://doi.org/10.1530/jrf.0.0250365
Weir BJ (1971b) The reproductive physiology of the plains viscacha, *Lagostomus maximus*. J Reprod Fertil 25:355–363. https://doi.org/10.1530/jrf.0.0250355
Wildt L, Häusler A, Marshall G, Hutchison JS, Plant TM, Belchetz PE, Knobil E (1981) Frequency and amplitude of gonadotropin releasing hormone stimulation and gonadotropin secretion in the rhesus monkey. Endocrinology 109:376–385. https://doi.org/10.1210/endo-109-2-376
Willingstorfer WJ, Burda H, Winckler J (1998) Ovarian growth and folliculogenesis in breeding and nonbreeding females of a social rodent, the Zambian common mole-rat, Cryptomys sp. J Morphol 237(1):33–41. https://doi.org/10.1002/(SICI)1097-4687(199807)237:1<33::AID-JMOR3>3.0.CO;2-P
Wintermantel TM, Campbell RE, Porteous R, Bock D, Gröne H, Todman MG, Korach KS, Greiner E, Pérez CA, Schütz G, Herbison AE (2006) Definition of estrogen receptor pathway critical for estrogen positive feedback to gonadotropin-releasing hormone neurons and fertility. Neuron 52:271–280. https://doi.org/10.1016/j.neuron.2006.07.023
Wray S, Zoeller RT, Gainer H (1989) Differential effects of estrogen on luteinizing hormone-releasing hormone gene expression in slice explant cultures prepared from specific rat forebrain regions. Mol Endocrinol 3(8):1197–1206. https://doi.org/10.1210/mend-3-8-1197
Wyrzykowski Z, Przybylska B, Wyrzykowska K (1987) Morphology of the pineal gland in pregnant and nursing sows. J Pineal Res 4:125–134. https://doi.org/10.1111/j.1600-079X.1987.tb00849.x
Yang C, Ran Z, Liu G, Hou R, He C, Liu Q, Chen Y, Liu Y, Wang X, Ling C, Fang F, Li X (2021) Melatonin administration accelerates puberty onset in mice by promoting FSH synthesis. Molecules 26(5):1474. https://doi.org/10.3390/molecules26051474
Yeo SH (2013) Neuronal circuits in the hypothalamus-controlling GnRH release: the neuroanatomical projections of kisspeptin neurons. Exp Physiol 98(11):1544–1549. https://doi.org/10.1113/expphysiol.2013.071944
Yip SH, Romano N, Gustafson P, Hodson DJ, Williams EJ, Kokay IC, Martin AO, Mollard P, Grattan DR, Bunn SJ (2019) Elevated prolactin during pregnancy drives a phenotypic switch in mouse hypothalamic dopaminergic neurons. Cell Rep 26:1787–1799. https://doi.org/10.1016/j.celrep.2019.01.067
Zhang L, Zhang Z, Wang F, Tian X, Ji P, Liu G (2017) Effects of melatonin administration on embryo implantation and offspring growth in mice under different schedules of photoperiodic exposure. Reprod Biol Endocrinol 15:78–87. https://doi.org/10.1186/s12958-017-0297-7
Zohar Y, Muñoz-Cueto JA, Elizur A, Kah O (2010) Neuroendocrinology of reproduction in teleost fish. Gen Comp Endocrinol 165:438–455. https://doi.org/10.1016/j.ygcen.2009.04.017

Chapter 17
Synchronization of the Pineal and Pituitary Morphophysiology by Photoperiodic Environmental Signals

Verónica P. Filippa, Fabricio I. Busolini, and Fabian H. Mohamed

Abstract Terrestrial environments have cyclical characteristics, and therefore, there must be a temporal harmonization between environmental cycles and biological rhythmicity. Thus, seasonal changes in lighting and temperature synchronize circannual rhythms to regulate the biological rhythms of organisms. This synchronization is carried out through biological clocks, which control the activity of numerous endocrine glands, such as the pineal gland and the secretion of its main hormone, melatonin. This chapter describes the morphophysiological variations of the pineal and pituitary glands of male vizcachas (*Lagostomus maximus maximus*), captured in San Luis, Argentina. It is a nocturnal rodent mammal that lives in underground caves in semiarid areas of central Argentina and, in its natural habitat, is a photoperiod-dependent seasonal breeder. The pineal gland is ovoid in shape, elongated, and connected by a stalk to the base of the brain. In it, two regions (proximal and distal) and three cell types are distinguished: pinealocytes (P), interstitial cells (IC), and pigmented cells. Seasonal morphological changes have been reported in the main parenchyma cells in adult male animals. At ultrastructural level, winter P showed an oval nucleus with dispersed chromatin and a large nucleolus. The cytoplasm exhibited numerous dense matrix tubular mitochondria, developed Golgi complex and RER, free polyribosomes, and dense bodies of various sizes suggesting active metabolism and increased cellular activity. Immunohistochemical studies made it possible to identify IC, through the expression of S-100, GFAP, and vimentin proteins. These cells were mainly located in the proximal region and exhibited an irregular cell body with abundant cytoplasmic processes surrounding the P and/or reaching blood vessels. Likewise, the existence of IC in different functional and maturation stages was demonstrated, probably in relation to the needs of the glan-

V. P. Filippa
Área de Morfología, Facultad de Química, Bioquímica y Farmacia, Universidad Nacionalde San Luis, San Luis, Argentina

Consejo Nacional de Investigaciones Científicas y Técnicas (CONICET), San Luis, Argentina

F. I. Busolini · F. H. Mohamed (✉)
Área de Morfología, Facultad de Química, Bioquímica y Farmacia, Universidad Nacionalde San Luis, San Luis, Argentina

L. L. Rasia et al. (eds.), *Plains Vizcachas*, https://doi.org/10.1007/978-3-031-49487-1_17

dular microenvironment or the presence of cell subpopulations with different functions. The expression of S-100 and GFAP was maximum during winter, in parallel to the P activity. This indicates a synchronized activity between both cell types through an intraglandular paracrine communication network that ensures optimal neuroendocrine regulation. In the pituitary gland of adult male vizcachas, specifically in the adenohypophysis, different cell populations and follicular structures that present seasonal morphological variations were described. The cell types identified in pars distalis (PD) were corticotrophs, thyrotrophs, gonadotrophs, lactotrophs, somatotrophs, and follicle-stellate cells. All these cell populations presented a defined regionalization in the PD parenchyma and histomorphometric variations in relation to cell activity during the reproductive cycle. In addition, they exhibited a specific cytological configuration that demonstrates the existence of an intrapituitary communication network to respond to the different physiological requirements of the animals. Besides, cell proliferation and androgen receptor expression exhibit seasonal changes according to the glandular activity. In conclusion, the morphological changes in these endocrine glands, central components in the neuroendocrine control system, demonstrate the synchronization of glandular activity with the main environmental signals, promoting reproductive and adaptive behaviors that guarantee the perpetuation of this species.

Keywords *Lagostomus* · Pineal gland · Pituitary gland · Photoperiod · Melatonin

17.1 Pineal Gland

Daily physiological rhythms are common to almost all life forms. The presence of light-sensitive structures provides a selective advantage that has evolved in almost all *phyla* (Darwin 1859; Gehring 2005; Peirson and Foster 2006). The ability to detect light was optimized with the development of an endogenous circadian oscillator that allows for the organism to anticipate twilight, dawn, feeding, and resting hours and to help to escape from predators and improve the survival of the species (Bell-Pedersen et al. 2005; Kumar and Sharma 2018; Meléndez-Fernández et al. 2023). Under this selective pressure, two specialized visual systems for handling light information have evolved in the diencephalon of all vertebrates: (1) lateral eyes, which form images of objects (Knudsen 2020), and (2) the pineal gland or epiphysis.

17.1.1 Historical Background: From "Seat of the Soul" to a Neuroendocrine Organ

The epiphysis in the pharaonic Egypt was morphologically associated with the eye of Horus. In Hinduism, it was considered the third eye, the Ajna chakra, involved in clairvoyance (Nichols 2018). Galen of Pergamon (130–200 B.C.) identified the pineal as a distinct organ within the brain. He characterized it as a gland, and due to its particular shape, he called it *konareion*, or conarium in Latin. In the Renaissance, the anatomical descriptions of Andreas Vesalius Bruxellensis (1515–1564) was the basis for the conceptualization of the pineal gland as the "seat of the soul" by René Descartes (1596–1650), or as the organ of psycho-physiological control (Gheban et al. 2019). According to the Cartesian theory, the pineal was in charge of extracting certain particles from the blood and converting them into animal spirits, which traveled to different parts of the ventricular system. Each change in the epiphyseal position would correspond to a different perception of the soul, and the latter, in turn, could move the gland by the mere fact of perceiving (López-Muñoz and Boya 1992; López-Muñoz and Álamo 2000). Niels Steensen (1638–1686) refuted Descartes' Cartesian theories by demonstrating that the pineal gland was not located in the center of the ventricular system but rather dorsally. This anatomical demonstration was later joined by the German philosopher Immanuel Kant (1724–1804), who attributed a spiritual character to the soul, so it could not be located in a delimited space (Kappers 1979). Throughout the eighteenth century, the pineal gland remained forgotten and was considered a vestigial organ. During the first decades of the twentieth century, the so-called Spanish Histological School gained great popularity, at the expense of Ramón y Cajal and his disciples: Achúcarro, Sacristán, and Rio-Hortega. The current era of pineal knowledge begins in 1954, when Kitay and Altschule published the book "The Pineal Gland." These authors attribute the intervention in the control of gonadal function, as well as in the dermal response to ambient light changes to the epiphysis (Guerrero et al. 2007). Undoubtedly, the most important finding in pineal research was the identification and isolation of melatonin, carried out by Lerner et al. (1958).

17.1.2 Pineal Gland: Anatomical, Histological, and Biochemical Features

The pineal gland is a small, centrally located neuroendocrine organ with irregular morphology. In most mammals, the pineal stalk connects the gland to the habenular commissure, usually attached to the third ventricle, inserting it into a region called the pineal recess (Rouvière and Delmas 2005). The shape and location of the gland exhibits great macroscopic variability. The pineal gland of vizcacha has an oval, elongated, and slightly flattened shape with a stalk connected to the brain. It lies deeply between the habenular and posterior commissures. According to the classification proposed by Vollrath (1981), the vizcacha pineal belongs to type AB (glandular length is slightly more than twice its greatest width; Domínguez 1990). Two pineal regions can be distinguished in relation to the pineal stalk. The closest region (proximal) is rounded and wider, while the farthest region (distal) is edged and thinner, resulting in a typical pine-cone shape (Busolini et al. 2017). The gland is submerged in cerebrospinal fluid and encapsulated by connective tissue. Several trabeculae of various sizes and lengths penetrate the parenchyma along with a large number of blood vessels (Fig. 17.1a). In males, the gland average size is 4 mm long and 2.5 mm wide, with an average weight of 4.23 mg (Cernuda-Cernuda et al. 2003).

The pineal gland originates from the neural tube and is one of the most vascularized endocrine organs (Møller and Baeres 2002). In mammals, the posterior cerebral arteries from which the posterior choroidal arteries arise provide the main vascular supply. These arteries surround and penetrate the pineal capsule (Stehle et al. 2011). Venous drainage comes to converge in a large vessel which, after joining the great cerebral vein, empties into the anterior portion of the straight sinus (Beltrán 2003). The pineal may present both "fenestrated" capillaries, with high vascular permeability, as well as non-fenestrated capillaries. The ratio between them varies between different species of mammals. In vizcacha, the lumina of the capillaries are invested by continuous endothelial cells. These capillaries exhibit an endothelium without fenestrations surrounded by a limited perivascular space where some pericytes are located (Cernuda-Cernuda et al. 2003). Additionally, collagen fibers, cellular processes, and nerve terminals containing vesicles can be seen. Numerous arterioles with open and closed lumen are also observed in the gland, suggesting the existence of thoroughfare vascular channels (Dominguez et al. 1987). The observations of continuous-type capillary endothelial cells surrounded by a basal lamina without fenestrations suggest the presence of a blood brain barrier in this gland, as described in human fetuses, cats, and other mammals (Moller 1974, 1976).

The main innervation of the gland is established by the sympathetic nerve fibers from the superior cervical ganglion. A multisynaptic pathway originating in the retina leads to the pineal light information processed in the form of nerve impulses (Reiter 1986). This pathway, called the retino-hypothalamic-pineal tract, is responsible for the circadian rhythm of melatonin biosynthesis and secretion (Vrang et al.

Fig. 17.1 (continued) pinealocytes constitute the main cell population and their electron density can be variable. Type II pinealocytes (P2) have more electron-dense nuclei with irregular outlines; red blood cell (rbc), endothelial cell (ec). (**f**) Light (lp) and dark (dp) pinealocyte nuclei with different electron densities. (**g**) Mitochondria with different shapes and sizes; vacuole (va). (**h**) Electrodense concentric lamellar structures of different sizes. (**i**) Phagolysosomal system (fg) with several inclusions of a homogeneous material, possibly lipids, wrapped in a simple membrane (white arrow) and numerous glycogen-like particles (black arrow). (**j**) Pigment granules in different stages of melanogenesis in a lysosomal-type dense body (arrowhead). (**k**) Unmyelinated nerve terminal, note the presence of numerous axons with abundant mitochondria and small clear vesicles. (**l**) Micrograph shows a typical axon (ax) in the vicinity of a perivascular space. Cilia (ci). (**m**) Interstitial cell with elongated nucleus, scattered chromatin with heterochromatin granules located beneath the nuclear envelope, and an electron-dense cytoplasm. (**n**) and (**o**) The S-100 protein is expressed in IC nuclei, perinuclear cytoplasm and cytoplasmic processes. (**p**) and (**q**) GFAP is expressed in IC cytoplasmic processes and perinuclear cytoplasm. The nuclei remain immunonegative. During winter, a significant increase of the immunostaining is observed for both proteins. Winter (N), (P); Summer (O), (Q). Scale bars: (**a, c**) 250μm; (**b, d, n–q**) 25μm; (**e**) 10μm; (**f**) 5μm; (**k, m**) 2μm; (**l**) 1μm; (**g–j**) 500 nm. License Number 5635061199104 (**a** and **b**)

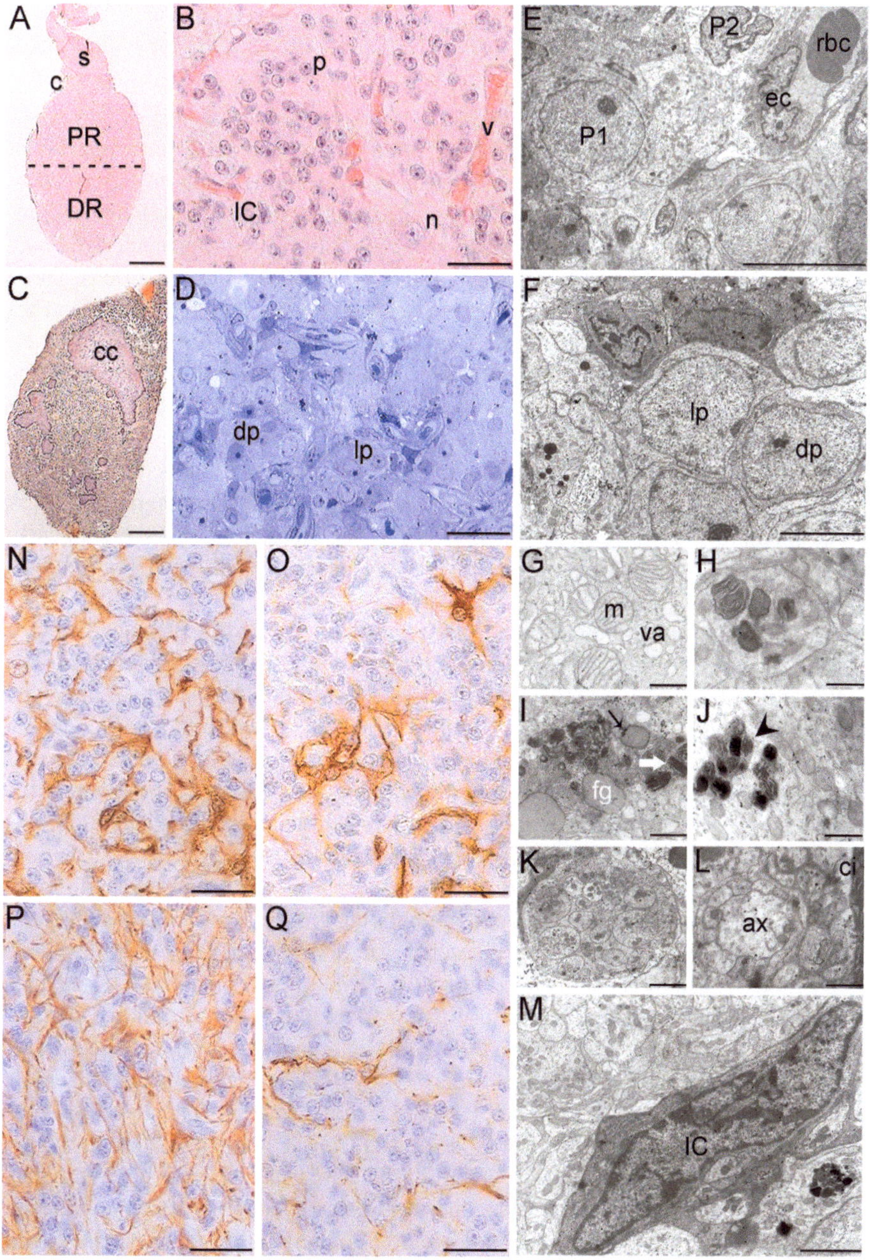

Fig. 17.1 Pineal gland of *Lagostomus maximus*. (**a**) The gland has a typical pine-cone shape covered by a connective tissue capsule (c), a stalk (s), and two regions: proximal (PR) and distal (DR). (**b**) The gland exhibits a homogeneous distribution with numerous pinealocytes (p), few interstitial cells (IC), and abundant blood vessels (v); neuron (n). Stain: hematoxylin-eosin. (**c**) Calcareous concretions (cc) of different size and location. Stain: Alizarin red. (**d**) Semi-thin sections show light (lp) and dark (dp) pinealocytes within the parenchyma. Stain: toluidine blue. (**e**) Type I (P1)

(continued)

1995; Larsen et al. 1998; Teclemariam-Mesbah et al. 1999; Munch et al. 2002). Parasympathetic innervation of the gland is carried out by pathways from the sphenopalatine and otic ganglia (Møller and Baeres 2002). Unmyelinated nerve fibers predominate in the vizcacha epiphysis. These fibers end in the vicinity of perivascular spaces and have club-shaped synaptic boutons that contain several vesicles of different sizes and shapes. However, two types of terminals can be described in detail. In one, small non-granulated spherical vesicles (50–60 nm in diameter) predominate, while in the other, granulated vesicles (120–150 nm) characterized by a dense core constitute the main accumulation (Cernuda-Cernuda et al. 2003). These vesicles have been reported to be found in nerve fibers of other vertebrates and act as storage sites for biogenic amines such as norepinephrine, serotonin, and their precursors. These nerves have been reported to play a key role in regulating the diameter of small blood vessels as well as vascular permeability. Furthermore, the release and diffusion of neurotransmitters in the perivascular space, after stimulation of these nerves, may be an important regulatory event in the synchronization of photoperiod and reproduction in the vizcacha pineal gland (Dominguez 1990). A predominant sympathetic "tone" may operate in the vizcacha during the long darkness of winter, stimulating the gland and inhibiting reproduction (Domínguez et al. 1987).

The pineal has great biochemical activity, as demonstrated by the abundant biogenic amines and multiple peptidergic substances synthesized within the gland. The melatonin isolated by Lerner is undoubtedly the most important indole of the gland and the most studied (Lerner et al. 1959; Arendt 1986; Arendt and Aulinas 2022; Ganguly et al. 2002; Stehle et al. 2002; Guerrero et al. 2007). Considering that melatonin biosynthesis is regulated by the photoperiod, it is not surprising that the enzymatic activities as well as its reaction products undergo circadian variations (Reuss 1996; Simmoneaux and Ribelayga 2003). Guzman et al. (1980) carried out studies on the biochemical activity of the vizcacha pineal gland. Under different light conditions, the activities of three methylating enzymes involved in melatonin synthesis were determined in vizcachas: hydroxyindole-O-methyltransferase (HIOMT), catechol-O-methyltransferase (COMT), and histamine-N-methyltransferase (HNMT), respectively. Results were compared to habitat animals. Under permanent illumination for 15 days, the HIOMT activity decreases while COMT activity increases; no changes are displayed by HNMT. Simultaneously, animals kept in the dark showed a decreased HIOMT activity. In other rodent species, illumination exerts a similar effect on HIOMT and COMT activities. Conversely, the decrease in HIOMT activity during the dark phase suggests that captivity may impair glandular function, probably due to reduced locomotor activity (Guzman et al. 1980). Seasonal changes in body weight, pineal HIOMT activity, and pineal gland weight were analyzed in adult male vizcacha by Pelzer et al. (1999). Based on this study, the highest HIOMT activity was observed in August (winter). Body weight did not show significant changes throughout the year, while glandular weight exhibited variations in some months. These findings suggest that the epiphysis responds to annual changes in photoperiod and probably other environmental factors. In all species, the production and secretion of melatonin is minimal during the day and increases sharply during dark (Forcada 2000). It is precisely this rhythmic and predictable behavior of environmental light the main signal that synchronizes many internal biological rhythms in photoperiod-dependent mammals, for example, reproductive and adaptive activity, metabolic rate, thermal insulation, mimicry, and camouflage, among others (Fuentes et al. 1991; Slominski et al. 2004; El Qandil et al. 2005; Guerrero et al. 2007). In its natural habitat, the adult male vizcacha captured in San Luis exhibits an annual reproductive cycle synchronized with the environmental photoperiod and modulated by the pineal gland and its main hormone, melatonin (Domínguez et al. 1987; Mohamed et al. 2000; Fuentes et al. 2003; Aguilera-Merlo et al. 2005; Perez-Romera et al. 2005; Filippa and Mohamed 2006a; Acosta et al. 2010; Chaves et al. 2012; Cruceño et al. 2012; Rosales et al. 2016; Busolini et al. 2017; Gallol et al. 2020). In San Luis, Argentina (33° 20′ south latitude, 760 m altitude), solar irradiation values reported as heliophany, precipitation, and temperature were minimum in winter (6.82 h, 11 mm and 12 °C) and maximum in summer (9.38 h, 90 mm and 19.5 °C), respectively. These average weather values were provided by the Servicio Meteorológico Nacional San Luis (www.smn.gov.ar).

The histoarchitecture of the epiphysis lacks a common pattern. Cell groups arranged in cords, follicles, and rosettes have been reported. Three cell types are distinguished in the gland parenchyma: pinealocytes, interstitial cells (IC), and pigmented cells (Vollrath 1981; Calvo et al. 1992). To a lesser extent, other cell types such as fibroblasts, macrophages, and neurons can be observed (Fig. 17.1b). It is important to highlight the presence of calcifications (cerebral grit; *corpora arenacea*) in the pineal glands, especially those that have aged (Fig. 17.1c; Beltrán 2003; Tan et al. 2018; Patel et al. 2020).

17.1.3 The Pinealocyte: King of the Pineal

As in other mammals, the pinealocyte is the most abundant parenchymal cell of the vizcacha pineal gland, comprising 75–90% of the pineal parenchyma. A clearly basophilic cytoplasm is observed on hematoxylin-eosin slides. The presence of cytoplasmic vacuoles is an occasional feature. The nucleus is large, rounded, or oval, with euchormatic and a prominent nucleolus (Busolini et al. 2017). The study of semi-thin sections stained with toluidine blue allows for a more comprehensive analysis within the gland. Most glandular parenchyma cells have a large, round, or slightly oval nucleus, with a visible nucleolus, and are associated with population I or type I (P1) pinealocytes. Some of these cells are low in density and appear clear, whereas others with higher density appear dark under a light microscope (Fig. 17.1d). In addition, some cells present irregular nuclei of variable size, with indentations, and are included within the population II or type II pinealocytes (P2; Cernuda-Cernuda et al. 2003; Gil et al. 2005). The predominant population within the gland is represented by P1. Under the electron microscope, the P1 population exhibits a lower electron density and larger cell size, which allows its identification (Fig. 17.1e). They show a large nucleus, obvious nucleolus, and peripherally arranged heterochromatin. The cytoplasm is slightly electron dense and rich in organelles, although, as described previously, it is sometimes more electron dense (Fig. 17.1f). Abundant mitochondria, well-developed smooth and granular endoplasmic reticulum, and a discrete Golgi complex are found. Multivesicular bodies, spherules, lamellar structures, phagolysosomes, pigment granules, and synaptic bars are seen with some frequency (Fig. 17.1g–j). Cytoplasmic processes of different lengths and variable numbers extend from the pinealocyte, ending in the vicinity of other pinealocytes, interstitial cells, blood vessels, or unmyelinated nerve terminals (Fig. 17.1k, l). In these processes, microtubules with small vesicles at their ends, sometimes dense in electrons, are observed (Domínguez et al. 1987; Domínguez 1990). Otherwise, the P2 population is randomly distributed and differs from P1 by the morphology of its nuclei. They present an irregular, electron-dense outline, with indentations and scattered heterochromatin. Its smaller cytoplasm, although similar to P1, exhibits several cellular processes. Numerous mitochondria, dilated endoplasmic reticulum cisterns, lysosomes of variable electron density, a well-developed Golgi complex, and some dense particles are observed (Cernuda-Cernuda et al. 2003).

Comparative electron microscopy studies between vizcacha pineal glands captured in summer and those captured in winter show seasonal variations. During the summer, the pinealocytes exhibit a large nucleus, and a defined cell outline can be observed. Its pale cytoplasm contains abundant ribosomes, either individually or in groups of polyribosomes, and numerous cisternae of smooth, flat-looking endoplasmic reticulum. The mitochondrial population is small and may be round or oval in shape with tubular cristae. Few dense bodies and opaque particles resembling glycogen inclusions are found (Dominguez et al. 1987; Dominguez 1990). On the other hand, some structural differences are displayed in winter pinealocytes. Cells appear elongated and the limiting membrane is corrugated. The nucleus is oval and voluminous and contains scattered chromatin with a visible nucleolus. The main feature of the cytoplasm is the presence of numerous dense mitochondria, mostly elongated or slightly oval in morphology. The internal membrane is invaginated into tubuliform cristae defining a dense matrix. The remaining cytoplasm has free ribosomes and dense bodies located close to the Golgi complex. This organelle is formed by abundant dilated sacs with parallel membranes that give it the classic concave-convex appearance (Dominguez et al. 1987; Dominguez 1990).

Cytological changes in vizcacha pinealocytes, mainly during winter, suggest an active metabolism that corresponds to greater cellular activity during this season. The hypothesis of increased secretory activity during this season is indirectly supported by the presence of numerous mitochondria and nearby dense bodies of a Golgi complex. These structural features are not seen in summer pinealocytes. They exhibit a pale-appearing cytoplasm, accompanied by a low population of oval mitochondria and few dense cytoplasmic bodies (Domínguez 1990). Furthermore, as has been suggested in different mammals during the short photoperiod, the pineal gland of the vizcacha can also induce sexual inactivity. This nocturnal rodent lives in deep burrows and only emerges to feed at dusk. Thus, winter darkness inhibits reproduction. During spring and early summer, the gonads appear to regenerate and glandular influence is minimal. Subsequently, during summer reproductive activity increases, because animals are exposed to a different photoperiod. In fact, during this season the vizcachas are gradually exposed to twilight, because the summer daylight phase is longer than the winter daylight phase (Domínguez et al. 1987). These changes in the light phase modifies the negative synchronization that the epiphysis exerts on the secretion of melatonin (reproductive inhibitory effect), affecting the biological clock that operates during winter. Other minor seasonal changes, such as annual differences in heliophany, precipitation, and temperature, might contribute to this effect.

17.1.4 The Interstitial Cell: A Supporting Cell or Something Else?

Using electron microscopy, IC are easily differentiated from pinealocytes by their shape, size, and electron density. They are scarce in number, and their location is close to perivascular spaces. The nucleus is elongated and irregular, with scattered chromatin and heterochromatin granules located on the nuclear envelope. They exhibit an electron-dense cytoplasm with abundant mitochondria, endoplasmic reticulum cisterns, glycogen granules, and microfilaments (Fig. 17.1m). The IC have been studied using immunohistochemistry (IHC) in several species such as rat (Møller et al. 1978; Schachner et al. 1984; López-Muñoz et al. 1992; Borregón et al. 1993; Suzuki and Kachi 1995), mouse (Schachner et al. 1984), golden hamster (Huang et al. 1984), monkey (Girod and Durand 1985), dog (Boya and Calvo 1993), cat (Boya and Calvo 1993; Boya et al. 1995), sheep (Franco et al. 1997; Redondo et al. 2001; Regodón et al. 2001), and donkey (Safwat 2012). The most commonly used markers include S-100 protein, glial fibrillary acidic protein (GFAP), and vimentin. The vertebrate central nervous system tissues are composed of several types of cells, including neural progenitors, neurons, astrocytes, oligodendrocytes, and microglia (Du et al. 2021). Expression of S-100 has been reported in these cells, as well as in other cell types including pituitary folliculostellate cells, chondrocytes, melanocytes, kidney epithelial cells, Langerhans cells, and myoblasts (Haimoto et al. 1987; Böni et al. 1997; Rambotti et al. 1999; Acosta et al. 2010; Acosta and Mohamed 2013). Intracellularly, the S-100 protein is involved in the regulation of proliferation, differentiation, inflammation and migration, energy metabolism, Ca^{2+} homeostasis, and apoptosis. These functions have been described through interactions with a variety of target proteins including receptors, cytoskeletal subunits, enzymes, transcription factors, and nucleic acids (Busolini et al. 2017). Extracellularly, the S-100 protein has an autocrine and/or paracrine function on cell proliferation, differentiation, migration, inflammation and tissue repair, and/or antimicrobial activity (Donato et al. 2009, 2013). GFAP and vimentin are members of the intermediate filament protein family Class III (Oshima 2007). GFAP is the main protein constituent of cytoskeleton intermediate filaments on mature astrocytes (Eng 1985; Eng et al. 2000). GFAP is a well-known marker of astrocytes. This protein provides structural stability to astrocytic processes, thereby modulating their shape and motility. Studies by Liedtke et al. (1996) have suggested a role for GFAP in the maintenance of normal white matter architecture and blood-brain barrier integrity. Some reports indicate that GFAP is also involved in the modulation of some neuronal functions (McCall et al. 1996; Shibuki et al. 1996). Vimentin is the main constituent of cytoskeleton intermediate filaments of immature glial cells (Dahl et al. 1981) and mesenchymal cells (Lehtonen et al. 1985). The role of organizer of critical proteins, involved in cell migration, attachment, and signaling, is carried out by vimentin. Besides maintaining the structural integrity of cells and tissues (Nieminen et al. 2006) and participating in adhesion and migration processes (Gonzales et al. 2001; Nieminen et al. 2006), vimentin is involved in signal transduction (Lopez-Egido et al. 2002; Runembert et al. 2002; Ivaska et al. 2007; Yang et al. 2005), apoptosis and immune defense (Yang et al. 2005; Schietke et al. 2006), and regulation of genomic DNA (Tolstonog et al. 2000, 2001). Immunolabeling for the S-100 protein and GFAP in male vizcachas reveals that the IC are located mainly in the proximal region, while in the distal region their presence is minimal or absent. These proteins are not detected in pinealocytes or endothelial cells. A nuclear and cytoplasmic immunolabeling pattern is observed with the S-100 protein. The cells have an irregular, stellate-like shaped body, with numerous cytoplasmic processes of great length, thickness, and irregular distribution within the parenchyma. Several of them surround the pinealocytes singly or in small groups, while others reach the blood vessels (Fig. 17.1n, o). GFAP expression is exclusively found in the IC cytoplasm and extends to its cytoplasmic processes. In this case, unlike what is observed with the S-100 protein, these processes are finer and more delicate (Fig. 17.1p, q). Some IC express colocalization for both analyzed proteins (S-100 and GFAP), while others express only one of them. This is revealed through the double IHC technique (Busolini et al. 2017). Qualitative analysis of vimentin shows a consistent expression pattern in both pineal regions because this protein not only is positive in the cytoplasmic processes of some IC but also is observed in endothelial cells and perivascular spaces. The morphometric analysis reveals that both the percentages of immunopositive area (%IA) for the S-100 protein (%IA-S-100) and GFAP (%IA-GFAP) in the proximal region are greater in rodents captured in winter compared to those caught in summer. On the contrary, both the %IA-S100 and %IA-GFAP in the distal region are lower in animals apprehend in summer if they are correlated with those sized in winter. Regarding the number of immunoreactive cells against the S-100 protein with a visible nucleus (number IC-S-100), no significant differences in the proximal region have been reported. However, there is a significant difference between the values reported in summer and winter for the distal region (Table 17.1).

Different components of the reproductive axis undergo biochemical and morphological changes during the short photoperiod. Gonadal and pituitary activities are inhibited as a result of increased pineal activity. In vizcacha, significant differences in the %IA-S-100 protein and GFAP may indicate that IC go through seasonal variations in their biochemical properties, in relation to seasonal changes in the pinealocyte activity (Busolini et al. 2017). In addition, the increased amount of

Table 17.1 Seasonal morphometric analysis of S-100 protein and GFAP in pineal gland

Parameters	Region	Summer	Winter
%IA S-100	PROXIMAL	3.44 ± 0.40	5.49 ± 0.12[a]
	DISTAL	1.45 ± 0.33	2.75 ± 0.44[b]
NUMBER IC-S-100	PROXIMAL	2.59 ± 0.22	3.31 ± 0.20
	DISTAL	1.07 ± 0.13	1.67 ± 0.06[c]
%IA-GFAP	PROXIMAL	2.80 ± 0.14	4.94 ± 0.22[d]
	DISTAL	1.33 ± 0.14	2.00 ± 0.54

Values are expressed as mean ± SEM ($n = 4$). Significant differences were determined by Kruskal Wallis test: % IA S-100: [a]$p < 0.01$; winter vs. summer; [b]$p < 0.01$; winter vs. summer. NUMBER IC-S-100: [c]$p < 0.01$; winter vs. summer. %IA GFAP: [d]$p < 0.01$; winter vs. summer. Comparison between regions showed significant differences for both proteins in each season ($p < 0.01$, Mann-Whitney test)

GFAP, a constituent of the intermediate filaments, is essential for the maintenance of cellular shape (Eng 1985; Inagaki et al. 1994). Some authors have reported a regionalized immunoexpression of these proteins in the pineal parenchyma, suggesting a relationship with the IC maturity stages (Lopez-Muñoz et al. 1992; Borregon et al. 1993; Suzuki and Kachi 1995; Safwat 2012). In the vizcacha epiphysis, IC have a regionalized distribution. A neuroectodermal origin could be demonstrated by the expression of the S-100 protein and GFAP. Moreover, the presence of few vimentin-positive cells suggests the existence of a population of immature glial cells, which probably constitutes a reserve population (Busolini et al. 2017). The existence of an extensive communication network within the gland could be evidenced by the presence of numerous long and branched cytoplasmic processes that come into contact with pinealocytes and other cell types within the pineal parenchyma. Therefore, it is likely that glandular regulation processes are carried out by these cells. Besides seasonal changes in the number, branching, and extension of the cytoplasmic processes, the increase in the number IC-S-100 during winter indicates an active proliferation of IC in relation to a higher glandular activity (Busolini et al. 2017). On the contrary, the lower glandular activity reported in summer is consistent with a decrease in the number IC-S-100. Findings over the years demonstrate that IC, despite being considered elements of support and nutrition in different species, have an active participation in the regulation of seasonal glandular activity.

17.2 Pituitary Gland

The pituitary gland (lat. Pituita: mucus) secretes different hormones that regulate, for example, reproduction, growth, and metabolism. Despite its small size, it is one of the most important organs in the body, produces at least nine hormones, and maintains reciprocal relationships with other endocrine glands. The pituitary gland establishes both neural and vascular connections with the brain, underscoring its pivotal role as an organ that serves as a mediator between the nervous and endocrine systems, the two primary integrative systems within the organism.

Pituitary is situated at the lower part of the brain, beneath the hypothalamus, to which it is connected as it extends along the diencephalon. Its location is within a small hollow in the sphenoid bone known as the sella turcica. Extending across it is a layer of protective tissue known as the dura mater, forming a diaphragm referred to as the diaphragma sellae or the sellar diaphragm. This diaphragm is also recognized as the pituitary tent. Within this structure, a central passage allows for the penetration of the pituitary stalk, also known as the pedicle, which connects the gland with the hypothalamus.

In *L. maximus*, the pituitary gland manifests as a planar anatomical structure with an approximate ovoid morphology. Within the context of mature individuals, the pituitary has average linear dimensions of 5 mm in length, 4.5 mm in width, and 1.5 mm in thickness, coupled with an estimated mass of approximately 28 mg (Mohamed 1995). Two regions can be recognized: an anterior one that is pink in color and soft in consistency, which represents the highest glandular percentage. This region constitutes the anterior lobe. The other is a rounded area, whitish, more rigid, and fibrous, which constitutes the posterior lobe. Among them is the pituitary cleft or Rathke infundibular recess, which is lined with epithelial cells and persists in adult specimens (Mohamed 1995).

The anterior lobe exhibits a well-defined and developed *pars distalis* (PD) with abundant blood supply and little connective tissue. Extensive blood vessels are found at the cephalic end, coursing throughout the parenchyma and anastomosing at the caudal end (Mohamed et al. 1995). The PD presents different regions called ventral, dorsal, medial, cephalic, and caudal. The parenchyma consists of granulated and non-granulated cells, organized in both cord-like and clustered arrangements in close proximity to well-established blood vessels. The cells are also organized, forming structures with follicular or pseudofollicular characteristics, which contain PAS-positive colloidal material (Mohamed et al. 1994, 2000). By electron microscopy, the cell types distinguished in PD are agranulated and granulated cells. Two types of agranulated cells are observed, type a and type b. The granulated cells have been classified based on their shape and size, organelle development, and characteristics of the secretory granules (Mohamed et al. 1995). Then, the hormonal composition of them enabled their identification by immunohistochemistry techniques.

In the pituitary gland, the presence of two distinct types of colloid-containing structures has been described: (1) follicles, constituted by folliculostellate cells (FSC), and (2) colloid extracellular accumulations, formed by various types of granulated cells. The follicular lumen exhibited copious electrodense colloid, along with numerous cellular debris encompassing vacuoles, altered membranes, mitochondria, and secretory granules likely indicative of the phagocytic activity of FSC (Mohamed et al. 2000). Follicles are randomly distributed in the medial region and PD extremes, preferentially in the peripheral region of the parenchyma. Their shapes vary, appearing round, ovoid, or irregularly elongated. These follicles are enveloped by one or two layers of endocrine cells and blood vessels. Additionally, certain follicles establish communication with others located in the *pars intermediate* (PI) and *pars nervosa* (PN). Besides, the number of colloidal accumulations varies during the year, showing the highest values in December and February (summer, reproductive period), but the lowest values in August (winter, gonadal regression period) (Mohamed et al. 2000; Acosta et al. 2010).

FSC are implicated in the processes of colloid secretion and absorption. They are characterized by an oval, elongated, or irregularly shaped nucleus that exhibits a relatively euchromatic nature and nuclear membrane invaginations. The cytoplasm contains considerable quantity of mitochondria, minimal granular endoplasmic reticulum, and. Extensions that come into contact with blood sinusoids and endocrine cells. In addition, junctional complexes and desmosomes connect the apical region's lateral membranes of adjacent cells. From the apical poles of FSC, a large number of long microvilli and cytoplasm processes protrude into the follicular lumen (Mohamed et al. 2000; Acosta et al. 2010).

Studies using immunohistochemical techniques and image analysis show that the proteins S-100, vimentin, and glial fibrillary acidic protein (GFAP) are expressed by FSC. These proteins are recognized, respectively, as hallmarks of glial, immature glial, and astrocyte cells. The different levels of those proteins and the immunolabeling profiles imply that FSC vary in their roles and derive from the neuroectoderm of the embryo. Besides, FSC that are positive for vimentin could be a minor subset of these cells that are not fully developed. The presence of S-100 protein in FSC and in the colloid indicates that this protein has several roles depending on where it is located. The FSC has a specific arrangement of distribution in the PD, and it is involved in the generation of an intercellular communication network for a paracrine regulation of the hormonal secretions in the adenohypophysis (Acosta et al. 2010). Only certain types of endocrine cells of the PD have a close association with the FSC and, consequently, with the follicles. The significant associations are presented by the lactotrophs, gonadotrophs, and corticotrophs (Filippa et al. 2012). This means that there probably exist between these cells and structures special mechanisms of regulation of the endocrine activity in the PD. In conclusion, FSC play pivotal and manifold roles within the pituitary gland parenchyma. These encompass crucial functions such as cellular support, phagocytosis, secretion of diverse mediators, and the establishment of a complex network of interconnected intercellular spaces (Mohamed et al. 2000; Acosta et al. 2010).

The follicular lumen is filled with a homogeneous colloidal appearing material. The size and content of the follicles vary, from small ones with little colloid to large ones with a lot of dense colloidal material. Some of the colloids have the hormones PRL (prolactin) and LH (luteinizing hormone), which might be from granulated cells that that were degraded or phagocytized by FSC. This may also explain the presence of the S-100 protein in the lumen of the follicle, which is released directly into from the FSC (Acosta et al. 2010). Studies involving the intrapituitary administration of lanthanum demonstrate its unimpeded diffusion between endocrine cells and the FSC. However, its course is interrupted beneath the junctional complexes of the FSC, impeding entry into the lumen of follicles. The lanthanum reaches endocrine cells in various stages of involution or degeneration, characterized by irregular and pycnotic nuclei. Notably, the tracer fails to permeate healthy cells. This observation suggests that endocrine cells undergoing degeneration and accessed by lanthanum may represent an incipient phase in the in the formation of colloid. The permeability to lanthanum in the PD parenchyma demonstrates the existence of an extensive and permeate network of intercommunicated intercellular spaces between the endocrine cells and FSC, and it supports the idea of a paracrine regulation between them (Acosta et al. 2010).

The PD cells exhibit a specific regionalization pattern within the parenchyma. These arrangements intricately connect with intercellular communication and vascular supply, both of which are important in the regulation of hormonal secretion dynamics. The gonadotrophs play a crucial role in the pituitary–gonadal axis, contributing significantly to the regulation of reproductive processes. Nevertheless, their activities are interconnected and not performed in isolation. Achieving pituitary homeostasis and adapting to reproductive demands require specific interactions among cell types for orchestrating these complex processes. For example, it is common to observe a proximity relationship between gonadotrophs and lactotrophs, as well as with somatotrophs, and with extensions of the FSC cytoplasm. These cellular associations vary with the environmental photoperiod probably in relation to the physiological changes of the gland (Filippa et al. 2012).

The study of each one of the populations of hormone-producing cells by immunohistochemical techniques and image analysis allows to relate the cellular activity with the variations of the environmental photoperiod.

The LH and FSH gonadotrophs are extensively localized throughout the PD, mainly in the ventromedial region. These cells are situated along the sinusoidal surface, in the interior cell cords, and near follicles. They are scattered or grouped in small clusters. The gonadotrophs are oval with an eccentric nucleus, usually located opposite the sinusoidal end of the cells (Filippa et al. 2005). The primary ultrastructural characteristics identified within this cellular subtype encompass the nucleus exhibiting finely granular chromatin and discernible nucleoli. Abundant cytoplasmic granules exhibit varying electron densities, showcasing a range in size from 250 to 750 nm, and are uniformly distributed throughout the cytoplasm. The cytoplasm is characterized by an abundance of elongated mitochondria, well-developed granular endoplasmic reticulum, and numerous Golgi vesicles (Mohamed et al. 1995).

Somatotrophs or GH cells exhibit a widespread distribution within the pituitary parenchyma. However, a small sector situated at the cephalic extremity, delineated by elongated blood vessels, is characterized by the absence of immunopositive cells. This distribution might depend on the variations in the vascularization that regulate PD endocrine activity. The somatotrophs are typically situated as constituents of the follicle in a basal position relative to the FSC, avoiding direct interaction with the colloidal lumen. They are distributed along the sinusoidal surface, occurring either in a scattered or clustered arrangement. These cells exhibit an oval, pyramidal, or round morphology, featuring a sizable nucleus that occupies the majority of the cytoplasm. The cytoplasm fine immunolabeling is more intense in cells localized at the cephalic end (Filippa and Mohamed 2006a). The electron microscopy examination reveals a centrally positioned nucleus distinguished by finely granular chromatin and prominent nucleoli. Spherical secretory granules (400–500 nm) exhibit uniform electron density and display polarization within the cytoplasm. These cells show a well-developed granular endoplasmic reticulum and Golgi apparatus (Mohamed et al. 1995).

Lactotrophs or PRL cells display considerable variation in their morphology, distribution, and function within the pituitary gland. They are primarily located in the ventromedial and caudal regions, with an intense concentration of cells near ventral areas defined by long blood vessels at the cephalic end. This irregular distribution might be attributed to blood flow dynamics and their responsiveness to secretagogues. Lactotrophs exhibit diverse shapes, ranging from oval to round, pyramidal, or cup-like, with eccentrically positioned nuclei. They are often found around blood vessels and adjacent to the colloidal lumen of follicular structures. Some lactotrophs form groups, while others are isolated, showing varied immunolabeling patterns in their cytoplasm. Additionally, they extend cytoplasmic prolongations towards blood vessels or neighboring cells (Filippa and Mohamed 2006a). Mohamed et al. (1995) described lactotrophs as having small eccentric nuclei with loosely arranged chromatin. These cells contain uniformly distributed electrodense secretory granules of varying sizes and a well-developed endoplasmic reticulum cisternae. In some cells, granules are concentrated in specific cytoplasmic regions near the Golgi complex.

The corticotrophs or ACTH cells are localized in all PD parenchyma, with a predominant concentration at the cephalic end and in the dorsal region. These cells exist either in isolation or in small clusters comprising three to five cells. Notably, they exhibit significant pleomorphism, manifesting as polygonal, round, or oval cells characterized by voluminous cytoplasm. Alternatively, some cells display a stellate morphology, featuring elongated cytoplasmic processes that typically extend towards blood vessels or encompass other neighboring endocrine cells. Certain corticotrophs, or their cytoplasmic extensions, demonstrate adjacency to the colloidal lumen, indicating a potential function of the colloid in either storing polypeptidic hormones or serving as a conduit for protein transportation. The cytoplasmic immunolabeling pattern is heterogeneous in the parenchyma. In cells situated at the cephalic end and dorsal region, secretory granules are present in both the perinuclear region and the cellular cytoplasm periphery. Conversely, cells located in the ventromedial and caudal regions display immunolabeling distributed uniformly across the cytoplasm. This observation implies the existence of distinct subpopulations that respond to diverse stressful stimuli. Thus, there could be hormone-secreting cells, others degranulated, and others in a standby situation, all of them in a different state of reactivity or in a different stage of their secretory cycle (Filippa and Mohamed 2006b). At the ultrastructural level, corticotrophs possess scant cytoplasm and exhibit underdeveloped organelles. The secretory granules are primarily situated in the peripheral regions. They display variations in size (250–350 nm) and maintain a homogeneous electron density. These granules are dispersed throughout the cellular periphery (Mohamed et al. 1995).

The thyrotrophs or TSH cells are predominantly situated in the ventromedial region and at the cephalic end of the PD. They are occasionally in the dorsal region next to Rathke's pouch. This regionalization within the parenchyma is constrained by the presence of blood vessels. Thyrotrophs are either isolated or form clusters adjacent to some follicles, creating a spatial arrangement along the surface of blood vessels. The predominant cell morphology is oval or pyramidal, with some cells featuring short cytoplasmic extensions that extend toward blood capillaries or envelop neighboring endocrine cells. These cells exhibit a spherical nucleus positioned either eccentrically or centrally. A finely distributed cytoplasmic immunolabeling, typically more evident, is present in the cellular extremity that comes into contact with the sinusoidal wall. Additionally, thyrotrophs present varying intensities of cytoplasmic immunolabeling, which may correspond to distinct cellular activity states. While certain thyrotrophs remain active in response to various stimuli, others remain in standby mode (Filippa and Mohamed 2008). The ultrastructure of these cells reveals a scarcity of elongated mitochondria and numerous secretory granules. These granules vary in size (150–300 nm) and are uniformly distributed throughout the cytoplasm, exhibiting moderate electron density (Mohamed et al. 1995) (Fig. 17.2a–o).

17.2.1 Seasonal Variations of PD Endocrine Cells

The LH and FSH gonadotrophs show seasonal changes in their arrangement, dimensions, and the %IA in male vizcachas. These cells display significant activity during distinct periods of their annual reproductive cycle, emphasizing their asynchronous secretion patterns. Besides, bihormonal gonadotrophs are present interspersed between monohormonal gonadotrophs, and their number increases in short photoperiod (Filippa et al. 2005). During the reproductive period (summer–early autumn), the LH cells exhibit heightened abundance, accompanied by an enlargement of cellular dimensions and increased %IA values compared to the gonadal regression phase (winter) (e.g., %IA 4.51 ± 0.23 and 2.48 ± 0.24, respectively). These changes in morphometric parameters reflect an elevated cellular engagement in LH biosynthesis and secretion, crucial for sustaining testicular steroidogenesis and spermatogenesis during the reproductive period (Filippa et al. 2005). This phenomenon is consistent with the peak serum LH and testosterone concentrations, alongside hypertrophic Leydig cells within the testes during this period (Fuentes et al. 1993; Muñoz et al. 1997; Filippa et al. 2015). In the reproductive and gonadal regression periods, the FSH cells are few and located mainly in the ventromedial region of PD; the values of %IA are 0.41 ± 0.06 and 1.02 ± 0.1, respectively. During the gonadal regression period, there is a notable reduction in the diameters of these cells. However, FSH cells exhibit high abundance and wide distribution in the PD during the gonadal recovery period, representing the highest %IA at 5.39 ± 0.25. This emphasizes their essential role in promoting the restoration of testicular activity in adult male vizcachas during the spring season. These findings suggest that these cells are synthesizing great quantities of the FSH hormone that stimulates Sertoli cells to initiate the spermatogenic processes (Filippa et al. 2005). The ability of the Sertoli cells to synthesize, pack, and secrete cellular products necessary in the spermatogenesis depends on the local hormone levels (Muñoz et al. 2001). The activity

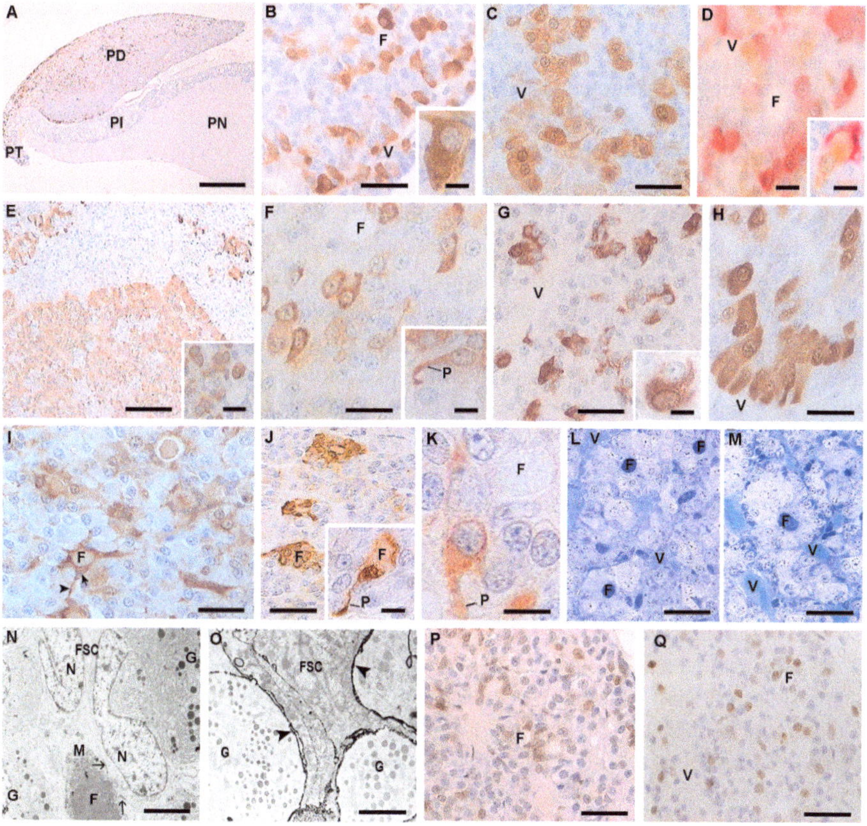

Fig. 17.2 Pituitary gland of *Lagostomus maximus.*(**a**) Histological section immunostained with anti-hFSHb. *PD* pars distalis, *PI* pars intermedia, *PN* pars nervosa. (**b**) and (**c**) LH and FSH cells, respectively. They are near follicular structures (F) and blood vessels (V). (**c**) Double immunohistochemistry image of LH-gonadotrophs (brown) and thyrotrophs (fuchsia). (**d**) LH-TSH associations are along the surfaces of blood vessels. Inset: Higher magnification of a FSC-S-100 positive cytoplasmic prolongation (fuchsia) involving a FSH-gonadotroph (brown). (**e**) In the cephalic extreme of PD, long blood vessels are delimiting a region where there are no immunopositive GH cells. Inset: Somatotrophs in basal position of the follicular structure, without contacting the colloidal lumen. Pleomorphism of lactotrophs (**f**) and corticotrophs (**g**) some of them with long cytoplasmic processes, extending towards blood vessels or surrounding non-immunolabeled cells. (**h**) Thyrotrophs forming fences along the blood vessels surface. (**i–j**) and inset: Immunostaining for S-100 in the follicular colloid and FSC. These cells show immunostaining in the nucleus and cytoplasm, and exhibit cytoplasmic processes (P). (**k**) Scarce FSC are immunostained with anti-GFAP in the cytoplasm around the nucleus. (**l–m**) Granulated cells are arranged around follicular structures, forming one or two cell layers, which are surrounded by blood vessels. Sections stained with Toluidine blue. (**n**) Electron micrographs of PD vizcachas. The FSC form follicles (F) have irregular nuclei (N), scarce cytoplasm with short cytoplasmic process (P) and absence of secretory granules. The different granulated cellular types (G) are closely associated with FSC, but without contacting the follicular lumen. Well developed junctional complex (arrows) are observed at the apical lateral surface of the FSC. These cells exhibit long microvilli (M) that protrude into the follicular lumen (F). (**o**) Transmission electron microscopy with lanthanum. The tracer freely penetrates into the intercellular spaces (arrowheads) between granulated endocrine cells (G) and FSC. (**p**) Numerous PCNA-ir cells during the reproductive period. (**q**) AR are expressed mainly in the nuclei of PD cells. Scale bar: (**a**) 500µm; (**b–d**), (**f–j**), (**l–m**), (**p–q**), and Inset of (**e**) 25 µm; (**e**) 100 μm; (**k**) and Inset of (**b, d, f, g, j**) 10µm; (**n**) 4µm; (**o**) 2µm. License Number 5635341352656 (**a** and **d**)

of both gonadotrophs fluctuates throughout the annual reproductive cycle, delineating distinct functions of FSH and LH in testicular steroidogenesis and spermatogenesis. These variations likely arise from intricate regulatory mechanisms involving the influence of melatonin, GnRH, GnIH, and testosterone. Moreover, peptides like activin, inhibin, and follistatin would be involved in complex autocrine and paracrine mechanisms that regulate the gonadotroph activity (Filippa et al. 2005).

The morphometric parameters of somatotrophs exhibit their peak values during the reproductive phase in comparison to the gonadal regression phase (e.g., %IA 12.66 ± 0.53 and 9.12 ± 0.31, respectively). These fluctuations may be attributed to the possibility that these cells store insufficient hormone quantities for their detectability or are in a state of standby within their secretory cycle. Furthermore, a notable rise in the morphometric parameters is observed during the gonadal recovery period, with the presence of immunolabeling within the intercellular space suggesting an augmented growth hormone secretion at this moment. As such, the somatotroph parameters undergo changes throughout the annual reproductive cycle of adult male vizcachas, likely influenced by shifts in gonadal activity (Filippa and Mohamed 2006a).

The arrangement of irregularly shaped groups of lactotrophs is frequently observed during the reproductive and gonadal recovery periods, while in the gonadal regression period, cells are primarily isolated or form small clusters. The %IA and the quantity of PRL cells experience a decline during the gonadal regression period (%IA 1.23 ± 0.13) compared to values in the reproductive period (%IA 2.62 ± 0.16), followed by an increase during the recovery period (%IA 2.81 ± 0.23). Lactotrophs exhibit diminished cellular activity during the gonadal regression phase, indicative of a reduced hormone reservoir within PD lactotrophs due to decreased hormonal synthesis. These findings suggest the possible presence of distinct subpopulations of lactotrophs that respond to varying stimuli or a division within lactotrophs, encompassing secreting cells and those in a standby state, exhibiting different levels of reactivity or stages of the secretory cycle (Filippa and Mohamed 2010).

The pituitary–adrenal cortex system is an integral part of the general adaptation mechanism observed in numerous animals under challenging environmental circumstances. During winter, vizcachas live under certain environmental conditions such as short photoperiod, low temperature, hydric restriction, and food variations (Bontti et al. 1999; Branch et al. 1994). These stressors are responsible for the increase in the glucocorticoids synthesis manifested by the high serum levels of corticosterone found in winter in adult male (Ribes et al. 1999). There are seasonal variations in the %IA and cellular size of the corticotrophs. During summer, the values reach their peak (%IA 3.23 ± 0.25), while in early winter, they reach the lowest levels (%IA 1.86 ± 0.16). The %IA increase during spring (3.41 ± 0.17%), indicating that the degranulated corticotrophs initiate the synthesis and storage of secretory granules containing ACTH (Filippa and Mohamed 2006b). This demonstrates that ACTH cells undergo morphological modifications in response to seasonal environmental conditions to contribute to the adaptive adjustments of vizcachas (Filippa and Mohamed 2006b). These cells persist in a state of total or partial degranulation post-ACTH secretion during winter and then begin to synthesize and release this hormone, which exerts its effects on the adrenal cortex to ultimately achieve the metabolic adjustments necessary for survival in an adverse environment. The intrapituitary regulation of corticotroph activity could be related to the elevated levels of melatonin determined during the winter months. The pineal hormone might inhibit the synthesis of ACTH, probably by affecting some secretagogue of this hypophysial hormone (Filippa and Mohamed 2006b).

The thyrotrophs make up a minor fraction of the cells in the PD. The values of morphometric parameters change throughout the year, %IA, and the number of thyrotrophs reaches their maximum in the last months of the reproductive period (%IA 3.35 ± 0.09). Subsequently, there is a substantial reduction during the gonadal regression period (%IA 1.43 ± 0.15), followed by an increase at the beginning of the gonadal recovery period (%IA 2.98 ± 0.21) (Filippa and Mohamed 2008). Hence, the seasonal fluctuations in thyrotroph activity can be directly associated with the modulation of specific secretory functions. In the summer (reproductive period), an elevated TSH storage in the pars distalis (PD) may result from either reduced hormonal secretion due to the negative feedback impact of thyroid hormones or a stimulatory influence on TSH synthesis exerted by testicular androgens. Morphometric parameters exhibit a decline in winter (gonadal regression period) and recover during the gonadal recovery period (spring), indicative of cellular activity being influenced by environmental cues that ultimately modulate the reactivation of testicular steroidogenic and spermatogenic processes (Filippa and Mohamed 2008). Hence, the challenges imposed by harsh winter environmental factors, including reduced photoperiod, cold temperatures, constraints in water and food availability, and modifications in social interactions, are likely to influence the functional dynamics of diverse cellular PD populations. These alterations manifest in concordance with the histophysiological transformations observed across the interrelated pineal–pituitary–gonadal, pineal–pituitary– adrenal, and pineal–pituitary–thyroid axes, as elucidated throughout the reproductive cycle of this rodent species.

17.2.2 Melatonin Effects

Given the pronounced seasonal fluctuations in the cellular compositions of the pituitary PD as a response to dynamic shifts in environmental conditions, particularly the photoperiodic cues, which subsequently induce variations in the physiological serum melatonin levels across different periods of the year, comprehensive investigations were undertaken involving the administration of melatonin in the vizcacha species. These studies were designed to elucidate the effects of this indol on the diverse cellular populations within the PD. In vizcacha PD, the melatonin administration induces a reduction in the number of colloidal accumulations. Therefore, it is possible to directly relate the heliophany values with the variations of the colloid deposits and suggest that pineal hormone mediates the effect of day length on PD follicles (Mohamed et al. 2000).

The administration of the pineal hormone results in a reduction in the dimensions of both gonadotrophs, along with a decrease solely in the %IA of LH gonadotrophs. This observation highlights the distinct effects of melatonin on the activity of both gonadotrophs, suggesting that melatonin may serve as the mediator of central photoperiodic control over LH cells by inhibiting cellular activity through the regulation of hypothalamic factors (Filippa et al. 2005).

After melatonin administration, PRL cells are predominantly isolated within the PD parenchyma. These cells with cytoplasmic prolongations are scarce, resulting in reduced cytoplasmic immunolabeling intensity and substantial morphometric parameter decrease. These morphological changes collectively indicate a decreased concentration of detectable PRL hormone within the pituitary gland. Hence, it is conceivable that the pineal hormone has suppressed PRL hormone production and secretion from these specific endocrine cells (Filippa and Mohamed 2010). Chronic melatonin treatment results in a reduction in the % IA of corticotrophs, implying a potential direct or indirect influence of the indol on cellular activity (Filippa and Mohamed 2006b). However, unlike the observed alterations in other morphometric parameters in the seasonal study, such changes are not apparent in this context. This distinction may be attributed to captivity-induced stress experienced by these wild rodent species, affecting both the experimental and control groups, as discussed by Filippa and Mohamed (2006b).

The administration of melatonin does not lead to morphometric parameter changes in somatotrophs and thyrotrophs. This lack of response is consistent with the idea that modifications observed in these cells during the reproductive cycle of adult male vizcachas are probably influenced by fluctuations in gonadal or thyroid gland activity, rather than being directly impacted by melatonin (Filippa and Mohamed 2006a, 2008). Finally, the findings of these studies effectively demonstrated that melatonin exerts a distinct and differential modulatory influence on the activities of synthesis and secretion within these cellular ensembles. This substantiates the role of melatonin as a key mediator in orchestrating the intricate interactions within the pituitary gland and sheds light on its central role in translating external environmental signals into hormonal responses that dynamically regulate cellular behavior and function.

17.2.3 Cell Proliferation

The proliferating cell nuclear antigen (PCNA) is a non-histone nuclear protein with a molecular weight of 36kD. It exhibits a high degree of conservation across various species and serves as a co-factor for DNA polymerase-delta, playing a crucial role in the initiation of cellular proliferation. Elevated expression of PCNA appears in the nucleus during late G1 phase before the onset of DNA synthesis, becomes maximal during S-phase, and declines during the G2 and M phases (Oishi et al. 1993). The usefulness of PCNA as a marker of pituitary proliferating cells has been supported by several studies (Lim et al. 2019; García-Barrado et al. 2017).

The proliferation of the pituitary cellular populations is regulated by different hypothalamic factors, hormones, and growth factors. In vizcacha PD, the proliferation was studied through the expression of PCNA by immunohistochemistry. The PCNA cells are primarily located at the caudal end, as well as in the ventral and dorsal regions. The immunolabeled nuclei frequently display a spherical morphology, while others present an oval shape. These nuclei are situated in proximity to follicular structures, indicating their potential involvement within these follicles. This suggests the likelihood of these structures exerting a regulatory impact on cellular proliferation (Rosales et al. 2016). During the annual reproductive cycle, there are variations in the PCNA expression in PD. The number of PCNA cells is highest in the reproductive period (1.40 ± 0.16%) compared to the gonadal regression (0.52 ± 0.06%) and gonadal recovery (0.36 ± 0.01%) periods. Moreover, there are noticeable changes in the distribution of PCNA-positive cells within the PD parenchyma across different seasons. These cells are particularly abundant at the caudal end and in the ventral region during the reproductive period. In these specific regions of the PD, androgen receptors (AR) are also found in abundance (Rosales et al. 2016). This suggests the possibility of androgens undergoing aromatization into estrogens, potentially stimulating the proliferation of certain cellular types. As a result, the seasonal fluctuations in cellular proliferation might be differentially regulated by both androgens and melatonin, contributing to the dynamic variations in hormonal secretion observed throughout the annual reproductive cycle (Rosales et al. 2016) (Fig. 17.2p).

17.2.4 Androgen Receptor

The investigation of androgen receptor (AR) expression within the pituitary gland of the vizcacha assumes a significant role in enhancing our comprehension of regulatory mechanisms, particularly due to the involvement of androgens in the reproductive functions of this seasonal breeder. The pituitary is an androgen-responsive tissue, and its potential to serve as a site for androgen-mediated feedback on hormone secretion has been extensively documented (Spady et al. 2004).

AR plays a central role within the pituitary gland, regulating responses to shifts in hormonal levels. AR-ir (immunoreactive: ir) cells are distributed across the entire PD, and immunochemical findings reveal their presence in the nuclei (ARn-ir) and in the cytoplasmic (ARc-ir). Throughout the reproductive cycle, AR expression predominantly localizes to the nucleus. The percentage of ARn-ir cells decreases during the gonadal regression period, demonstrating a direct correlation with low testosterone serum concentration (Filippa et al. 2015). In addition, changes in the regionalization and immunolabeling pattern of AR expression in PD occur throughout the annual reproductive cycle. During the reproductive period, numerous immunopositive nuclei are situated at the caudal end and dorsal region, as well as along the dorsal and ventral PD borders. In the medial region, ARc-ir cells are predominantly located. The cytoplasmic immunolabeling pattern is generally uniform and occasionally intense. During the gonadal regression period, there are more ARn-ir cells in the medial region. During the gonadal recovery period, there is an augmentation of ARn-ir cells across the PD parenchyma, particularly concentrated at the caudal end. The Arc-ir cells exhibit intense cytoplasms compared to the gonadal regression period (Filippa et al. 2015). These seasonal variations in the quantity of pituitary AR-ir cells throughout the annual reproductive cycle may be attributed to fluctuations in hormonal supply through blood vessels. Peak values of AR expression are observed during the reproductive period, while minimum values coincide with the regression period, in concordance with testosterone serum levels. The AR plays a regulatory role in the activity of pituitary cells responsible for maintaining secondary male characteristics and initiating and progressing spermatogenesis. Thus, variations in the androgen serum concentration and their receptors in different tissues are responsible for biological effects throughout the life cycle of adult animals (Filippa et al. 2015) (Fig. 17.2q).

17.3 Conclusions

The pineal and pituitary glands have a remarkable ability to dynamically adapt their morphology in response to fluctuating environmental conditions. Through these seasonal morphological changes, these glands finely regulate hormone production and release, optimizing reproductive strategies and essential adaptive behaviors crucial for the vizcacha survival and successful reproduction. In conclusion, these endocrine glands, which serve as central components of the intricate neuroendocrine control system, exemplify the remarkable synchronization of glandular activity with essential environmental cues.

References

Acosta M, Mohamed F (2013) Immunohistochemical and morphometric study of pituitary pars distalis folliculostellate cells of nonpregnant and pregnant viscachas. Biotech Histochem 88(3–4):161–169. https://doi.org/10.3109/10520295.2012.750737. Epub 2012 Dec 17. PMID: 23244236

Acosta M, Filippa V, Mohamed F (2010) Folliculostellate cells in pituitary pars distalis of male viscacha: immunohistochemical, morphometric and ultrastructural study. Eur J Histochem 54(1):e1. https://doi.org/10.4081/ejh.2010.e1. PMID: 20353904; PMCID: PMC3167288

Aguilera-Merlo C, Muñoz E, Dominguez S, Scardapane L, Piezzi R (2005) Epididymis of viscacha (*Lagostomus maximus maximus*): morphological changes during the annual reproductive cycle. Anat Rec A Discov Mol Cell Evol Biol 282(1):83–92. https://doi.org/10.1002/ar.a.20144. PMID: 15622516

Arendt J (1986) Role of the pineal gland and melatonin in seasonal reproductive function in mammals. Oxf Rev Reprod Biol 8:266–320. PMID: 3540805

Arendt J, Aulinas A (2022) Physiology of the pineal gland and melatonin. In: Feingold KR, Anawalt B, Blackman MR, Boyce A, Chrousos G, Corpas E, de Herder WW, Dhatariya K,

Dungan K, Hofland J, Kalra S, Kaltsas G, Kapoor N, Koch C, Kopp P, Korbonits M, Kovacs CS, Kuohung W, Laferrère B, Levy M, McGee EA, McLachlan R, New M, Purnell J, Sahay R, Shah AS, Singer F, Sperling MA, Stratakis CA, Trence DL, Wilson DP (eds) Endotext [Internet]. MDText.com, Inc, South Dartmouth. 2000. PMID: 31841296

Bell-Pedersen D, Cassone VM, Earnest DJ, Golden SS, Hardin PE, Thomas TL, Zoran MJ (2005) Circadian rhythms from multiple oscillators: lessons from diverse organisms. Nat Rev Genet 6(7):544–556. https://doi.org/10.1038/nrg1633. PMID: 15951747; PMCID: PMC2735866

Beltrán D (2003) Estudio comparativo de la organización estructural y funcional de la pineal en la especie ovina. Departamento de Ciencias Morfológicas, Universidad de Valencia, Valencia

Böni R, Burg G, Doguoglu A, Ilg EC, Schäfer BW, Müller B, Heizmann CW (1997) Immunohistochemical localization of the Ca2+ binding S100 proteins in normal human skin and melanocytic lesions. Br J Dermatol 137(1):39–43. PMID: 9274623

Bontti EE, Boo RM, Lindstrom LI, Elia OR (1999) Botanical composition of cattle and vizcacha diets in central Argentina. J Range Manag 52:370–377

Borregón A, Boya J, Calvo JL, López-Muñoz F (1993) Immunohistochemical study of the pineal glial cells in the postnatal development of the rat pineal gland. J Pineal Res 14(2):78–83. https://doi.org/10.1111/j.1600-079x.1993.tb00489.x. PMID: 8320635

Boya J, Calvo JL (1993) Immunohistochemical study of the pineal astrocytes in the postnatal development of the cat and dog pineal gland. J Pineal Res 15(1):13–20. https://doi.org/10.1111/j.1600-079x.1993.tb00504.x. PMID: 8229641

Boya J, Calvo JL, Rancaño D (1995) Structure of the pineal gland in the adult cat. J Pineal Res 18(2):112–118. https://doi.org/10.1111/j.1600-079x.1995.tb00148.x. PMID: 7629690

Branch LC, Villarreal D, Pelliza Shriller A, Sosa RA (1994) Diet selection of plains vizcacha (Lagostomus maximus, Family Chinchillidae) in relation to resource abundance in semi-arid scrub. Can J Zool 72:2210–2216

Busolini FI, Rosales GJ, Filippa VP, Mohamed FH (2017) A seasonal and age-related study of interstitial cells in the pineal gland of male viscacha (*Lagostomus maximus maximus*). Anat Rec (Hoboken) 300(10):1847–1857. https://doi.org/10.1002/ar.23621. Epub 2017 Jun 27. PMID: 28622452

Calvo JL, Boya J, García-Mauriño JE, Rancaño D (1992) Presence of melanin in the cat pineal gland. Acta Anat (Basel) 145(1):73–78. https://doi.org/10.1159/000147345. PMID: 1414215

Cernuda-Cernuda R, Piezzi RS, Domínguez S, Alvarez-Uría M (2003) Cell populations in the pineal gland of the viscacha (*Lagostomus maximus*). Seasonal variations. Histol Histopathol 18(3):827–836. https://doi.org/10.14670/HH-18.827. PMID: 12792895

Chaves EM, Aguilera-Merlo C, Cruceño A, Fogal T, Piezzi R, Scardapane L, Domínguez S (2012) Seasonal morphological variations and age-related changes of the seminal vesicle of viscacha (*Lagostomus maximus maximus*): an ultrastructural and immunohistochemical study. Anat Rec (Hoboken) 295(5):886–895. https://doi.org/10.1002/ar.22434. Epub 2012 Mar 2. PMID: 22383422

Cruceño AM, de Rosas JC, Fóscolo M, Chaves EM, Scardapane L, Dominguez S, Aguilera-Merlo C (2012) Effects of photoperiod on epididymal and sperm morphology in a wild rodent, the viscacha (*Lagostomus maximus maximus*). ISRN Anat 2013:128921. https://doi.org/10.5402/2013/128921. PMID: 25969823; PMCID: PMC4403560

Dahl D, Rueger DC, Bignami A, Weber K, Osborn M (1981) Vimentin, the 57 000 molecular weight protein of fibroblast filaments, is the major cytoskeletal component in immature glia. Eur J Cell Biol 24(2):191–196. PMID: 7285936

Darwin C (1859) On the origin of species (On the origin of species by means of natural selection, or the preservation of favoured races in the struggle for life). J. Murray, London

Domínguez Nora Susana (1990) Aspectos morfológicos y ultraestructurales de la glándula pineal del *Lagostomus maximus maximus*. Director: Dr. Ramón Salvador Piezzi. Tesis doctoral para optar al grado de Doctora en Bioquímica. Universidad Nacional de San Luis, Argentina

Domínguez S, Piezzi RS, Scardapane L, Guzmán JA (1987) A light and electron microscopic study of the pineal gland of the viscacha (*Lagostomus maximus maximus*). J Pineal Res 4(2):211–219. https://doi.org/10.1111/j.1600-079x.1987.tb00858.x. PMID: 3598856

Donato R, Sorci G, Riuzzi F, Arcuri C, Bianchi R, Brozzi F, Tubaro C, Giambanco I (2009) S100B's double life: intracellular regulator and extracellular signal. Biochim Biophys Acta 1793(6):1008–1022. https://doi.org/10.1016/j.bbamcr.2008.11.009. Epub 2008 Dec 7. PMID: 19110011

Donato R, Cannon BR, Sorci G, Riuzzi F, Hsu K, Weber DJ, Geczy CL (2013) Functions of S100 proteins. Curr Mol Med 13(1):24–57. PMID: 22834835; PMCID: PMC3707951

Du J, Yi M, Zhou F, He W, Yang A, Qiu M, Huang H (2021) S100B is selectively expressed by gray matter protoplasmic astrocytes and myelinating oligodendrocytes in the developing CNS. Mol Brain 14(1):154. https://doi.org/10.1186/s13041-021-00865-9. PMID: 34615523; PMCID: PMC8496084

El Qandil S, Chakir J, El Moussaouiti R, Oukouchoud R, Rami N, Benjelloun WA, Lakhdar-Ghazal N (2005) Role of the pineal gland and melatonin in the photoperiodic control of hypothalamic gonadotropin-releasing hormone in the male jerboa (*Jaculus orientalis*), a desert rodent. Brain Res Bull 64(5):371–380. https://doi.org/10.1016/j.brainresbull.2004.06.010. PMID: 15607825

Eng LF (1985) Glial fibrillary acidic protein (GFAP): the major protein of glial intermediate filaments in differentiated astrocytes. J Neuroimmunol 8(4–6):203–214. https://doi.org/10.1016/s0165-5728(85)80063-1. PMID: 2409105

Eng LF, Ghirnikar RS, Lee YL (2000) Glial fibrillary acidic protein: GFAP-thirty-one years (1969-2000). Neurochem Res 25(9–10):1439–1451. https://doi.org/10.1023/a:1007677003387. PMID: 11059815

Filippa V, Mohamed F (2006a) ACTH cells of pituitary pars distalis of viscacha (*Lagostomus maximus maximus*): immunohistochemical study in relation to season, sex, and growth. Gen Comp Endocrinol 146(3):217–225. https://doi.org/10.1016/j.ygcen.2005.11.012. Epub 2006 Feb 8. PMID: 16466724

Filippa V, Mohamed F (2006b) Immunohistochemical study of somatotrophs in pituitary pars distalis of male viscacha (*Lagostomus maximus maximus*) in relation to the gonadal activity. Cells Tissues Organs 184(3–4):188–197. https://doi.org/10.1159/000099626. PMID: 17409745

Filippa V, Mohamed F (2008) Immunohistochemical and morphometric study of pituitary pars distalis thyrotrophs of male viscacha (*Lagostomus maximus maximus*): seasonal variations and effect of melatonin and castration. Anat Rec (Hoboken) 291(4):400–409. https://doi.org/10.1002/ar.20671. PMID: 18286616

Filippa V, Mohamed F (2010) Morphological and morphometric changes of pituitary lactotrophs of viscacha *(Lagostomus maximus maximus)* in relation to reproductive cycle, age, and sex. Anat Rec (Hoboken) 293(1):150–161. https://doi.org/10.1002/ar.21013. PMID: 20039438

Filippa V, Penissi A, Mohamed F (2005) Seasonal variations of gonadotropins in the pars distalis male viscacha pituitary. Effect of chronic melatonin treatment. Eur J Histochem 49(3):291–300. https://doi.org/10.4081/956. PMID: 16216815

Filippa V, Acosta M, Mohamed F (2012) Cellular associations of pituitary gonadotrophs in a rodent *(Lagostomus maximus maximus)* with photoperiod-dependent reproduction. Tissue Cell 44(6):351–357. https://doi.org/10.1016/j.tice.2012.05.003. Epub 2012 Jun 28. PMID: 22749373

Filippa VP, Rosales GJ, Cruceño AA, Mohamed FH (2015) Androgen receptors expression in pituitary of male viscacha in relation to growth and reproductive cycle. Int. J Endocrinol 2015:168047. https://doi.org/10.1155/2015/168047. Epub 2015 Apr 7. PMID: 25945090; PMCID: PMC4405020

Forcada F (2000) El papel del fotoperiodo y la melatonina en la actividad reproductora. Monografíaovis n° 71:1–86

Franco A, Regodon S, Masot AJ, Redondo E (1997) A combined immunohistochemical and electron microscopic study of the second cell type in the developing sheep pineal gland. J Pineal Res 22(3):130–136. https://doi.org/10.1111/j.1600-079x.1997.tb00314.x. PMID: 9213266

Fuentes LB, Caravaca N, Pelzer LE, Scardapane LA, Piezzi RS, Guzmán JA (1991) Seasonal variations in the testis and epididymis of vizcacha (*Lagostomus maximus maximus*). Biol Reprod 45(3):493–497. https://doi.org/10.1095/biolreprod45.3.493. PMID: 1782299

Fuentes LB, Calvo JC, Charreau EH, Guzmán JA (1993) Seasonal variations in testicular LH, FSH, and PRL receptors; in vitro testosterone production; and serum testosterone concentration in adult male vizcacha (*Lagostomus maximus maximus*). Gen Comp Endocrinol 90(2):133–141. https://doi.org/10.1006/gcen.1993.1068. PMID: 8391497

Fuentes LB, Møller M, Muñoz E, Calderón C, Pelzer L (2003) Seasonal variations in the expression of the mRNA encoding $\beta 1$-adrenoceptor and AA-NAT enzyme, and in the AA-NAT activity in the pineal gland of vizcacha (*Lagostomus maximus maximus*) – correlation with serum melatonin. Biol Rhythm Res 34(2):193–206. https://doi.org/10.1076/brhm.34.2.193.14488

Gallol LE, Busolini FI, Mohamed FH (2020) Influence of melatonin and sexual hormones on the expression of proliferating cell nuclear antigen in the adrenal cortex of a seasonal breeder (*Lagostomus maximus*). Anat Rec (Hoboken) 303(12):3052–3067. https://doi.org/10.1002/ar.24457. Epub 2020 Jun 18. PMID: 32445505

Ganguly S, Coon SL, Klein DC (2002) Control of melatonin synthesis in the mammalian pineal gland: the critical role of serotonin acetylation. Cell Tissue Res 309(1):127–137. https://doi.org/10.1007/s00441-002-0579-y. Epub 2002 May 29. PMID: 12111543

García-Barrado MJ, Blanco EJ, Iglesias-Osma MC, Carretero-Hernández M, Catalano-Iniesta L, Sanchez-Robledo V, Carretero M, Herrero JJ, Carrero S, Carretero J (2017) Relation among aromatase P450 and tumoral growth in human Prolactinomas. Int J Mol Sci 18(11):2299. https://doi.org/10.3390/ijms18112299. PMID: 29104246; PMCID: PMC5713269

Gehring WJ (2005) New perspectives on eye development and the evolution of eyes and photoreceptors. J Hered 96(3):171–184. https://doi.org/10.1093/jhered/esi027. Epub 2005 Jan 13. PMID: 15653558

Gheban BA, Rosca IA, Crisan M (2019) The morphological and functional characteristics of the pineal gland. Med Pharm Rep 92(3):226–234. https://doi.org/10.15386/mpr-1235. Epub 2019 Jul 31. PMID: 31460502; PMCID: PMC6709953

Gil E, Calderón C, Pelzer L, Domínguez S, Fogal T, Scardapane L, Piezzi RS (2005) Morphological and biochemical study of the pineal gland of pregnant and non-pregnant female vizcachas (*Lagostomus maximus maximus*). Neuro Endocrinol Lett 26(5):575–580. PMID: 16264398

Girod C, Durand N (1985) Immunocytochemical localization of S-100 protein in interstitial cells of the monkey *Macaca irus* pineal gland. Neurosci Lett 57(2):181–184. https://doi.org/10.1016/0304-3940(85)90060-6. PMID: 3897913

Gonzales M, Weksler B, Tsuruta D, Goldman RD, Yoon KJ, Hopkinson SB, Flitney FW, Jones JC (2001) Structure and function of a vimentin-associated matrix adhesion in endothelial cells. Mol Biol Cell 12(1):85–100. https://doi.org/10.1091/mbc.12.1.85. PMID: 11160825; PMCID: PMC30570

Guerrero JM, Carrillo-Vico A, Lardone PJ (2007) La Melatonina. Investigación y Ciencia 373:30–38

Guzmán JA, Pelzer LE, Piezzi RS, Scardapane L, Dominguez S (1980) Hydroxyindole-O-methyltransferase (HIOMT), catechol-O-methyltransferase (COMT) and histamine-N-methyltransferase (HNMT) in the pineal gland of the vizcacha (*Lagostomus maximus maximus*). J Neural Transm 47(1):23–28. https://doi.org/10.1007/BF01256636. PMID: 7359120

Haimoto H, Hosoda S, Kato K (1987) Differential distribution of immunoreactive S100-alpha and S100-beta proteins in normal nonnervous human tissues. Lab Investig 57(5):489–498. PMID: 3316838

Huang SK, Nobiling R, Schachner M, Taugner R (1984) Interstitial and parenchymal cells in the pineal gland of the golden hamster. A combined thin-section, freeze-fracture and immunofluorescence study. Cell Tissue Res 235(2):327–337. https://doi.org/10.1007/BF00217857. PMID: 6367998

Inagaki M, Nakamura Y, Takeda M, Nishimura T, Inagaki N (1994) Glial fibrillary acidic protein: dynamic property and regulation by phosphorylation. Brain Pathol 4(3):239–243. https://doi.org/10.1111/j.1750-3639.1994.tb00839.x. PMID: 7952265

Ivaska J, Pallari HM, Nevo J, Eriksson JE (2007) Novel functions of vimentin in cell adhesion, migration, and signaling. Exp Cell Res 313(10):2050–2062. https://doi.org/10.1016/j.yexcr.2007.03.040. Epub 2007 Apr 14. PMID: 17512929

Kappers JA (1979) Short history of pineal discovery and research. Prog Brain Res 52:3–22. https://doi.org/10.1016/S0079-6123(08)62908-8. PMID: 398533
Knudsen EI (2020) Evolution of neural processing for visual perception in vertebrates. J Comp Neurol 528(17):2888–2901. https://doi.org/10.1002/cne.24871. Epub 2020 Feb 13. PMID: 32003466; PMCID: PMC7586818
Kumar V, Sharma A (2018) Common features of circadian timekeeping in diverse organisms. Curr Opin Physio 5:58–67. https://doi.org/10.1016/j.cophys.2018.07.004
Larsen PJ, Enquist LW, Card JP (1998) Characterization of the multisynaptic neuronal control of the rat pineal gland using viral transneuronal tracing. Eur J Neurosci 10(1):128–145. https://doi.org/10.1046/j.1460-9568.1998.00003.x. PMID: 9753120
Lehtonen E, Virtanen I, Saxén L (1985) Reorganization of intermediate filament cytoskeleton in induced metanephric mesenchyme cells is independent of tubule morphogenesis. Dev Biol 108(2):481–490. https://doi.org/10.1016/0012-1606(85)90051-x. PMID: 2416612
Lerner AB, Case JD, Takahashi Y, Lee TH, Mori W (1958) Isolation of melatonin, the pineal gland factor that lightens melanocytes. J Am Chem Soc 80:2537. https://doi.org/10.1021/ja01543a060
Lerner AB, Case JD, Heinzelman RV (1959) Structure of melatonin. J Am Chem Soc 81:6084–6085. https://doi.org/10.1021/ja01531a060
Liedtke W, Edelmann W, Bieri PL, Chiu FC, Cowan NJ, Kucherlapati R, Raine CS (1996) GFAP is necessary for the integrity of CNS white matter architecture and long-term maintenance of myelination. Neuron 17(4):607–615. https://doi.org/10.1016/s0896-6273(00)80194-4. PMID: 8893019
Lim JS, Eom YW, Lee ES, Kwon HJ, Kwon JY, Choi J, Chung CH, Jo YS, Lee EJ (2019) Effects of oxytocin on cell proliferation in a corticotroph adenoma cell line. Endocrinol Metab (Seoul) 34(3):302–313. https://doi.org/10.3803/EnM.2019.34.3.302. PMID: 31565883; PMCID: PMC6769335
Lopez-Egido J, Cunningham J, Berg M, Oberg K, Bongcam-Rudloff E, Gobl A (2002) Menin's interaction with glial fibrillary acidic protein and vimentin suggests a role for the intermediate filament network in regulating menin activity. Exp Cell Res 278(2):175–183. https://doi.org/10.1006/excr.2002.5575. PMID: 12169273
López-Muñoz F, Álamo C (2000) El Tratado del Hombre: Interpretación cartesiana de la neurofisiología del dolor. Asclepio Rev Hist Med Ciencia 52:239–267
López-Muñoz F, Boya J (1992) El papel de la glándula pineal en la doctrina psicofisiológica cartesiana [The role of the pineal gland in cartesianpsychophysiological doctrine]. Acta Physiol Pharmacol Ther Latinoam 42(4):205–216. Spanish. PMID: 1343979
López-Muñoz F, Calvo JL, Boya J, Carbonell AL (1992) Coexpression of vimentin and glial fibrillary acidic protein in glial cells of the adult rat pineal gland. J Pineal Res 12(4):145–148. https://doi.org/10.1111/j.1600-079x.1992.tb00041.x. PMID: 1403607
McCall MA, Gregg RG, Behringer RR, Brenner M, Delaney CL, Galbreath EJ, Zhang CL, Pearce RA, Chiu SY, Messing A (1996) Targeted deletion in astrocyte intermediate filament (Gfap) alters neuronal physiology. Proc Natl Acad Sci U S A 93(13):6361–6366. https://doi.org/10.1073/pnas.93.13.6361. PMID: 8692820; PMCID: PMC39027
Meléndez-Fernández OH, Liu JA, Nelson RJ (2023) Circadian rhythms disrupted by light at night and mistimed food intake alter hormonal rhythms and metabolism. Int J Mol Sci 24(4):3392. https://doi.org/10.3390/ijms24043392. PMID: 36834801; PMCID: PMC9963929
Mohamed F (1995) Tesis Doctoral: "Aspectos morfológicos y ultraestructurales de la parsdistalis hipofisaria del Lagostomus maximus maximus". Biblioteca Central de la Universidad Nacional de San Luis, Argentina
Mohamed F, Fogal T, Dominguez S, Scardapane L, Guzmán J, Piezzi RS (1994) Seasonal variations and ultrastructure of colloid accumulation in the pituitary pars distalis of viscacha (L.m.m.). Biocell 18:23–31
Mohamed F, Fogal T, Dominguez S, Scardapane L, Guzmán J, Piezzi RS (1995) Ultrastructure of pars distalis of viscacha (*Lagostomus maximus maximus*). Pituitary cellular types. Biocell 19:133–140

Mohamed F, Fogal T, Dominguez S, Scardapane L, Guzmán J, Piezzi RS (2000) Colloid in the pituitary pars distalis of viscacha (*Lagostomus maximus maximus*): ultrastructure and occurrence in relation to season, sex, and growth. Anat Rec 258(3):252–261. https://doi.org/10.1002/(SICI)1097-0185(20000301)258:3<252::AID-AR4>3.0.CO;2-D. PMID: 10705345

Møller M (1974) The ultrastructure of the human fetal pineal gland. I. Cell types and blood vessels. Cell Tissue Res 152(1):13–30. https://doi.org/10.1007/BF00224208. PMID: 4374305

Møller M (1976) The ultrastructure of the human fetal pineal gland. II. Innervation and cell junctions. Cell Tissue Res 169(1):7–21. https://doi.org/10.1007/BF00219303.0. PMID: 1277287

Møller M, Baeres FM (2002) The anatomy and innervation of the mammalian pineal gland. Cell Tissue Res 309(1):139–150. https://doi.org/10.1007/s00441-002-0580-5. Epub 2002 May 18. PMID: 12111544

Møller M, Ingild A, Bock E (1978) Immunohistochemical demonstration of S-100 protein and GFA protein in interstitial cells of rat pineal gland. Brain Res 140(1):1–13. https://doi.org/10.1016/0006-8993(78)90234-2. PMID: 342071

Munch IC, Møller M, Larsen PJ, Vrang N (2002) Light-induced c-Fos expression in suprachiasmatic nuclei neurons targeting the paraventricular nucleus of the hamster hypothalamus: phase dependence and immunochemical identification. J Comp Neurol 442(1):48–62. https://doi.org/10.1002/cne.1421. PMID: 11754366

Muñoz E, Fogal T, Dominguez S, Scardapane L, Guzmán J, Piezzi RS (1997) Seasonal changes of the Leydig cells of viscacha (*Lagostomus maximus maximus*). A light and electron microscopy study. Tissue Cell 29(1):119–128. https://doi.org/10.1016/s0040-8166(97)80078-x. PMID: 18627814

Muñoz EM, Fogal T, Dominguez S, Scardapane L, Piezzi RS (2001) Ultrastructural and morphometric study of the Sertoli cell of the viscacha (*Lagostomus maximus maximus*) during the annual reproductive cycle. Anat Rec 262(2):176–185. https://doi.org/10.1002/1097-0185(20010201)262:2<176::AID-AR1022>3.0.CO;2-I. PMID: 11169912

Nichols DE (2018) N,N-dimethyltryptamine and the pineal gland: separating fact from myth. J Psychopharmacol 32(1):30–36. https://doi.org/10.1177/0269881117736919. Epub 2017 Nov 2. PMID: 29095071

Nieminen M, Henttinen T, Merinen M, Marttila-Ichihara F, Eriksson JE, Jalkanen S (2006) Vimentin function in lymphocyte adhesion and transcellular migration. Nat Cell Biol 8(2):156–162. https://doi.org/10.1038/ncb1355. Epub 2006 Jan 22. PMID: 16429129

Oishi Y, Okuda M, Takahashi H, Fujii T, Morii S (1993) Cellular proliferation in the anterior pituitary gland of normal adult rats: influences of sex, estrous cycle, and circadian change. Anat Rec 235:111–120

Oshima RG (2007) Intermediate filaments: a historical perspective. Exp Cell Res 313(10):1981–1994. https://doi.org/10.1016/j.yexcr.2007.04.007. Epub 2007 Apr 11. PMID: 17493611; PMCID: PMC1950476

Patel S, Rahmani B, Gandhi J, Seyam O, Joshi G, Reid I, Smith NL, Waltzer WC, Khan SA (2020) Revisiting the pineal gland: a review of calcification, masses, precocious puberty, and melatonin functions. Int J Neurosci 130(5):464–475. https://doi.org/10.1080/00207454.2019.1692838. Epub 2020 Feb 25. PMID: 31714865

Peirson S, Foster RG (2006) Melanopsin: another way of signaling light. Neuron 49(3):331–339. https://doi.org/10.1016/j.neuron.2006.01.006. PMID: 16446137

Pelzer LE, Calderón CP, Guzmán J (1999) Changes in weight and hydroxyindole-O-methyltransferase activity of pineal gland of the plains viscacha (*Lagostomus maximus maximus*). Mastozool Neotrop 6:31–38

Perez Romera E, Mohamed F, Filippa V, Fogal T, Dominguez S, Scardapane L, Piezzi R (2005) Ultrastructural and immunocytochemical studies of the viscacha(*Lagostomus maximus maximus*) pituitary pars tuberalis. Anat Rec A Discov Mol Cell Evol Biol 284(1):431–438. https://doi.org/10.1002/ar.a.20163. PMID: 15791582

Rambotti MG, Giambanco I, Spreca A, Donato R (1999) S100B and S100A1 proteins in bovine retina: their calcium-dependent stimulation of a membrane-bound guanylate cyclase activity

as investigated by ultracytochemistry. Neuroscience 92(3):1089–1101. https://doi.org/10.1016/s0306-4522(99)00074-3. PMID: 10426548
Redondo E, Franco A, Masot AJ, Regodón S (2001) Ultrastructural and immunocytochemical characterization of interstitial cells in pre- and postnatal developing sheep pineal gland. Eur J Histochem 45(3):249–258. https://doi.org/10.4081/1635. PMID: 11759811
Regodón S, Franco A, Masot AJ, Redondo E (2001) Immunohistochemical and ultrastructural study of interstitial cells during postnatal development of the sheep pineal gland. Int J Dev Biol 45:87–88
Reiter RJ (1986) Pineal melatonin production. Photoperiodic and hormonal influences. In: Reiter RJ (ed) Advances in pineal research Londres. John Libbey Co Ltd, pp 77–78
Reuss S (1996) Components and connections of the circadian timing system in mammals. Cell Tissue Res 285(3):353–378. https://doi.org/10.1007/s004410050652. PMID: 8772150
Ribes AC, Mohamed F, Dominguez S, Delgado M, Scardapane L, Guzman J, Piezzi R (1999) Probable effect of photoperiod on seasonal variation in the nuclear volume of the adrenal cortex of viscacha *(Lagostomus maximus maximus)*. Braz J Med Biol Res 32(9):1115–1120. https://doi.org/10.1590/S0100-879X1999000900010. PMID: 10464388
Rosales GJ, Busolini FI, Mohamed FH, Filippa VP (2016) Effects of melatonin and gonadal androgens on cell proliferation in the pituitary of viscachas (*Lagostomus maximus maximus*). Cell Prolif 49(5):644–653. https://doi.org/10.1111/cpr.12280. Epub 2016 Aug 3. PMID: 27484731; PMCID: PMC6496599
Rouvière H, Delmas A (2005) Anatomía Humana; Descriptiva, Topográfica y Funcional, 11ª edn. Ed. Masson
Runembert I, Queffeulou G, Federici P, Vrtovsnik F, Colucci-Guyon E, Babinet C, Briand P, Trugnan G, Friedlander G, Terzi F (2002) Vimentin affects localization and activity of sodium-glucose cotransporter SGLT1 in membrane rafts. J Cell Sci 115(Pt 4):713–724. https://doi.org/10.1242/jcs.115.4.713. PMID: 11865027
Safwat E (2012) Morphological and immunohistochemical studies on the pineal gland of the donkey (*Equus asinus*). J Vet Anat 5:47–74. https://doi.org/10.21608/jva.2012.44883
Schachner M, Huang SK, Ziegelmüller P, Bizzini B, Taugner R (1984) Glial cells in the pineal gland of mice and rats. A combined immunofluorescence and electron-microscopic study. Cell Tissue Res 237(2):245–252. https://doi.org/10.1007/BF00217142. PMID: 6478492
Schietke R, Bröhl D, Wedig T, Mücke N, Herrmann H, Magin TM (2006) Mutations in vimentin disrupt the cytoskeleton in fibroblasts and delay execution of apoptosis. Eur J Cell Biol 85(1):1–10. https://doi.org/10.1016/j.ejcb.2005.09.019. Epub 2005 Nov 8. PMID: 16373170
Shibuki K, Gomi H, Chen L, Bao S, Kim JJ, Wakatsuki H, Fujisaki T, Fujimoto K, Katoh A, Ikeda T, Chen C, Thompson RF, Itohara S (1996) Deficient cerebellar long-term depression, impaired eyeblink conditioning, and normal motor coordination in GFAP mutant mice. Neuron 16(3):587–599. https://doi.org/10.1016/s0896-6273(00)80078-1. PMID: 8785056
Simonneaux V, Ribelayga C (2003) Generation of the melatonin endocrine message in mammals: a review of the complex regulation of melatonin synthesis by norepinephrine, peptides, and other pineal transmitters. Pharmacol Rev 55(2):325–395. https://doi.org/10.1124/pr.55.2.2. PMID: 12773631
Slominski A, Tobin DJ, Shibahara S, Wortsman J (2004) Melanin pigmentation in mammalian skin and its hormonal regulation. Physiol Rev 84(4):1155–1228. https://doi.org/10.1152/physrev.00044.2003. PMID: 15383650
Spady TJ, Shayya R, Thackray VG, Ehrensberger L, Bailey JS, Mellon PL (2004) Androgen regulates follicle-stimulating hormone β gene expression in an activin-dependent manner in immortalized gonadotropes. Mol Endocrinol 18(4):925–940. https://doi.org/10.1210/me.2003-0115. PMID: 14701939
Stehle JH, von Gall C, Korf HW (2002) Organisation of the circadian system in melatonin-proficient C3H and melatonin-deficient C57BL mice: a comparative investigation. Cell Tissue Res 309(1):173–182. https://doi.org/10.1007/s00441-002-0583-2. Epub 2002 May 25. PMID: 12111547

Stehle JH, Saade A, Rawashdeh O, Ackermann K, Jilg A, Sebestény T, Maronde E (2011) A survey of molecular details in the human pineal gland in the light of phylogeny, structure, function and chronobiological diseases. J Pineal Res 51(1):17–43. https://doi.org/10.1111/j.1600-079X.2011.00856.x. Epub 2011 Apr 26. PMID: 21517957.

Suzuki T, Kachi T (1995) Immunohistochemical studies on supporting cells in the adrenal medulla and pineal gland of adult rat, especially on S-100 protein, glial fibrillary acidic protein and vimentin. Kaibogaku Zasshi 70(2):130–139. PMID: 7785412

Tan DX, Xu B, Zhou X, Reiter RJ (2018) Pineal calcification, melatonin production, aging, associated health consequences and rejuvenation of the pineal gland. Molecules 23(2):301. https://doi.org/10.3390/molecules23020301. PMID: 29385085; PMCID: PMC6017004

Teclemariam-Mesbah R, Ter Horst GJ, Postema F, Wortel J, Buijs RM (1999) Anatomical demonstration of the suprachiasmatic nucleus-pineal pathway. J Comp Neurol 406(2):171–182. PMID: 10096604

Tolstonog GV, Wang X, Shoeman R, Traub P (2000) Intermediate filaments reconstituted from vimentin, desmin, and glial fibrillary acidic protein selectively bind repetitive and mobile DNA sequences from a mixture of mouse genomic DNA fragments. DNA Cell Biol 19(11):647–677. https://doi.org/10.1089/10445490050199054. PMID: 11098216

Tolstonog GV, Mothes E, Shoeman RL, Traub P (2001) Isolation of SDS-stable complexes of the intermediate filament protein vimentin with repetitive, mobile, nuclear matrix attachment region, and mitochondrial DNA sequence elements from cultured mouse and human fibroblasts. DNA Cell Biol 20(9):531–554. https://doi.org/10.1089/104454901317094954. PMID: 11747605

Vollrath L (1981) The pineal organ. Berlin Heidlberg 6(7):28–29. Springer, New York

Vrang N, Larsen PJ, Møller M, Mikkelsen JD (1995) Topographical organization of the rat suprachiasmatic-paraventricular projection. J Comp Neurol 353(4):585–603. https://doi.org/10.1002/cne.903530409. PMID: 7759617

Yang X, Wang J, Liu C, Grizzle WE, Yu S, Zhang S, Barnes S, Koopman WJ, Mountz JD, Kimberly RP, Zhang HG (2005) Cleavage of p53-vimentin complex enhances tumor necrosis factor-related apoptosis-inducing ligand-mediated apoptosis of rheumatoid arthritis synovial fibroblasts. Am J Pathol 167(3):705–719. https://doi.org/10.1016/S0002-9440(10)62045-7. PMID: 16127151; PMCID: PMC1698724

Chapter 18
Histological Changes in the Male Reproductive System During the Annual Reproductive Cycle

Albana A. M. Cruceño, Eduardo M. Chaves, Claudia I. Aguilera Merlo, and Fabian H. Mohamed

Abstract Abundant evidence obtained during more than four decades of research has demonstrated that adult male plains vizcacha who inhabit west-central Argentina exhibit a seasonal reproductive cycle synchronized by environmental signals, mainly the natural photoperiod. The cycle has three well characterized gonadal stages: activity (early summer-fall), regression (late fall-winter), and recovery (spring). During regression, the testis has a significant reduction in the number of seminiferous tubules sections as well as in their diameter, until their transformation into cords. Sperm was not observed within the tubular lumen. Sertoli and Leydig cells showed marked signs of involution: declined in organelles content such as mitochondria, smooth endoplasmic reticulum, and Golgi complex and increased content of dense bodies. In the epididymis, corpus and cauda segments showed greater sensitivity to seasonal variations: a decrease in luminal diameter and an increase in both the epithelium height and the thickness of the lamina propria, especially in cauda, during the regression period. In both epididymal segments, principal and clear cells showed signs of involution with respect to the activity period: irregular nuclei, small cytoplasms, large vacuoles, altered mitochondria, and glycogen deposits. All cellular populations of the epididymal epithelium in regression exhibited abundant lysosome-like dense bodies. Finally, in seminal vesicles and prostate, the epithelial height and the luminal diameter decrease. The muscle tunic increased in the seminal vesicles and decreased in the prostate. In both glands, the main epithelial cell populations exhibit morphological characteristics consistent with a reduced functional activity. The seasonal morphological changes observed in the male reproductive system of vizcacha show a high degree of synchrony between the different organs along the seasonal reproductive cycle. These morphological variations are regulated by pineal melatonin, whose serum concentrations are high during winter while the pituitary gland activity decreases in this season.

A. A. M. Cruceño · E. M. Chaves · C. I. Aguilera Merlo · F. H. Mohamed (✉)
Área de Morfología, Facultad de Química, Bioquímica y Farmacia, Universidad Nacional de San Luis, San Luis, Argentina

L. L. Rasia et al. (eds.), *Plains Vizcachas*, https://doi.org/10.1007/978-3-031-49487-1_18

Keywords Male reproductive system · Vizcacha · Morphology · Seasonal reproduction

18.1 Introduction

The survival of the species constitutes one of the essential objectives of reproduction in wild animals, which develops under the influence of the external environment. The reproductive processes occur in a synchronized manner under the control of the environment that determines the genetic potential of the individuals, establishing the reproduction periods and their intensity throughout the year. The initiation and cessation of sexual activity in small wild mammals is conditioned by diverse factors such as photoperiod, temperature, food availability, water restriction, and social behavior (Bronson 1985; Bustos Obregón and Torres Díaz 2012).

Lagostomus maximus is one of the largest rodents of the Argentine territory. Adult males weigh between 4000 and 8000 g and measure between 66 and 68 cm, while females are smaller and weigh between 2000 and 5000 g. Considering the weight and body length of males, three age categories have been established: prepubescent, pubescent, and adult. This species reaches sexual maturity at the age of a year and a half and consequently begins its preparation for reproduction well in advance, thus ensuring gonadal development, gametes production, and the establishment of territories for copulation and gestation moments. Under natural conditions, environmental signals determine the precise moment to start the reproductive process, which manifests itself at the beginning of autumn, when the females come into heat while the males face each other in violent combats. Copulation occurs inside the vizcacheras, giving rise to the gestation period beginning with an approximate duration of 5 months. Then, with the beginning of spring, two or three pups per litter are born with a weigh of approximately 200 g (Llanos and Crespo 1954; Quintanilla et al. 1973). The newborn pups will remain in the vizcacheras during their lactating stage, which ranges from birth until 2 or 3 months after birth. Then, they begin to go out in search of solid food.

18.2 Anatomical and Histophysiological Characteristics of the Male Vizcacha Reproductive System Organs: Seasonal Reproductive Cycle

18.2.1 Testicle

The adult male vizcacha testicle is an ovoid organ whose weight and dimensions vary throughout the annual reproductive cycle (Fuentes et al. 1991). As in other mammals, the vizcacha testis is covered with a fibrous connective tissue capsule, called tunica albuginea, which penetrates into the gonad dividing it into two compartments: tubular and interstitial. The tubular compartment contains the tortuous seminiferous tubules where spermatogenesis is developed, while the interstitial compartment contains the androgen-producing Leydig cells. These cells are arranged in groups close to the blood vessels and are associated with fibroblasts, macrophages, and collagen fibers (Muñoz 1998).

The vizcacha seminiferous tubules are lined with a complex stratified epithelium composed of germ cells and Sertoli cells and coated with a single layer of myoid cells which constitutes the tubular limiting membrane. Surrounding this, the parietal lymphatic endothelium is observed, which is a component of the extensive lymphatic sinusoids wall.

Sertoli cells are located in the testis tubular compartment and together with germ cells constitute the seminiferous epithelium. In addition to the multiple functions attributed to Sertoli cells, this cell type represents the point of convergence of hormones and local testicular factors that regulate and modulate spermatogenic activity. This is possible because this cell type constitutes a nexus between the interstitial compartment and the developing germ cells (Pescovitz et al. 1994; Muñoz 1998).

Studies carried out by Muñoz et al. (1998) described the different phases of spermatogenesis of the adult vizcacha which is produced through organized and synchronized processes that determine nine well-defined phases during the active period. The identification of the different cellular associations of the seminiferous epithelium was carried out based on the acrosomal system morphogenesis and the elongation and condensation degree of the spermatic head. These studies revealed that spermiation occurs during late phase III and early phase IV, contrary to what was observed in rats, where spermiation occurs between phases VIII and IX (Clermont and Perey 1957; Russell et al. 1990).

Reproductive activity period. The seminiferous tubules have greater diameter and spermatozoa within the tubular lumen. Multiple concentric layers of germ cells at different stages are observed in the seminiferous epithelium, indicating complete spermatogenic activity (Fig. 18.1a). Sertoli cells distribution within the seminiferous tubule is variable throughout the reproductive cycle of the vizcacha, showing a greater distribution in the active period (15.24 ± 0.15) when compared to the regression period (9.53 ± 0.15). These values are due to an evident seminiferous tubule retraction that occurs during the gonadal regression period (Fuentes et al. 1991; Muñoz et al. 2001). Regarding their morphological characteristics, during the active period, Sertoli cells have a large and irregular nucleus in basal position. The cytoplasm has extensive cytoplasmic projections surrounding the associated germ cells (Fig. 18.1b). The most developed organelles are the endoplasmic reticulum, Golgi complex, mitochondria, polyribosomes, and single ribosomes. The smooth endoplasmic reticulum is predominant over the rough endoplasmic reticulum. In addition, well-developed Golgi complexes associated with abundant vesicles are observed. All these ultrastructural observations allow us to confirm that the Sertoli cell has a great capacity for the synthesis and secretion of peptides and proteins (Muñoz et al. 2001). Another particular characteristic of the Sertoli cell is the presence of extensive intersertolian junction complexes that correspond to a highly

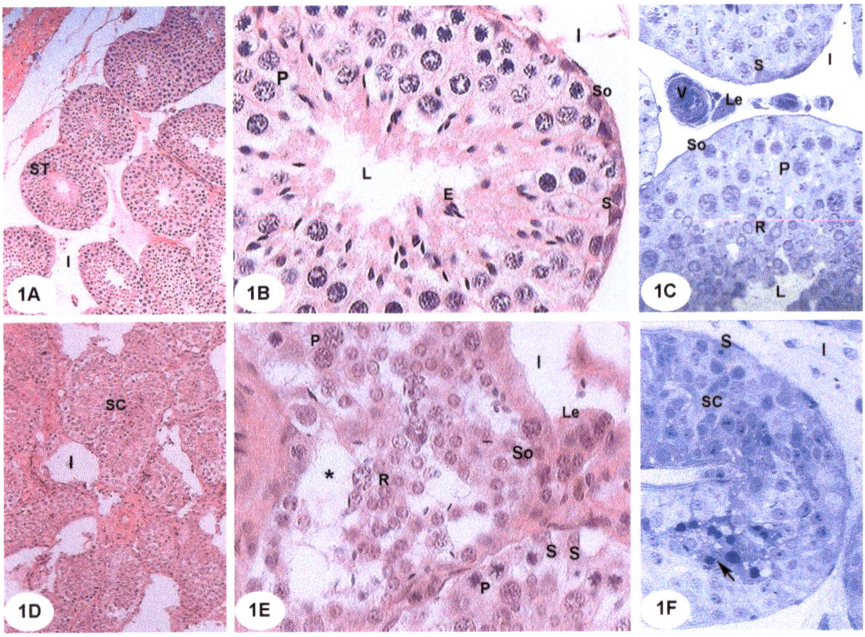

Fig. 18.1 (**a–c**) Testicular activity. (**a**) Seminiferous tubules (ST) highly organized. *I* interstitial space. H-E ×100. (**b**) Multiple concentric layers of germ cells at different stages of maturation. *So* spermatogonia, *P* primary spermatocyte, *R* round spermatid, *E* elongated spermatid, *S* Sertoli cell, *L* lumen, *I* interstitial space. Hematoxylin-Eosin (H-E) ×400. (**c**) Semi-thin section of a seminiferous tubule showing the relationship of Leydig cells (Le) close to blood vessels (V) in the interstitial space (I). *So* spermatogonia, *P* primary spermatocyte, *R* round spermatid. *S* Sertoli cell, *L* lumen. Toluidine Blue (TB) ×400. (**d–f**) Testicular regression. (**d**) Seminiferous epithelium highly disorganized Seminiferous cords (SC). *I* interstitial space. H-E ×100. (**e**) Seminiferous epithelium showing a drastic change in spermatogenic activity. Note that the seminiferous tubules are organized in vacuolated cords (asterisk) and there is no sperm content in the lumen. Populations of Sertoli and germ cells morphologically altered are observed. Sertoli cells (S) exhibit an irregular nucleus. In the interstitium space (I), Leydig cells (Le) constitute irregular groups with scarce cytoplasm. *So* spermatogonia, *P* primary spermatocyte, *R* round spermatid. H-E ×400. (**f**) Semi-thin section of a seminiferous cord (SC) showing altered and degenerated germinal cells (arrow). *S* Sertoli cell, *I* interstitial space. TB ×400

functional blood-testis barrier during the active period (Morales and Cavicchia 1993). In the testicular interstitium, Leydig cells are organized constituting groups of variable size around blood vessels associated with fibroblasts and macrophages (Fig. 18.1c). It is among these cell types that desmosome-like and communicating junctions are observed. During the active period, the cell nucleus has an eccentric position with oval or round morphology, and the wide cytoplasm shows regional distribution of organelles. The smooth endoplasmic reticulum and mitochondria are the predominant organelles, both involved in steroid hormone biosynthesis. It is important to note that the smooth endoplasmic reticulum is constituted by a network of interconnected cisterns, vesicles, and tubules, closely related to mitochondria, lipid inclusions, and dense bodies. Mitochondria varied in shapes and sizes with abundant tubular cristae in their inner membrane. Numerous lipid inclusions are distributed irregularly or in groups in this period (Muñoz et al. 1997).

Regression period. The seasonal study of the vizcacha seminiferous epithelium reveals a drastic drop in spermatogenic activity during the period of testicular regression. The seminiferous tubules show a gradual morphological transformation until they become seminiferous cords without sperm content in the tubular lumen which is lined only with spermatogonia and Sertoli cells. Germ cell populations show a noticeable decrease in all their stages and a significant increase in altered and degenerated cells (Fuentes et al. 1991) (Fig. 18.1d–f). These cellular changes are accompanied by a lower number of spermatozoa retrieved from the epididymal tail (Aguilera Merlo 2006). Sertoli cells exhibit a marked reduction in their size and organelle content. The nuclei are irregular, round, or oval, and their chromatin shows a dense granular texture. The nucleolus is not often observable. The cytoplasm of Sertoli cells usually contains numerous clear vacuoles. The degree of cell vacuolization is related to seminiferous tubule damage. The endoplasmic reticulum is constituted mainly of short tubules and cisterns and a few vesicles of smooth membrane. The Golgi complex is rudimentary, and the number of associated vesicles is low. The content of ribosome, rough endoplasmic reticulum, and mitochondria decreases. The spermatogenic arrest is accompanied by a drastic increase, in the basal cytoplasm, of lipid inclusions and membrane-limited dense bodies (Muñoz et al. 2001). In the tubular interstitium, the Leydig cells show significant morphological changes with small and dense nuclei and reduced cytoplasm (Fuentes et al. 1991). During this period, scarce dense bodies and lipid inclusions are observed in the cytoplasm of these cells. In relation to the serum concentration of testosterone, the maximum level is reached during the period of maximum gonadal activity, while the minimum level occurs during the regression period (Fuentes et al. 1991).

18.2.2 Epididymis

The epididymis is a single, elongated, and coiled duct. This organ is folded on itself into a short structure (5–7 cm) and located behind the testis. The epididymis is divided into four segments: the initial segment, located in the upper pole of the testis, the caput, the corpus, and the cauda in the lower pole. Each epididymal segment is constituted by three histological tunics, namely, mucosa, muscular, and adventitia. The mucosa has a pseudostratified columnar epithelium with stereocilia constituted by principal, basal, clear, halo, and narrow cells which lines the epididymal lumen and is surrounded by a lamina propria. This lamina is constituted by smooth muscle fibers and connective tissue. The second tunic is a very thin muscle layer which runs along most of the duct whose thickness increases near the vas deferens. Finally, the adventitia surrounds and envelopes the entire duct and is constituted by loose connective tissue. The tissue organization is similar to that seen in the interstitial tissue present in other organs which is observed in the cross sections of the organ. In this structure numerous blood vessels, lymphatic vessels and nerves are observed. A particular characteristic of vizcacha epididymis is that the cauda segment presents an enveloping fibromuscular capsule formed by striated muscle fibers and collagen fibers, positioned concentrically around the entire epididymal segment. This capsule maintains its histoarchitecture along the annual reproductive cycle of *Lagostomus* (Aguilera Merlo et al. 2005a).

The epididymis has the greatest length during the period of maximum reproductive activity (7.41 ± 0.063 cm) compared to the regression period (5.27 ± 0.075 cm). The epididymal corpus is the segment that has the greatest changes in length along the annual reproductive cycle. During the active period (4.3 ± 0.15 μm), the values are the highest, while during the regression period are minimum (3.25 ± 0.13 μm) (Aguilera Merlo et al. 2005a).

As regards the structural parameters, epithelial height, luminal diameter, and thickness of the lamina propria show significant variations in the different segments during the periods of activity and regression. *In the period of maximum gonadal activity*, the epididymal sections have a wide luminal diameter. The height of the epithelium decreases throughout the epididymis, being higher in the initial segment, while the thickness of the lamina propria shows a greater development in the epididymal cauda. *During the regression period*, the parameters change significantly in comparison to the active period. The luminal diameter in each of the studied segments is lower than that observed in the period of reproductive activity. The epithelium height decreases in the initial segment and the caput but increases in corpus and cauda, with respect to that observed in the active period. The thickness of the lamina propria is higher during gonadal regression, mainly in the epididymal cauda. Calvo et al. (1997, 1999) report similar results in hamsters subjected to artificial short photoperiods. These changes are likely due to a decrease in testosterone levels. These results seem to indicate that the cauda is the epididymal segment most sensitive to the changes induced by the photoperiod (Aguilera Merlo et al. 2005a).

The epididymis epithelium of vizcacha has the same cell populations which are observed in rats (Robaire and Hermo 1988) and hamsters (Calvo et al. 1999). However, their distribution, frequency, and structural characteristics differ from those described in rats (Reid and Cleland 1957; Sun and Flickinger 1979) and hamsters (Flickinger et al. 1978). In vizcacha, narrow cells are only present in the initial segment of the epididymal epithelium during the active period, suggesting that narrow cells could be precursors of clear cells. This is in agreement with findings described for other species such as rats (Sun and Flickinger 1980), hamsters (Flickinger et al. 1978), and mice (Abou-Haila and Fain-Maurel 1984).

In the morphological study of vizcacha epididymis, corpus and cauda segments show the greatest morphological changes during the regression compared to the characteristics observed during the activity period.

During the reproductive cycle in the epididymal corpus, only the principal and clear cells exhibit significant morphological changes. The principal cells have elongated and large nuclei with lax chromatin and one or more evident nucleoli. The cytoplasm is abundant with numerous dense bodies, rough endoplasmic reticulum cisternae, and abundant elongated mitochondria located in the peripheral region

(Fig. 18.2a, c). *During the regression period*, the principal cells have small nuclei with irregular shapes and sizes. Their cytoplasm have abundant dense bodies, multivesicular bodies, large vacuoles, and structures like phagolysosome and absence of endoplasmic reticulum. Studies performed in other animal species such as rabbits (Cran and Jones 1980) and rats (Serre and Robaire 1998) described the presence of large vacuoles in the principal cells located in the epididymal corpus of elderly animals. In our animal model, we also observe in the principal cells that the dense bodies merge with vacuoles which remain in the medial region and neither reach the epithelium nor touch the basement membrane (Fig. 18.2d–f). According to Serre and Robaire (1998), the presence of vacuoles that neither reach the epithelium nor touch the basement membrane suggests an autophagocytosis in the vacuolated principal cells that might produce a temporary disruption of the endosomal lysosomal cycling system causing functional variations in principal cells during the seasonal reproductive cycle of the vizcacha.

As regards clear cells, they have large, round, or elongated nuclei located in basal position during the period of activity. In the cytoplasm, elongated and rounded mitochondria are observed in the middle and basal regions of the cytoplasm, which is clear and less dense than that of the principal cells. The rest of the cytoplasmic organelles are scarcely developed. In the perinuclear region, small vesicles and few dense bodies of different sizes and intensities are observed (Fig. 18.2b). However, during the regression period, the organelles of the clear cells show a higher degree of development. The apical cytoplasm have abundant lysosomal-like electrodense material, numerous mitochondria, and vesicles. This cell population has irregularly shaped nuclei with dense clumps of heterochromatin (Fig. 18.2e). Similar findings are reported in other animal species, such as hamsters, exposed to long and short photoperiods (Calvo et al. 1997). Clear cells perform different functions such as secretion of L-α-glycerophosphorylcholine (Martan 1969), absorption of luminal proteins (Lea et al. 1978), absorption of cytoplasmic droplets of sperm (Hermo et al. 1985), and also participation in the acidification of the luminal microenvironment (Shum et al. 2008, 2009). In the vizcacha epididymal corpus during the regression period, the clear cells might be especially important in the absorption of sperm remnants and luminal components found in the lumen of this segment.

In the epididymal cauda, during the active period, a higher number of halo cells are observed. In plastic-embedded sections, this epididymal segment has the two types of halo cells, classical and mature, both located in the base of the epithelium. Classic halo cells have a round nuclei with lax chromatin and a weakly stained cytoplasm. Mature halo cells have a small nucleus with numerous dense bodies in the cytoplasm. In addition, abundant spermatozoids and early germinal cells constitute the luminal dense spermatic mass. *In the regression period, the cauda* shows a strong reduction of the lumen, a significant increment of epithelial height compared to the active period and low sperm content within the lumen. A decrease in halo cells is observed compared to the period of gonadal activity. The lamina propria in this segment is more developed than in the active period (Fig. 18.2g–l).

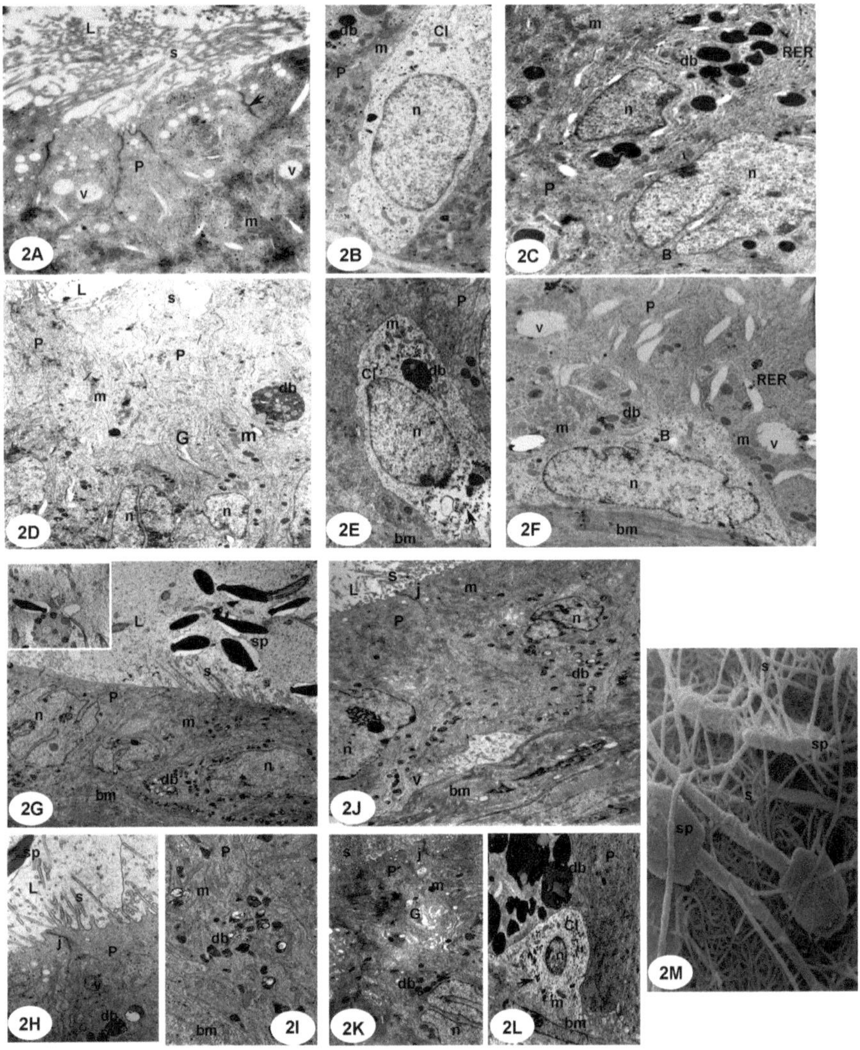

Fig. 18.2 (**a**–**c**) Epididymal Corpus – Activity. (**a**) Principal cell apical region (P) showing at the base of the stereocilia (s) numerous vesicles (v) of variable sizes and mitochondria (m). *L* lumen. Transmission Electron Microscopy (TEM) ×3000. (**b**) Basal region of clear cell (Cl) elongated nucleus (n) with lax chromatin and pale cytoplasm. Observe part of the cytoplasm of a neighboring principal cell (P) with mitochondria (m) and dense phagolysosome-like bodies (db) of variable sizes. TEM ×5000. (**c**) Central region of a principal cell (P) showing a particular arrangement of the granular endoplasmic reticulum (RER) in relation to dense phagolysosome-like bodies (db). *B* basal cell, *n* nucleus, *m* mitochondria. TEM ×5000. (**d**–**f**) Epididymal Corpus – Regression. (**d**) Apical and central regions of a principal cell (P) exhibit few secretion vesicles and dilated Golgi complex (G). Note the presence of dense phagolysosome-like bodies (db) with variable sizes and pleomorphic mitochondria (m). *L* lumen, *s* stereocilia, *n* nucleus. TEM ×3000. (**e**) Basal region of clear cell (Cl), showing an organelle-rich cytoplasm. *n* nucleus, *m* mitochondria, *db* dense phagolysosome-like bodies. Arrow: glycogen granules. *P* principal cell, *bm* basement membrane.

18.2.2.1 Enzymatic Biochemistry in Vizcacha Epididymis

We analyzed four of the numerous epididymal lysosomal enzymes from the epididymal cauda using perfusion techniques: *β*-D-Galactosidase (*β*-GAL), N-Acetyl---D-Glucosaminidase (*β*-NAG), *α*-D-Mannosidase (*α*-MAN), and *β*-D-Glucuronidase (*β*-GLU). In the studied enzymes, the values of total enzyme activity/mg of tissue suggest that the capacity for enzymatic synthesis of epididymal epithelial cells varies throughout the seasonal reproductive cycle. This functional variation in epididymal cells of vizcacha might occur as a consequence of endosome-lysosomal cyclic system temporary disruption. The compartmentalization of lysosomal enzymes in epididymis of vizcacha shows that the highest activity of the studied enzymes, except *β*-NAG and *α*-MAN, is found in the tissue during periods of activity and regression (Aguilera Merlo et al. 2009). Similar findings have been observed in other species (Serre and Robaire 1998; O'Donnell et al. 2017).

β-GAL has slightly higher total activity, mainly in the tissue, during the active period. Inside the epididymal lumen, and in association with sperm, this enzyme has greater distribution during the regression period. This is similar to what is observed in rats and rabbits. In rat epididymis, it has been demonstrated that *β*-GAL binds to the surface of spermatozoa through high affinity binding sites (Belmonte et al. 2002). In rabbit, sperm have been found high affinity binding sites for *β*-GAL in acrosomal and post-acrosomal regions (Nizolajczyk and O'Rand 1992).

Fig. 18.2 (continued) TEM ×3000. (**f**) Basal cell (B) showing an indented nucleus (n) and cytoplasm with poor organelle content. Cytoplasmic areas of neighboring principal cells (P) showing highly vacuolated (v) and dense phagolysosome-like bodies (db). *RER* granular endoplasmic reticulum, *m* mitochondria, *bm* basal membrane. TEM ×4000. (**g–i**) Epididymal Cauda – Activity. (**g**) Low epididymal epithelium showing principal cells (P) with nuclei (n) of variable sizes and shapes. A high content of dense phagolysosome-like bodies (db) distributed in the cytoplasm is observed. Note the presence of spermatozoa (sp) anchored between the epithelium stereocilia (s). *m* mitochondria, *bm* basal membrane, *L* lumen. TEM ×2500. Inset: Detail of a spermatozoon in the lumen with excess residual cytoplasm located behind the head and around the midpiece. TEM ×10000. (**h**) Principal cell (P) apical region with numerous vesicles (v) and dense phagolysosome-like bodies (db) of variable sizes. *j* cell junctional complex, *L* lumen, *s* stereocilia, *sp.* spermatozoa. TEM ×10000. (**i**) Observe in the cytoplasm of the Principal cells (P) clumps of dense phagolysosome-like bodies (db). *m* mitochondria, *bm* basal membrane. TEM ×10000. (**j–l**) Epididymal Cauda – Regression. (**j**) Note the tight junction below the stereocilia (s), followed by adhesion and gap junctions. Principal cells (P) present irregular nuclei (n) with pronounced invaginations. Associations of dense phagolysosome-like bodies (db) and vacuoles (v) are observed in the basal region. *L* lumen, *s* stereocilia, *bm* basement membrane. TEM ×3000. (**k**) Apical and central surface of principal cells (P) showing dilated Golgi cisterns (G) and altered mitochondria (m). Dense phagolysosome-like bodies (db) located around the nucleus (n) are observed. *j* cell junctional complex, *s* stereocilia. TEM ×5000. (**l**) Clear cell (Cl) with small nucleus (n) and cytoplasm with few mitochondria (m) and abundant glycogen granules (arrow). In the adjacent principal cell (P), note the heterogeneity of the dense phagolysosome-like bodies (db) and organelle disorganization. *bm* basement membrane. TEM ×2500. (**m**) Vizcacha Spermatozoa. Mature gametes (sp) of vizcacha located in the cauda of the epididymis anchored between the stereocilia (s). These gametes have a flattened spoon-shaped sperm head. Scanning Electron Microscopy (SEM) ×8000

β-GLU is located mainly in the epididymal tissue of vizcacha, showing higher total activity during the active period and exhibiting similar distribution in the fluid or in association with sperm in both periods. Inside the lumen, the greater distribution of β-GLU is observed during the regression period. The main intracellular function of β-GLU might be related to lysosomal acid hydrolysis (Abou-Aila et al. 1996).

The enzymes β-GAL and β-GLU have also been located in the acrosome of the sperm. In vitro immunolocation experiments demonstrate that these enzymes change their fluorescence pattern during the capacitation process, demonstrating their key participation in sperm maturation (Reid et al. 2011).

α-MAN has slightly higher total activity during the active period and is located mainly in the fluid in both periods. Inside the lumen, and in association with sperm, the enzyme is located mainly during the regression period. Studies performed by Okamura et al. (1995) demonstrated α-MAN participation in the interaction process of sperm and oocytes. This seems to suggest that in vizcacha α-MAN might participate in the maintenance of the sperm-oocyte interaction throughout the annual reproductive cycle.

β-NAG is located mainly in epididymal fluid during the active and regression periods and in association with sperm, mainly during the regression period, in opposition to that observed in other rodents (Belmonte et al. 2002). This enzyme has the highest total activity during the regression period, which might indicate a higher synthesis or a decrease of other soluble proteins into epididymal fluid. β-NAG might hydrolyze the membrane of abnormal sperm in the epididymal lumen, collaborating in sperm concentration decrease during the regression period.

Our results demonstrate that the epididymal enzymes β-GAL, α-MAN, and β-GLU show a primarily luminal distribution during the regression period. Both the activity and the distribution of lysosomal enzymes in vizcacha epididymis might collaborate in the maintenance of the microenvironment along the annual reproductive cycle.

18.2.2.2 Epididymal Luminal Compartment and Spermatozoa

The epididymal epithelium through its absorption and secretion functions plays a key role in maintaining the intraluminal microenvironment where sperm maturation occurs. Several investigations indicate that interaction between spermatozoa and epididymal epithelium secretion is necessary for gametes to acquire fertilizing capacity. Functional studies have confirmed that each epididymal segment contributes to maturation in a particular way, and to the protection and storage of sperm, depending on the cell populations present in each epididymal segment and the control provided by androgens and testicular factors (Martínez-Fresneda et al. 2019).

Considering the morphological and functional complexity of the epididymis, the main morphological and physiological characteristics of the spermatozoa extracted from epididymal cauda of vizcacha were evaluated. Functionally competent gametes are stored in this epididymal segment which is one of the segments most sensitive to natural and artificial photoperiod (Yanagimachi 1994; Calvo et al. 1997; Aguilera Merlo et al. 2005a). Mature gametes located in the epididymal cauda were obtained from animals captured during the period from summer to winter. The obtained spermatozoa were released into a culture medium for quantitative, qualitative, and functional morphological evaluation throughout the seasonal reproductive cycle of this rodent. The vizcacha gametes are characterized by a flattened, spoon-shaped sperm head (length 6.25 ± 0.10 mm and width 4.2 ± 0.03 mm) and an acrosomal region located in the anterior third of the head, separated from post acrosomal region by an equatorial segment visible under light microscopy (Fig. 18.2m). The total length of flagellum varies from 43.8 ± 0.93 to 48.67 ± 0.42 mm (Aguilera Merlo 2006).

The sperm quality of spermatozoa stored into vizcacha epididymal cauda is evaluated by measuring the seminal parameters of concentration, vitality, motility, and morphology. These studies show a significant variation in sperm quantification between the active and regression periods. During the period of maximum gonadal activity (summer-autumn), the sperm concentration/ml is 470 × 106 (±12.5 × 106) with an approximate sperm vitality of 58%, while in the regression period there is a decrease in sperm count, reaching a value of 70.8 × 106 (±6.27 × 106)/ml and a spermatic vitality of 44%. However, sperm motility is progressively fast in both periods (Aguilera Merlo et al. 2005b). Regarding sperm morphology, it was analyzed using stained smears with modified Papanicolaou staining for sperm, finding a low percentage of morphological anomalies along the *Lagostomus* seasonal reproductive cycle. The incidence of morphological defects in epididymal spermatozoa is higher during the regression period, mainly in the sperm head and tail. The most observed morphological alterations are microcephaly, pyriform heads, coiled, and poorly implanted or broken tails (Cruceño et al. 2013). In addition, functional assays for sperm capacitation and acrosomal reaction, carried out along the vizcacha reproductive cycle, showed seasonal changes in protein patterns obtained from capacitated and reacted sperm pellets. These results demonstrate that vizcacha spermatozoa undergo remodeling processes in their membranes, both plasmatic and acrosomal, in order to achieve their fertilizing capacity (Aguilera Merlo 2006).

Our results suggest that in our model the epididymis retains the ability to maintain a special microenvironment for subsistence of stored sperm during the seasonal cycle. Several studies performed in other mammals demonstrated a prolonged survival of spermatozoa in the epididymal cauda during several days to weeks depending on the species (Setchell et al. 1993; Jones and Murdoch 1996; Zhou et al. 2018).

The final destination of nonviable sperm into epididymal lumen raises questions about the mechanisms of the epididymis for their elimination. In this process, the stored sperm in epididymal lumen might not be valid to postulate a unique mechanism of phagocytosis performed by the epididymal epithelium because millions of spermatozoa produced daily by the testes must be eliminated.

18.2.3 Male Accessory Glands

The morphological characteristics of the accessory glands (seminal vesicles, prostate, coagulant and bulbourethral glands) of the male vizcacha have important differences and similarities with other mammalian species in relation shape, size, function, and development (Luke and Coffey 1994; Setchell et al. 1994; Adebayo et al. 2009).

Some morphological and functional differences between species have been attributed to the great diversity of environments and reproductive habitats that determine activity and inactivity periods of these glands. In different studies, it has been observed that size and function of accessory glands are seasonally regulated (Luke and Coffey 1994; Orostegui et al. 2000). On the other hand, different structures of the male genital system are present in some animal species and absent in other species, such as the vas deferens ampulla, which is absent in vizcachas and in most hystricomorph rodents as well as in wild African rats, cats, and pigs (Adebayo et al. 2009; Sisson 1975).

18.2.4 Seminal Vesicles

In *L. maximus*, as in most mammals, the seminal vesicles are located in a retroperitoneal position. The vizcacha seminal vesicle is a paired organ of whitish coloration, elongated, tortuous in shape whose glands release their content independently into the urethral lumen. Histologically, the seminal vesicles are constituted by three radially arranged histological tunics: mucosa, muscular, and adventitia. The mucosa forms folds of variable length (between 56 ± 12.2 μm and 96 ± 10 μm) that extend into the glandular lumen. These folds branch and anastomose to form secondary folds. The glandular epithelium is a pseudostratified columnar epithelium and its height is 6.6 ± 1.6 μm. The epithelium has columnar cells with elongated nuclei, moderately condensed chromatin, and evident nucleoli. At the base of the epithelium are located basal cells with round or oval nuclei. The glandular lumen is large and contains abundant secretion with a homogeneous appearance. Both the epithelium and the glandular secretion are negative for periodic acid-Schiff's (PAS) reactivity. The muscular thickness is 59.2 ± 18.8 μm, and it is formed, mainly, by smooth muscle fibers and collagen fibers. Surrounding these structures is an adventitia of loose connective tissue (Chaves et al. 2011).

The seminal vesicles structure undergoes important changes during the annual reproductive cycle. Specifically, parameters such as lumen diameter and epithelial height show an important increase during the period of maximum gonadal activity compared to what is observed in the regression period. These changes are probably due to the high levels of testosterone, observed during the period of activity, which would stimulate the epithelial cells secretory activity and a greater secretion storage in the glandular lumen. The secretory epithelium of the seminal vesicles is formed by three types of cells: principal, clear, and basal, of which the principal cell population exhibits greater ultrastructural changes during the annual reproductive cycle of vizcacha. *During the period of gonadal activity*, principal cells display abundant rough endoplasmic reticulum and ribosomes (Fig. 18.3a), which suggests a greater activity of protein synthesis (Chaves et al. 2012). This activity consumes at least four high-energy phosphate bonds, which represents a higher ATP requirement (Alberts et al. 2002). This large increase in energy requirements might explain the high abundance of mitochondria observed in principal cell cytoplasm during the period of activity (Chaves et al. 2012). On the other hand, *during the period of gonadal regression*, the principal cells show signs of decreased cellular activity characterized by reduced cytoplasmic size, irregularly shaped nuclei, dilated rough endoplasmic reticulum cisterns, and alteration of mitochondria morphology (Fig. 18.3b). These changes correlate with a decrease in both the luminal diameter of the gland and serum testosterone levels (Chaves et al. 2012). These results have allowed us to postulate a decrease in protein synthesis in seminal vesicles, in agreement with what has been observed in other species such as the Saharan rodent (Gernigon et al. 1994).

Principal cells have been previously reported in different animal species such as bulls (Amselgruber and Feder 1986), goats (Wrobel 1970), and humans (Riva and Aümuller 1994), while clear cells, present in vizcacha seminal vesicles, have only been observed in wild boar seminal vesicles (Badia et al. 2006). In relation to these cell populations' functions, in our animal model, we postulate that principal cells observed in seminal vesicles might be functionally differentiated cells, while clear cells, which do not have secretory granules in their cytoplasm and have little endoplasmic reticulum development, might constitute an initial state of differentiated columnar cells. On the other hand, basal cells are scarce and characterized by their poor development of cytoplasmic organelles and the absence of secretory granules (Chaves et al. 2012). The exact function of this cell population within the glandular epithelium is unclear, but it could act as a potential mediator of luminal and stromal cells activity, regulating the secretion of material (El-Alfy et al. 2000). In addition, basal cells could also play a structural role (Hayward et al. 1996) and constitute a columnar cell reserve (Verhagen et al. 1988; Riva and Aumüller 1994).

In the immunohistochemical determination of androgen receptor (isoform B) in seminal vesicle during the regression period, in both epithelium and stroma, higher percentages of cells immunolabeled for androgen receptor with a weak labeling intensity are observed, compared to what is observed during the period of activity (Chaves et al. 2012). Considering that, during the regression period, the vizcachas have the lowest serum testosterone levels, our immunohistochemical results show a significant inverse correlation between percentage of androgen receptor-immunopositive cells and serum testosterone concentration. The higher expression of androgen receptor when testosterone levels are lower would allow the maintenance of glandular function during the period of gonadal regression (Chaves et al. 2012). Our results agree with previous studies carried out in other animal species where a decrease in serum testosterone levels and an increase in androgen receptor expression in different organs of the male reproductive system were observed, after performing surgical and chemical castration (Li-Xin et al. 1990; Blok et al. 1992).

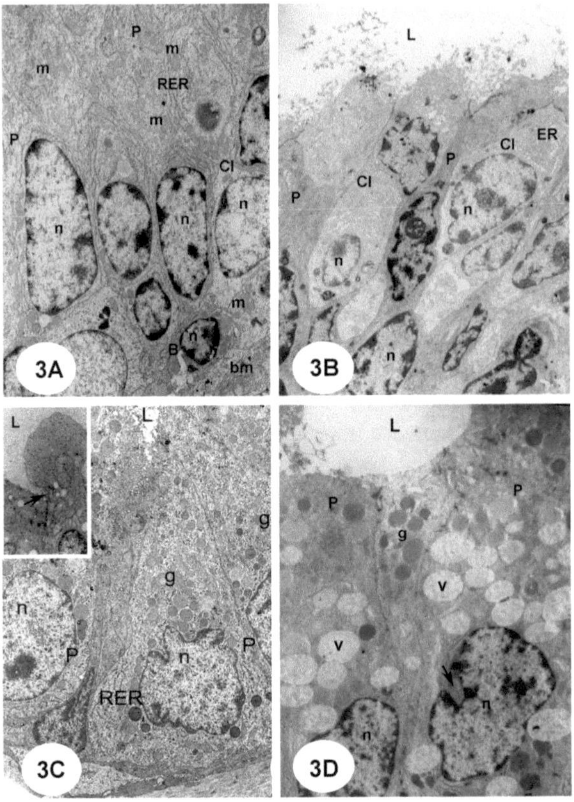

Fig. 18.3 (**a**) Seminal vesicles. Period of activity. Glandular epithelium pseudostratified columnar with principal (P), basal (B), and clear (Cl) cells showing nuclei (n) with variable shapes and sizes. Observe on the apical surface of principal cells a large development of mitochondria (m) and rough endoplasmic reticulum (RER). *bm* basal membrane. TEM ×3000. (**b**) Seminal vesicles. Period of regression. Principal (P) cells show a scarce cytoplasm and irregular nuclei (n) with variable sizes. Clear cells (Cl) present dilated endoplasmic reticulum (ER). *L* lumen. TEM ×3000. (**c**) Prostate gland. Period of activity. Principal cells (P) showing irregular shaped nuclei (n) with moderately condensed chromatin. Observe abundant secretory granules (g) in the supranuclear region. *RER* rough endoplasmic reticulum, *L* lumen. TEM ×3000. Inset: Note a cytoplasmic protrusion (arrow) into the lumen (L), showing the cellular secretory activity. TEM ×7000. (**d**) Prostate gland. Period of regression. Principal cells (P) with irregular shapes and variable sizes are observed. The nuclei (n) show irregular edges, pronounced invaginations (arrows), and electron-dense chromatin forming clumps next to the nuclear envelope. Abundant vacuoles (v) and a few secretory granules (g) are present in the cytoplasm. *L* lumen. TEM ×3000

18.2.5 Prostate Gland

The prostate gland activity depends on several factors such as androgen stimulation, pathological conditions, and environmental signals. Environmental stimuli such as photoperiod induce seasonal variations during the year, thus determining activity and inactivity periods for this gland.

In vizcacha, the macroscopic morphology of the prostate gland was similar to that described in other hystricomorphs such as *Chinchilla lanigera* (Cepeda et al. 2006) and guinea pigs (Vásquez and del Sol 2002) or in African wild rats (Adebayo et al. 2009). The vizcacha prostate gland is a non-paired gland of ochre color and soft consistency, located in caudal position with respect to the bladder, above the floor of the perineum, ventrally with respect to the rectum, and dorsally with respect to the coxal bones.

The prostate epithelium in the vizcacha has two cell populations as in other rodents: principal and basal cells. A cell subpopulation of principal cells, called dense cells, was also identified (Chaves et al. 2015). These cells have nuclear and cytoplasmic features typical of cells with low secretory activity. Dense cells were also described in the prostate epithelium of the gray squirrel and in the disseminate prostate of pigs (Siwela and Tam 1984; Badia Brea 2000). In agreement with these authors, we postulate that different subpopulations of principal cells found in vizcacha could be part of different stages of the same cell in relation to its secretory activity (Chaves et al. 2015).

Histologically, the prostate is made up of two well-differentiated zones, which according to the distance between glandular adenomeres and the urethra and also their morphological and histochemical differences can be classified as a peripheral zone and a central zone. Both zones possess an organized parenchyma with abundant adenomeres and the fibromuscular stroma (Chaves et al. 2011). A thin fibrous capsule surrounds the gland structure.

Throughout the seasonal cycle, this gland shows significant changes in its weight, indicating a greater glandular activity during the period of maximum reproductive activity. In relation to glandular morphological organization, both the central and the peripheral zone show significant changes during the periods of maximum and minimum gonadal activity.

During the period of maximal reproductive activity, the peripheral zone, the most distant from the urethra, has adenomeres with large lumen (luminal diameter: 775 ± 43.43 μm) and scarce folds (fold length: 87.2 ± 3.44 μm). Around the adenomeres, a thin layer of smooth muscle (thickness of the muscle layer: 40 ± 2.8 μm) and collagen fibers can be observed. The adenomeres epithelium is low cubic (epithelial height: 11.2 ± 0.44 μm) with nuclei in central position, loosely chromatin and evident nucleolus. The PAS histochemical technique reveals that epithelial cells are negative for neutral glycoproteins, while the secretion contained in the adenomeres lumen is PAS positive (Chaves et al. 2011). *During the active period, the central zone*, the region closest to the urethral duct, has adenomeres with a reduced lumen (435 ± 50.32 μm), surrounded by more developed fibromuscular tissue than that observed in the peripheral zone. The glandular epithelium is cylindrical pseudostratified (epithelial height: 13.6 ± 0.64 μm) and forms abundant long folds (fold length: 165 ± 6.40 μm) that project to the glandular lumen. Cylindrical epithelial cells with elongated nuclei are predominant, with lax chromatin and evident nucleolus. A higher secretory activity of these cells is demonstrated by the presence of cytoplasmic organelles such as rough endoplasmic reticulum, ribosomes, and secretory granules (Fig. 18.3c). Furthermore, in the apical region, cytoplasmic protrusions suggestive of an apocrine secretion mechanism are observed, similar to that described in other species such as rats (Whalqvist 2000), hamsters (Toma and Buzzel 1988), and humans (Biagini et al. 1982). In agreement with this, a histochemical study revealed scarce PAS-positive cells, and the secretion contained in the glandular lumen is intensely stained (Chaves et al. 2011).

During the period of minimal reproductive activity, the mucosa in both the central and peripheral zones have adenomeres with a small number of folds and glandular epithelium with a lower height in relation to that described in the active period and the glandular lumen, which becomes smaller. In addition to this, it is observed that the principal cells of both zones have small nuclei of irregular shapes and varying sizes, cytoplasms with dilated cisternae in the Golgi complex, scarce organelles with altered morphology, and fewer apical cytoplasmic protrusions (Fig. 18.3d). Variations in glandular morphological organization throughout the seasonal reproductive cycle show increased secretory activity of the glandular epithelium during the period of maximum gonadal activity and decreased synthesis and secretion activity during the regression period. The lower secretory activity in epithelial cells of the prostate during the minimal reproductive activity of *Lagostomus* could be directly related to low serum testosterone levels found in this species (Chaves et al. 2015). Androgen receptor immunohistochemical determination in the muscle layer of the prostate central zone shows intense labeling during the regression period (Chaves et al. 2015), in agreement with what was observed in rats (Nishino et al. 2004) and in humans (Cunha 1994). This suggests that the prostate gland can increase receptor expression in the stroma to facilitate the uptake of low levels of circulating testosterone. On the other hand, muscle fibers could modulate the glandular activity through their androgen receptors by means of paracrine effectors interacting on epithelial cells (Chaves et al. 2015).

18.3 Conclusions

According to the previously described histological characteristics of the testis, epididymis, seminal vesicles, and prostate of *Lagostomus maximus*, it demonstrates that these organs exhibit seasonal morphological changes, with maximum activity in summer and minimum in winter. These modifications are synchronized by the natural photoperiod, modulated by melatonin, and mainly regulated by pituitary hormones. In this way, the vizcacha coordinates its reproductive activity with environmental conditions, guaranteeing survival.

References

Abou-Haila A, Fain-Maurel MA (1984) Regional differences of the proximal part of mouse epididymis: morphological and histochemical characterization. Anat Rec 209:197–208

Abou-Haila A, Tulsiani DR, Skudlarek MD, Orgebin-Crist MC (1996) Androgen regulation of molecular forms of Beta-D-glu-glucuronidase in the mouse epididymis: comparison with liver and kidney. J Androl 17:194–207

Adebayo AO, Oke BO, Akinloye AK (2009) The gross morphometry and histology of the male accessory sex glands in the greater cane rat (*Thryonomys swinderianus*, Temminck). J Vet Anat 2:41–51

Aguilera Merlo C (2006) Morphological and biochemical study of the epididymis of *Lagostomus maximus maximus*. Effect of the photoperiod and potential fertilizing capacity of the spermatozoon. Doctoral thesis, National University of San Luis

Aguilera Merlo C, Muñoz E, Dominguez S, Scardapane L, Piezzi R (2005a) Epididymis of viscacha (*Lagostomus maximus maximus*). Morphological changes during the annual reproductive cycle. Anat Rec 282:83–92

Aguilera Merlo C, Muñoz E, Dominguez S, Scardapane L, de Rosas JC (2005b) Seasonal variations in the heterologous binding of viscacha spermatozoa. A scanning electron microscopy study. Biocell 29:243–251

Aguilera Merlo C, Fogal T, Sator T, Dominguez S, Sosa M, Scardapane L, Piezzi R (2009) Ultrastructural and biochemical seasonal changes in epididymal corpus and cauda of viscacha (*Lagostomus maximus maximus*). J Morphol 270:805–814

Alberts B, Bray D, Lewis J, Raff M, Roberts K, Watson JD (2002) Molecular biology of the cell. Garland Publishing, New York, p 1463

Amselgruber W, Feder FH (1986) Licht- und elektronenmikroskopi-sche untersuchugen der samenblasendru ̈se (Glandula vesicularis) desbullen. Anat Histol Embryol 15:361–379

Badia Brea ME (2000) Estudi structural, ultraestructural i histoquimic de les glándules sexuals accessóries del mascle reproductor porcı (Sus domesticus). Thesis doctoral, Biblioteca Universidad de Girona, España

Badia E, Briz MD, Pinart E, Garcia-Gil N, Bassols J, Pruneda A, Bussalleu E, Yeste M, Casas I, Bonet S (2006) Structural and ultrastructural features of boar seminal vesicles. Tissue Cell 38:79–91

Belmonte SA, Romano P, Sartor T, Sosa MA (2002) Compartmentalization of lysosomal enzymes in cauda epididymis of normal and castrated rats. Arch Androl 48:193–201

Biagini G, Preda P, Lo Cigno M, Soli M, Bercovich E (1982) Ultrastructural aspects of human prostate in benign prostatic hyperplasia. Am J Anat 181:132–140

Blok LJ, Bartlett JM, Bolt de Vries J, Themmen AP, Brinkmann AO, Weinbauer GF, Nieschlag E, Grootegoed JA (1992) Effect of testosterone deprivation on expression of the androgen receptor in rat prostate, epididymis and testis. Int J Androl 15:182–198

Bronson FH (1985) Mammalian reproduction: an ecological perspective. Biol Reprod 32(1):1–26. https://doi.org/10.1095/biolreprod32.1.1. PMID: 3882162

Bustos Obregón E, Torres-Díaz L (2012) Seasonal reproduction in the male. Int J Morphol 30(4):1266–1279. https://doi.org/10.4067/S0717-95022012000400004

Calvo A, Bustos-Obregon E, Pastor LM (1997) Morphological and histochemical changes in the epididymis of hamsters (*Mesocricetus auratus*) subjected to short photoperiod. J Anat 191:77–88

Calvo A, Pastor LM, Martínez E, Vázquez JM, Roca J (1999) Age-related changes in the hamster epididymis. Anat Rec 256:335–346

Cepeda R, Adaro L, Peñailillo G (2006) Morphometric variations of *Chinchilla laniger* prostate and plasmatic testosterone concentration during its annual reproductive cycle. Int J Morphol 24:89–97

Chaves EM, Aguilera-Merlo C, Filippa V, Mohamed F, Dominguez S, Scardapane L (2011) Anatomical, histological and immunohistochemical study of the reproductive system accessory glands in male viscacha (*Lagostomus maximus maximus*). Anat Histol Embryol 40:11–20

Chaves EM, Aguilera-Merlo C, Cruceño A, Fogal T, Piezzi R, Scardapane L, Dominguez S (2012) Seasonal morphological variations and age-related changes of the seminal vesicle of viscacha (*Lagostomus maximus maximus*): an ultrastructural and immunohistochemical study. Anat Rec (Hoboken) 295(5):886–895. https://doi.org/10.1002/ar.22434. Epub 2012 Mar 2. PMID: 22383422

Chaves M, Aguilera-Merlo C, Cruceño A, Fogal T, Mohamed F (2015) Morphological study of the prostate gland in viscacha (*Lagostomus maximus maximus*) during periods of maximal and minimal reproductive activity. Anat Rec (Hoboken) 298(11):1919–1931. https://doi.org/10.1002/ar.23261. Epub 2015. PMID: 26290122

Clermont Y, Perey B (1957) Quantitative study of the cell population of the seminiferous tubules in immature rats. Am J Anat 100:241–266

Cran DC, Jones R (1980) Aging male reproductive system: changes in the epididymis. Exp Gerontol 15:93–101

Cruceño AM, de Rosas JC, Fóscolo M, Chaves EM, Scardapane L, Dominguez S, Aguilera-Merlo C (2013) Effects of photoperiod on epididymal and sperm morphology in a wild rodent, the viscacha (*Lagostomus maximus maximus*). ISRN Anat 2:128921. https://doi.org/10.5402/2013/128921. PMID: 25969823; PMCID: PMC4403560

Cunha G (1994) Chapter 2: role of mesenchymal-epithelial interactions in nor-mal and abnormal development of male urogenital glands. In: Riva A, Testa-Riva F, Motta PM (eds) Ultrastructure of the male urogenital glands: prostate, seminal vesicles, urethral, and bulbourethral glands. Kluwer Academic Publisher, New York, pp 15–34

El-Alfy M, Pelletier G, Hermo LSF (2000) Unique features of the basal cells of human prostate epithelium. Microsc Res Tech 51:436–446

Flickinger CJ, Howards SS, English HF (1978) Ultrastructural differences in the efferent ducts in several regions of the epididymis of the hamster. Am J Anat 152:557–586

Fuentes L, Caravaca N, Pelzer L, Scardapane L, Piezzi R, Guzmán R (1991) Seasonal variations in the testis and epididymis of viscacha (*L.m.m.*). Biol Reprod 45:493–497

Gernigon T, Berger M, Lecher P (1994) Seasonal variations in the ultrastructure and production of androgen-dependent proteins in the seminal vesicles of a Saharian rodent (*Psammomys obesus*). J Endocrinol 142:37–46

Hayward SW, Brody JR, Cunha GR (1996) An edgewise look at basal epithelial cells: three-dimensional views of the rat prostate, mammary gland and salivary gland. Differentiation 60:219–227

Hermo L, Clermont Y, Morales C (1985) Fluid-phase and ad- sorptive endocytosis in ciliated epithelial cells of the rat ductuli efferentes. Anat Rec 211t285–294

Jones RC, Murdoch RN (1996) Regulation of the motility and metabolism of spermatozoa for storage in the epididymis of eutherian and marsupial mammals. Reprod Fertil Dev 8(4):553–568. https://doi.org/10.1071/rd9960553. PMID: 8870080

Lea OA, Petrusz P, French FS (1978) Purification and localization of acidic epididymal glycoprotein (AEG): a sperm coating protein secreted by the rat epididymis. Int J Androl 2:592–607

Li-Xin S, Rodriguez MC, Jeanne OA (1990) Regulation of androgen receptor protein and mRNA concentrations by androgens in rat ventral prostate and seminal vesicles and in human hepatoma cells. Mol Endocrinol 4:1636–1646

Llanos A, Crespo J (1954) Ecología de la vizcacha (*Lagostomus maximus maximus*) en el nordeste de la provincia de Entre Ríos. Revista de Investigaciones Agrícolas. Extra Nueva Serie n° 10, 5–95

Luke M, Coffey D (1994) The male sex accessory tissue. Structure, androgen, action, and physiology. In: Knobil E, Neil J (eds) The physiology of reproduction. Raven Press, New York

Martan J (1969) Epididymal histochemistry and physiology. Biol Reprod 1:134–154

Martínez-Fresneda L, Castaño C, Bóveda P et al (2019) Epididymal and ejaculated sperm differ on their response to the cryopreservation and capacitation processes in mouflon (*Ovis musimon*). Sci Rep 9:15659

Morales A, Cavicchia JC (1993) Seasonal changes of the blood-testis barrier in viscacha (*Lagostomus maximus maximus*): a freeze-fracture and lanthanum tracer study. Anat Rec 236(3):459–464

Muñoz E (1998) Seasonal study of tubular and interstitial compartments of the viscacha (*Lagostomus maximus maximus*) testis. Role of melatonin in reproductive cyclicity and its relationship with photoperiod. Doctoral thesis, Central library National University of San Luis

Muñoz E, Fogal T, Dominguez S, Scardapane L, Guzman J, Piezzi R (1997) Seasonal changes of the Leydig cells of viscacha (*Lagostomus maximus maximus*). A light and electron microscopy study. Tissue Cell 29:119–128

Muñoz E, Fogal T, Dominguez S, Scardapane L, Guzman J, Piezzi R (1998) Stages of the cycle of the seminiferous epithelium of the viscacha (*Lagostomus maximus maximus*). Anat Rec 252:8–16

Muñoz E, Fogal T, Dominguez S, Scardapane L, Guzmán J (2001) Ultrastructural and morphometric study of the Sertoli cell of the viscacha (*Lagostomus maximus maximus*) during the annual reproductive cycle. Anat Rec 262:176–185

Nikolajczyk BS, O'Rand MG (1992) Characterization of rabbit testis beta-galactosidase and arylsulfatase A: purification and localization in spermatozoa during the acrosome reaction. Biol Reprod 46(3):366–378. https://doi.org/10.1095/biolreprod46.3.366. PMID: 1352147

Nishino T, Wedel T, Schmitt O, Bühlmeyer K, Schönfelder M, Hirtreiter C, Schulz T, Kühmel W, Michna H (2004) Androgen-dependent morphology of prostate and seminal vesicles in the Hershberger Assay: evaluation of immunohistochemical and morphometric parameters. Ann Anat 186:247–253

O'Donnell L, Stanton P, de Kretser DM (2017) Endocrinology of the male reproductive system and spermatogenesis. In: Feingold KR, Anawalt B, Boyce A et al (eds) Endotext [Internet]. MDText.com, Inc, South Dartmouth. Available from https://www.ncbi.nlm.nih.gov/books/NBK279031/

Okamura N, Tamba M, Liao HJ, Onoe S, Sugita Y, Dacheux F, Dacheux JL (1995) Cloning of complementary DNA encoding a 135-kilodalton protein secreted from porcine corpus epididymis and its identification as an epididymis-specific alpha-mannosidase. Mol Reprod Dev 42(2):141–148. https://doi.org/10.1002/mrd.1080420203. PMID: 8562059

Oróstegui P, Parraguez G, Adaro A, Peñailillo G, Cepeda R (2000) Histological and morphometric changes of the seminal vesicles of *Chinchilla laniger* (grey) in captivity induce by seasonal variations. Rev Chil Anat 18:89–96

Pescovitz OH, Srivastava CH, Breyer PR, Monts BA (1994) Paracrine control of spermatogenesis. TEM 5:126–131

Quintanilla RH, Rizzo HR, Fraga CP (1973) Rodents harmful to agriculture in the Argentine Republic. Buenos Aires University Press, p 110

Reid BL, Cleland KW (1957) The structure and function of the epididymis: I, the histology of the rat epididymis. Aust J Zool 5:223–246

Reid A, Redgrove K, Aitken J, Nixon B (2011) Cellular mechanisms regulating sperm-zona pellucida interaction. Asian J Androl 13(1):88–96. https://doi.org/10.1038/aja.2010.74

Riva A, Aümuller G (1994) Epithelium of the distal portion of the human spermatic pathway: seminal vesicle, ampulla ductus deferentis, and ejaculatory duct. In: Riva A, Riva FT (eds) Ultrastructure of male urogenital glands, vol 1. Raven Press, New York, pp 1435–1487

Robaire B, Hermo L (1988) Efferent ducts, epididymis and vas deferens: structure, functions and their regulation. In: Knobil E, Neil J (eds) The physiology of reproduction. Raven Press, New York, pp 999–1080

Russell LD, Ettlin RA, Sinha Hikim AM, Clegg ED (1990) Histological and histopathological evaluation of the testis. Cache River Press

Serre V, Robaire B (1998) Segment-specific morphological changes in aging Brown Norway rat epididymis. Biol Reprod 58:497–495

Setchell BP, Sánchez-Partida LG, Chairussyuhur A (1993) Epididymal constituents and related substances in the storage of spermatozoa: a review. Reprod Fertil Dev 5(6):601–612. https://doi.org/10.1071/rd9930601. PMID: 9627723

Setchell BP, Maddocks S, Brooks DE (1994) Anatomy, vasculature, innervations, and fluids of the male reproductive tract. In: Knobil E, Neill JD (eds) The physiology of reproduction. Raven Press, New York, pp 1063–1175

Shum WW, Da Silva N, McKee M, Smith PJ, Brown D, Breton S (2008) Transepithelial projections from basal cells are luminal sensors in pseudostratified epithelia. Cell 135(6):1108–1117. https://doi.org/10.1016/j.cell.2008.10.020. PMID: 19070580; PMCID: PMC264

Shum WW, Da Silva N, Brown D, Breton S (2009) Regulation of luminal acidification in the male reproductive tract via cell-cell crosstalk. J Exp Biol 212(Pt 11):1753–1761. https://doi.org/10.1242/jeb.027284. PMID: 19448084; PMCID: PMC26830156085

Sisson S (1975) The anatomy of domestic animals. In: Geity R (ed) Sisson & Grossman, vol 1, 5th edn. Saunders Co, Philadelphia, pp 768–769

Siwela A, Tam WH (1984) Ultrastructural changes in the prostate gland of a seasonally breeding mammal, the grey squirrel (*Sciurus carolinensis Gmelin*). J Anat 138:153–162

Sun EL, Flickinger CJ (1979) Development of cell types and regional differences in the postnatal rat epididymis. Am J Anat 154:27–55

Sun EL, Flickinger CJ (1980) Morphological characteristics of cells with apical nuclei in the initial segment of the adult rat epididymis. Anat Rec 196:285–293

Toma JG, Buzzell GR (1988) Fine structure of the ventral and dor-sal lobes of the prostate in the young adult Syrian hamster, *Mesocricetus auratus*. Am J Anat 181:132–140

Vásquez B, del Sol M (2002) Prostatic complex in the rabbit (*Oryctolagus cuniculus*). Rev Chil Anat 20:175–180

Verhagen AP, Aalders TW, Ramaekers FC, Debruyne FM, Schalken JA (1988) Differential expression of keratins in the basal adluminal compartment of rat prostatic epithelium during degeneration. Prostate 13:25–38

Wahlqvist R (2000) Ultrastructure of the accessory sex glands of the male rat under various conditions. Holstad Graphics, Oslo. 131 p

Wrobel KH (1970) Zur morphologie des samenblasen epithels derziege. Zentralbl Veterinarmed A 17:634–643

Yanagimachi R (1994) Fertility of mammalian spermatozoa: its development and relativity. Zygote 2(4):371–372. https://doi.org/10.1017/S0967199400002240

Zhou W, De Iuliis GN, Dun MD, Nixon B (2018) Characteristics of the epididymal luminal environment responsible for sperm maturation and storage. Front Endocrinol (Lausanne) 28(9):59. https://doi.org/10.3389/fendo.2018.00059. PMID: 29541061; PMCID: PMC5835514

Chapter 19
Histological and Histochemical Characteristics of the Intestinal Tract: Morphofunctional Specializations to Herbivory

María Florencia Tano de la Hoz, Stefanía Cohen, Mirta Alicia Flamini, and Alcira Ofelia Díaz

Abstract In recent years, the plains vizcacha (*Lagostomus maximus*) has been used as an animal model due to its zootechnical importance and because it has unusual morphological-physiological characteristics among eutherian mammals. The available information on the digestive physiology of *L. maximus* makes this rodent an interesting model to evaluate morphofunctional adaptations of the intestinal tract to herbivory. The current chapter provides a general overview of the morphology, ultrastructure, and glycosylation pattern of different segments of the small and large intestines of *L. maximus* and discusses the role of mucins in the intestinal tract functioning. In particular, this chapter focuses on the morphology and histochemical profile of an anatomical adaptation to herbivory, called the longitudinal colonic

M. F. Tano de la Hoz (✉)
Department of Biology, School of Exact and Natural Sciences (FCEyN), National University of Mar del Plata (UNMdP), Mar del Plata, Argentina

Institute of Materials Science and Technology Research (INTEMA) UNMdP–CONICET, Applied Electrochemical Division, Mar del Plata, Argentina
e-mail: mftano@mdp.edu.ar

S. Cohen
Fisheries Research and Development Institute (INIDEP), Mar del Plata, Argentina

Institute of Marine and Coastal Research (IIMyC), FCEyN, CONICET-UNMdP, Mar del Plata, Argentina

M. A. Flamini
Laboratory of Histology and Descriptive, Experimental and Comparative Embryology, Department of Basic Sciences, School of Veterinary Sciences, National University of La Plata, La Plata, Argentina

A. O. Díaz
Department of Biology, School of Exact and Natural Sciences (FCEyN), National University of Mar del Plata (UNMdP), Mar del Plata, Argentina

Institute of Marine and Coastal Research (IIMyC), FCEyN, CONICET-UNMdP, Mar del Plata, Argentina

L. L. Rasia et al. (eds.), *Plains Vizcachas*, https://doi.org/10.1007/978-3-031-49487-1_19

groove, which is present in the ascending colon of *L. maximus* and functions as a route of retrograde transport of a mixture of bacteria and mucus. In addition, the distribution pattern of the telocytes formerly interstitial cells of Cajal (ICC) is described, assessing its regulatory influence on gut motility.

Keywords Plains vizcacha · Intestinal tract · Colonic groove · Mucins · Morphology

The small intestine of *Lagostomus maximus* is divided into three anatomical portions: duodenum, jejunum, and ileum (Tano de la Hoz et al. 2014, 2016). The large intestine begins at the level of the ileocecal valve and comprises the cecum, colon, rectum, and anal canal (Tano de la Hoz et al. 2019). The cecum is the largest organ in the intestinal tract, lacks a vermiform appendix, and has a proximal portion composed of the base and body and an elongated distal region that terminates at the apex. The external surface of the cecum and ascending colon is characterized by numerous sacculations and projections of the serosa replete with adipose tissue, commonly referred to as epiploic appendages. Throughout its entire length, the wall of the intestinal tract is formed by the four typical tunics of the digestive tube (mucosa, submucosa, muscular, serosa, or adventitia).

The ascending colon of *L. maximus* has a specialized anatomical modification for herbivory known as the longitudinal colonic groove (LCG) along the mesenteric side (Fig. 19.1) (Tano de la Hoz et al. 2017). As described in other small herbivores, the LCG works as a retrograde transport route to the cecum for a mixture of bacteria and mucus (Björnhag and Snipes 1999; Takahashi and Sakaguchi 2000, 2006; Sakaguchi 2003; Franz et al. 2011). Because of this process, known as the colonic separation mechanism (CSM), bacteria are not released in feces but are accumulated in the cecum, where bacterial fermentation occurs. The material separated by the CSM is excreted in the feces through cecal contractions and re-ingested by the vizcacha. Unlike hard, fiber-rich feces, the CSM generates soft feces, or cecotrophs, with high protein content. The presence of the LCG is considered a prerequisite for the practice of coprophagy, demonstrating that *L. maximus* utilizes this digestive strategy. Furthermore, it has been shown through the use of digestibility marker tests that *L. maximus*, similar to other hystricomorph rodents, re-ingests a portion of its own feces. Studies based on ingesta passage have revealed that the selective transport of bacteria to the cecum is promoted by mucins secreted in the colonic groove, classifying it as a mucus-dependent CSM (Clauss et al. 2007; Hagen et al. 2015).

In *L. maximus*, the LCG originates near the cecocolic junction and is delimited at both edges by non-papillated ridges (Fig. 19.1a, b) that divide the lumen of the ascending colon into two compartments: the groove lumen and the main lumen, with the latter being larger in dimension (Fig. 19.1a). Through the length of the ascending colon, there is a gradual reduction in the spacing between ridges. This results in a constriction of the groove towards the distal portion of the organ, culminating in its eventual disappearance (Fig. 19.1c and inset of c). Each ridge presents

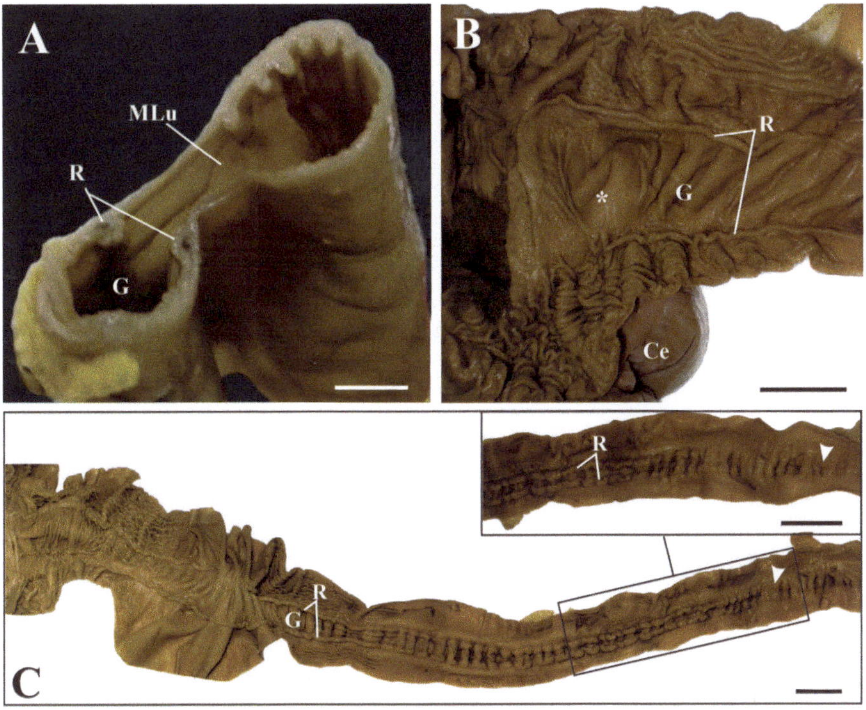

Fig. 19.1 Morphological characterization of the longitudinal colonic groove. (**a**) Cross section of the ascending colon. (**b**) Macroscopic anatomy of the groove's origin. (**c**) Macroscopic view of the inner surface of the ascending colon. Inset of **c**. Macroscopic view of the caudal region of the ascending colon. Arrowhead, caudal region of the ascending colon where the groove disappeared; asterisk, cecocolic junction; Ce, cecum; G, groove; MLu, main lumen; R, ridge. Scale bar: 1 cm (**a**); 1.5 cm (**b**, **c**, inset of **c**). (Reproduced with permission of John Wiley and Sons and Copyright Clearance Center (License Number 5494711266574))

two lateral edges known as "colonic luminal side" (CLS), which lines the main lumen of the organ, and "groove side" (GS), which borders the colonic groove (Fig. 19.2a). While the histological and anatomical arrangement of the colonic groove in *L. maximus* shares similarities with that of other hystricomorph rodents, it also possesses unique characteristics specific to the species. Like all studied rodent species, the mucosal tunic is composed of a superficial lining epithelium, a proper layer of connective tissue with Lieberkühn's glands, and a layer of smooth muscle cells called the muscularis mucosae. The center of the ridges consists of an extension of the submucosa tunic and contains large-caliber veins (Fig. 19.2b, c) (Snipes et al. 1988; Kotzé et al. 2009; Tano de la Hoz et al. 2017). As proposed by Kotzé et al. (2009), it is possible that the presence of these large veins leads to swelling of the ridges due to increased blood irrigation, resulting in greater compartmentalization of the lumen. At the height of the ridges, the structural characteristics of the tunica muscularis of the ascending colon undergo a sudden change. The outer longitudinal layer, which is present in most regions of the intestinal tube, gradually

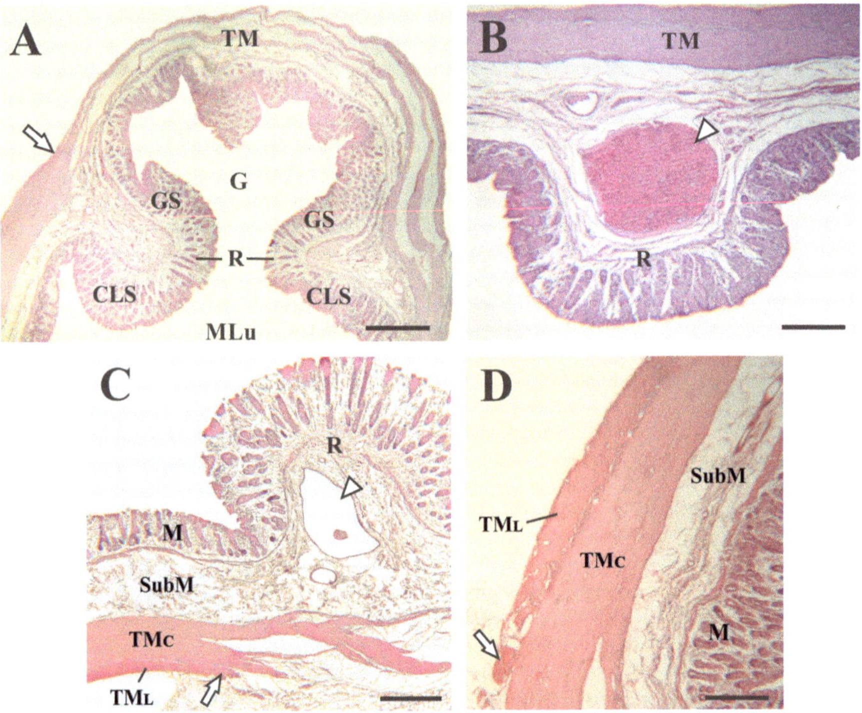

Fig. 19.2 Histological characterization of the longitudinal colonic groove, H-E. (**a**) Microphotography of the groove. (**b**) Non-papillated ridge. (**c**) Detail of the abrupt change of the tunica muscularis morphology. (**d**) Detail of the tunica muscularis. Arrowhead, large caliber vein; block arrow, morphological change of the tunica muscularis; CLS, lateral border of the ridge limiting with the main lumen; G, groove; GS, lateral border of the ridge limiting the groove; M, tunica mucosa; MLu, main lumen; R, ridge; SubM, tunica submucosa; TM, tunica muscularis; TMC, inner circular layer of the tunica muscularis; TML, outer longitudinal layer of the tunica muscularis. Scale bar: 800 mm (**a**); 400 mm (**b**); 250 mm (**c**); 100 mm (**d**). (Reproduced with permission of John Wiley and Sons and Copyright Clearance Center (License Number 5494711266574))

reduces in thickness until it finally disappears within the groove (Fig. 19.2c, d). Consequently, the tunica muscularis of the groove exclusively comprises a circular layer composed of smooth muscle organized into bundles separated by partitions of connective tissue (Fig. 19.2a, c). Despite having only one circular layer, the tunica muscularis of the groove is thicker than the muscular layer of the rest of the organ. Similar histological characteristics have been described in the nutria *Myocastor coypus*, although in this rodent, the muscular bundles of the circular layer have a different organization than in *L. maximus* (Snipes et al. 1988). In contrast, in six Species of African mole-rats (Rodentia, Bathyergidae), Kotzé et al. (2009) observed that precisely below the ridges that delimit the groove, the thickness of the outer longitudinal layer increases. Although notable differences exist between species, changes in the histological characteristics of the tunica muscularis have been described at the level of the ridgests in all rodents studied. This highly specialized structural adaptation of the muscular tissue is possibly related to the antiperistaltic movements that occur in the colonic groove to transport the luminal content in a retrograde way (Takahashi and Sakaguchi 2006).

Advancements made in recent decades have revealed that the motor activity of the intestinal tract is a complex physiological process involving the interaction of three cell types: enteric neurons, telocytes formerly interstitial cells of Cajal (ICC), and smooth muscle cells (Mazzone and Farrugia 2007; Mazet 2015; Foong et al. 2020). Within the intestinal tract of *L. maximus*, different subpopulations of ICC are identified based on their morphology and anatomical locations (Fig. 19.3) (Tano de la Hoz et al. 2017). In particular, there is a subpopulation of ICC located exclusively in the colonic groove, known as submucosal ICC (ICC-SM) (Fig. 19.4). These ICC-SM cells reside in the submucosal region bordering the tunica muscularis and extend from ridge to ridge, forming a continuous band of multipolar cells with thin, long, and interdigitating processes. While there are currently no similar studies in the LCG of other herbivores, ICC-SM cells are known to generate slow waves in the colon of other mammals and provide pacemaker activity in colonic muscles (Mazzone and Farrugia 2007; Blair et al. 2014).

The mucosa of *L. maximus*'s intestinal tract is entirely covered by a simple columnar epithelium comprising different types of mature cells: enterocytes, enteroendocrine cells, and goblet cells. Additionally, in the small intestine, a fourth type of cell called Paneth cells is found. This epithelium composed of highly specialized cells is related to the diverse digestive, absorptive, endocrine, and immunological functions performed by the intestinal mucosa. Like in other vertebrates, enterocytes in *L. maximus* present distinct cellular polarity, with uniform microvilli on the apical region, and complexes of junctions and interdigitating cytoplasmic extensions on their lateral surfaces (Fig. 19.5a and inset of a). In the ileum, enterocytes can be distinguished based on the electron density of their cytoplasm; the majority of these absorptive cells have electron-lucid (EL) cytoplasm, while a smaller number of enterocytes possess electron-dense (ED) cytoplasm (Fig. 19.5b). Similar ultrastructural findings have been reported in some bird and mammal species (Anwar et al. 2021; Roshdy et al. 2021). Additionally, molecular studies have identified two types of enterocytes in the human small intestine, with differences in gene expression profiles and in distribution patterns along the crypt-villus axis (Gassler et al. 2006). In *L. maximus*, the spatial distribution of these two enterocyte types within the intestinal epithelium is not clearly defined. Nevertheless, the observed morphological differences between these cell types could potentially indicate different physiological states (Tano de la Hoz et al. 2016). Lectin-histochemical studies have demonstrated the presence of glycoconjugates (GCs) in the supranuclear region of enterocytes in most intestinal segments of *L. maximus* (Tano de la Hoz et al. 2014, 2016). This labeling pattern is likely indicative of the glycosylation process that GCs undergo within the endomembrane system (Gimeno and Barbeito 2004; Gabrielli and Tomassoni 2017). Morphologically, these absorptive cells exhibit well-developed Golgi cisternae in the supranuclear region and a highly developed rough endoplasmic reticulum (RER).

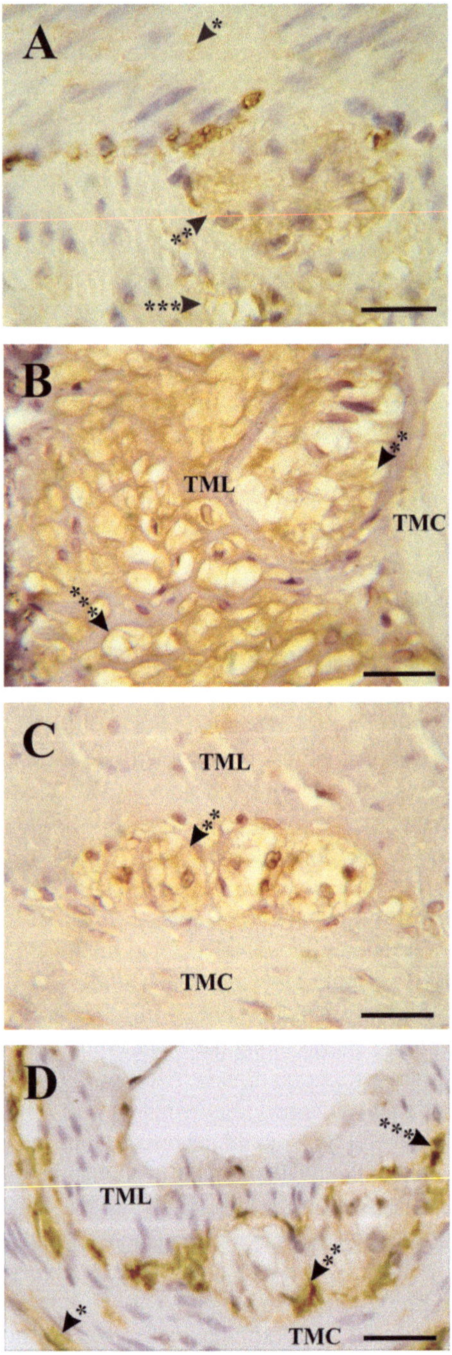

Fig. 19.3 Immunohistochemistry anti-c-kit to detect interstitial cells of Cajal (ICC). (**a**) Jejunum. (**b**) Descending colon. (**c**, **d**) Rectum. ◄* positive cells in the circular layer of the tunica muscularis; ◄** positive cells in the myenteric plexus region. ◄*** positive cells in the longitudinal layer of the tunica muscularis. TMC, inner circular layer of the tunica muscularis; TML, outer longitudinal layer of the tunica muscularis. Scale bar: 20 μm (**a**, **c**, **d**); 15 μm (**b**)

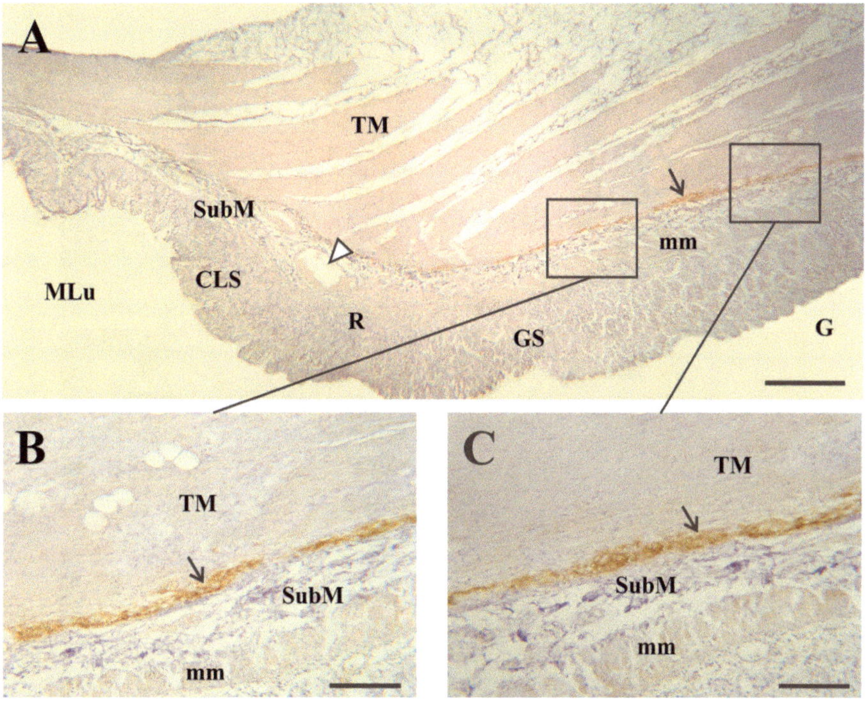

Fig. 19.4 Immunohistochemistry anti-c-kit to evidentiate the interstitial cells of Cajal (ICC) of the colonic groove. (**a**) Microphotography of a ridge showing a continuous band of c-kit positive cells in the submucosa region limiting the tunica muscularis. (**b**, **c**) Detail of the tunica submucosa with c-kit positive cells. Arrow, ICC-SM; arrowhead, large caliber vein; CLS, lateral border of the ridge limiting the main lumen; G, groove; Gs, lateral border of the ridge limiting the groove; MLu, main lumen; mm, muscularis mucosa; R, ridge; SubM, tunica submucosa; TM, tunica muscularis. Scale bar: 500 mm (**a**); 100 mm (**b**, **c**). (Reproduced with permission of John Wiley and Sons and Copyright Clearance Center (License Number 5503170007766))

Paneth cells are located at the base of the intestinal or Lieberkühn crypts (Fig. 19.6f and inset of f) in all three portions of the small intestine of *L. maximus*, showing a distribution pattern similar to that found in mice, rats, guinea pigs, horses, and humans (Satoh et al. 1990; Takehana et al. 1998; Vásquez Cachay et al. 2014). In *L. maximus*, these cells share morphological similarities with Paneth cells from other species, such as their pyramidal shape and supranuclear acidophilic granules (Zanuzzi et al. 2008; Vásquez Cachay et al. 2014; Tano de la Hoz et al. 2016). At the ultrastructural level, most of the granules in *L. maximus* Paneth cells have a homogeneous matrix, while others display a bipartite structure with an electron-dense center and a less dense peripheral region (Fig. 19.5d). This bipartite structure described in some Paneth cell granules in *L. maximus* has also been observed in rodents of the suborder Myomorpha and in humans (Satoh et al. 1990).

Enteroendocrine cells are the least predominant cell type present in the intestinal mucosa. In the small intestine of *L. maximus*, these cells are mainly located at the

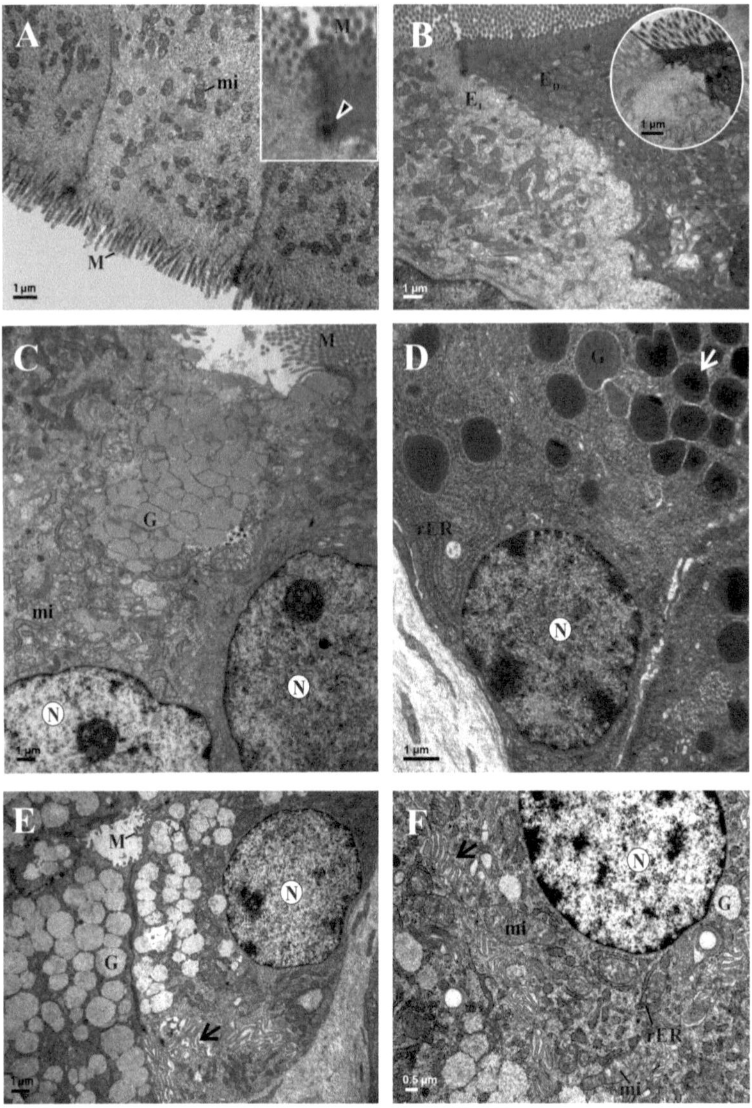

Fig. 19.5 Transmission electron micrographs of intestinal epithelial cells. (**a**) Duodenal enterocytes. Inset of **a**. Detail of the apicolateral region of enterocytes. (**b**) Ileal enterocytes with different cytoplasmic electron densities. Inset of **b**. Detail of the apicolateral region of both types of enterocytes. (**c**) Goblet cell with numerous electron-lucid mucous granules. (**d**) Paneth cell with numerous cytoplasmatic granules, jejunum. Note the secretory granules with a bipartite substructure (white arrow). (**e**, **f**) Brunner's glands (duodenal glands). Arrowhead, junctional complex; black arrow, folds of the lateral membrane; E_D, electron-dense enterocyte; E_L, electron-lucid enterocyte; G, secretory granules; M, microvilli; mi, mitochondria; N, nucleus; RER, rough endoplasmic reticulum

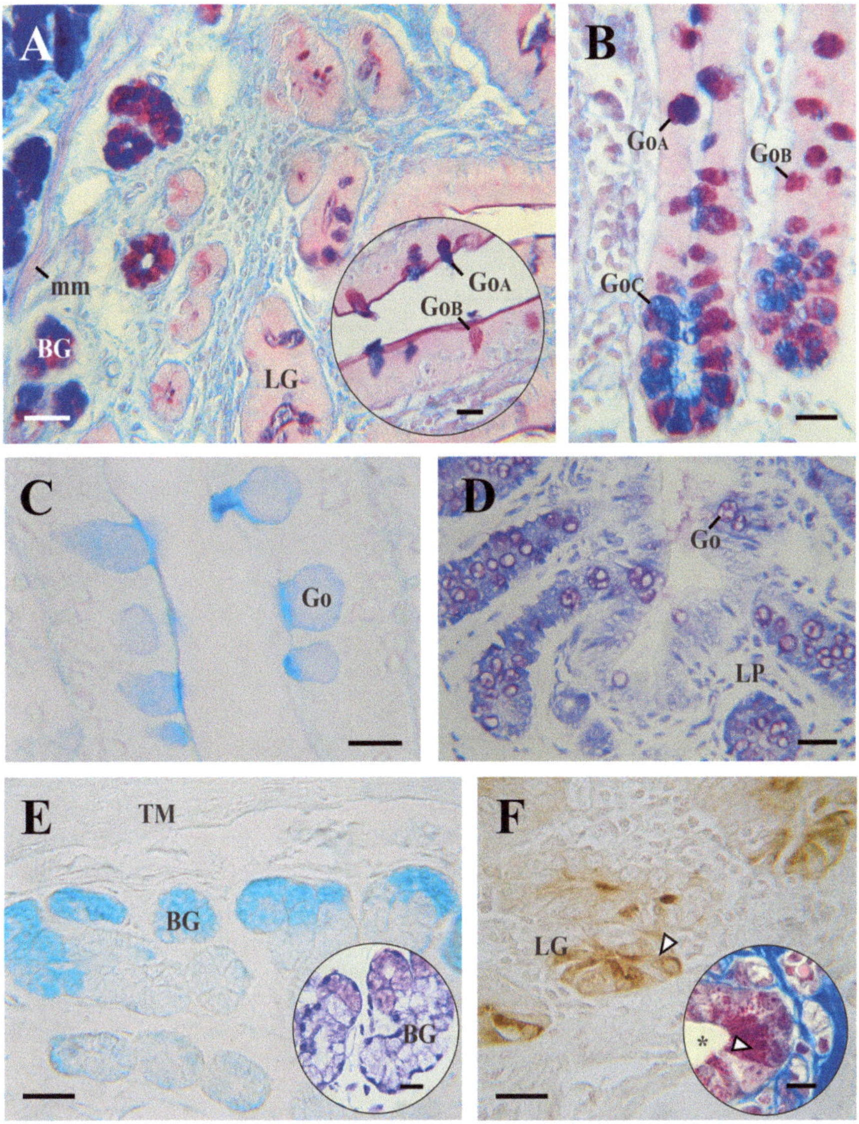

Fig. 19.6 Histochemical characterization of the intestinal tract. (**a**) Brunner's glands in the duodenal mucosa. Inset of **a**. Duodenal goblet cells AB/PAS positive (GoA; mixed mucins) and PAS positive (GoB; neutral mucins), AB pH 2.5/PAS. (**b**) Goblet cells with three glycosylation patterns (GoA; mixed mucins; GoB; neutral mucins; GoC, acidic mucins) in the cecum, AB pH 2.5/PAS. (**c**) Jejunal goblet cells slightly AB positive, AB pH 2.5. (**d**) Metachromatic goblet cells in the distal colon showing high concentrations of anionic groups, Toluidine blue technique pH 5.6. (**e**) Brunner's glands from the deep area of the submucosa, AB pH 1.0. Inset of **e**. Toluidine blue technique pH 5.6. (**f**) Jejunal Paneth cells, UEA-I lectin. Inset of **f**. Masson's trichrome. Arrowhead, Paneth cells; asterisk, lumen; BG, Brunner's glands; Go, goblet cell; LG, Lieberkühn gland; LP, lamina propria; mm, muscularis mucosae; SubM, submucosa; TM, tunica muscularis. Scale bar = 20 µm (**a**, **b**, **d**, **f**); 10 µm (Inset of **a**, **c**, inset of **f**); 40 µm (**e**). (Reproduced with permission of John Wiley and Sons and Copyright Clearance Center (License Number 5494711266574))

base of Lieberkühn crypts and contain an abundance of electron-dense secretory granules with diverse shapes and sizes in their basal cytoplasm (Tano de la Hoz et al. 2016). The population of enteroendocrine cells in the large intestine of most mammals is generally less diverse than in the small intestine (Rindi et al. 2004; Fakhry et al. 2017). However, despite comprising only about 1% of the cells lining the intestinal lumen, in mammals, they represent the largest population of hormone-producing cells (Schonhoff et al. 2004). The various peptide hormones produced by these cells regulate important physiological functions of the digestive system, as well as electrogenic ion secretion and intestinal motility (Gunawardene et al. 2011).

In the intestinal tract of *L. maximus*, mucins are synthesized and secreted by mucous branched tubular glands limited to the duodenum, called Brunner's glands, and by unicellular glands distributed along the entire epithelial lining, known as goblet cells. The high carbohydrate content of mucins is the main factor determining their high molecular weight and many of their physicochemical properties. Their functions include mucosal lubrication, modulation of water and electrolyte absorption, and protection of the underlying epithelium from physical and chemical abrasion. Moreover, mucins have been shown to act as binding sites for commensal and pathogenic bacteria (Robbe et al. 2004; Kim and Ho 2010). Similar to other species of hystricomorph and myomorph rodents, Brunner's glands in *L. maximus* possess adenomers consisting of mucous cells and excretory ducts that open into the intestinal lumen. Mucous cells exhibit a pyramidal shape with few short microvilli on their luminal surface and numerous lateral membrane folds (Fig. 19.5e, f). They also contain electron-lucent granules in their apical cytoplasm, abundant mitochondria, extensive RER, and a supranuclear Golgi complex (Fig. 19.5f). While in most mammals, Brunner's glands are restricted to the submucosa, in *L. maximus*, besides being present throughout the submucosal layer, some adenomers are found in the lamina propria (Fig. 19.6a) (Tano de la Hoz et al. 2014). This distribution has also been described in humans and canines by Skutelsky et al. (1989). Histochemical and lectin-histochemical studies in *L. maximus* have demonstrated that the adenomers of Brunner's glands in the deep area of the submucosa synthesize sulfated GCs, displaying a different glycosylation pattern compared to those in the superficial area (Fig. 19.6e and inset of e). This histochemical profile of Brunner's glands agrees with that described by Verdiglione et al. (2002) in cattle. In the bovine duodenum, different degrees of sulfation have been described among the adenomers of Brunner's glands, being the adenomers of the deepest region of the submucosa the only ones that contain sulfomucins. Similar findings have also been described in the duodenum of guinea pigs (Scillitani and Mentino 2015). This pattern may be attributed to the presence of adenomers in different secretory stages or the existence of two subpopulations of glands with different glycosylation patterns.

Goblet cells play a fundamental role in the synthesis of intestinal mucins, which has led to significant interest in recent years in studying their morphology, function, distribution, and glycosylation pattern (Fekete and Buret 2023). Like all mammals, the goblet cells in *L. maximus* are distributed throughout the epithelium of the intestinal tract, with higher abundance in the large intestine (Boonzaier et al. 2013a, 2013b; Pawlina 2020). Transmission electron microscope (TEM) reveals a substantial accumulation of mucinogen granules in the apical cytoplasm, causing distension in this region of the cell (Fig. 19.5c). The presence of abundant mucinogen granules is an ultrastructural feature used as a criterion to identify cells in a mature secretory stage (Birchenough et al. 2015). In the large intestine of *L. maximus*, mature goblet cells are found in the lower third of the Lieberkühn glands. These observations corroborate that, like other vertebrates, goblet cells in this region of *L. maximus* intestine begin to mature in the replication cell zone.

Throughout the intestinal tract of *L. maximus*, there are goblet cells with different glycosylation patterns. In the large intestine, three distinct histochemical profiles can be identified, neutral, acidic, and mixed (Fig. 19.6b), while in the small intestine, only two types of secretion, neutral and mixed, are characterized along the crypt-villus axis (Fig. 19.6, inset of a). Among all these glycosylation patterns, the predominant cells throughout the intestinal tract of *L. maximus* are those with mixed secretion. This same type of secretion was described by Boonzaier et al. (2013b) in the gastrointestinal tract of three species of insectivorous mammals. These similarities between taxonomically distant species suggest that mixed mucins are essential for most mammals, regardless of their diet or taxonomic location. On the other hand, the existence of goblet cells with different glycosylation patterns demonstrates that functionally, there are distinct subclasses of goblet cells within a morphologically homogeneous cell population, both at the histological and ultrastructural levels. As described in other vertebrate species, the glycosylation pattern remains highly conserved among individuals of the same species and is independent of individuals' sex. Currently, it is proposed that the glycosylation pattern of intestinal mucins could be the result of a co-evolutionary process between the microbiota and the host to maintain an optimal symbiotic relationship (Johansson et al. 2011a, b; Fekete and Buret 2023). A growing gradient of sulfated mucins (sulfomucins), associated with a decreasing gradient of neutral mucins, is observed along the intestinal tract of *L. maximus* (Fig. 19.6c, d) (Tano de la Hoz et al. 2016, 2019). Studies in different mammalian species have described a glycosylation pattern similar to that of *L. maximus*. Several authors have shown that acidic glycoconjugates, especially sulfated glycoconjugates, act as a defensive barrier, protecting the intestine from potential pathogens and digestive enzymes (Robertson and Wright 1997; Beyaz and Liman 2009). Additionally, it has been demonstrated that sulfomucins are more resistant to degradation by bacterial glycosidases and have a stronger inhibitory effect on bacterial growth compared to non-sulfated mucins (Conour et al. 2002). Lectin-histochemical studies using Wheat Germ Agglutinin (WGA) revealed terminal residues of N-acetylglucosamine (GlcNAc) in goblet cells throughout the intestinal tract of *L. maximus*. Moreover, an increase in the labeling pattern was observed towards the large intestine. In mammals, the presence of GlcNAc is particularly associated with the degree of mucin acidification since sulfate groups are generally attached to this monosaccharide (Liquori et al. 2012). In *L. maximus*, there is a correspondence between the acidic gradient of secretory mucins and the distribution of terminal residues of GlcNAc, consistent with the findings described by Liquori et al. (2012) in mice.

In *L. maximus*, abrupt changes in the histochemical profile of goblet cells are identified at the level of the longitudinal colonic groove throughout the ascending colon. In all examined animals, the mucosa lining the luminal colonic side of the ridge (CLS) shows a distinct glycosylation pattern when compared to the lateral ridge bordering the groove (groove side, GS) (Fig. 19.7). Both the groove and the main lumen of the organ have goblet cells containing neutral, acidic, and mixed mucins, but their distribution pattern is considerably different (Fig. 19.7). The crypts outside the groove present a gradient with neutral and mixed GCs in the upper zone and mainly acidic GCs in the basal zone (Fig. 19.7a–c). In contrast, within the groove, the three types of GCs are located throughout the axis of the crypts (Fig. 19.7d). This variation highlights the crucial role of mucus glycosylation pattern in the functioning of this anatomical adaptation. As described by Hansson (2012), mucus can trap bacteria in the intestinal lumen in several ways. Microorganisms can be trapped in pores formed by the arrangement of different mucin types, or bacteria can attach to a diverse range of mucus glycans using both fimbrial and afimbrial adhesins, each selectively recognizing distinct carbohydrate residues. Given the importance of mucus composition in bacterial aggregation, it is possible that the particular histochemical profile of the colonic groove generates specific anchoring sites for the high density of bacteria present in this region. Current evidence demonstrates that mucins interact and dynamically respond to the intestinal microflora (Corfield 2015). Freitas et al. (2002) have shown that the microbiota induces modifications in the glycosylation of intestinal mucins, leading to an increase in sulfated mucin secretion. Similarly, the amount of sulfomucins is higher in the mucus secreted within the groove compared to that secreted in the remainder of the ascending colon (Fig. 19.7e, f). Some studies have demonstrated that sulfate groups provide mucus with greater resistance to degradation by bacterial glycosidases and host proteases (Robertson and Wright 1997; McGuckin et al. 2011). In vitro experiments have also determined that highly sulfated mucins can reduce the growth of certain types of bacteria compared to non-sulfated mucins (Chance and Mawhinney 2000). Considering these findings and given that the colonic groove lumen transports a higher concentration of bacteria than the main lumen of the ascending colon (Takahashi and Sakaguchi 2006), it is possible that sulfomucins are involved in protecting the groove mucosa by inhibiting the proliferation of pathogenic microorganisms. Moreover, there are differences between the colonic groove and the main lumen regarding the types of sialomucins secreted by goblet cells in both regions (Tano de la Hoz et al. 2017). Mucous cells secrete GCs with sialic acid residues throughout the ascending colon; however, GCs O-acyl substituted at C7, C8, or C9 are secreted only in the main lumen. Similarly, other studies have described a high proportion of O-acetylated sialomucins in the colon of humans and rodent mammals (Accili et al. 2008; Mastrodonato et al. 2013). As previously proposed (Angata and Varki 2002), the acetylation of sialic acid residues can substantially modify their functional role in different biological processes. The higher degree of acetylation of sialic acid residues secreted by goblet cells in the main lumen may alter the viscoelastic and protective properties of mucus, providing greater resistance to bacterial neuraminidases (Mastrodonato et al. 2013).

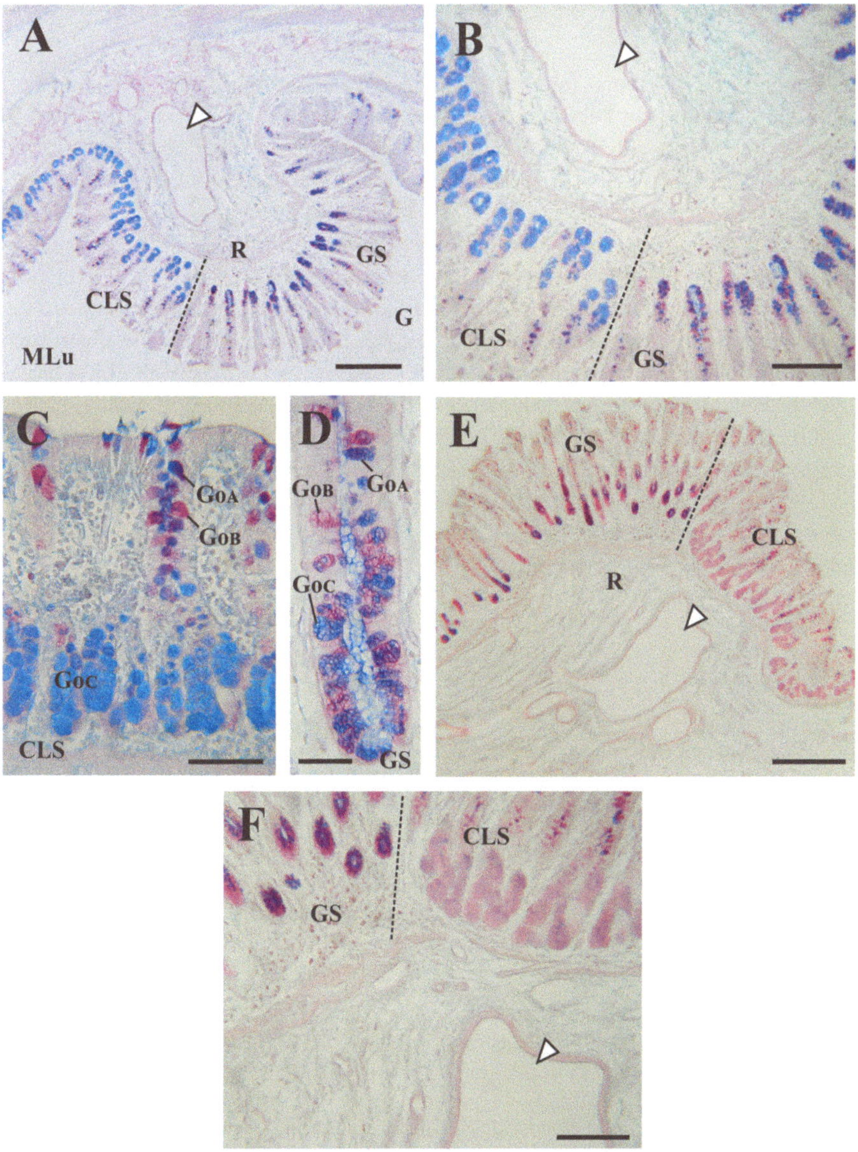

Fig. 19.7 In situ characterization of the glycosylation pattern of the ascending colon. (**a**) Microphotography of a colonic ridge, AB pH 2.5/PAS. (**b**) Detail of figure **a** showing the abrupt change in the histochemical profile of goblet cells (see dotted line), AB pH 2.5/PAS. (**c**) Microphotography of the lateral ridge limiting the main lumen of the colon (CLS), AB pH 2.5/PAS. (**d**) Microphotography of an intestinal crypt in the side of the ridge limiting the colonic groove (GS), AB pH 2.5/PAS. (**e**) Non-papillated ridge showing an abrupt change in the histochemical pattern of goblet cells (see dotted line), AB pH 1.0/PAS. (**f**) Detail of figure **e**, AB pH 1.0/PAS. Arrowhead, large caliber vein; CLS, lateral border of the ridge limiting the main lumen; G, groove; GoA, goblet cell AB/PAS positive; GoB, goblet cell PAS positive; GoC, goblet cell AB positive; GS, lateral border of the ridge limiting the groove; MLu, main lumen; R, ridge. Scale bar: 500 mm (**a**, **e**); 250 mm (**b**, **f**); 75 mm (**c**); 50 mm (**d**)

19.1 Conclusions

In summary, the intestinal tract of *L. maximus* has anatomical adaptations to herbivory. In particular, the cecum is a voluminous organ that, like in other herbivorous rodents, functions as a fermentation chamber, and the ascending colon has a longitudinal colonic groove (LCG) in the mesenteric region. Given that these characteristics are considered a prerequisite for the practice of coprophagy, it is possible to affirm that *L. maximus* uses this digestive strategy. Furthermore, histochemical analysis reveals that the composition of mucosubstances produced by goblet cells varies throughout the intestinal tract, with an increase in sulfated mucins toward the colon. It is possible that acidic GCs increase the negative charge of the mucus and confer greater resistance to degradation by bacterial glycosidases and host proteases. In particular, the abrupt changes described in the histochemical profile of goblet cells at the level of the LCG suggest that the mucus glycosylation pattern plays a key role in the functioning of the colonic separation mechanism. In conclusion, the integrated analysis of the morphological and histochemical characteristics of the digestive tract of *L. maximus* provided a better understanding of the morphofunctional adaptations used by this species to achieve greater digestive efficiency.

References

Accili D, Menghi G, Gabrielli MG (2008) Lectin histochemistry for in situ profiling of rat colon sialoglycoconjugates. Histol Histopathol 23:863–875

Angata T, Varki A (2002) Chemical diversity in the sialic acids and related α-keto acids: an evolutionary perspective. Chem Rev 102:439–469

Anwar SM, Abd-Elhafeez HH, Abdel-maksoud FM, Abdalla KE (2021) Morph-anatomic and histochemical study of ileum of goose (*Alopochen aegyptiacus*) with special references to immune cells, mucous and serous goblet cells, telocytes, and dark and light smooth muscle fibers. Microsc Res Tech 84(6):1328–1347

Beyaz F, Liman N (2009) The prenatal development and histochemistry of the ileal mucins in the bovine fetuses. Anat Histol Embryol 38:436–442

Birchenough GMH, Johansson MEV, Gustafsson JK, Bergström JH, Hansson GC (2015) New developments in goblet cell mucus secretion and function. Mucosal Immunol 8:712–719

Björnhag G, Snipes RL (1999) Colonic separation mechanism in lagomorph and rodent species – a comparison. Zoosyst Evol 75:275–281

Blair PJ, Rhee PL, Sanders KM, Ward SM (2014) The significance of interstitial cells in neurogastroenterology. J Neurogastroenterol Motil 20(3):294–317

Boonzaier J, Van der Merwe EL, Bennett NC, Kotzé SH (2013a) Comparative gastrointestinal morphology of three small mammalian insectivores: *Acomys spinosissimus* (Rodentia), *Crocidura cyanea* (Eulipotyphla), and *Amblysomus hottentotus* (Afrosoricida). J Morphol 274:615–626

Boonzaier J, Van der Merwe EL, Bennett NC, Kotzé SH (2013b) A comparative histochemical study of the distribution of mucins in the gastrointestinal tracts of three insectivorous mammals. Acta Histochem 115:549–556

Chance DL, Mawhinney TP (2000) Carbohydrate sulfation effects on growth of *Pseudomonas aeruginosa*. Microbiology 146:1717–1725

Clauss M, Besselmann D, Schwarm A, Ortmann S, Hatt JM (2007) Demonstrating coprophagy with passage markers? The example of the plains viscacha (*Lagostomus maximus*). Comp Biochem Physiol A Mol Integr Physiol 147(2):453–459

Conour JE, Ganessunker D, Tappenden KA, Donovan SM, Gaskins HR (2002) Acidomucin goblet cell expansion induced by parenteral nutrition in the small intestine of piglets. Am J Physiol Gastrointest Liver Physiol 283:G1185–G1196

Corfield AP (2015) Mucins: a biologically relevant glycan barrier inmucosal protection. Biochim Biophys Acta 1850:236–252

Fakhry J, Wang J, Martins P, Fothergill LJ, Hunne B (2017) Distribution and characterisation of CCK containing enteroendocrine cells of the mouse small and large intestine. Cell Tissue Res. https://doi.org/10.1007/s00441-017-2612-1

Fekete E, Buret AG (2023) The role of mucin O-glycans in microbiota dysbiosis, intestinal homeostasis, and host-pathogen interactions. Am J Physiol Gastrointest Liver Physiol 324(6):G452–G465

Foong D, Zhou J, Zarrouk A, Ho V, O'Connor MD (2020) Understanding the biology of human interstitial cells of Cajal in gastrointestinal motility. Int J Mol Sci 21(12):4540

Franz R, Kreuzer M, Hummel J, Hatt JM, Clauss M (2011) Intake, selection, digesta retention, digestion and gut fill of two coprophagous species, rabbits (*Oryctolagus cuniculus*) and guinea pigs (*Cavia porcellus*), on a hay-only diet. J Anim Physiol Anim Nutr 95:564–570

Freitas M, Axelsson LG, Cayuela C, Midtvedt T, Trugnan G (2002) Microbial–host interactions specifically control the glycosylation pattern in intestinal mouse mucosa. Histochem Cell Biol 118:149–161

Gabrielli MG, Tomassoni D (2017) Starch-enriched diet modulates the glucidic profile in the rat colonic mucosa. Eur J Nutr. https://doi.org/10.1007/s00394-017-1393-3

Gassler N, Newrzella D, Böhm C, Lyer S, Li L, Sorgenfrei O, van Laer L, Sido B, Mollenhauer J, Poustka A, Schirmacher P, Gretz N (2006) Molecular characterisation of non-absorptive and absorptive enterocytes in human small intestine. Gut 55:1084–1089

Gimeno EJ, Barbeito CG (2004) Glicobiología, una nueva dimensión para el estudio de la biología y la patología. Academia Nacional de Agronomía y Veterinaria LVIII:5–34

Gunawardene AR, Corfe BM, Staton CA (2011) Classification and functions of enteroendocrine cells of the lower gastrointestinal tract. Int J Exp Pathol 92:219–231

Hagen KB, Besselmann D, Cyrus-Eulenberger U, Vendl C, Ortmann S, Zingg R, Kienzle E, Kreuzer M, Jean-M H, Clauss M (2015) Digestive physiology of the plains viscacha (*Lagostomus maximus*): a large herbivorous Hystricomorph rodent. Zoo Biol 34:345–359

Hansson GC (2012) Role of mucus layers in gut infection and inflammation. Curr Opin Microbiol 15:57–62

Johansson MEV, Ambort D, Pelaseyed T, Schitte A, Gustafsson JK, Ermund A, Subramani DB, Holmén-Larsson JM, Thomsson KA, Bergström JH et al (2011a) Composition and functional role of the mucus layers in the intestine. Cell Mol Life Sci 68:3635–3641

Johansson ME, Larsson JM, Hansson GC (2011b) The two mucus layers of colon are organized by the MUC2 mucin, whereas the outer layer is a legislator of host-microbial interactions. Proc Natl Acad Sci 108:4659–4665

Kim YS, Ho SB (2010) Intestinal goblet cells and mucins in health and disease: recent insights and progress. Curr Gastroenterol Rep 12:319–330

Kotzé SH, van der Merwe EL, Ndou R, O'Riain MJ, Bennett NC (2009) The colonic groove or furrow: a comparative morphological study of six species of African mole-rats (Rodentia, Bathyergidae). J Morphol 270:966–975

Liquori GE, Mastrodonato M, Mentino D, Scillitani G, Desantis S, Portincasa P, Ferri D (2012) In situ characterization of O-linked glycans of Muc2 in mouse colon. Acta Histochem 114:723–732

Mastrodonato M, Mentino D, Liquori GE, Ferri D (2013) Histochemical characterization of the sialic acid residues in mouse colon mucins. Microsc Res Tech 76:156–162

Mazet B (2015) Gastrointestinal motility and its enteric actors in mechanosensitivity: past and present. Pflugers Arch – Eur J Physiol 467:191–200

Mazzone A, Farrugia G (2007) Evolving concepts in the cellular control of gastrointestinal motility: neurogastroenterology and enteric sciences. Gastroenterol Clin N Am 36:499–513

McGuckin MA, Lindén SK, Sutton P, Florin TH (2011) Mucin dynamics and enteric pathogens. Nat Rev Microbiol 9:265–278

Pawlina W (2020) Ross: Histología texto y atlas, correlación con biología molecular y celular, 8° edición. Wolters Kluwer Health, 1036 p

Rindi G, Leiter AB, Kopin AS, Bordi C, Solcia E (2004) The "normal" endocrine cell of the gut changing concepts and new evidences. Ann N Y Acad Sci 1014:1–12

Robbe C, Capon C, Coddeville B, Michalski JC (2004) Structural diversity and specific distribution of O-glycans in normal human mucins along the intestinal tract. Biochem J 384:307–316

Robertson AM, Wright DP (1997) Bacterial glycosulphatases and sulphomucin degradation. Can J Gastroenterol Hepatol 11:361–366

Roshdy K, Alsafy M, El-Gendy S, Elghoul M (2021) Microscopic characterization of the jejunal epithelial cells of the pig. Ann Rom Soc Cell Biol 25(6):18527–18537

Sakaguchi E (2003) Digestive strategies of small hindgut fermenters. Anim Sci J 74:327–337

Satoh Y, Yamano M, Matsuda M, Ono K (1990) Ultrastructure of Paneth cells in the intestine of various mammals. J Electron Microsc Tech 16:69–80

Schonhoff SE, Giel-Moloney M, Leiter AB (2004) Minireview: development and differentiation of gut endocrine cells. Endocrinology 145:2639–2644

Scillitani G, Mentino D (2015) Comparative glycopattern analysis of mucins in the Brunner's glands of the guinea-pig and the house mouse (Rodentia). Acta Histochem 117:612–623

Skutelsky E, Moore RP, Alroy J (1989) Lectin histochemistry of mammalian Brunner's glands. Histochemistry 90:383–390

Snipes RL, Hörnicke H, Björnhag G, Stahl W (1988) Regional differences in hindgut structure and function in the nutria, *Myocastor* coypus. Cell Tissue Res 252:435–447

Takahashi T, Sakaguchi E (2000) Rol of the furrow of the proximal colon in the production of soft and hard feces in nutrias, *Myocastor coypus*. J Comp Physiol B 170:531–535

Takahashi T, Sakaguchi E (2006) Transport of bacteria across and along the large intestinal lumen of guinea pigs. J Comp Physiol B 176:173–178

Takehana K, Masty J, Yamaguchi M, Kobayashi A, Yamada O, Kuroda M, Park YS, Iwasa K, Abe M (1998) Fine structural and histochemical study of equine Paneth cells. Anat Histol Embryol 27:125–129

Tano de la Hoz MF, Flamini MA, Díaz AO (2014) Histological and histochemical study of the duodenum of the plains viscacha (*Lagostomus maximus*) at different stages of its ontogenetic development. Acta Zool (Stockholm) 95:21–31

Tano de la Hoz MF, Flamini MA, Díaz AO (2016) Comparative analysis of the morphology, ultrastructure, and glycosylation pattern of the jejunum and ileum of the wild rodent *Lagostomus maximus*. Anat Rec 299(5):630–642

Tano de la Hoz MF, Flamini MA, Zanuzzi CN, Díaz AO (2017) The colonic groove of the plains viscacha (*Lagostomus maximus*): histochemical evidence of an abrupt change in the glycosylation pattern of goblet cells. J Morphol 278:1606–1618

Tano de la Hoz MF, Flamini MA, Portiansky EL, Díaz AO (2019) Analysis of glycoconjugates and morphological characterization of the descending colon and rectum of the plains viscacha, *Lagostomus maximus*. Zoology 135:125691

Vásquez Cachay ME, Pebe Gomez E, Rodriguez Gutierrez JL, Lira Mejia B, Falcon Perez N, Zanuzzi CN, Barbeito C (2014) Paneth cell identification in the small intestine of guinea pig offsprings (*Cavia porcellus*). Anat Rec 297:856–863

Verdiglione R, Mamola CL, Filotto U (2002) Glycoconjugate histochemistry of bovine Brunner glands. Ann Anat 184:61–69

Zanuzzi CN, Fontana PA, Barbeito CG, Portiansky EL, Gimeno EJ (2008) Paneth cells: histochemical and morphometric study in control and *Solanum glaucophyllum* intoxicated rabbits. Eur J Histochem 52:93–100

Index

L. L. Rasia et al. (eds.), *Plains Vizcachas*,
https://doi.org/10.1007/978-3-031-49487-1

MIX
Papier aus verantwortungsvollen Quellen
Paper from responsible sources
FSC® C105338

If you have any concerns about our products,
you can contact us on
ProductSafety@springernature.com

In case Publisher is established outside the EU,
the EU authorized representative is:
Springer Nature Customer Service Center GmbH
Europaplatz 3, 69115 Heidelberg, Germany

Printed by Libri Plureos GmbH
in Hamburg, Germany